Neue Nummer Fam. Bauer

06634/205015

Rebecca Handy

0172/4955838 → Billig Nummer vorwählen!!!!

01782/5493744 D.W

D1692675

de Gruyter Studies in Mathematics 1

Editors: Heinz Bauer · Peter Gabriel

Wilhelm Klingenberg

Riemannian Geometry

Walter de Gruyter · Berlin · New York · 1982

Author:

Dr. phil. Wilhelm Klingenberg
Professor of Mathematics
Rheinische Friedrich-Wilhelms-Universität, Bonn

For my wife

Library of Congress Cataloging in Publication Data

Klingenberg, Wilhelm, 1924–
Riemannian geometry.
(De Gruyter studies in mathematics ; 1)
Includes bibliographical references and index.
1. Geometry, Riemannian. I. Title. II. Series.
QA649.K544 1982 516.3'73 82-9772
ISBN 3-11-008673-5 AACR2

CIP-Kurztitelaufnahme der Deutschen Bibliothek

Klingenberg, Wilhelm:
Riemannian geometry / Wilhelm Klingenberg. –
Berlin ; New York : de Gruyter, 1982.
(De Gruyter studies in mathematics ; 1)
ISBN 3-11-008673-5
NE: GT

© Copyright 1982 by Walter de Gruyter & Co., Berlin. All rights reserved, including those of translation into foreign languages. No part of this book may be reproduced in any form – by photoprint, microfilm, or any other means – nor transmitted nor translated into a machine language without written permission from the publisher. Typesetting and Printing: Fa. Tutte, Salzweg. – Binding: Lüderitz & Bauer Buchgewerbe GmbH, Berlin. – Cover design: Rudolf Hübler, Berlin. **Printed in Germany.**

Preface

The present book is an outcome of my course on Riemannian Geometry. Its origin can be traced back to a series of special lectures which I gave during the summer semester 1961 in Bonn. At that time, D. Gromoll and W. Meyer were among my students and in 1967 we jointly published in the Lecture Notes Series our "Riemannsche Geometrie im Großen".

These lectures have met with great interest, because for the first time a concise introduction into Riemannian Geometry was combined with global methods culminating in the so-called Sphere Theorem, which states that the underlying topological manifold of a simply connected Riemannian manifold with suitably restricted positive curvature is a sphere.

Over the past twenty years, global Riemannian Geometry has experienced considerable growth in various areas. Here I wish to mention in particular the work of Gromov [1], [2], [3] on manifolds with restrictions on the curvature and the numerous results on the eigenvalues of the Laplace operator. Also – and this is the field in which I have been most active myself – a great number of new results on the existence and on the properties of closed geodesics have been obtained. For an excellent survey of the present state of research cf. Yau [1].

It is only natural that in my course I have chosen topics which are close to my own areas of research. But I have always begun with a full exposition of the classical, local Riemannian Geometry.

Thus, chapter 1 is devoted to the foundations. What is unusual here is that from the very beginning I have allowed manifolds to be modelled on separable Hilbert spaces. This presents no difficulties when compared with the case of finite dimensional manifolds and it has the advantage of yielding the necessary framework for later applications in chapter 2. Of course, there are some differences between Hilbert manifolds and finite dimensional manifolds, which appear for the first time when considering the tensor product. However, for most of the basic results, the step from finite dimensions to Hilbert space is no bigger than the step from 2 dimensions to n dimensions. Whenever the restriction to the finite dimensional case brings about some simplification, I have pointed this out clearly.

In chapter 1 – and the same is true for the later chapters – the first part of a section is usually more basic than the rest. At least this is the case when the sections are longer than 10 pages or so. While the expert will have no difficulty in constructing a 'basic course' from our material, the beginner should keep in mind that it might be wise to switch to the next section when he reaches subjects like general vector bundles, submersions or focal points, to mention a few. He can always return to the previous sections when the need arises.

In chapter 2, entitled *Curvature and Topology*, I restrict myself to finite dimensional manifolds because the local compactness of the manifold is needed. Complete manifolds are studied and there follows a rather complete account of the theory of symmetric spaces from the point of view of Riemannian Geometry which differs from the usual, more algebraic approach. After this, I develop in three sections the basic theory of the manifold of curves. Here, I take advantage of the fact that in chapter 1 Riemannian manifolds modelled on Hilbert Spaces were allowed.

When it comes to the critical point theory I only develop the Lusternik-Schnirelmann approach. I do not enter into the much more delicate Morse Theory. The full power of this approach has not been sufficiently recognized. Among other things I show that one can give a completely elementary proof of the fundamental estimate on the injectivity radius of 1/4-pinched manifolds.

The chapter continues with a simplified proof of the Alexandrov-Toponogov Comparison Theorem. This is an essential tool in the proof of the Sphere Theorem, given in the next section. I conclude with the basic constructions on non-compact manifolds of positive curvature.

With the end of chapter 2 I have covered all the material contained in the "Riemannsche Geometrie im Großen" and indeed much more; e.g., symmetric spaces and the manifold of curves with its various important submanifolds.

In chapter 3, entitled *Structure of the Geodesic Flow*, I deal with a subject which, traditionally, is not presented in a course on Riemannian Geometry. I feel, however, that this field should not be left to specialists in ergodic theory or Hamiltonian systems. Rather, it should be tied more closely to Riemannian Geometry proper. In fact, it is one of the oldest fields of research. Thus, e.g., the geodesic flow on the ellipsoid was even studied by C.G.J. Jacobi and the problem of stability for periodic orbits plays a fundamental rôle in Poincaré's investigations on Celestial Mechanics. I present many of the classical results together with numerous examples. Among them, there are theorems for periodic orbits with elementary proofs employing only the Lusternik-Schnirelmann theory developed in chapter 2. The last two sections deal with manifolds of non-positive curvature. Here, in particular the case of strictly negative curvature is treated for the first time in a monograph, with elementary proofs of many of the basic results in this important area.

After this brief description of the contents, it would certainly take more space to describe what has been omitted. To have an idea of topics not covered in this book see e.g. Chern [2] and de Rham [2]. Most notable is the absence of integration methods. It is clear that on Hilbert manifolds differential forms are bound to play a lesser rôle than on manifolds of finite dimension. But the deeper explanation for this and for most of the other omissions is simply that a book on mathematics, like any other literary work, is necessarily prejudiced by the personal experiences of the author and thus reveals strong autobiographical traits. As a matter of fact, I have composed this monograph around my own area of research in Riemannian geometry over the past 25 years, thereby including the work of younger colleagues who I had the privilege of meeting to mutual advantage. I have no other excuse to proffer for the selection of the contents, except that I am convinced that my choice represents a lasting contribution to the field and that future fruitful developments seem most likely.

Thus, I hope that my efforts in writing this book over many years will not just be a record of results and methods but will also serve as an impetus towards further research.

It only remains to express my gratitude to the people who helped me with the manuscript, by reading whole sections. I wish to mention in particular W. Ballmann, V. Bangert, J. Eschenburg, H. Matthias, A. Thimm, G. Thorbergsson and F. Wolter. Finally, I wish to thank Walter de Gruyter & Co. for accepting my manuscript in their new series 'Studies in Mathematics'.

Bonn, 1982 Wilhelm Klingenberg

Table of Contents

Chapter 1: Foundations

This chapter contains the basic definitions and results on differentiable manifolds, vector and tensor bundles over such manifolds and Riemannian metrics. The material presented here differs little from that in other well-known text books, except that we consider manifolds modelled on Hilbert spaces rather than on finite dimensional spaces. This will be useful in Chapter 2 and presents no conceptual difficulties anyway, as was demonstrated by Lang [1].

Not quite standard in our chapter on the Foundations is the discussion of submersions (see 1.11) and Jacobi fields (see 1.12). This constitutes a first step towards global geometry, which is the subject of the remainder of the book.

1.0 Review of Differential Calculus and Topology

In this section we set forth some notation and recall some basic properties of differentiable maps between Banach spaces. For details we refer to Dieudonné [1] and Lang [1]. We shall conclude with some facts on topological spaces. Reference will be made to Bourbaki [1].

1. We denote by $\mathbb{E}, \mathbb{E}', \mathbb{E}_i, \ldots, \mathbb{F}, \mathbb{F}', \mathbb{F}_i, \ldots$ real Banach spaces. In fact, most of the time these will actually be separable (complete) Hilbert spaces. Subspaces are always assumed to be closed and linear mappings are assumed to be continuous.

Note. Subspaces of finite dimension or finite codimension are always closed.

We say that a closed (linear) subspace $\mathbb{E}'$ of $\mathbb{E}$ *splits* if there exists a closed complement $\mathbb{E}''$ such that $\mathbb{E}$ is isomorphic to $\mathbb{E}' \times \mathbb{E}''$. Note that for a Hilbert space $\mathbb{E}$, every subspace $\mathbb{E}'$ splits: Take for $\mathbb{E}''$ the orthogonal complement of $\mathbb{E}'$.

Let $F\colon \mathbb{E} \to \mathbb{F}$ be an injective linear mapping whose image is a closed subspace $\mathbb{F}'$. F is called a *splitting* mapping if $\mathbb{F}'$ splits, i.e., if $\mathbb{F} \cong \mathbb{F}' \times \mathbb{F}''$. More generally, a linear mapping $F\colon \mathbb{E} \to \mathbb{F}$ with closed image is called a splitting mapping if the induced injection $\mathbb{E}/\mathit{ker}\,F \to \mathbb{F}$ splits. Again, for Hilbert spaces, any closed linear mapping splits – closed means that the image is a closed subspace.

Let us denote by $L(\mathbb{E}; \mathbb{F})$ the vector space of linear mappings $F\colon \mathbb{E} \to \mathbb{F}$. $L(\mathbb{E}; \mathbb{F})$ becomes a Banach space by taking as norm $|F|$ of an $F \in L(\mathbb{E}; \mathbb{F})$ the greatest lower bound of all numbers k such that

$$|F.X| \leqslant k|X|, \quad \text{for all } X \in \mathbb{E}.$$

If $\mathbb{E}$ and $\mathbb{F}$ are finite dimensional, one can define a scalar product on $L(\mathbb{E}, \mathbb{F})$ so as to make it into a (finite dimensional) Hilbert space, see (1.0.2).

More generally, we define a norm on the space

$$L(\mathbb{F}_1, \ldots, \mathbb{F}_r; \mathbb{G})$$

of r-linear mappings from $\mathbb{F}_1 \times \ldots \times \mathbb{F}_r$ into $\mathbb{G}$ by taking for $|F|$ the greatest lower bound of real numbers k satisfying

$$|F(X_1, \ldots, X_r)| \leqslant k|X_1| \ldots |X_r|,$$

where $(X_1, \ldots, X_r) \in \mathbb{F}_1 \times \ldots \times \mathbb{F}_r$.

With this, the canonical mapping

$$L(\mathbb{F}_1; L(\mathbb{F}_2; \ldots; L(\mathbb{F}_r; \mathbb{G})) \to L(\mathbb{F}_1, \mathbb{F}_2, \ldots, \mathbb{F}_r; \mathbb{G})$$

from the space of iterated linear maps into the space of multilinear maps becomes a Banach space isomorphism.

Of particular importance are the various tensor spaces associated to a Hilbert space $\mathbb{E}$. Let $\mathbb{E}^*$ denote the dual of $\mathbb{E}$. Then

$$T_s^r \mathbb{E} \equiv L(\underbrace{\mathbb{E}^*, \ldots, \mathbb{E}^*}_{r}, \underbrace{\mathbb{E}, \ldots, \mathbb{E}}_{s}; \mathbb{R})$$

is called the *space of r-fold contravariant and s-fold covariant tensors.*

We also use $T_s^r \mathbb{E}$ to denote any of the $(r+s)!/r!s!$ spaces $L(\mathbb{E}_1, \ldots, \mathbb{E}_{r+s}; \mathbb{R})$, where r of the $\mathbb{E}_i$ are equal to $\mathbb{E}^*$ and the remaining s of the $\mathbb{E}_i$ are equal to $\mathbb{E}$.

Since $L(\mathbb{E}^*; \mathbb{R}) = \mathbb{E}$; $L(\mathbb{E}; \mathbb{R}) = \mathbb{E}^*$ we have for $rs > 0$

$$\begin{aligned} T_s^r \mathbb{E} = L(\underbrace{\mathbb{E}^*, \ldots, \mathbb{E}^*}_{r}, \underbrace{\mathbb{E}, \ldots, \mathbb{E}}_{s}; \mathbb{R}) &\cong L(\underbrace{\mathbb{E}^*, \ldots, \mathbb{E}^*}_{r-1}, \underbrace{\mathbb{E}, \ldots, \mathbb{E}}_{s}; \mathbb{E}) \\ &\cong L(\underbrace{\mathbb{E}^*, \ldots, \mathbb{E}^*}_{r}, \underbrace{\mathbb{E}, \ldots, \mathbb{E}}_{s-1}; \mathbb{E}^*) \end{aligned}$$

and

$$T_1^0 \mathbb{E} \cong \mathbb{E}^*; \; T_0^1 \mathbb{E} \cong \mathbb{E}.$$

A word to explain the terminology: Take e.g. $X \in T_0^1 \mathbb{E} = \mathbb{E}$, a 1-fold contravariant tensor. Let F: $\mathbb{E} \to \mathbb{E}$ be an automorphism. Choose an (orthonormal) Hilbert basis $\{e_i\}$ and its dual $\{e^i\}$. Then the i-th coordinate of X ist given by $X^i = \langle e^i, X\rangle$ where we denote by $\langle\, ,\rangle$ the *canonical pairing* $\mathbb{E}^* \times \mathbb{E} \to \mathbb{R}$. The i-th coordinate of FX is given by $X'^i = \langle e^i, FX\rangle = \langle {}^tFe^i, X,\rangle$, ${}^tF$: $\mathbb{E}^* \to \mathbb{E}^*$ being the transpose of F. That is, $X'^i = \sum_k ({}^tF)^i_k X^k$ where the $({}^tF)^i_k$ are the elements of the transposed matrix ${}^tF$ of F. Thus, the coordinates of a vector are transformed with the transposed matrix which is why X is called contravariant.

Let $\mathbb{F}$ be a Banach space. By $GL(\mathbb{F})$ we mean the group of (linear) automorphisms of $\mathbb{F}$. Consider a tensor space $T^r_s\,\mathbb{E}$ of the Hilbert space $\mathbb{E}$. Then we have a canonical group morphism

$$T^r_s\colon GL(\mathbb{E}) \to GL(T^r_s\,\mathbb{E})$$

given by associating with $\{F\colon \mathbb{E} \to \mathbb{E}\} \in GL(\mathbb{E})$ the mapping

$$T^r_s F\colon T^r_s\,\mathbb{E} \to T^r_s\,\mathbb{E}\,;$$

$$X^r_s \mapsto X^r_s \circ (\underbrace{{}^tF \times \ldots \times {}^tF}_{r} \times \underbrace{F^{-1} \times \ldots \times F^{-1}}_{s}),$$

where ${}^tF\colon \mathbb{E}^* \to \mathbb{E}^*$ is the transpose of F.

Indeed, one verifies at once that $T^r_s F$ is linear and that $T^r_s(F_2 \circ F_1) = T^r_s F_2 \circ T^r_s F_1$. Moreover, $T^r_s(id|\,\mathbb{E}) = id|\,\mathbb{E}^r_s$. Thus T^r_s is what is called in category theory a covariant functor.

A sub[illegible] under the subgroup $T^r_s GL(\mathbb{E})$ of $GL(T^r_s\,\mathbb{E})$ is called a [illegible]
We giv[illegible] of such general tensor spaces in $T^0_s\,\mathbb{E}$
$= L(\mathbb{E},$ [illegible]

(i) Th[illegible] and symmetric tensors consists of those elements [illegible]

[illegible] $\ldots, X_s)$,

for all [illegible] $s\}$.

(ii) T[illegible]nt antisymmetric tensors consisting of the $Z^0_s \in T^0_s$ [illegible]

[illegible] $(X_1, \ldots, X_s)$,

for all [illegible]

We s[illegible]

[illegible] 1)! whereas

[illegible] r $1 \leqslant s \leqslant n$

[illegible] for $s > n$

We conclude this section by indicating some canonical isomorphisms between spaces of linear maps *when all vector spaces have finite dimension.*

Recall that the *tensor product* $\mathbb{E} \otimes \mathbb{F}$ of two vector spaces $\mathbb{E}$ and $\mathbb{F}$ is characterized by the following properties:

(i) There exists a bilinear mapping

$$\Phi\colon \mathbb{E} \times \mathbb{F} \to \mathbb{E} \otimes \mathbb{F};\ (X, Y) \mapsto X \otimes Y$$

such that the image generates $\mathbb{E} \otimes \mathbb{F}$ as a vector space.

(ii) Given any bilinear map $F \in L(\mathbb{E}, \mathbb{F}; \mathbb{G})$, there exists a unique $G \in L(\mathbb{E} \otimes \mathbb{F}; \mathbb{G})$ with $F = G \circ \Phi$.

In particular, if $\{e_i\}$, $\{f_j\}$ are bases of $\mathbb{E}$ resp. $\mathbb{F}$, then $\{e_i \otimes f_j\}$ is a basis for $\mathbb{E} \otimes \mathbb{F}$. One has the canonical isomorphisms:

$$L(\mathbb{E} \otimes \mathbb{F}; \mathbb{G}) \cong L(\mathbb{E}, \mathbb{F}; \mathbb{G}) \cong L(\mathbb{E}; L(\mathbb{F}; \mathbb{G}))$$

$$L(\mathbb{E}; \mathbb{F}^*) \cong \mathbb{E}^* \otimes \mathbb{F}^*$$

For instance, $F \in L(\mathbb{E}; \mathbb{F}^*)$ corresponds to $\sum_{i,j} \langle f_j, F(e_i)\rangle\, e^i \otimes f^j$ where $\{e^i\}$, $\{f^j\}$ are the dual bases of the bases $\{e_i\}$, $\{f_j\}$ of $\mathbb{E}$ and $\mathbb{F}$. Combining these isomorphisms, we get the

1.0.1 Proposition. *Let* $\dim \mathbb{E} < \infty$. *Then*

$$T_s^r\, \mathbb{E} \cong \underbrace{\mathbb{E} \otimes \ldots \otimes \mathbb{E}}_{r} \otimes \underbrace{\mathbb{E}^* \otimes \ldots \otimes \mathbb{E}^*}_{s}$$

and also

$$T_s^r\, \mathbb{E} \cong L(\underbrace{\mathbb{E}, \ldots, \mathbb{E}}_{s}; \underbrace{\mathbb{E} \otimes \ldots \otimes \mathbb{E}}_{r}) =$$

$$L(\underbrace{\mathbb{E}, \ldots, \mathbb{E}}_{s-1}; \mathbb{E}^* \otimes \underbrace{\mathbb{E} \otimes \ldots \otimes \mathbb{E}}_{r}) \quad etc. \quad \square$$

Note. The concept of the projective tensor product for Banach spaces allows one to extend these isomorphisms to the case of infinite dimensions. See Schatten [1].

For later use we point out another feature of vector spaces $\mathbb{E}$ of finite dimension.

1.0.2 Proposition. *Let* $\dim E = n < \infty$. *Let* $\langle , \rangle$ *denote the scalar product on* $\mathbb{E}$. *Then on* $T_s^r\, \mathbb{E}$ *this determines intrinsically a scalar product as follows: Let* $\{e_i\}$, $1 \leqslant i \leqslant n$, *be an orthonormal basis for* $\mathbb{E}$. *Together with the dual basis* $\{e^j\}$, $i \leqslant j \leqslant n$, *the*

$$e_{i_1 \ldots i_r}^{j_1 \ldots j_s} = e_{i_1} \otimes \ldots \otimes e_{i_r} \otimes e^{j_1} \otimes \ldots \otimes e^{j_s}$$

$$\textit{for } 1 \leqslant i_1, \ldots, i_r \leqslant n;\ 1 \leqslant j_1, \ldots, j_s \leqslant n$$

form a basis for $T_s^r\, \mathbb{E}$. *Now define the metric on* $T_s^r\, \mathbb{E}$ *by letting this basis be orthonormal.*

This definition is independent of the choice of the orthonormal bases $\{e_i\}$.

Proof. Any two orthonormal bases $\{e_i\}$, $\{e_i'\}$ are related by an orthogonal transformation $A = (a_i^k)$:

$$e_i' = \sum_k a_i^k e_k, \quad \sum_k a_i^k a_j^k = \delta_{ij}.$$

The corresponding bases

$$\{e_{i_1 \ldots i_r}^{j_1 \ldots j_s}\}; \quad \{e'^{j_1 \ldots j_s}_{i_1 \ldots i_r}\}$$

then are related by the linear transformation $T_s^r A$, i.e.,

$$e'^{j_1 \dots j_s}_{i_1 \dots i_r} = \sum_{\substack{k_1, \dots, k_r \\ l_1, \dots, l_s}} a^{k_1}_{i_1} \dots a^{k_r}_{i_r} b^{j_1}_{l_1} \dots b^{j_s}_{l_s} e^{l_1 \dots l_s}_{k_1 \dots k_r}$$

Here $e'^j = \sum_l b^j_l e^l$, i.e., (b^j_l) ist the transposed inverse or contragradient ${}^t A^{-1}$ of A.

From $A^t A = E$, $({}^t A^{-1})^t ({}^t A^{-1}) = E$ we see that the $e'^{j_1 \dots j_s}_{i_1 \dots i_r}$ also form an orthonormal basis. That is to say, $A \in \mathbb{O}(\mathbb{E})$ implies $T_s^r A \in \mathbb{O}(T_s^r \mathbb{E})$. □

2. Let $U \subset \mathbb{F}$, $U' \subset \mathbb{F}'$ be open sets. Let $F\colon U \to U'$ be a mapping. F is called *differentiable at* $u_0 \in U$ if there exists a $DF(u_0) \in L(\mathbb{F}; \mathbb{F}')$ such that

$$F(u) - F(u_0) - DF(u_0) \,.\, (u - u_0) = o(|u - u_0|).$$

Here, $o(r)$ satisfies $\lim_{r \to 0; r > 0} \|o(r)/r\| = 0$. F is called *differentiable of class* C^1 if it is differentiable for all $u \in U$ and $u \in U \mapsto DF(u) \in L(\mathbb{F}; \mathbb{F}')$ is continuous.

That $F\colon U \to U'$ is *differentiable of class* C^r is defined by induction. Assume we have defined $D^{r-1}F$ as a mapping from U into $L(\mathbb{F}; L(\mathbb{F}; \dots L(\mathbb{F}; \mathbb{F}')))$ which we can identify with $L(\mathbb{F}, \mathbb{F}, \dots; \mathbb{F}')$, with $(r-1)$ times $\mathbb{F}$. If $D^{r-1}F$ is differentiable of class C^1, put $D(D^{r-1}F) = D^r F$ and call F differentiable of class C^r.

Finally, we call $F\colon U \to U'$ *differentiable*, if it is differentiable of class C^r for all r.

Sometimes we will find it convenient to use the language of categories and functors. Thus we may speak of the category formed by the open subsets of Banach spaces as objects and the differentiable mappings between them as morphisms. This means that with

$$F_1\colon U_1 \subset \mathbb{E}_1 \to U_2 \subset \mathbb{E}_2;\ F_2\colon U_2 \subset \mathbb{E}_2 \to U_3 \subset \mathbb{E}_3,$$

being differentiable the composition

$$F_2 \circ F_1\colon U_1 \to U_3$$

is also differentiable. Moreover, $id_U\colon U \subset \mathbb{E} \to U \subset \mathbb{E}$ is differentiable.

Let $U \subset \mathbb{E}$ be open. For every $u_0 \in U$ we define the *tangent space* $T_{u_0} U$ *of* U *at* u_0 as the set $\{(u_0, X); X \in \mathbb{E}\}$, endowed with the vector space structure arising from the canonical mapping

$$pr_2\colon (u_0, X) \in T_{u_0} U \mapsto X \in \mathbb{E}$$

The collection of the tangent spaces $T_{u_0} U$, $u_0 \in U$, is denoted by TU. The canonical isomorphism

$$TU \cong U \times \mathbb{E}$$

makes TU into an open subset of $\mathbb{E} \times \mathbb{E}$.

The projection $pr_1\colon U \times \mathbb{E} \to U$ onto the first factor will also be written as

$$\tau \equiv \tau_U\colon TU \to U;\ (u_0, X) \mapsto u_0 .$$

τ_U is called *tangent bundle of* U. TU is called the *total tangent space of* U and τ ist called the *projection* of the tangent bundle.

For a differentiable

$$F\colon U \subset \mathbb{E} \to V \subset \mathbb{F}$$

we define the *tangential of F*,

$$TF\colon TU \to TV,$$

by $(u, X) \mapsto \bigl(F(u), DF(u) \,.\, X\bigr)$.

Note that, for each $u \in U$, the restriction $T_u F = TF | T_u U$ is a linear mapping which is completely determined by the differential $DF(u)\colon \mathbb{E} \to \mathbb{F}$. For this reason, $DF(u)$ and $T_u F$ sometimes are identified. But basically, $DF(u)$ is a mapping from $\mathbb{E}$ to $\mathbb{F}$ while $T_u F$ is a mapping between tangent spaces of $\mathbb{E}$ and $\mathbb{F}$.

Associating with $F\colon U \to V$ its tangential $TF\colon U \times \mathbb{E} \to V \times \mathbb{F}$ constitutes a covariant functor from the category of differentiable mappings between open sets into the same category. Indeed, let

$$F_1\colon U_1 \subset \mathbb{E}_1 \to U_2 \subset \mathbb{E}_2;\ F_2\colon U_2 \subset \mathbb{E}_2 \to U_3 \subset \mathbb{E}_3$$

be morphisms. Then the morphisms

$$T(F_2 \circ F_1)\colon T_1 U_1 \to TU_3;\ TF_2 \circ TF_1\colon TU_1 \to TU_3$$

are the same. And the tangential Tid_U of the identity mapping $id_U\colon U \to U$ is the identity mapping $id_{TU}\colon TU \to TU$.

Actually, the tangential is a special sort of morphism; it preserves the product structure $U \times \mathbb{E}$ of the objects TU. This amounts to the commutativity of the diagram

$$\begin{array}{ccc} U \times \mathbb{E} = TU & \xrightarrow{\quad TF \quad} & TV = V \times \mathbb{E} \\ \downarrow & & \downarrow \\ U & \xrightarrow{\quad F \quad} & V \end{array}$$

Moreover, the restrictions $T_u F = TF | T_u U$ are linear. Therefore we may say that the pair (F, TF) becomes a morphism in the category of tangent bundles of the open sets U of Banach spaces.

3. We continue with the inverse mapping theorem and two corollaries concerning locally injective and locally surjective differentiable mappings. For our later applications it suffices to consider the case that all spaces are Hilbert spaces.

1.0.3 Theorem. *Let U be an open neighborhood of $0 \in \mathbb{E}$. Let*

(*) $\quad F\colon U \to \mathbb{F},\ F(0) = 0$

be differentiable such that $DF(0)\colon \mathbb{E} \to \mathbb{F}$ is a (bijective) isomorphism. Then F is a local *diffeomorphism. That is to say, there exist open neighborhoods* U', V' of $0 \in \mathbb{E}$, $0 \in \mathbb{F}$, $U' \subset U$, *such that* $F|U'\colon U' \to V'$ *is a diffeomorphism.*

A *diffeomorphism* is a differentiable homeomorphism such that also the inverse is differentiable.

1.0.4 Corollary 1. *Assume that the mapping* F, (*), *has the property that* $DF(0)\colon \mathbb{E} \to \mathbb{F}$ *is an isomorphism with a closed subspace* $\mathbb{F}_1$ *of* $\mathbb{F}$. *Write* $\mathbb{F} = \mathbb{F}_1 \times \mathbb{F}_2$. *Then there exists a local diffeomorphism*

$$g\colon \mathbb{F} \to \mathbb{F}_1 \times \mathbb{F}_2;\ g(0) = 0$$

and an open neighborhood $U_1 \subset U$ *of* $0 \in \mathbb{E}$ *such that*

$$g \circ (F|U_1)\colon U_1 \to U_1 \times \{0\} \subset \mathbb{E} \times \{0\} \cong \mathbb{F}_1 \times \{0\} \subset \mathbb{F}$$

is the canonical linear injection.

1.0.5 Corollary 2. *Assume that the mapping* F, (*), *has the property that* $DF(0)\colon \mathbb{E} \to \mathbb{F}$ *is surjective. Write* $\mathbb{E} = \mathbb{E}_1 \times \mathbb{E}_2$ *with* $DF(0)|\mathbb{E}_2\colon \mathbb{E}_2 \to \mathbb{F}$ *bijective, i.e.,* $\mathbb{E}_2 \cong \mathbb{F}$ *via* $DF(0)|\mathbb{E}_2$. *Then there exists a local diffeomorphism*

$$h\colon (U_1 \times U_2, 0) \subset (\mathbb{E}_1 \times \mathbb{E}_2, 0) \to (\mathbb{E}, 0)$$

with U_i *an open neighborhood of* $0 \in \mathbb{E}_i$ *such that*

$$F \circ h\colon U_1 \times U_2 \to U_2 \subset \mathbb{E}_2 \cong \mathbb{F}$$

is the projection pr_2 *onto the second factor.*

Note. If $\mathbb{E}$, $\mathbb{F}$ are Banach spaces, one must assume that ker $DF(0)$ splits.

Proof. For the proof of (1.0.3) one uses the contraction lemma. For details we refer to the literature, cf. Dieudonné [1], Lang [1].

Corollary 1 is deduced from the theorem by extending $F\colon U \to \mathbb{F}$ to a locally invertible mapping

$$\Phi\colon U \times \mathbb{F}_2 \subset \mathbb{E} \times \mathbb{F}_2 \cong \mathbb{F} \to \mathbb{F}_1 \times \mathbb{F}_2; \quad (u, v_2) \mapsto F(u) + (0, v_2).$$

Indeed, $D\Phi(0, 0) = DF(0) + (0, id|\mathbb{F}_2)$. Taking g as the local inverse of Φ we prove our claim.

Similarly, for the proof of Corollary 2, we consider

$$\Phi\colon U \subset \mathbb{E}_1 \times \mathbb{E}_2 \cong \mathbb{E}_1 \times \mathbb{F} \to \mathbb{E}_1 \times \mathbb{E}_2; \quad (u_1, u_2) \mapsto (u_1, F(u_1, u_2)).$$

Then $D\Phi(0) = (id|\mathbb{E}_1, 0) + (DF(0)|\mathbb{E}_1, DF(0)|\mathbb{E}_2)$, i.e., $D\Phi(0)$ is bijective. Taking for h the local inverse of Φ we get a mapping satisfying our requirements.

4. A topological space M is called *metrizable* if there exists a metric on M which induces the given topology.

M is called *separable* if it possesses a countable base for the open sets. For metric spaces this is equivalent to saying that there exists a countable dense set of points in M.

Let $(M_\alpha)_{\alpha\in A}$ be an open covering of M. That is, all M_α are open and every $p\in M$ is contained in some M_α. An open covering $(\tilde{M}_\beta)_{\beta\in B}$ of M will be called *refinement* of the covering $(M_\alpha)_{\alpha\in A}$ if there exists a mapping $\sigma\colon B\to A$ such that $\tilde{M}_\beta\subset M_{\sigma(\beta)}$.

An open covering $(M_\alpha)_{\alpha\in A}$ of M is called *locally finite* if every point $p\in M$ possesses a neighborhood U such that $U\cap M_\alpha\neq\emptyset$ for finitely many α only.

A topological space M is called *paracompact* if every open covering of M possesses a locally finite refinement. Clearly, a compact space is paracompact. So are finite dimensional Banach spaces. An important sufficient condition for a space to be paracompact is that it is metrizable.

The property of a topological space to be metrizable is preserved under the operations of forming products and taking subsets. Our interest in this property stems from the fact that we will be considering Riemannian manifolds on which the Riemannian structure defines a metric which induces the given topology.

A *partition of unity* on a topological space M consists of a family $(\phi_\beta, \tilde{M}_\beta)_{\beta\in B}$. Here, $(\tilde{M}_\beta)_{\beta\in B}$ forms a locally finite open covering of M and $\phi_\beta\colon M\to\mathbb{R}$ is continuous $\geqslant 0$ with $\{\phi_\beta>0\}\subset\tilde{M}_\beta$ and $\sum_\beta \phi_\beta(p)=1$, for all $p\in M$.

A topological space M is said to *admit partitions of unity* if, for every open covering $(M_\alpha)_{\alpha\in A}$, there exists a partition of unity $(\phi_\beta, \tilde{M}_\beta)_{\beta\in B}$ with $(\tilde{M}_\beta)_{\beta\in B}$ being a refinement of $(M_\alpha)_{\alpha\in A}$. A paracompact separable space admits partitions of unity.

If M is a differentiable manifold in the sense of definition (1.1.2) then M even admits *differentiable partitions of unity*, i.e., the functions $\phi_\beta\colon M\to\mathbb{R}$ are of class C^∞. Cf. Lang [1]. For the finite dimensional case cf. also Hirsch [1] and Sulanke und Wintgen [1].

1.1 Differentiable Manifolds

In this section we introduce the concept of a differentiable manifold M, modelled on a (separable) Hilbert space $\mathbb{E}$. Essentially, this is a topological space which locally looks like an open set U of $\mathbb{E}$. Such a local representation of M is called a chart. So far, M is only a topological manifold. What makes M into a differentiable manifold is that the transition mappings, determined by the overlap of two charts, are diffeomorphisms, cf. (1.1.2).

The morphisms, i.e., the structure preserving mappings between differentiable manifolds, are introduced in (1.1.4). With this, we get the category of differentiable manifolds and mappings, see (1.1.5).

We conclude by showing that differentiable mappings can be localized. This means that every differentiable mapping, defined on some open set of a manifold M, when restricted to a suitable open neighborhood of a point $p\in M$, can be viewed as the restriction of a differentiable mapping defined on all of M. Thus, for local properties, there is no difference between local and global morphisms.

It would also have been possible to introduce the concept of a manifold modelled on a Banach space. However, for Riemannian geometry this is not the right object since in general the norm on a Banach space is not equivalent to the norm of a Hilbert space structure. Still, when we come in (1.2) to the concept of vector bundles over manifolds we will be forced to allow as fibres general Banach spaces. Thus, we will get spaces (the total spaces of such vector bundles) which are modelled on a product $\mathbb{E} \times \mathbb{F}$ of a Hilbert space $\mathbb{E}$ and a Banach space $\mathbb{F}$. These manifolds possess atlases, however, with special charts having as image open sets of the form $U \times \mathbb{F} \subset \mathbb{E} \times \mathbb{F}$. Moreover, the transition mappings are linear when restricted to a 'fibre' $\{u\} \times \mathbb{F}$. Therefore, we do not deem it necessary to introduce the concept of a Banach manifold in its full generality.

1.1.1 Definition. *A topological manifold M, modelled on the Hilbert space* $\mathbb{E}$, is a separable, metrizable space which is locally homeomorphic to $\mathbb{E}$. That is to say, every point of M has an open neighborhood which is homeomorphic to $\mathbb{E}$.

1.1.2 Definition. (i) Let M be a topological manifold modelled on $\mathbb{E}$. A *differentiable atlas* for M is a family

$$(u_\alpha, M_\alpha)_{\alpha \in A}$$

of *charts*, having the following properties: $(M_\alpha)_{\alpha \in A}$ is a covering of M by open sets. For each $\alpha \in A$, (u_α, M_α) stands for a homeomorphism

$$u_\alpha \colon M_\alpha \to U_\alpha$$

of M_α onto an open set $U_\alpha \subset \mathbb{E}$. Finally, if we put $M_\alpha \cap M_\beta = M_{\alpha\beta}$ and $u_\alpha(M_{\alpha\beta}) = U_{\alpha\beta}$, *the transition mapping*

$$u_\beta \circ u_\alpha^{-1} \colon U_{\alpha\beta} \to U_{\beta\alpha}$$

is a diffeomorphism.

(ii) We call two differentiable atlases $(u_\alpha, M_\alpha)_{\alpha \in A}$, $(u_{\alpha'}, M_{\alpha'})_{\alpha' \in A'}$ for the topological manifold M *equivalent* if their union gives a differentiable atlas.

(iii) *A differentiable structure* on a topological manifold is an equivalence class of differentiable atlases. A *differentiable manifold* – or simply *manifold* – is a topological manifold, endowed with a differentiable structure.

Notes. 1. If the model space $\mathbb{E}$ of the topological manifold M has dimension n then M is called *n-dimensional*. In this case the hypothesis metrizable can be replaced by Hausdorff.

2. Let (u, M') be a chart of the manifold M. That is for each $p \in M'$, $u(p)$ is an element of $U \subset \mathbb{E}$. If we take an orthonormal Hilbert basis $\{e_i\}$ in $\mathbb{E}$ then u^i, the projection of u on the i-th component, is called *i-th coordinate function.*

1.1.3 Examples. (i) The Hilbert space $\mathbb{E}$ with the atlas $(id, \mathbb{E})$. More generally, every open set $U \subset \mathbb{E}$ with (id, U) as atlas.

(ii) *The sphere of radius* $\varrho > 0$. Put

$$S_\varrho^{\mathbb{E}} = \{(x_0, x) \in \mathbb{R} \times \mathbb{E} = \mathbb{E}', x_0^2 + |x|^2 = \varrho^2\}.$$

Let p_+, p_- be the points on $S_\varrho^{\mathbb{E}}$ with $x_0 = +\varrho$ or $-\varrho$. We define two charts

$$u_\pm : S_\varrho^{\mathbb{E}} - \{p_\pm\} \to \mathbb{E};\, p = (x_0, x) \mapsto \varrho x/(\varrho \mp x_0).$$

Here, $\mathbb{E}$ is the subspace $\{x_0 = 0\}$ of codimension 1 in $\mathbb{E}'$. Since

$$|u_+(x, x_0)|^2 = \varrho^2(\varrho + x_0)/(\varrho - x_0)$$

we find

$$u_- \circ u_+^{-1} : u \in \mathbb{E} - \{0\} \mapsto \varrho^2 u/|u|^2 \in \mathbb{E} - \{0\}.$$

That is, the two charts $(u_+, S_\varrho^{\mathbb{E}} - \{p_+\})$, $(u_-, S_\varrho^{\mathbb{E}} - \{p_-\})$ define a differentiable atlas for $S_\varrho^{\mathbb{E}}$. These charts are called the *stereographic projections from* p_+ and p_-, respectively.

(iii) Any open subset M' of a differentiable manifold M is a differentiable manifold. Indeed, if $(u_\alpha, M_\alpha)_{\alpha \in A}$ is a differentiable atlas for M, its restriction $(u_\alpha | M_\alpha \cap M', M_\alpha \cap M')_{\alpha \in A}$ to M' gives a differentiable atlas for M'.

(iv) As a particular example for (iii) consider for M the *space* $M(n; \mathbb{R})$ *of all real* (n,n)*-matrices*. Take as differentiable atlas the isomorphism with $\mathbb{R}^{n^2}$, where the $(n(i-1)+j)-th$ coordinate of $A \in M(n; \mathbb{R})$ is given by the element a_{ij} in the i-th row and j-th column.

(v) *The general linear group* $GL(n; \mathbb{R})$ is the open subset M' of $M(n; \mathbb{R})$ defined by $\det A \neq 0$. Thus, $GL(n, \mathbb{R})$ is a manifold of dimension n^2.

(vi) If M and M' are manifolds, modelled on $\mathbb{E}$ and $\mathbb{E}'$ respectively, the *product manifold* $M \times M'$, modelled on $\mathbb{E} \times \mathbb{E}'$, is defined by taking as atlas the product atlas $(u_\alpha \times u_{\alpha'}, M_\alpha \times M'_{\alpha'})_{(\alpha, \alpha') \in A \times A'}$.

The structure preserving mappings, briefly morphisms, between differentiable manifolds are the differentiable mappings:

1.1.4 Definition. (i) A mapping $F: M \to N$ from a manifold M into a manifold N is called *differentiable* if F is continuous and if, for some atlas $(u_\alpha, M_\alpha)_{\alpha \in A}$ of M and some atlas $(v_\beta, N_\beta)_{\beta \in B}$ of N, the mappings

$$v_\beta \circ F \circ (u_\alpha | (M_\alpha \cap F^{-1} N_\beta))^{-1} : u_\alpha(M_\alpha \cap F^{-1} N_\beta) \to v_\beta(N_\beta)$$

are differentiable, for all $(\alpha, \beta) \in A \times B$.

(ii) If in particular N is a Hilbert space $\mathbb{F}$ then, using the canonical chart $(id, \mathbb{F})$, $F: M \to \mathbb{F}$ being differentiable means that

$$F \circ u_\alpha^{-1} : u_\alpha(M_\alpha) \to \mathbb{F}$$

is differentiable, for all $\alpha \in A$. In the case $\mathbb{F} = \mathbb{R}$, the real line, we also call $F: M \to \mathbb{R}$ a *differentiable function*.

(iii) If $F: M \to N$ is a homeomorphism such that F as well as F^{-1} is differentiable we also call F a *diffeomorphism*.

We show that this definition does not depend on the atlas.

1.1.5 Proposition. *Whether or not $F: M \to N$ is differentiable does not depend on the particular choices of the atlases on M and N. Moreover, if*

$$F_1: M_1 \to M_2 \quad \textit{and} \quad F_2: M_2 \to M_3$$

are differentiable, the composition

$$F_2 \circ F_1: M_1 \to M_3$$

is also differentiable.

Note. This allows us to speak of the category of differentiable manifolds whose morphisms are the differentiable mappings.

Proof. Let $(u_{\alpha'}, M_{\alpha'})_{\alpha' \in A'}$ and $(v_{\beta'}, N_{\beta'})_{\beta' \in B'}$ be a second pair of atlases for M and N. Then the mapping (with suitably restricted domains of definition)

$$v_{\beta'} \circ F \circ u_{\alpha'}^{-1} = (v_{\beta'} \circ v_{\beta}^{-1}) \circ (v_{\beta} \circ F \circ u_{\alpha}^{-1}) \circ (u_{\alpha'} \circ u_{\alpha}^{-1})^{-1}$$

is differentiable.

As for the claim about the composition $F_2 \circ F_1$ of F_1 and F_2 we reduce its proof to the local representation. Here it reads

$$u_3 \circ F_2 \circ F_1 \circ u_1^{-1} = (u_3 \circ F_2 \circ u_2^{-1}) \circ (u_2 \circ F_1 \circ u_1^{-1})$$

which is obviously true. □

Example. Given a chart (u, M') on $M, u^i: M' \to \mathbb{R}$, the i-th coordinate function, is a differentiable function on the manifold M'. M' is a manifold according to (1.1.3, iii). Actually, its differentiable structure is determined by the atlas with the single chart (u, M').

We add a result on differentiable functions on M.

1.1.6 Proposition. *The set $\mathcal{F}M$ of differentiable functions on a manifold M is an $\mathbb{R}$-algebra under the natural compositions*

$$(f+g)(p) = f(p) + g(p); \; (af)(p) = af(p); \; (fg)(p) = f(p)g(p)$$

with f, g in $\mathcal{F}M$, $a \in \mathbb{R}$.

Moreover, if $F: M \to N$ is a differentiable mapping, we obtain an algebra morphism

$$F^*: \mathcal{F}N \to \mathcal{F}M; \; g \mapsto g \circ F.$$

Proof. Evident from the definitions. □

1.1.7 Definition. Let M, N be manifolds. A *local diffeomorphism at $p \in M$* is a bijective mapping

$$F:(M'(p),p)\to(N'(q),q)$$

of some open neighborhood $M'(p)$ of p onto an open neighborhood $N'(q)$ of the image $q=F(p)$ of p such that F and F^{-1} are differentiable.

Remark. The concept of a diffeomorphism $F\colon M\to N$ between two manifolds clearly defines an equivalence relation. If M and N have the same underlying topological manifold (which we denote for the moment by M_0) then it may happen that an atlas on M is not equivalent to an atlas on N. Nevertheless, there might exist a diffeomorphism $F\colon M\to N$.

Example. $M_0=\mathbb{R}$, $M=(id,\mathbb{R})$; $N=(u,\mathbb{R})$ with $u(x)=x^3$. The two atlases on M_0 are not equivalent. But $F\colon M\to N;\ x\mapsto\sqrt[3]{x}$ is a diffeomorphism.

In this connection one can ask whether there exist several non-isomorphic differentiable structures on a topological manifold or perhaps none at all. As for the first question, Milnor discovered that the 7-sphere admits non-isomorphic differentiable structures. Actually, there are 28 such structures on S^7. See Milnor [1]. After this it soon bedame clear that in general a topological manifold, if it admits a differentiable structure, will also admit more than one, provided the dimension is at least 5. But there also exist topological manifolds which admit no differentiable structure at all. The first example was discovered by Kervaire [1], having dimension 10.

We conclude by describing a procedure for extending a locally defined mapping to a globally defined one.

1.1.8 Lemma. *Let (u,M') be a chart on M such that $U=u(M')$ is a neighborhood of $0\in\mathbb{E}$. Let r be >0 such that $B_r(0)\subset U$, with $B_r(0)=\{x\in\mathbb{E},|x|<r\}$. If we have a differentiable function $f_0\colon B_r(0)\to\mathbb{R}$ and r_1,r_2 with $0<r_1<r_2<r$, then there exists a differentiable $f\colon M\to\mathbb{R}$ such that*

$$f\circ u^{-1}(u)=f_0(u),\quad\text{for } u\in B_{r_1}(0)$$
$$f(q)=0,\qquad\qquad\text{for } q\notin u^{-1}B_{r_2}(0).$$

Proof. Assume we have a differentiable function $\phi\colon\mathbb{R}\to\mathbb{R}$, $0\leqslant\phi\leqslant 1$, with $\phi(x)=1$ for $|x|\leqslant r_1$, $\phi(x)=0$ for $|x|\geqslant r_2$. Then we define $f\circ u^{-1}|B_{r_2}(0)$ by $\phi(|u|)f_0(u)$ and put $f(q)=0$ for the $q\in M$, not in $u^{-1}B_{r_2}(0)$.

To prove the existence of such a ϕ we proceed in three steps:

$$\lambda(x)=\begin{cases}e^{1/(x+r_1)(x+r_2)}, & -r_2<x<-r_1\\ 0 & \text{, otherwise}\end{cases};$$

$$\mu(x)=\int_{-r_2}^{-x}\lambda(y)\,dy\Big/\int_{-r_2}^{-r_1}\lambda(y)\,dy;\qquad \phi(x)=\begin{cases}\mu(x) & ,\ x>0\\ \mu(-x), & x\leqslant 0\end{cases}.\qquad\square$$

1.2 Tensor Bundles

In (1.0) 2 we associated with an open set $U \subset \mathbb{E}$ its tangent bundle $\tau_U: TU \to U$. We want to extend this construction to an arbitrary manifold M modelled on $\mathbb{E}$. Our procedure is to consider first the 'local tangent bundle' $\tau_U: TU = U \times \mathbb{E} \to U$ over the range $U = u(M')$ of a chart (u, M'). If (u', M'') is another chart and $\tau_{U'}$ the tangent bundle of its range $U' = u'(M'')$, then we identify those tangent vectors of $Tu(M' \cap M'') \subset TU$ and $Tu'(M'' \cap M') \subset TU'$ which correspond to each other under the tangential $T(u' \circ u^{-1})$.

After having thus constructed in (1.2.4) the tangent bundle $\tau_M: TM \to M$ we use a similar procedure for the construction of the co-tangent bundle $\tau_M^*: T^*M \to M$.

We complement the construction $M \rightsquigarrow TM$ by associating with $\{F: M \to N\}$ the so-called tangential $\{TF: TM \to TN\}$. This generalizes the local constructions carried out in (1.0) 2. We thus get the tangent functor T, cf. (1.2.7).

Actually, the tangential $TF: TM \to TN$ is special in so far as it respects the projection mappings, i.e., $\tau_N \circ TF = F \circ \tau_M$. This leads us to the important concept of a vector bundle $\pi: P \to M$ over M with fibre $\pi^{-1}(p)$ modelled on a Banach space $\mathbb{F}$. $\tau_M: TM \to M$ is a special case where the fibre is a Hilbert space, i.e., the model space of M. However, applying to Hilbert spaces operations like forming the space of linear mappings, leads us out of the category of Hilbert spaces into the wider category of Banach spaces, cf. (1.0) 1. For that reason we allow as fibres general Banach spaces.

We conclude with the definition of various tensor bundles, associated to a manifold M. The fact that the model $\mathbb{E}$ of M might have infinite dimension implies that we have to be careful in choosing our definitions. Some definitions which are equivalent for $\dim \mathbb{E} < \infty$ cease to be so in the case of infinite dimension, cf. (1.0) 1.

We begin with the definition of the tangent space T_pM of M at a point $p \in M$. T_pM will be a vector space isomorphic (although not canonically isomorphic) to the model Hilbert space $\mathbb{E}$ of M.

1.2.1 Definition. Let M be a differentiable manifold. (i) Let (u, M') be a chart covering the point $p \in M$. Put $u(M') = U \subset \mathbb{E}$. Then $T_{u(p)}U$ is said to be *representative of the tangent space of M at p (given by the chart (u, M')).*

The elements of $T_{u(p)}U$ are also denoted by $(u(p), X_{u(p)})$ with $X_{u(p)} \in \mathbb{E}$, cf. (1.0) 2. $X_{u(p)}$ is called the *principal part* of the vector $(u(p), X_{u(p)})$.

(ii) Let (u, M'), (u', M'') be two charts covering $p \in M$. Put $u(M') = U$, $u'(M'') = U'$. Then the transition mapping

$$u' \circ u^{-1}: u(M' \cap M'') \to u'(M'' \cap M')$$

determines the linear isomorphism

$$T_{u(p)}(u' \circ u^{-1}): T_{u(p)}U \to T_{u'(p)}U'.$$

We say that $(u(p), X_{u(p)})$ and $(u'(p), X_{u'(p)})$ represent the same tangent vector of M at p if the latter is the image of the first under $T_{u(p)}(u' \circ u^{-1})$.

A tangent vector to M at p is defined by the family of its representatives, given by the charts covering p.

(iii) The *tangent space* $T_p M$ is the set of tangent vectors to M at p. $T_p M$ has the structure of a vector space isomorphic to $\mathbb{E}$ by letting the representation mapping $T_p u: T_p M \to T_{u(p)} U$ be a linear isomorphism.

(iv) Using the isomorphism $pr_2: T_{u(p)} U = \{u(p)\} \times \mathbb{E} \to \mathbb{E}$, we put $pr_2 \circ T_p u = Du(p)$ and call this the *differential of u at p*.

Remark. We have to show that the vector space structure of $T_p M$ is independent of the choice of the representation mapping. Indeed, if (u, M') and (u', M'') are charts covering $p \in M$, then

$$T_p u' \circ T_p u^{-1} = T_{u(p)} (u' \circ u^{-1}) : T_{u(p)} U \to T_{u'(p)} U'$$

is a linear isomorphism, given by

$$\big(u(p), X_u\big) \mapsto \big(u'(p), D(u' \circ u^{-1})(u(p)) \,.\, X_u\big).$$

Note: $Du'(p) \circ Du(p)^{-1} = D\big(u' \circ u^{-1}\big)\big(u(p)\big)$.

1.2.2 Definition. Let M be a manifold. By TM we denote the collection of tangent spaces $T_p M, p \in M$. Define $\tau \equiv \tau_M: TM \to M$ by associating with a vector $X \in T_p M$ its so-called *base point* $p \in M$. τ_M is called *projection.*

1.2.3 Proposition. *TM is a differentiable manifold, modelled on $\mathbb{E} \times \mathbb{E}$. The projection $\tau: TM \to M$ is differentiable.*

More precisely, every atlas $(u_\alpha, M_\alpha)_{\alpha \in A}$ of M determines an atlas $(Tu_\alpha, TM_\alpha)_{\alpha \in A}$ of TM as follows: Put $u_\alpha(M_\alpha) = U_\alpha$. Let

$$TM_\alpha = \bigcup_{p \in M_\alpha} T_p M$$

and

$$Tu_\alpha: TM_\alpha \to TU_\alpha = U_\alpha \times \mathbb{E};\ X \mapsto T_{\tau X} u_\alpha(X) = \big(u_\alpha(\tau X), X_{u_\alpha(\tau X)}\big).$$

The differentiable structure thus determined on TM is independent of the choice of the atlas of M.

1.2.4 Definition. The *tangent bundle of a manifold M* is the mapping $\tau: TM \to M$. TM, considered as differentiable manifold, is called *total tangent space of M.*

Proof of 1.2.3. The Tu_α are obviously bijections between subsets of TM and open sets of $\mathbb{E} \times \mathbb{E}$. The transition mappings

$$Tu_\beta \circ Tu_\alpha^{-1} = T(u_\beta \circ u_\alpha^{-1})$$

are differentiable.

The local representative of the projection mapping τ reads

$$u_\alpha \circ \tau \circ Tu_\alpha^{-1}: (u_\alpha, X_{u_\alpha}) \in TU_\alpha \mapsto u_\alpha \in U_\alpha.$$

Thus, τ is differentiable.

Finally, the differentiable structure on TM is independent of the choice of the atlas $(u_\alpha, M_\alpha)_{\alpha\in A}$ of M. If $(u_{\alpha'}, M_{\alpha'})_{\alpha'\in A'}$ is another atlas, equivalent to $(u_\alpha, M_\alpha)_{\alpha\in A}$, then the diffeomorphism $u_{\alpha'} \circ u_\alpha^{-1}$ determines the diffeomorphism $T(u_{\alpha'} \circ u_\alpha^{-1})$, the latter being the mapping $Tu_{\alpha'} \circ Tu_\alpha^{-1}$. □

Besides the tangent bundle $\tau: TM \to M$ of M, we have as another important object the co-tangent bundle $\tau^*: T^*M \to M$ associated to M. For each $p \in M$ let T_p^*M be the dual of T_pM. Denote by T^*M the collection of the T_p^*M, $p \in M$. Define $\tau^*: T^*M \to M$ by associating to a $X^* \in T_p^*M$ its *base point* $p \in M$.

Recall that any isomorphism $F: \mathbb{E} \to \mathbb{E}'$ determines the contragradient (i.e., inverse transposed) isomorphism $F^\vee: \mathbb{E}^* \to \mathbb{E}'^*$ between the duals. Thus, in particular we have from $T_pu: T_pM \to T_{u(p)}U$ the mapping $T_pu^\vee: T_p^*M \to T_{u(p)}^*U$.

1.2.5 Proposition. *T^*M is a differentiable manifold modelled on $\mathbb{E} \times \mathbb{E}^* \cong \mathbb{E} \times \mathbb{E}$. $\tau^*: T^*M \to M$ is differentiable.*

More precisely, every atlas $(u_\alpha, M_\alpha)_{\alpha\in A}$ of M determines an atlas $(T^*u_\alpha, T^*M_\alpha)_{\alpha\in A}$ of T^*M as follows: With $u_\alpha(M_\alpha) = U_\alpha$ put

$$T^*M_\alpha = \bigcup_{p\in M_\alpha} T_p^*M \quad \textit{and}$$

$$T^*u_\alpha: T^*M_\alpha \to T^*U_\alpha = U_\alpha \times \mathbb{E}^*;$$

$$X^* \mapsto (T_{\tau^*X^*}u_\alpha)\ (X^*) = (u_\alpha(\tau^*X^*), X^*_{u_\alpha(\tau^*X^*)})$$

1.2.6 Definition. The *co-tangent bundle* of a manifold M is the mapping $\tau^*: T^*M \to M$. T^*M is called the *total space of the co-tangent bundle*. We also write τ_M^* instead of τ^*.

Proof of 1.2.5. As in the proof of (1.2.3) we see that

$$T^*u_\beta \circ T^*u_\alpha^{-1} = T(u_\beta \circ u_\alpha^{-1})^\vee$$

Thus, the transition mappings are differentiable. The remainder of the proof is like the proof of (1.2.3). □

We want to show that the construction of the total space TM of the tangent bundle can be supplemented by constructing the tangential $TF: TM \to TN$ for a differentiable mapping $F: M \to N$. Here, if $F: M \to M$ is the identity, then $TF: TM \to TM$ will also be the identity. Moreover, if

$$F_1: M_1 \to M_2;\ F_2: M_2 \to M_3$$

are differentiable mappings, and thus

$$F_2 \circ F_1 : M_1 \to M_3$$

is differentiable, then

$$T(F_2 \circ F_1) : TM_1 \to TM_3 \quad \text{and} \quad TF_2 \circ TF_1 : TM_1 \to TM_3$$

are the same.

In the language of categories, this means that T is a covariant functor from the category $\{M, F\}$ of differentiable manifolds M and differentiable mappings F between them into even this category.

Actually, the image $\{TM, TF\}$ is a subcategory, its objects TM having the local product structure given by the bundle charts and its morphisms respecting this local product structure, cf. (1.2.7). In (1.2.8), (1.2.10) we will introduce the category of vector bundles over a manifold with its vector bundle morphisms as the appropriate setting for this structure.

1.2.7 Proposition. *Let $F\colon M \to N$ be a differentiable map. Then we define the* tangential *of F*

(*) $\qquad TF\colon TM \to TN,$

as follows. For each $p \in M$ we define

$$T_pF\colon T_pM \to T_{F(p)}N,$$

the tangential of F at p, *by taking the tangential*

$$T_{u(p)}(v \circ F \circ u^{-1})\colon T_{u(p)}U \to T_{v(F(p))}V$$

of a local representative of F, cf. (1.1.4). *This definition is then independent of the choice of the representation of F. TF is differentiable and commutes with the projections; i.e., the diagram*

$$\begin{array}{ccc} TM & \xrightarrow{\quad TF \quad} & TN \\ \big\downarrow{\scriptstyle \tau_M} & & \big\downarrow{\scriptstyle \tau_N} \\ M & \xrightarrow{\quad F \quad} & N \end{array}$$

is commutative.

Finally, if $F_1\colon M_1 \to M_2$, $F_2\ M_2 \to M_3$ are differentiable, $T(F_2 \circ F_1) = TF \circ TF_1$.

Remarks. 1. If in particular N is a Hilbert space $\mathbb{F}$ we define the *differential of F at $p \in M$* to be the linear mapping

$$DF(p)\colon T_pM \to \mathbb{F}$$

given by composing $T_pF\colon T_pM \to T_{F(p)}\mathbb{F} = \{F(p)\} \times \mathbb{F}$ with the projection $pr_2\colon \{F(p)\} \times \mathbb{F} \to \mathbb{F}$.

2. In (1.2.1, iv) we defined for a chart $u: M' \to u(M') \subset \mathbb{E}$ the differential $Du(p)$: $T_p M' \equiv T_p M \to \mathbb{E}$. This obviously is a special case of the differential $DF(p)$ since u is a differentiable mapping defined on the manifold M'.

Proof. Let $v' \circ F \circ u'^{-1}$ be a second representative of F. Then the identity

$$T(v' \circ F \circ u'^{-1}) = T(v' \circ v^{-1}) \circ T(v \circ F \circ u^{-1}) \circ T(u' \circ u^{-1})^{-1},$$

defined on an appropriately small neighborhood of $u'(p) \in u'(M'')$, shows that $T_p F$ is independent of the choice of the representative.

The differentiability of TF is evident from its local representation

$$Tv \circ TF \circ Tu^{-1} = T(v \circ F \circ u^{-1}).$$

$F \circ \tau_M = \tau_N \circ TF$ is clear from the definitions. Finally, $T(F_2 \circ F_1) = TF_2 \circ TF_1$ follows from the corresponding fact for the local representatives:

$$u_3 \circ F_2 \circ F_1 \circ u_1^{-1} = (u_3 \circ F_2 \circ u_2^{-1}) \circ (u_2 \circ F_1 \circ u_1^{-1})$$

implies

$$T(u_3 \circ F_2 \circ F_1 \circ u_1^{-1}) = T(u_3 \circ F_2 \circ u_2^{-1}) \circ T(u_2 \circ F_1 \circ u_1^{-1}). \quad \square$$

We now come to the concept of a vector bundle over a manifold which generalizes the bundles τ_M and τ_M^*.

1.2.8 Definition. (i) Let M be manifold, modelled on the Hilbert space $\mathbb{E}$, and $\mathbb{F}$ a Banach space. Let there be given a set P and a surjective mapping

$$\pi: P \to M$$

(called a *projection*) such that, for every $p \in M$, $\pi^{-1}(p)$ *is a Banach space isomorphic to* $\mathbb{F}$. We also write $\mathbb{F}_p$ or P_p for $\pi^{-1}(p)$ and call it the *fibre* of π over p.

In addition, let M possess an atlas such that for each chart (u, M') of this atlas there is given a *bundle chart* (Pu, u, M') consisting of a commutative diagram of mappings

$$\begin{array}{ccc} PM' \equiv \pi^{-1} M' & \xrightarrow{\quad Pu \quad} & U \times \mathbb{F} \\ \Big\downarrow \pi | PM' & & \Big\downarrow pr_1 \\ M' & \xrightarrow{\quad u \quad} & U \end{array} \quad ; \quad \begin{array}{ccc} \xi & \longmapsto & (u(p), \xi_u) \\ \Big\downarrow & & \Big\downarrow \\ p & \longmapsto & u(p) \end{array}$$

with the following properties:

(a) Pu is bijective and $P_p u \equiv Pu | \mathbb{F}_p$: $\mathbb{F}_p \to \{u(p)\} \times \mathbb{F}$ is a Banach space isomorphism. Here the Banach space structure on $\{u(p)\} \times \mathbb{F}$ is defined by the canonical identification pr_2: $\{u(p)\} \times \mathbb{F} \to \mathbb{F}$.

(b) If (Pu, u, M'), (Pu', u', M'') are two bundle charts then

$$(*) \qquad Pu' \circ (Pu \,|\, P(M' \cap M''))^{-1} : u(M' \cap M'') \times \mathbb{F} \to u'(M'' \cap M') \times \mathbb{F}$$

is a diffeomorphism. We also denote this mapping briefly by $Pu' \circ Pu^{-1}$.

The object thus defined is called a *vector bundle* (over M) *with bundle atlas*.

(ii) Two bundle atlases $(Pu_\alpha, u_\alpha, M_\alpha)_{\alpha \in A}$, $(Pu_{\alpha'}, u_{\alpha'}, M_{\alpha'})_{\alpha' \in A'}$

of the same $\pi: P \to M$, associated with two equivalent atlases $(u_\alpha, M_\alpha)_{\alpha \in A}$, $(u_{\alpha'}, M_{\alpha'})_{\alpha' \in A'}$ are called *equivalent* if their union is a bundle atlas associated with the union of $(u_\alpha, M_\alpha)_{\alpha \in A}$ and $(u_{\alpha'}, M_{\alpha'})_{\alpha' \in A'}$.

An equivalence class of vector bundles over M with atlases is called a *vector bundle over M with fibre modelled on* $\mathbb{F}$. If $\pi: P \to M$, then P is called the *total space of* π and M the *base space of* π.

Remark. Some authors find it preferable to introduce the bundle of frames. At least in the case of finite dimensional manifolds, cf. Kobayashi and Nomizu [1], Sulanke und Wintgen [1], Bishop and Crittenden [1]. For our purposes, it seems more appropriate to restrict ourselves to vector bundles.

Note. The fibre $\mathbb{F}_p = \pi^{-1}(p)$ over p is isomorphic to $\mathbb{F}$, but there is no canonical isomorphism. Rather, for each bundle chart (Pu, u, M') covering $\mathbb{F}_p$, we obtain the isomorphism

$$Pu(p) = pr_2 \circ P_p u : \mathbb{F}_p \to \mathbb{F}.$$

Here we have introduced the notation $Pu(p)$ in analogy to the differential $Du(p)$ for the tangent bundle.

Let (Pu, u, M'), (Pu', u', M'') be two bundle charts. Then we have the transition mapping $Pu' \circ Pu^{-1}$ introduced under (b) above. Let us define, for $p \in M' \cap M''$, the isomorphism

$$P(u' \circ u^{-1})(u(p)) = Pu'(p) \circ Pu(p)^{-1} : \mathbb{F} \to \mathbb{F}.$$

$Pu' \circ Pu^{-1}$ may then be written as

$$(u, \xi_u) \mapsto ((u' \circ u^{-1})(u), P(u' \circ u^{-1})(u).\, \xi_u).$$

Since the transition mapping is differentiable so is the mapping

$$u \in u(M' \cap M'') \mapsto P(u' \circ u^{-1})(u) \in GL(\mathbb{F}).$$

Example. The tangent bundle $\tau_M : TM \to M$ of a manifold M is a vector bundle over M with fibre modelled on $\mathbb{E}$, the model for M. The atlas $(Tu_\alpha, TM_\alpha)_{\alpha \in A}$ of TM, associated with an atlas $(u_\alpha, M_\alpha)_{\alpha \in A}$ of M, cf. (1.2.3), is a bundle atlas; as such it is denoted by $(Tu_\alpha, u_\alpha, M_\alpha)_{\alpha \in A}$.

From (1.2.8) we immediately obtain the

1.2.9 Proposition. *Let $\pi: P \to M$ be a vector bundle over M with fibre modelled on $\mathbb{F}$.*

Let $\mathbb{E}$ *be the model of* M. *Then the bundle atlas* $(Pu_\alpha, u_\alpha, M_\alpha)_{\alpha \in A}$ *determines an atlas* $(Pu_\alpha, PM_\alpha)_{\alpha \in A}$ *of the total space* P *making it into a manifold modelled on* $\mathbb{E} \times \mathbb{F}$.

Proof. Clear from the definition. □

Remark. If $\mathbb{F}$ is not a Hilbert space, then we have used a slight generalization of the concept of a differentiable manifold: Instead of the model space being a Hilbert space as in (1.1.1), in our case it is a Banach space. However, this is not the full picture if we observe that in our case the transition mappings, determined by the overlap of two bundle charts (Pu, u, M'), (Pu', u', M''), have the special feature of being linear isomorphisms on the second factor $\mathbb{F}$. More precisely, the transition mapping $Pu' \circ Pu^{-1}$ may be written in the form $u' \circ u^{-1} \times P(u' \circ u^{-1})$: $u(M' \cap M'') \times \mathbb{F} \to u'(M'' \cap M') \times \mathbb{F}$ where

$$P(u' \circ u^{-1}): u(M' \cap M'') \to GL(\mathbb{F})$$

is a differentiable mapping into the group $GL(\mathbb{F})$ of linear isomorphisms. In this way, the transition mapping preserves the product structure of the model $\mathbb{E} \times \mathbb{F}$, i.e., we obtain the commutative diagram

$$\begin{array}{ccc} u(M' \cap M'') \times \mathbb{F} & \xrightarrow{Pu' \circ Pu^{-1}} & u'(M'' \cap M') \times \mathbb{F} \\ \downarrow{\scriptstyle pr_1} & & \downarrow{\scriptstyle pr_1} \\ u(M' \cap M'') & \xrightarrow{u' \circ u^{-1}} & u'(M'' \cap M') \end{array}$$

We complement the definition of vector bundles over manifolds by introducing the corresponding structure preserving mappings or, briefly, morphisms. This yields the *category of vector bundles over manifolds*. Cf. Lang [1] and (in the case of finite dimensions) Sulanke und Wintgen [1].

1.2.10 Definition. Let

$$\pi: P \to M; \pi^*: P^* \to M^*$$

be vector bundles. Denote by $\mathbb{F}$, $\mathbb{F}^*$ the model spaces for the fibres of π and π^*, respectively. A *morphism* from π into π^* consists of a pair of differentiable maps $F: M \to M^*$; $PF: P \to P^*$ such as to make the diagram

$$\begin{array}{ccc} P & \xrightarrow{PF} & P^* \\ \downarrow{\scriptstyle \pi} & & \downarrow{\scriptstyle \pi^*} \\ M & \xrightarrow{F} & M^* \end{array}$$

commutative. Moreover, the restriction $P_p F$ of PF to the fibre $\mathbb{F}_p$ over p shall be a linear mapping

$$P_p F\colon \mathbb{F}_p \to \mathbb{F}^*_{F(p)}$$

where $\mathbb{F}_p$, $\mathbb{F}^*_{F(p)} = \pi^{*-1}(F(p))$ are endowed with their Banach space structure.

Remark. Let (Pu, u, M'), $(Pu', u^*, M^{*\prime})$ be bundle charts for π and π^*. Then the above diagram possesses the local representation

$$\begin{array}{ccc}
U \times \mathbb{F} & \xrightarrow{\;Pu^* \circ PF \circ Pu^{-1}\;} & U^* \times \mathbb{F}^* \\
\Big\downarrow{\scriptstyle pr_1} & & \Big\downarrow{\scriptstyle pr_1} \\
U & \xrightarrow{\;u^* \circ F \circ u^{-1}\;} & U^*
\end{array}$$

Here we have written U for $u(M' \cap F^{-1} M^{*\prime})$ and U^* for $u^*(M^{*\prime})$.
From $Pu(p)\colon \mathbb{F}_p \to \mathbb{F}$; $P_p F\colon \mathbb{F}_p \to \mathbb{F}^*_{F(p)}$; $Pu^*(F(p))\colon \mathbb{F}^*_{F(p)} \to \mathbb{F}^*$ we form

$$P(u^* \circ F \circ u^{-1})(u(p)) \equiv Pu^*(F(p)) \circ P_p F \circ Pu(p)^{-1}\colon \mathbb{F} \to \mathbb{F}^*.$$

Then

$$u \in U \mapsto P(u^* \circ F \circ u^{-1})(u) \in L(\mathbb{F}; \mathbb{F}^*)$$

is differentiable. Actually, this characterizes the differentiability of the mapping $Pu^* \circ PF \circ Pu^{-1}$ in the above diagram.

For the case of the tangent bundle we see that the tangential $TF\colon TM \to TN$ of a morphism $F\colon M \to N$ is a vector bundle morphism, cf. (1.2.7).

In (1.0) 1 we have constructed from Banach spaces $\mathbb{F}_1, \ldots, \mathbb{F}_r$; $\mathbb{G}$ the Banach space $L(\mathbb{F}_1, \ldots \mathbb{F}_r; \mathbb{G})$. We complement this by observing that this construction yields a canonical inclusion

$$(*) \qquad L(\;,\ldots,\;;\;)\colon GL(\mathbb{F}_1) \times \ldots \times GL(\mathbb{F}_r) \times GL(\mathbb{G}) \to GL(L(\mathbb{F}_1, \ldots, \mathbb{F}_r; \mathbb{G}))$$

as follows: If $X \in L(\mathbb{F}_1, \ldots \mathbb{F}_r; \mathbb{G})$, $A_i \in GL(\mathbb{F}_i)$, $B \in GL(\mathbb{G})$, then $L(A_1, \ldots, A_r; B)(X)$ shall be the element $B \circ X \circ (A_1^{-1} \times \ldots \times A_r^{-1})$. Compare this with the special case T_s^r in (1.0) 1.

1.2.11 Definition. Let $\alpha_i\colon A_i \to M$, and $\beta\colon B \to M$ be vector bundles over M, where $\mathbb{F}_i$ and $\mathbb{G}$ are the model spaces for the fibres of α_i and β, respectively, $i = 1, \ldots, r$. We may assume – by taking, if necessary, intersections of charts – that we have an atlas for M such that for each chart (u, M') of that atlas we simultanously have for all bundles α_i, β bundle charts $(A_i u, u, M')$, (Bu, u, M').

For each $p \in M$ we define

$$L(\mathbb{F}_{1,p}, \ldots, \mathbb{F}_{r,p}; \mathbb{G}_p) = \text{(briefly) } \mathbb{H}_p.$$

By $L(A_1, \dots, A_r; B) =$ (briefly) C we denote the union of the $\mathbb{H}_p$, $p \in M$. Let

$$L(\alpha_1, \dots, \alpha_r; \beta): L(A_1, \dots, A_r; B) \to M$$

or briefly

$$\gamma: C \to M$$

be the projection mapping defined by $\gamma^{-1}(p) = \mathbb{H}_p$.

For each chart (u, M') of M we define the bundle chart (Cu, u, M') as follows. For each $p \in M'$ let $Cu(p): \mathbb{H}_p \to \mathbb{H}$ be given by

$$L(A_1 u(p), \dots, A_r u(p); Bu(p): L(\mathbb{F}_{1,p}, \dots, \mathbb{F}_{r,p}; \mathbb{G}_p) \to L(\mathbb{F}_1, \dots, \mathbb{F}_r; \mathbb{G});$$
$$X_p \mapsto Bu(p) \circ X_p \circ (A_1 u(p)^{-1} \times \dots \times A_r u(p)^{-1}).$$

1.2.12 Lemma.

$$L(\alpha_1, \dots, \alpha_r; \beta): L(A_1, \dots, A_r; B) \to M$$

is a vector bundle over M, with the bundle charts defined in (1.2.11).

Proof. We only need to show that the transition mappings

$$Cu' \circ Cu^{-1}: u(M' \cap M'') \times \mathbb{H} \to u'(M'' \cap M') \times \mathbb{H}$$

for the above defined charts are differentiable. Observe that we may write $Cu' \circ Cu^{-1} \equiv u' \circ u^{-1} \times C(u' \circ u^{-1})$ with

$$C(u' \circ u^{-1}) = L\big(A_1(u' \circ u^{-1}), \dots, A_r(u' \circ u^{-1}); B(u' \circ u^{-1})\big).$$

Since the inclusion $(*)$ $L(\ ,\dots,\ ;\ )$ before (1.2.11) is differentiable, we see that

$$u \in u(M' \cap M'') \mapsto C(u' \circ u^{-1}) \in GL(\mathbb{H})$$

is differentiable. □

1.2.13 Definition. *The bundle of r-fold contravariant and s-fold covariant tensors over M* is defined by

$$L(\underbrace{\tau^*, \dots, \tau^*}_{r}, \underbrace{\tau, \dots, \tau}_{s}; \varrho): L(T^*M, \dots, T^*M, TM, \dots, TM; \mathbb{R}M) \to M$$

or briefly

$$\tau_s^r: T_s^r M \to M.$$

Here, $\varrho: \mathbb{R}M \equiv M \times \mathbb{R} \to M$ denotes the *trivial* $\mathbb{R}$*-bundle* over M.

Remarks. 1. One also denotes by the same name any of the $(r+s)!/r!s!$ bundles

$$L(\tau_1, \dots, \tau_{r+s}; \varrho): L(T_1 M, \dots, T_{r+s} M; \mathbb{R}M) \to M$$

where r of the τ_i are equal to τ^* and s of the τ_i are equal to τ. All these bundles are canonically isomorphic to each other. Moreover, since $L(\mathbb{E}^*; \mathbb{R}) \cong \mathbb{E}$, we have

$$L(\underbrace{\mathbb{E}^*, \ldots, \mathbb{E}^*}_{r}, \underbrace{\mathbb{E}, \ldots, \mathbb{E}}_{s}; \mathbb{R}) \cong L(\underbrace{\mathbb{E}^*, \ldots, \mathbb{E}^*}_{r-1}, \underbrace{\mathbb{E}, \ldots, \mathbb{E}}_{s}; \mathbb{E}).$$

Therefore we also denote by τ_s^r, $r > 0$, the bundle $L(\underbrace{\tau^*, \ldots \tau^*}_{r-1}, \underbrace{\tau, \ldots \tau}_{s}; \tau)$.

2. Since $L(\mathbb{E}^*; \mathbb{R}) = \mathbb{E}$ and $L(\mathbb{E}; \mathbb{R}) = \mathbb{E}^*$ we see that $\tau_0^1 = \tau$ and $\tau_1^0 = \tau^*$. We also define τ_0^0 to be the trivial bundle.

3. The transition functions for τ_s^r,

$$T_s^r u' \circ (T_s^r u)^{-1}: u(M' \cap M'') \times L(\mathbb{E}^*, \ldots, \mathbb{E}; \mathbb{R}) \to \\ u'(M'' \cap M') \times L(\mathbb{E}^*, \ldots, \mathbb{E}; \mathbb{R}),$$

determine for each $u \in u(M' \times M'')$ the linear mapping

$$T_s^r(u' \circ u^{-1}): T_s^r \mathbb{E} \to T_s^r \mathbb{E} \text{ with}$$

$$X_s^r \mapsto X_s^r \circ (\underbrace{{}^t D(u' \circ u^{-1}) \times \ldots \times {}^t D(u' \circ u^{-1})}_{r} \times \underbrace{D(u' \circ u^{-1})^{-1} \times \ldots \times D(u' \circ u^{-1})^{-1}}_{s})$$

In (1.0) 1 we defined the concept of a general tensor space as a subspace of $T_s^r \mathbb{E}$ which is invariant under $T_s^r F$, $F \in GL(\mathbb{E})$. The construction of $\tau_s^r: T_s^r M \to M$ shows that for each such general tensor space we obtain a subbundle for τ_s^r. In particular, we get from the s-fold covariant symmetric {antisymmetric} spaces $S_s \mathbb{E} \{A_s \mathbb{E}\}$ the corresponding bundles.

1.2.14 Definition. The *bundle of s-fold covariant symmetric {antisymmetric} tensors,*

$$\sigma_r: S_r M \to M \ \{\alpha_r: A_r M \to M\},$$

is the bundle obtained from restricting the bundle τ_r^0 to the subspace $S_r T_p M \{A_r T_p M\}$.

1.2.15 Remark: We have constructed various tensor bundles from the tangent bundle $\tau_M: TM \to M$ over M, Most of these constructions can also be carried out with a general vector bundle $\pi: P \to M$ over M. At the basis of these constructions is Lemma (1.2.12). For details see Bourbaki [2], § 7.

Here we mention only the bundle

$$S_2(\pi): S_2 P \to M$$

of symmetric bilinear forms with fibre isomorphic to $S_2 \mathbb{F}$, if the fibres of π are isomorphic to $\mathbb{F}$.

1.2.16 Remark. We conclude with an alternative description of the tensor bundle $\tau_s^r: T_s^r M \to M$ in the case where M is modelled on a space $\mathbb{E}$ of finite dimension.

From (1.0.1) we then have that

$$T_s^r \mathbb{E} = \underbrace{\mathbb{E} \otimes \ldots \otimes \mathbb{E}}_{r} \otimes \underbrace{\mathbb{E}^* \otimes \ldots \otimes \mathbb{E}^*}_{s}.$$

An $F \in GL(\mathbb{E})$ determines

$$T_s^r F = F \otimes \ldots \otimes F \otimes F^\vee \otimes \ldots \otimes F^\vee \in GL(T_s^r \mathbb{E})$$

where $F^\vee = {}^tF^{-1}$ is the contragradient of F. With this we put

$$T_{s,p}^r M = \underbrace{T_p M \otimes \ldots \otimes T_p M}_{r} \otimes \underbrace{T_p{}^* M \otimes \ldots \otimes T_p{}^* M}_{s}.$$

$T_s^r M$ is defined as the union of the $T_{s,p}^r M$, $p \in M$, and τ_s^r: $T_s^r M \to M$ is given by $(\tau_s^r)^{-1}(p) = T_{s,p}^r M$. Finally, from an atlas $(u_\alpha, M_\alpha)_{\alpha \in A}$ of M we construct the bundle atlas $(T_s^r u_\alpha, u_\alpha, M_\alpha)_{\alpha \in A}$ with

$$T_s^r u_\alpha \colon (\tau_s^r)^{-1} M_\alpha \to u_\alpha(M_\alpha) \times T_s^r \mathbb{E}$$

given by

$$X_{s,p}^r \in T_{s,p}^r M \mapsto \left(u_\alpha(p), T_s^r(T_p u_\alpha)(X_{s,p}^r)\right).$$

1.3 Immersions and Submersions

In (1.0) 3 we characterized locally injective (surjective) differentiable mappings by their tangential at a point. Carrying this over to manifolds we obtain local immersions and submersions, cf. (1.3.3). Also of interest are the global versions of this property. In particular, we have the concept of a submanifold which appeared in the very beginning of Differential Geometry when Gauss developed the theory of surfaces in 3-dimensional Euklidian space $\mathbb{R}^3$.

Submanifolds possess a local characterization, cf. (1.3.5). They occur as counter image of submanifolds for a transversal (regular) differentiable mapping $F\colon M \to N$, cf. (1.3.6), (1.3.8). Quadratic hypersurfaces and groups of matrices can, in this way, be seen to have a natural structure of a differentiable manifold, cf. (1.3.9).

In (1.3.10) we introduce the important concept of an induced bundle. A first example is the normal bundle of an immersion, see (1.3.11). We conclude with a look at the structure of the tangent bundle of the total space of a vector bundle $\pi: P \to M$ over a manifold M. Here we point out that each tangent space $T_\xi P$ contains a distinguished subspace, called the vertical space, i. e., the tangent space to the fibre containing ξ.

We begin by reformulating the inverse mapping theorem (1.0.3).

1.3.1 Lemma. *Let $F\colon M \to N$ be differentiable. If, for $p \in M$, $T_p F$ is a linear bijection from $T_p M$ onto $T_q N$, $q = F(p)$, then F is a local diffeomorphism.* □

1.3.2 Definition. *Let* $F: M \to N$ be differentiable.

(i) F is called an *immersion at* $p \in M$ if $T_p F$ is injective and closed. It is called an *immersion* if $T_p F$ is injective for all $p \in M$.

(ii) F is called a *submersion at* p if $T_p F$ is surjective. If $T_p F$ is surjective for all $p \in M$, then F is called a *submersion*.

With this notation the corollaries (1.0.4) and (1.0.5) read:

1.3.3 Theorem. *Let* $F: M \to N$ *be differentiable. Consider* $p \in M$ *and put* $F(p) = q$. *If* F *is an immersion* {*submersion*} *at* p, *then there exist charts* (u, M') *around* p *and* (v, N') *around* q *such that*

$$v \circ F \circ u^{-1}: U = u(M') \to V = v(N') = U \times V_2$$

is given by

$$u \mapsto (u, 0)$$

$$\{v \circ F \circ u^{-1}: U = u(M') = U_1 \times V \to V = v(N')$$

is given by $$u = (u_1, v) \mapsto v\}. \quad \square$$

Among the immersions, the following class is particularly important:

1.3.4 Definition. (i) $F: M \to N$ is called an *embedding* if it is an immersion and if $F: M \to F(M)$ is a homeomorphism, where $F(M) \subset N$ is endowed with the induced topology.

(ii) If a subset M of a manifold N can be given the structure of a differentiable manifold such that the inclusion $i: M \to N$ is an embedding, then M is called a *submanifold of* N.

For submanifolds we have the following characterization. Note that here our notation differs from the one used in (1.3.3), (1.3.4).

1.3.5 Theorem. *A subset* N *of a manifold* M *is a submanifold if and only if the following holds:*

Let $\mathbb{E}$ *be the model of* M. *There exists a splitting* $\mathbb{E} = \mathbb{E}_1 \times \mathbb{E}_2$ *and, for every* $p \in N$, *there exists a chart* (u, M') *of* M *around* p *of the form*

$$u: (M', p) \to (U_1 \times U_2, (0,0)).$$

Here, U_j *is an open neighborhood of* $0 \in \mathbb{E}_j, j = 1,2,$ *and* $u|(M' \cap N)$ *is given by*

$$u|(M' \cap N): M' \cap N \to U_1 \times \{0\}.$$

Proof. Let N be submanifold, $i: N \to M$ the inclusion. According to (1.3.3) there exists around $p \in N$ a chart (u, M') such that $U = u(M') = U_1 \times U_2$, $u \circ i \circ (u|M' \cap N)^{-1}: U_1 \to U_1 \times \{0\}$ a linear injection. Conversely, if we have around each $p \in N$ such a chart, then i is an immersion. The topology of $N \subset M$ is generated by the open subsets of $M' \cap N$ which are homeomorphic to the open subsets $U_1 \subset \mathbb{E}_1$ via $u|M' \cap N$. □

1.3.6 Definition. Let $F\colon M \to N$ be differentiable. Let N_1 be a submanifold of N. F is called *transversal to* N_1 *at* p if either $q = F(p) \notin N_1$ or else $T_q N$ is generated by $T_q N_1$ and $T_p F(T_p M)$. Notation: $F \pitchfork_p N_1$.

If $F \pitchfork_p N_1$ for all $p \in M$, then F is called *transversal to* N_1. Notation: $F \pitchfork N_1$.

Example. $N_1 = \{q\}$. Then $F \pitchfork N_1$ is equivalent to saying that, for every p with $F(p) = q$, $T_p F$ is surjective. In this case q is called a *regular value of* F.

Remark. The set of mappings $F\colon M \to N$, transversal to a given submanifold N_1 of N, plays an important rôle since they are "typical" in the family of all mappings F from M to N. Intuitively, this is clear: If F is not transversal to N_1, arbitrary small modifications of F will suffice to make F transversal. Moreover, if F is transversal, sufficiently small perturbations of F will not destroy this property. One therefore calls the property $F \pitchfork N_1$ *generic.*

We shall not enter into the details of this theory. It was created by R. Thom and has found many applications, see Abraham and Robbin [1], Thom [1].

The transversality at a point possesses the following characterization:

1.3.7 Proposition. *Let* N_1 *be a submanifold of* N *and* $F\colon M \to N$ *be a morphism. Assume* $q = F(p) \in N_1$.

Then $F \pitchfork_p N_1$ *is equivalent to the existence of charts* (u, M') *around* p *and* (v, N') *around* q *of the following form: The model* $\mathbb{E}$ *of* M *splits into* $\mathbb{E}_1 \times \mathbb{F}_2$ *and* $u(M') = U = U_1 \times V_2 \subset \mathbb{E}_1 \times \mathbb{F}_2$ *with* $u(p) = (0,0)$. *The chart* (v, N') *is adapted to the submanifold* N_1 *in the point* q, *in the sense of* (1.3.5). *That is, the model* $\mathbb{F}$ *of* N *splits into* $\mathbb{F}_1 \times \mathbb{F}_2$ *and* $v(N') = V = V_1 \times V_2 \subset \mathbb{F}_1 \times \mathbb{F}_2$ *with* $v(q) = (0,0)$, $v(N' \cap N_1) = V_1 \times \{0\}$. *And finally, the mapping* $F|M'$ *has its image in* N' *and possesses the representation*

$$v \circ F \circ u^{-1}\colon U_1 \times V_2 \to V_1 \times V_2;\ (u_1, v_2) \mapsto (a(u_1, v_2), v_2)$$

with $a(0,0) = 0$.

Proof. Obviously, the local representation implies $F \pitchfork_p N_1$. Conversely, with (v, N') adapted to N_1 as above and (u, M') such that $u(p) = 0$, $F(M') \subset N'$, we can consider the mappings

$$U = u(M') \xrightarrow{u^{-1}} M' \xrightarrow{F} N' \xrightarrow{v} V_1 \times V_2 \xrightarrow{pr_2} V_2.$$

$F \pitchfork_p N_1$ means that the differential at $0 \in U$ of the composition of these mappings is surjective. According to (1.3.3, ii) we can therefore choose u such that $u(M') = U_1 \times V_2$ and the composition is given by the projection on the second factor. But this means that $v \circ F \circ u^{-1}$ is of the form given above. □

1.3.8 Theorem. *Let N_1 be a submanifold of N and $F: M \to N$ be transversal to N_1. If $M_1 = F^{-1}(N_1) \neq \emptyset$, M_1 is a submanifold with* $\operatorname{codim} M_1 = \operatorname{codim} N_1$. $T_p M_1 = T_p F^{-1}(T_q N_1)$, *for $p = F(q)$.*

Remark. $\operatorname{codim} M_1 = \operatorname{codim} N_1$ means that in the splitting $\mathbb{E} = \mathbb{E}_1 \times \mathbb{E}_2$ of the model of M and the splitting $\mathbb{F} = \mathbb{F}_1 \times \mathbb{F}_2$ in the model of $\mathbb{F}$, associated to the submanifolds $M_1 \subset M$ and $N_1 \subset N$ according to (1.3.5), $\mathbb{E}_2$ is isomorphic to $\mathbb{F}_2$.

Proof. We have from (1.3.7): For $p \in M_1$ there exists a chart (u, M') with $u: (M', p) \to (U_1 \times V_2, (0,0))$ such that $u(M' \cap F^{-1}(N_1)) = u \circ F^{-1} \circ v^{-1}(V_1 \times \{0\}) = U_1 \times \{0\}$. That is, we can apply (1.3.5). □

1.3.9 Examples. (i) Let q be a regular value for $F: M \to N, F^{-1}(q) \neq \emptyset$. Then $M_1 = F^{-1}(q)$ is a submanifold, $\operatorname{codim} M_1 = \dim N$. $T_p M_1 = \ker T_p F$.

For instance, consider

$$F: (x_0, x) \in \mathbb{R} \times \mathbb{E} = \mathbb{E}' \mapsto x_0^2 + |x|^2 \in \mathbb{R}.$$

Every $\varrho^2 > 0$ is a regular value. $F^{-1}(\varrho^2) = S_\varrho^{\mathbb{E}}$.

More generally, denote by

$$F: x \in \mathbb{E}' \mapsto Q(x)$$

a non-degenerate quadratic form. If $x^* \in \mathbb{E}'$ is a such that $Q(x^*) \neq 0$ then $Q(x^*)$ is a regular value for F, since for $x \in F^{-1}(Q(x^*))$

$$T_x F . x = 2Q(x) = 2Q(x^*) \neq 0.$$

$F^{-1}(Q(x^*))$ is called a *quadratic hypersurface* in $\mathbb{E}'$.

(ii) $M = M(n; \mathbb{R})$, the space of all real (n, n)-matrices, cf. (1.1.3, iv). $N = S(n, \mathbb{R})$, the submanifold of all *symmetric matrices* C, i. e., $C = {}^tC$. Consider

$$F: M \to N; \; A \mapsto A\,{}^tA.$$

We claim that the unit matrix E is a regular value.

To see this we use the canonical coordinates $u_{ik}(A) = a_{ik}$ of $A \in S(n, \mathbb{R})$, considered as a linear subspace of $M(n; \mathbb{R}) \cong \mathbb{R}^{n^2}$. Then $u_{ik}(F(A)) = \sum_j u_{ij}(A) u_{jk}({}^tA) = \sum_j a_{ij} a_{kj}$.

We determine the differential $Du_{ik}(F(A))$ applied to the principal part B_u of a representative tangent vector with coordinates b_{ij}:

$$Du_{ik}(F(A)).B_u = \sum_{r,s} \frac{\partial}{\partial a_{rs}} \Big(\sum_j a_{ij} a_{kj}\Big) b_{rs} =$$

$$\sum_j b_{ij} a_{kj} + \sum_j a_{ij} b_{kj},$$

i. e., $\quad T_A F . B = B\,{}^tA + A\,{}^tB.$

If $A \in F^{-1}(E)$, $T_A F$ is surjective. Indeed every $C = {}^tC \in T_E N \cong S(n, \mathbb{R})$ can be written as $B\,{}^tA + {}^tA\,B$ with $B = CA/2$.

Thus, the *orthogonal group* $\mathbb{O}(n) = \{A \in M(n; \mathbb{R});\, A\,{}^tA = E\}$ is a submanifold of $M(n, \mathbb{R})$ of dimension $n^2 - \dim S(n, \mathbb{R}) = n(n-1)/2$. Its tangent space at the neutral element E is the kernel of $T_E F$,

$$T_E \mathbb{O}(n) = \{B \in M(n; \mathbb{R});\, B + {}^tB = 0\}.$$

(iii) The inclusion

$$i: M \to TM;\, p \mapsto 0_p = \text{Nullvector in } T_p M$$

is an embedding. If we identify M with its image under i, M becomes a submanifold. This can be seen from (1.3.5), since in the chart

$$Tu: TM' \to TU \cong U \times \mathbb{E}$$

$Tu|(TM' \cap M)$ is given by $TM' \to U \times \{0\}$.

(iv) For each $p \in M$, $T_p M$ is a submanifold of TM. Indeed, $\tau: TM \to M$ has $p \in M$ as a regular value.

(v) The *diagonal* $\Delta = \Delta_M = \{(p, q) \in M \times M;\, p = q\}$ is a submanifold of $M \times M$ of dimension = codimension = dim $\mathbb{E}$. To see this consider an atlas $(u_\alpha, M_\alpha)_{\alpha \in A}$ of M. It determines an atlas, the product atlas, for $M \times M$. The subfamily of this atlas consisting of $(u_\alpha \times u_\alpha, M_\alpha \times M_\alpha)_{\alpha \in A}$ covers Δ. For each α, $u_\alpha|(M_\alpha \times M_\alpha) \cap \Delta_M$ has as image the linear subspace $\Delta_{u_\alpha} = \{(u_\alpha, u_\alpha) \in U_\alpha \times U_\alpha\}$ of $U_\alpha \times U_\gamma$.

Vector bundles over a manifold M occur naturally when we start with a morphism $F: M \to N$, where N is the base space N of a vector bundle $\pi: P \to N$. In the case where π is the product bundle $pr_1: N \times \mathbb{F} \to N$, this is the left hand side of the diagram

$$\begin{array}{ccc} M \times \mathbb{F} & \xrightarrow{(F \times id)} & N \times \mathbb{F} \\ \downarrow{\scriptstyle pr_1} & & \downarrow{\scriptstyle pr_1} \\ M & \xrightarrow{F} & N \end{array}$$

The full diagram gives the natural bundle morphism between the new and the old bundle.

1.3.10 Lemma. *Let $\pi: P \to N$ be a vector bundle over N with fibre modelled on $\mathbb{F}$. Let $F: M \to N$ be a differentiable map. Then the* vector bundle over M induced by F,

$$F^*\pi: F^*P \to M,$$

is defined as follows: Consider the morphism

$$(*) \qquad F \times \pi: M \times P \to N \times N.$$

Take for F^*P the counter image of the diagonal Δ_N of $N \times N$. The fibre $F^*P_p = (F^*\pi)^{-1}(p)$ over $p \in M$ shall be the subset $\{p\} \times P_{F(p)} \equiv \{p\} \times \mathbb{F}_p$ of F^*P, belong-

ing to the counter image of $(F(p), F(p)) \in \Delta_N$. $\{p\} \times P_{F(p)}$ *gets the structure of a Banach space, isomorphic to* $\mathbb{F}$, *by the canonical identification of this set with* $P_{F(p)} = \pi^{-1}(F(p))$. *One obtains a canonical bundle morphism*

$$\begin{array}{ccc} F^*P & \xrightarrow{\quad \pi^*F \quad} & P \\ \Big\downarrow{\scriptstyle F^*\pi} & & \Big\downarrow{\scriptstyle \pi} \\ M & \xrightarrow{\quad F \quad} & N \end{array}$$

by mapping the fibre $\{p\} \times P_{F(p)}$ *into* $P_{F(p)} \equiv \mathbb{F}_{F(p)}$.

Note. The mapping (*) is transversal to the submanifold Δ_N of $N \times N$. Indeed, $T_{(q,q)}\Delta_N$ consists of the pairs $(Y, Y) \in T_qN \times T_qN$. Let $(p, Y) \in M \times P$ with $(F(p), \pi Y) = (q, q)$. Since $\pi: P \to N$ is surjective, $T_{(p,Y)}(F \times \pi): T_{(p,Y)}(M \times P) \to T_{(q,q)}(N \times N)$ contains the subspace $\{0_q\} \times T_qN$ wich is complementary to $T_{(q,q)}\Delta_N$ in $T_{(q,q)}(N \times N)$. Hence, according to (1.3.8), $F^*P = (F \times \pi)^{-1}\Delta_N$ is a submanifold of $M \times P$. Here we have tacitly assumed that (1.3.8) is true also if P is not modelled on a Hilbert manifold. Otherwise, we would have to restrict ourselves to the case when the fibre model $\mathbb{F}$ of π is a Hilbert space.

Proof. All that has to be shown is the existence of a bundle atlas for $F^*\pi$ and the existence of the bundle morphism. This latter needs to be verified only for a pair of bundle charts.

Now, let (Pv, v, N') be a bundle chart for π. Let (u, M') be a chart for M with $F(M') \subset N'$. Then we define the bundle chart

$$F^*Pu: (F^*\pi)^{-1}(M') \to u(M') \times \mathbb{F}$$

by $(p, Y) \mapsto (u(p), Pv(F(p)).Y)$. Hence,

$$F^*Pu|F^*P_p: F^*P_p \to \{u(p)\} \times \mathbb{F}$$

is a linear isomorphism which we also denote by F^*P_pu.

With this, the bundle morphism (π^*F, F) is represented by

$$\begin{array}{ccc} u(M') \times \mathbb{F} & \xrightarrow{\quad (v \circ F \circ u^{-1}, id) \quad} & v(N') \times \mathbb{F} \\ \Big\downarrow{\scriptstyle pr_1} & & \Big\downarrow{\scriptstyle pr_1} \\ u(M') & \xrightarrow{\quad v \circ F \circ u^{-1} \quad} & v(N') \end{array} \quad \square$$

We apply the construction of the induced bundle to obtain the normal bundle of an immersion.

1.3.11 Proposition. *Let*

$$F: M \to N$$

be an immersion. The normal bundle associated with F

$$\nu_F : T^{\perp} F \to M$$

is obtained from the induced bundle $F^*\tau_N$ by taking on each fibre $\{p\} \times T_{F(p)}N, p \in M$, the quotient with respect to the subspace $\{p\} \times T_p F(T_p M)$.

Proof. To construct a bundle atlas, we consider pairs (u, M'), (v, N') of charts on M and N adapted to the immersion $F: M \to N$ in the sense of (1.3.3). Thus, the model space $\mathbb{F}$ of N has a splitting $\mathbb{F} = \mathbb{E}_1 \times \mathbb{F}_2$ with $\mathbb{E}_1 \cong \mathbb{E} =$ the model space of M such that $v \circ F \circ u^{-1}$ coincides locally with the inclusion $\mathbb{E} \to \mathbb{E}_1 \times \mathbb{F}_2$.

For the induced bundle $F^*\tau_N$ we have the bundle charts

$$F^* Tu: (p, Y) \in (F^*\tau_N)^{-1}(M') \mapsto (u(p), Dv(F(p)).Y) \in U \times \mathbb{F}.$$

The subspace $\{p\} \times T_p F(T_p M)$ of $(F^*\tau_N)^{-1}(p)$ then has the representation

$$\{u(p)\} \times Dv(F(p)) \circ T_p F(T_p M) = \{u(p)\} \times \mathbb{E}_1 \subset \{u(p)\} \times \mathbb{F}.$$

If (u', M''), (v, N'') is a second pair of charts adapted to F, $\{p\} \times T_p F(T_p M)$ is represented by

$$\{u'(p)\} \times Dv'(F(p)) \circ T_p F(T_p M) = \{u'(p)\} \times \mathbb{E}_1 \subset \{u'(p)\} \times \mathbb{F}.$$

Put $F(p) = q$. The transition mapping

$$Dv'(q) \circ Dv(q)^{-1} = D(v' \circ v^{-1})(v(q)) : \mathbb{F} \to \mathbb{F}$$

carries the subspace $\mathbb{E}_1$ of $\mathbb{E}_1 \times \mathbb{F}_2 = \mathbb{F}$ into itself. We therefore can construct a bundle atlas for ν_F by going for a bundle chart

$$F^* Tu: F^*\tau_N^{-1}(M') \to u(M') \times \mathbb{F}.$$

of $F^*\tau_N$ to the quotient mapping

$$F^* T_p u: F^* T_p / \{p\} \times T_p F(T_p M) \to$$

$$\{u(p)\} \times \mathbb{F} / (\{u(p)\} \times \mathbb{E}_1) \cong \{u(p)\} \times \mathbb{F}_2$$

1.3.12 Example. Let

$$F: \Delta_M \to M \times M$$

be the inclusion of the diagonal. Then the normal bundle of Δ_M is isomorphic to the tangent bundle of M.

Indeed, the fibre over $(p, p) \in \Delta_M$ in the induced bundle $F^*\tau_{M \times M}$ consists of $T_{(p,p)} M \times M = T_p M \times T_p M$, whereas the fibre of the tangent bundle $\tau_{\Delta_M}: T\Delta_M \to \Delta_M$ consists of the pairs $(X, X) \in T_p M \times T_p M$. The mapping

$$X \in T_p M \mapsto (X, -X) \in T_p M \times T_p M$$

is a linear isomorphism of $T_p M$ onto a complement of $T_{(p,p)} \Delta_M$ in $T_{(p,p)} M \times M$ for each $p \in M$. We thus get a diffeomorphism of TM with the total space of the normal bundle of the diagonal Δ_M in $M \times M$ which commutes with the respective projection maps.

We conclude with an observation on the structure of the tangent bundle of a vector bundle.

1.3.13 Proposition. *Let $\pi: P \to M$ be a vector bundle over M, modelled on $\mathbb{E} \times \mathbb{F}$. Then the total tangent space TP of P can be viewed also as the total space of the bundle over TM induced from $\tau_M: TM \to M$ by π,*

$$\begin{array}{ccc} TP = \tau_M^* P & \xrightarrow{\tau_P = \pi^* \tau_M} & P \\ \Big\downarrow {\scriptstyle T\pi = \tau_M^* \pi} & & \Big\downarrow {\scriptstyle \pi} \\ TM & \xrightarrow{\tau_M} & M \end{array}$$

As always for the tangent space of a manifold, a chart of the manifold yields a chart for the tangent space. In this case, a bundle chart (Pu, u, M') of π gives rise to the bundle chart (TPu, Tu, TM'), i.e.,

$$\begin{array}{ccc} TPM' & \xrightarrow{TPu} & (U \times \mathbb{F}) \times (\mathbb{E} \times \mathbb{F}) \\ \Big\downarrow {\scriptstyle \tau_{PM'}} & & \Big\downarrow {\scriptstyle pr_1} \\ PM' & \xrightarrow{Pu} & U \times \mathbb{F} \end{array}$$

With this, the above diagram reads, with $u(M') = U$,

$$\begin{array}{ccc} (U \times \mathbb{F}) \times (\mathbb{E} \times \mathbb{F}) & \xrightarrow{(\tau_P)_U} & U \times \mathbb{F} \\ \Big\downarrow {\scriptstyle (T\pi)_U} & & \Big\downarrow {\scriptstyle pr_1} \\ U \times \mathbb{E} & \xrightarrow[pr_1]{} & U \end{array} \; ; \qquad \begin{array}{ccc} (u, \xi, X, \eta) & \longmapsto & (u, \xi) \\ \Big\downarrow & & \Big\downarrow \\ (u, X) & \longmapsto & (u) \end{array}$$

Proof. Everything is clear from the definitions in (1.3.10) and (1.2.1). □

1.3.14 Corollary. *In each tangent space $T_\xi P$ of the total space P of a vector bundle π there exists a distinguished subspace, the* vertical space *$T_{\xi v} P$. This is the subspace tangent to the fibre $\mathbb{F}_p, p = \pi\xi$, containing ξ. Also,*

$$T_{\xi v} P = ker\{T_\xi \pi: T_\xi PM \to T_{\pi\xi} M\}.$$

In a bundle chart (TPu, Tu, TPM') of τ_P, covering $\xi \in P$, $T_{\xi v} P$ is represented by $(u, \xi) \times \{0\} \times \mathbb{F}$.

Proof. According to (1.3.13), $T\pi|T_\xi P = T_\xi \pi$ has in the chart (TPu, TPM') the representation

$$(u, \xi, X, \eta) \in T_{(u,\xi)}(U \times \mathbb{F}) \mapsto (u, X) \in T_u U.$$

Thus, the elements with $X = 0$ form the kernel. □

Remark. One may ask whether there exists a distinguished complementary space for the distinguished subspace $T_{\xi v} P$ in $T_\xi P$. This is in general not the case. Rather, the existence of distinguished so-called horizontal spaces amounts to having an additional structure for $\pi: P \to M$. We shall come back to this in (1.5.9) where we introduce the concept of a linear connection.

1.4 Vector Fields and Tensor Fields

Having introduced various bundles over a manifold, we now consider sections in these bundles. As a special case, we obtain the vector fields; they yield the first natural example of a Lie algebra, cf. (1.4.7).

Combining the concepts of an induced bundle and a section in a bundle we obtain vector fields (or more generally, sections) along a mapping, see (1.4.8).

Other examples of sections are the differential forms. They have the particular property of behaving contravariantly under morphisms $F: M \to N$ of the base manifolds, see (1.4.12).

For finite dimensions, tensor fields can be characterized as $\mathscr{F}M$-linear mappings, see (1.4.14). In the treatises where only finite dimensional manifolds are considered, this fact is often used to define tensor fields.

We begin with the sections in the tangent bundle of a manifold.

1.4.1 Definition. A *vector field on* M is a differentiable map $X: M \to TM$ such that $\tau \circ X = id$.

Remark. It is useful to represent a vector field by the commutative diagram

$$\begin{array}{ccc} & & TM \\ & \nearrow X & \downarrow \tau \\ M & \xrightarrow{id} & M \end{array}$$

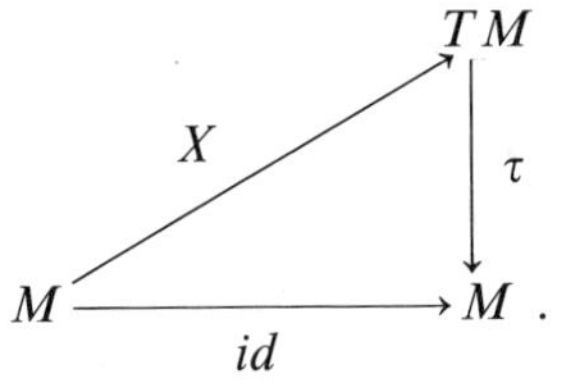

That is to say, the value $X(p)$ of X at $p \in M$ lies in the fibre $\tau^{-1}(p) = T_p M$ of the tangent bundle. More generally, one calls mappings from the base space of a bundle

into the total space of the bundle such that the above diagram is commutative, *sections* of the bundle.

1.4.2 Proposition. *The set* $\mathfrak{V}M$ *of vector fields on* M *is an* $\mathbb{R}$*-vector space and a* $\mathcal{F}M$*-module under the natural compositions, i. e.,*

$$(aX+bY)(p) = aX(p)+bY(p)$$

$$(fX)(p) = f(p)X(p).$$

Proof. Clear. $\square$

1.4.3 Proposition. *Given a vector field* $X\colon M \to TM$*, we get for every chart* (u, M') *of* M *a differentiable mapping*

$$(*) \qquad X_U\colon U = u(M') \to \mathbb{E};\, u = u(p) \mapsto Du(p).X(p),$$

with $Du(p)\colon T_pM \mapsto \mathbb{E}$ *defined as in* (1.2.1, iv). *Conversely, a differentiable mapping* (*) *determines a vector field on* M' *by*

$$(**) \quad X\colon p \in M' \mapsto Du(p)^{-1}.X_U(u(p)) \in TM'.$$

Proof. With $X\colon M \to TM$ differentiable,

$$pr_2 \circ Tu \circ X \circ u^{-1}\colon U \to \mathbb{E}$$

is also differentiable, and so therefore is (*). Conversely, if (*) is differentiable, the mapping X, (**), has the local representation $Tu \circ X \circ u^{-1}\colon (u(p) \mapsto (u(p), X_U(u(p)))$, i. e., it is differentiable and $\tau \circ X = id$. $\square$

1.4.4 Example. *The Gauss frames.* Let (u, M') be a chart. On the model space $\mathbb{E}$ let $\{e_i\}$ be an orthonormal basis. Then we have, for each i, the vector field $e_i\colon u \in U \mapsto (u, e_i) \in TU = U \times \mathbb{E}$. Via the isomorphism $(T_pu)^{-1}\colon T_uU \to T_pM$ we get a vector field on M' which we briefly denote by $e_i(p), p \in M'$. In this notation we do not indicate that this vector field depends on the chart (u, M'). We also call $e_i(p)$ the *i-th basis vector field* (w. r. t. the chart (u, M')).

For each $p \in M'$, $\{e_i(p)\}$ is a basis of $T_pM' \cong T_pM$. This field of bases over M' is called *Gauss frame* (w. r. t. the chart (u, M')).

1.4.5 Definition. A *derivation* on M is a mapping

$$\delta\colon \mathcal{F}M \to \mathcal{F}M$$

on the $\mathbb{R}$-Algebra $\mathcal{F}M$ of differentiable functions which is $\mathbb{R}$-linear and satisfies

$$(*) \qquad \delta(fg) = (\delta f)g + f(\delta f).$$

Example. Let (u, M') be a chart. Then we get for every basis vector field $e_i(p)$ on M' a derivation on $\mathcal{F}M'$ by associating to $f = \{p \mapsto f(p)\}$ the function

$\delta_{e_i} f = \{p \mapsto Df(p).e_i(p) = \frac{\partial f \circ u^{-1}}{\partial u^i}(u(p))\}$. Clearly, $\delta_{e_i}(fg) = (\delta_{e_i} f)g + f(\delta_{e_i} g)$. In particular, $\delta_{e_i} u^j = \delta^j_i$. For this reason one sometimes denotes the vector field $e_i(p)$ on M' also by $\frac{\partial}{\partial u^i}(p)$.

Actually, we can let every vector field X on M operate as a derivation on $\mathscr{F}M$.

1.4.6 Lemma. *Let X be a vector field on M. Then we obtain a derivation, denoted by δ_X, by associating with $f \in \mathscr{F}M$ the function $\delta_X f = \{p \in M \mapsto Df(p).X(p)\}$. We also write $Df.X$ or $X(f)$ or simply Xf instead of $\delta_X f$.*

$\delta_X = \delta_{X'}$ implies $X = X'$. If $\dim M < \infty$ then every derivation δ is of the form δ_X, X some vector field.

Proof. δ_X is clearly $\mathbb{R}$-*linear*. The relation

$$D(fg)(p).X(p) = Df(p).X(p)g(p) + f(p)Dg(p).X(p)$$

is the derivation property (*).

Assume $X \neq X'$. That is, there is $p \in M$ where $Y(p) = X(p) - X'(p) \neq 0$. Let (u, M') be a chart around p with $u(p) = 0$. Then $Y(0)$ (the principal part of the representative of $Y(p)$) is $\neq 0$. There exists a $f_0 \in \mathbb{E}^* = L(\mathbb{E}; \mathbb{R})$ with $Y(0)(f_0) = Df_0.Y(0) \neq 0$. From (1.1.8) we get the existence of a $f \in \mathscr{F}M$ with $f = f_0 \circ u$ near p. Hence, $Y(f)(p) \neq 0$.

Let now $\dim M = n < \infty$. For a chart (u, M') we consider the n coordinate functions $u^i = pr_i \circ u : M \to \mathbb{R}, i = 1 \ldots, n$. Given a derivation δ, we define on M' a vector field X by

$$X(p) = \sum_j \delta u^j(p) e_j(p).$$

We claim: For all $f' \in \mathscr{F}M', \delta_X f' = \delta f'$, which will complete the proof of (1.4.6).

Now, for the $u^i \in \mathscr{F}M'$ we clearly have from $\delta_{e_j} u^i = \delta^i_j : \delta_X u^i = \delta u^i$. For an arbitrary $f' \in \mathscr{F}M'$ we consider $f'(u) \equiv (f' \circ u^{-1})(u)$ as function on $U = u(M')$. Near a given $u_0 \in U$ we can write

$$f'(u) = f'(u_0) + \sum_i (u^i - u^i_0) g_i(u)$$

with $$g_i(u) = \int_0^1 \frac{\partial f'}{\partial u^i}(t(u - u_0) + u_0)\, dt.$$

Thus, f' can be written for p near $p_0 \in U$ as

$$f'(p) = f'(p_0) + \sum_i (u^i - u^i_0) g_i(p).$$

Using the derivation property (*) we prove our claim. □

Remark. Fix a $p \in M$. On the $\mathbb{R}$-algebra $\mathcal{F}M$ of real valued functions on M consider the relation $f \underset{p}{\sim} g$ which means $f \equiv g$ on some neighborhood of p on M. This is an equivalence relation. The equivalence classes are called *germs (of functions at p)*. The set $\mathcal{F}_p$ of germs inherits the structure of an $\mathbb{R}$-algebra from $\mathcal{F}M$.

With $\tilde{f} \in \mathcal{F}_p$ we associate the value $\tilde{f}(p)$ given by $f(p)$, f being a representative of $\tilde{f}$. The set $\mathcal{M}_p = \{\tilde{f}; \tilde{f}(p) = 0\}$ forms a maximal ideal in $\mathcal{F}_p$.

We have a linear mapping

$$\mathcal{M}_p/\mathcal{M}_p^2 \to T_p^* M; \tilde{f} \mapsto \{Df(p): T_p M \to \mathbb{R}\}$$

where $f \in \tilde{f}$. This is an isomorphism.

In the last part of the proof of (1.4.6) we constructed, from a basis of $T_p M$, linearly independent elements in $\mathcal{M}_p/\mathcal{M}_p^2 \cong T_p^* M$. Only for $\dim T_p M = n < \infty$ this is a basis for $T_p^* M$.

$\mathcal{M}_p/\mathcal{M}_p^2$ yields a representation of $T_p^* M$, the space of differentials at p. One gets a natural representation of higher order differentials or *jets*, by $\mathcal{M}_p/\mathcal{M}_p^{k+1}$, $k > 1$. See Warner [1] and Bourbaki [2] for details.

Jets play an essential rôle in the study of singularities of maps, see Arnold [1].

1.4.7 Proposition. *With $X, Y \in \mathfrak{V}M$ there is associated a vector field $[X, Y]$ characterized by*

$$[X, Y](f) = X(Y(f)) - Y(X(f)),$$

for all $f \in \mathcal{F}M$.

$[X, Y]$ is called the Lie bracket of the pair (X, Y). *It has the properties*

(i) $(X, Y) \in \mathfrak{V}M \times \mathfrak{V}M \mapsto [X, Y] \in \mathfrak{V}M$ *is $\mathbb{R}$-bilinear,*

(ii) $[X, Y] = -[Y, X]$,

(iii) $[[X, Y], Z] + [[Z, X], Y] + [[Y, Z], X] = 0$,
(Jacobi identity)

(iv) $[X, fY] = f[X, Y] + X(f)\, Y$.

Remarks. 1. A *Lie algebra* is a vector space on which there is defined a bilinear mapping $(X, Y) \mapsto [X, Y]$ into itself, satisfying (ii) and (iii).

2. As we will see in the proof, if $\dim M < \infty$ and X, Y have the principal parts $X(u) = \sum_i x^i(u) e_i$ and $Y(u) = \sum_j y^j(u) e_j$ then the principal part of $[X, Y]$ reads

$$[X, Y](u) = \sum_j \left(\sum_i x^i \frac{\partial y^j}{\partial u^i} - \sum_i y^i \frac{\partial x^j}{\partial u^i} \right)(u)\, e_j.$$

3. In (1.4.6) we associated with a vector field a derivation which characterized the vector field. If $\dim M = \infty$, not all derivations can be described in this way by vector

fields. However, it is possible to define a Lie bracket on the set of all derivations so as to make it a Lie algebra which extends the Lie algebra of vector fields, cf. Lang [1].

Proof. The principal part $[X, Y](u)$ is given, using the principal parts $X(u)$ and $Y(u)$ of X and Y, by

$$[X, Y](u) = DY(u).X(u) - DX(u).Y(u).$$

Indeed, for any given function $f(u)$ on U we then have

$$[X, Y](f) = Df.(DY.X) - Df.(DX.Y) = D(Df.Y).X - D(Df.X).Y =$$
$$= X(Y(f)) - Y(X(f)),$$

since $D^2f.(X, Y) = D^2f.(Y, X)$.

From (1.4.6) it follows that $[X, Y]$ is uniquely determined.

The validity of (i) to (iv) is an easy consequence of the above expression for the principal part. □

1.4.8 Definition. Let $F: M \to N$ be a morphism. *A vector field V along F* is a morphism

$$V: M \to TN \text{ with } \tau_N \circ X = F.$$

Note. The last property is the commutativity of the diagram

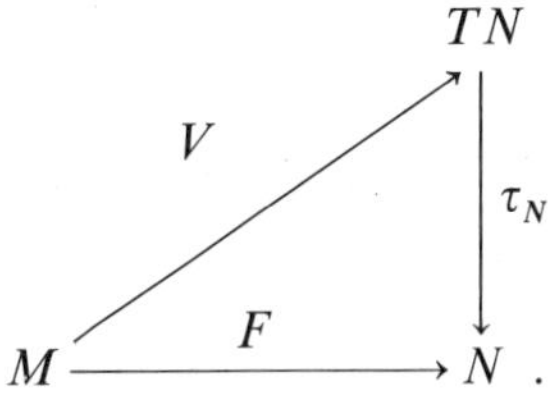

For $F = id: M \to M$ we get back (1.4.1).

Another way of describing a vector field V along F is to consider the bundle $F^*\tau_N$, induced by F from τ_N, cf. (1.3.10). Then a section V^* in $F^*\tau_N$ gives the vector field $V(p) = \tau_N^* F \circ V^*(p)$ along F. Conversely, if V is a vector field F, define the section V^* in $F^*\tau_N$ by $V^*(p) = (\tau_N^* F | (\{p\} \times \mathbb{F}_p))^{-1} V(p)$.

Examples. Let $c: \mathbb{R} \to M$ be a *curve*. Then a vector field along c is a mapping $t \in \mathbb{R} \mapsto V(t) \in T_{c(t)}M$ depending differentiably on t.

More generally consider a *k-dimensional singular surface* in M, i. e., a differentiable mapping

$$F: t \in \mathbb{R}^k \to F(t) \in M.$$

The basis vector fields $(t, e_i), 1 \leqslant i \leqslant k$, determine the vector fields

$$t \in \mathbb{R}^k \mapsto \frac{\partial F}{\partial t^i}(t) \in TM \text{ along } F \text{ with } \frac{\partial F}{\partial t^i}(t) = T_t F(t, e_i).$$

1.4.9 Proposition. *Let $F: M \to N$ be a morphism.*

(i) *The set $\mathfrak{V}_F M$ of vector fields along F is a $\mathscr{F}M$-module under the pointwise composition*

$$(fV + gW)(p) = f(p)\,V(p) + g(p)\,W(p)$$

(ii) *The tangential TF of F determines a $\mathscr{F}M$-module morphism*

$$TF: \mathfrak{V}M \to \mathfrak{V}_F;\; X \mapsto TF \circ X$$

The image $\mathfrak{V}_F^T$ consists of the so-called tangential vector fields along F.

(iii) *F determines an $\mathbb{R}$-algebra morphism*

$$F^*: \mathscr{F}N \to \mathscr{F}M;\; g \mapsto g \circ F$$

which extends to a $\mathscr{F}N$-module morphism

$$F^*: \mathfrak{V}N \to \mathfrak{V}_F;\; Y \mapsto Y \circ F.$$

Proof. Everything is clear from the definition. □

Note. Let $F: M \to N$ be a morphism. In general, for a vector field X on M there does not exist a vector field Y on N such that $T_p F . X(p) = Y(F(p))$. If there should exist such a Y we say that *Y is related to X by F.*

1.4.10 Proposition. *Let $F: M \to N$ be a morphism and $Y_i \in \mathfrak{V}N$ be related to $X_i \in \mathfrak{V}M$ by F, $i = 1, 2$. Then $[Y_1, Y_2]$ is related to $[X_1, X_2]$ by F, i.e.,*

$$TF \circ [X_1, X_2] = [Y_1, Y_2].$$

Proof. Using charts (u, M') on M and (v, N') on N around p and $q = F(p)$, respectively, we employ the formula for the principal parts $[X_1, X_2](u)$ and $[Y_1, Y_2](F(u))$ from the proof of (1.4.7). Here we have written F instead of $v \circ F \circ u^{-1}$.

$$[Y_1, Y_2](F(u)) = DY_2(F(u)) \,.\, Y_1(F(u)) - DY_1(F(u)) \,.\, Y_2(F(u)) =$$

$$DY_2(F(u)) \circ DF(u) \,.\, X_1(u) - DY_1(F(u)) \circ DF(u) \,.\, X_2(u) =$$

$$D(Y_2 \circ F(u)) \,.\, X_1(u) - D(Y_1 \circ F(u)) \,.\, X_2(u) =$$

$$D(DF(u) \,.\, X_2(u)) \,.\, X_1(u) - D(DF(u) \,.\, X_1(u)) \,.\, X_2(u) =$$

$$DF(u) \,.\, (DX_2(u) \,.\, X_1(u) - DX_1(u) \,.\, X_2(u)). \quad \square$$

We can now generalize the concept of a vector field.

1.4.11 Definition. Denote by $\mu(\tau): \mu TM \to M$ one of the various tensor bundles over M. A *tensor field (of type μ)* is a section in $\mu(\tau)$, i.e., a differentiable mapping $Z: M \to \mu TM$ satisfying $\mu(\tau) \circ Z = id$.

Denote by $\Gamma(\mu(\tau))$ the set of tensor fields of type μ. Then this is in a canonical way a $\mathscr{F}M$-module.

In the special case $\mu(\tau) = \alpha_s$, cf. (1.2.14), a tensor field also is called a (*differential*) *form of degree s* on M. The $\mathscr{F}M$-module of such forms is denoted by $\Omega_s M$.

Note. Tensor fields of type $\binom{1}{0}$ are the same as vector fields. Tensor fields of type $\binom{0}{1}$ are sometimes also called *co-vector fields*. Functions on M are tensor fields of type $\binom{0}{0}$; one can interpret them also as *differential forms of degree 0*.

We want to point out an important feature of purely covariant tensor fields which is not shared by general tensor fields. Recall from (1.2.12, 3) that the transition mappings of the general tensor space $\tau_s^r : T_s^r M \to M$ involve, besides $T_p(u' \circ u^{-1}) = T_p u' \circ T_p u^{-1}$, its contragradient $(T_p u' \circ T_p u^{-1})\check{\ }$. Compare also the definition of the morphism $T_s^r : GL(\mathbb{E}) \to GL(T_s^r \mathbb{E})$ in (1.0) 2.

If, however, $r = 0$, only $T_p(u' \circ u^{-1})$ is involved. Thus, T_s^0 (as well as A_s, S_s) can be extended to contravariant functors of the category of vector spaces:

With $\mathbb{E}$ we associate $T_s^0 \mathbb{E} = L(\underbrace{\mathbb{E}, \ldots, \mathbb{E}}_{s}; \mathbb{R})$ and with $F: \mathbb{E} \to \mathbb{F}$ we associate

$$T_s^0 F : T_s^0 \mathbb{F} \to T_s^0 \mathbb{E}; \quad Y_s^0 \mapsto Y_s^0 \circ (F \times \ldots \times F).$$

In the following Lemma we use the language of functors. It suffices to have in mind one of the special cases T_s^0, S_s, A_s. Cf. also (1.2.11).

1.4.12 Lemma. (i) *Let λ be one of the previously considered contravariant functors in the category of vector spaces. Consider the bundle $\lambda(\tau_M)$ over M,*

$$\lambda(\tau_M) : \lambda TM \to M$$

by taking as fibre over $p \in M$ the space $\lambda T_p M$. The transition functions are given by

$$\lambda(T_p u' \circ T_p u^{-1}) : \lambda T_{u(p)} U \to \lambda T_{u'(p)} U'.$$

Now let $F: M \to N$ be a morphism from the manifold M to the manifold N. Then any section ω_N in $\lambda(\tau_N)$ determines a section ω_M in $\lambda(\tau_M)$ by $\omega_M(p) = \lambda T_p F \circ \omega_N \circ F(p)$. In other words, by writing λTF for the morphism which on each fibre is given by $\lambda T_p F : \lambda T_{F(p)} N \to \lambda T_p M$, i.e., by the application of the contravariant functor λ to $T_p F : T_p M \to T_{F(p)} M$, we have the commutative diagram

$$\begin{array}{ccc} \lambda TM & \xleftarrow{\lambda TF} & \lambda TN \\ \omega_M \updownarrow \lambda(\tau_M) & & \omega_N \updownarrow \lambda(\tau_N) \\ M & \xrightarrow{F} & N \end{array}$$

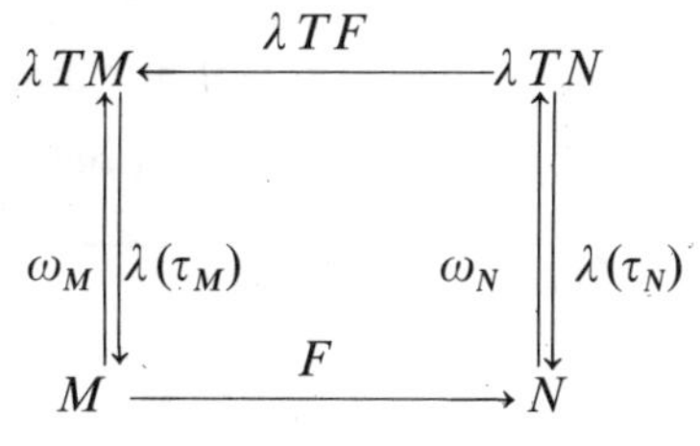

Proof. Clear from the definition. □

Remark. Note that $\lambda(\tau_M)$ in general is not the bundle induced by F from $\lambda(\tau_N)$.

Example. Let $F\colon M \to N$ be a morphism. A differential form ω_N on N determines a differential form ω_M on M by

$$\omega_M(p) = \omega_N(F(p)) \circ (T_p F \times \ldots \times T_p F).$$

We conclude with a useful characterization of tensor fields for manifolds of finite dimension.

If $\dim M = n < \infty$, a tensor field can be described over the domain M' of a chart (u, M') by its components in the canonical base. Take e. g. a tensor field of type $\binom{r}{s}$. Then we have on M' and $U = u(M')$ the n^{r+s} basis vector fields

$$(u, e_{i_1} \otimes \ldots \otimes e_{i_r} \otimes e^{j_1} \otimes \ldots \otimes e^{j_s})$$

or briefly

$$(u, e^{j_1 \ldots j_s}_{i_1 \ldots i_r}).$$

1.4.13 Proposition. *Let* $\dim M = n < \infty$. *Consider a tensor field* Z *on* M *of type* $\binom{r}{s}$. *Then the principal part* $Z(u)$ *of the representative of* Z *with respect to a chart* (u, M') *can be written as*

$$Z(u) = \sum_{\substack{i_1 \ldots i_r \\ j_1 \ldots j_s}} Z^{i_1 \ldots i_r}_{j_1 \ldots j_s}(u)\, e^{j_1 \ldots j_s}_{i_1 \ldots j_s}$$

where the n^{r+s} *coefficients* $Z^{i_1 \ldots i_r}_{j_1 \ldots j_s}(u)$ *are differentiable functions on* $U = u(M')$.

Conversely, given n^{r+s} *differentiable functions* $Z^{i_1 \ldots i_r}_{j_1 \ldots j_s}(u)$, *the above formula is the principal part of a tensor of type* $\binom{r}{s}$ *on the domain* M' *of the chart* (u, M').

Proof. Given a tensor field Z we define

$$Z^{i_1 \ldots i_r}_{j_1 \ldots j_s}(u) = Z(u, e^{i_1 \ldots i_r}_{j_1 \ldots j_s})$$

These are the components of Z with respect to the base $(u, e^{j_1 \ldots j_s}_{i_1 \ldots i_r})$. And they depend differentiably on u since Z was supposed to be differentiable. Conversely, the above formula clearly defines a differentiable section for $\tau^r_s | M'$. □

For $\dim M < \infty$ it is customary to define a tensor field as a $\mathscr{F}M$-multilinear function on vector fields and co-vector fields, having as value a differentiable function or a vector field again, as the case may be. Such a definition is not possible for $\dim M = \infty$.

We conclude this paragraph by showing in a special case how our definition and the customary definition are related.

Recall from (1.0) 1 that for $\dim \mathbb{E} < \infty$

$$T^r_s \mathbb{E} \cong L(\underbrace{\mathbb{E}, \ldots, \mathbb{E}}_{s}; \underbrace{\mathbb{E} \otimes \ldots \otimes \mathbb{E}}_{r}).$$

1.4.14 Proposition. *Let* $\dim M < \infty$. *Then there is a* 1 : 1 *correspondence between the* *tensor fields* Z *of type* $\binom{r}{s}$ *and the* $(r+s)$*-fold* $\mathcal{F}M$*-linear mappings*

$$Z: (\mathfrak{V}M)^s \times (\Omega_1 M)^r \to \mathcal{F}M;$$

$$(X_1, \ldots, X_s, \omega_1, \ldots, \omega_r) \mapsto Z(X_1, \ldots X_s, \omega_1, \ldots \omega_r).$$

In the same way one can – for $\dim M < \infty$ *– characterize a tensor field* Z *of type* $\binom{r}{s}$ *as an s-fold* $\mathcal{F}M$*-linear mapping*

$$Z: \mathfrak{V}M^s \to \Gamma(\tau_0^r);$$

$$(X_1, \ldots, X_s) \mapsto Z(X_1, \ldots X_s).$$

Here, $\Gamma(\tau_0^r)$ *is the* $\mathcal{F}M$*-module of r-fold contravariant tensors.*

Proof. The composition of differentiable sections clearly shows that a tensor field Z of type $\binom{r}{s}$ determines such a mapping. This is true without restrictions on the dimension of M.

The converse uses the characterization of tensor fields given in (1.4.13). Define the components of the principal part $Z(u)$ of a tensor Z of type $\binom{r}{s}$ by the formula given there. □

Remark. Most of the concepts and constructions in this section can be generalized from the tangent bundle of a manifold M to an arbitrary vector bundle $\pi: P \to M$ over M. We only write down the generalization of (1.4.8).

1.4.15 Definition. Let $\pi: P \to N$ be a vector bundle and consider a morphism $F: M \to N$. A *section along* F is a morphism $\eta: M \to P$ with $\pi \circ \eta = F$, i.e.,

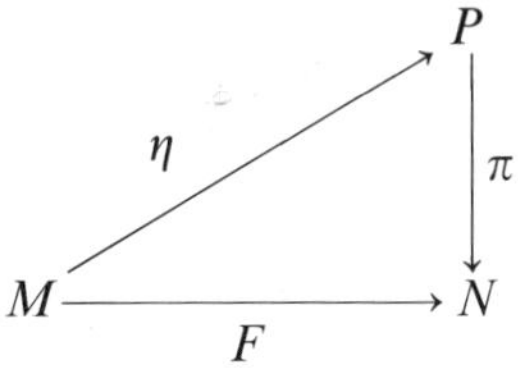

is commutative.

Remark. Obviously, a section along F may be viewed also as section in the induced bundle $F^*\pi: F^*P \to M$, cf. (1.3.10), (1.4.8).

1.5 Covariant Derivation

Let X be a vector field on an open set U of a Hilbert space. Then its differential DX operates as a derivation on the $\mathcal{F}U$-module $\mathfrak{V}U$ of vector fields on U. For a vector field X on a manifold M we cannot define such an operator on the $\mathscr{F}M$-module $\mathfrak{V}M$ of vector fields on M since here we only have the tangential $TX: TM \to TTM$. Hence, TX, applied to a vector field Y, belongs to TTM and not to TM.

To be able to define a 'covariant derivation' ∇X of a vector field X which acts as a derivation on $\mathfrak{V}M$ we need an additional structure on M (or its tangent bundle). In (1.5.1) we describe this structure with the help of the so-called Christoffel symbols. Associated to a covariant derivation are its torsion and its curvature tensor, see (1.5.3).

In (1.5.6) we show how a covariant derivation of vector fields along a morphism $F: M \to N$ can be defined if N possesses a covariant derivation.

A covariant derivation on the tangent bundle τ_M of a manifold M falls under the general concept of a connection for a vector bundle $\pi: P \to M$ over M, cf. (1.5.9). Such a connection is given by a differentiable mapping $K: TP \to P$, which on each fibre $T_\xi P$ of $\tau_P: TP \to P$ is a linear surjective morphism. Its kernel, the so-called horizontal space in ξ, is a complementary space to the vertical space in ξ.

An induced bundle (cf. (1.3.10)) inherits a connection from the inducing bundle, see (1.5.13).

There always exist linear connections, see (1.5.15). We conclude with the observation that – for finite dimensions – a covariant derivation on a manifold M, i.e., a linear connection on the tangent bundle $\tau_M: TM \to M$, determines in a canonical way a covariant derivation on all tensor bundles associated with τ_M, see (1.5.17).

For a vector field $Y: U \subset \mathbb{E} \to TU \cong U \times \mathbb{E}$ we can define the differential by interpreting its principal part $Y(u)$ as mapping from U into $\mathbb{E}$. Thus, we get a $\mathbb{R}$-bilinear mapping

$$(X, Y) \in \mathfrak{V}U \times \mathfrak{V}U \mapsto DY.X \in \mathfrak{V}U$$

which is $\mathscr{F}U$-linear in its first argument and a derivation in its second. In particular, if we write $D_X Y$ instead $DY.X$ we have

$$D_{fX}Y = fD_X Y;\ D_X(fY) = fD_X Y + X(f)\,Y.$$

We want to generalize this by defining for vector fields $(X, Y) \in \mathfrak{V}M \times \mathfrak{V}M$ a covariant derivation $\nabla_X Y \equiv \nabla Y.X \in \mathfrak{V}M$ of Y in the direction X, which is $\mathscr{F}M$-linear in X and a $\mathscr{F}M$-derivation in Y, cf. (1.5.2).

In a first approach we might take the tangential $TY: TM \to TTM$. If X is a vector field, then $TY(X)$ is a mapping from M into TTM. We therefore need some mapping $K: TTM \to TM$ which brings us back into TM.

Before we introduce the concept of such a connection map K (even for general vector bundles) we present a different approach to the solution of our problem by using charts. Let (u, M') be a chart and $(X(u), Y(u))$ the principle parts of the local representatives of vector fields X, Y w.r.t. this chart. Define $\nabla Y.X$ on M' by letting $(u, DY(u).X(u))$ be its local representative. If now $(u', Y(u')), (u', X(u'))$ are the local representatives of Y, X with respect to a chart (u', M''), then we have from

$$X(u') = D(u' \circ u^{-1}) \,.\, X(u); \; Y(u') = D(u' \circ u^{-1}) \,.\, Y(u):$$

$$(*) \qquad \begin{cases} DY(u') \,.\, X(u') = D\big(D(u' \circ u^{-1})(u) \,.\, Y(u)\big) \,.\, X(u) \\ = D^2(u' \circ u^{-1})(u) \,.\, \big(Y(u), X(u)\big) + D(u' \circ u^{-1})(u) \,.\, \big(DY(u) \,.\, X(u)\big) \,. \end{cases}$$

Thus, for general $(u' \circ u^{-1})(u)$, $DY(u') \,.\, X(u')$ will not be equal to $D(u' \circ u^{-1})(u) \,.\, DY(u) \,.\, X(u)$, i.e., the definition of ∇Y by using $DY(u)$ in local coordinates does not work.

In fact, the possibility of defining a 'differential' for vector fields on a manifold M requires from M the presence of an additional structure.

1.5.1 Definition. Let M be a manifold. (i) Assume that for every chart (u, M') of an atlas of M there is given a differentiable mapping

$$\Gamma : u \in u(M') \mapsto \Gamma(u) \in L(\mathbb{E}, \mathbb{E}, \mathbb{E}^*; \mathbb{R}) = L(\mathbb{E}, \mathbb{E}; \mathbb{E})$$

with the following property: If (u', M'') is a second chart with mapping

$$\Gamma' : u' \in u'(M'') \mapsto \Gamma'(u') \in L(\mathbb{E}, \mathbb{E}, \mathbb{E}^*; \mathbb{R}) = L(\mathbb{E}, \mathbb{E}; \mathbb{E})$$

then $\Gamma | u(M' \cap M'')$ and $\Gamma' | u'(M'' \cap M')$ are related by

$$(\dagger) \qquad D(u' \circ u^{-1}) \,.\, \Gamma(u) = D^2(u' \circ u^{-1}) + \Gamma'(u') \circ \big(D(u' \circ u^{-1}) \times D(u' \circ u^{-1})\big) \,.$$

The Γ's are called *Christoffel symbols.*

(ii) Given a family of Christoffel symbols on M we define a *covariant derivation*

$$(X, Y) \in \mathfrak{V}M \times \mathfrak{V}M \mapsto \nabla_X Y \in \mathfrak{V}M$$

by taking for the principal part $\nabla_X Y(u)$ of $\nabla_X Y$ w.r.t. a chart (u, M') the expression

$$(\dagger\dagger) \qquad \nabla_X Y(u) = DY(u) \,.\, X(u) + \Gamma(u)\big(X(u), Y(u)\big) \,.$$

Here, $X(u)$, $Y(u)$ are the principle parts of X, Y w.r.t. (u, M').

(iii) When M is endowed with an atlas for which we have Christoffel symbols we also say *M possesses a covariant derivation.*

Remarks. In this definition, we have assumed already that $\nabla_X Y$ is well-defined, i.e., that the principle parts $\nabla_X Y(u)$, $\nabla_X Y(u')$ of $\nabla_X Y$ w.r.t. two charts (u, M'), (u', M'') are related by

$$\nabla_X Y(u') = D(u' \circ u^{-1}) \,.\, \nabla_X Y(u) \,.$$

But this is easily verified by substituting $(\dagger)$ in $(\dagger\dagger)$ and using $(*)$:

$$\nabla_X Y(u') = DY(u') \,.\, X(u') + \Gamma'(u')\big(X(u'), Y(u')\big) =$$
$$= D^2(u' \circ u^{-1}) \,.\, \big(Y(u), X(u)\big) + D(u' \circ u^{-1}) \,.\, \big(DY(u) \,.\, X(u)\big)$$
$$+ D(u' \circ u^{-1}) \,.\, \Gamma(u)\big(Y(u), X(u)\big) - D^2(u' \circ u^{-1}) \,.\, \big(Y(u), X(u)\big)$$
$$= D(u' \circ u^{-1}) \,.\, \nabla_X Y(u) \,.$$

1.5.2 Proposition. (i) *Let there be given a covariant derivation on* M. *The mapping thus determined*

(§) $\qquad (X, Y) \in \mathfrak{V}M \times \mathfrak{V}M \mapsto \nabla_X Y \in \mathfrak{V}M$

is $\mathscr{F}M$*-linear in the first argument and a* $\mathscr{F}M$*-derivation in the second. By the latter we mean*

$$\nabla_X(aY + a'Y') = a\nabla_X Y + a'\nabla_X Y', \ \nabla_X(fY) = X(f)\,Y + f\nabla_X Y,$$

with a, a' *in* $\mathbb{R}$, f *in* $\mathscr{F}M$

For a given $Y \in \mathfrak{V}M$, *the value* $\nabla_X Y(p)$ *of* $\nabla_X Y$ *in* p *depends on* $X(p)$ *only.*

(ii) *Assume* $\dim M < \infty$. *Then a mapping* (§) *with the above properties in its two arguments determines a covariant derivation on* M.

(iii) *Again let* $\dim M < \infty$ *and* M *be endowed with a covariant derivation. Consider a chart* (u, M'). *We consider the coordinates* $X^i(u)$, $Y^j(u)$ *of the principal parts* $X(u)$, $Y(u)$ *of vector fields* X, Y *and the coordinates* $\Gamma_{ij}^k(u)$ *of* $\Gamma(u)$ *with respect to the canonical basis vector fields* $u \mapsto (u, e_i)$ *on* $U = u(M')$, *i.e.,*

$$X(u) = \sum_i X^i(u)\,e_i; \ Y(u) = \sum_j Y^j(u)\,e_j; \quad \Gamma(u) = \sum_{i,j,k} \Gamma_{ij}^k(u)\,e^i \otimes e^j \otimes e_k \,.$$

Then we get the coordinates of $\nabla_X Y(u)$ *from*

(§§) $$\nabla_X Y(u) = \sum_k \left\{ \sum_i X^i(u) \frac{\partial Y^k}{\partial u^i}(u) + \sum_{i,j} X^i(u)\,Y^j(u)\,\Gamma_{ij}^k(u) \right\} e_k$$

Proof. (i) is an immediate consequence of the definition of $\nabla_X Y$ by its local representative $\nabla_X Y(u)$, cf. (1.5.1) (††).

To prove (ii) we show how the Christoffel symbols are determined. Let (u, M') be a chart. Denote by e_i: $M' \to TM'$ the i-th basis vector field on M', given by $e_i(p) = T_p u^{-1} \,.\, e_i(u)$. Then we express the principal part $\nabla_{e_i} e_j(u)$ of $\nabla_{e_i} e_j$ w.r.t. the basis vector fields on $U = u(M')$:

$$\nabla_{e_i} e_j(u) = \sum_k \Gamma_{ij}^k(u)\,e_k \,.$$

We define now $\Gamma(u)$ by

$$\Gamma(u) = \sum_{i,j,k} \Gamma_{ij}^k(u)\,e^i \otimes e^j \otimes e_k \,.$$

We thus get a differentiable mapping $\Gamma: U \to L(\mathbb{E}, \mathbb{E}, \mathbb{E}^*; \mathbb{R}) = L(\mathbb{E}, \mathbb{E}; \mathbb{E})$. It remains to show that the principal part $\nabla_X Y(u)$ of $\nabla_X Y$ is completely determined by $\Gamma(u)$ (and by the principal parts $X(u)$, $Y(u)$ of X, Y).

But this will follow when we establish the formula (§§) for $\nabla_X Y(u)$. Using the fact that the mapping (§) is $\mathscr{F}M$-linear in the first argument and a $\mathscr{F}M$-derivation in the second we find

$$(\nabla_X Y)(u) = \nabla_{\sum_i X^i(u) e_i} \Big(\sum_j Y^j(u) e_j\Big)(u) =$$

$$= \sum_i X^i(u) \nabla_{e_i} \Big(\sum_j Y^j(u) e_j(u)\Big) = \sum_{i,j} X^i(u) \frac{\partial Y^j(u)}{\partial u^i} e_j(u) +$$

$$+ \sum_{i,j} X^i(u) Y^j(u) \nabla_{e_i} e_j(u) = \text{right hand side of (§§).}$$

With this, we also have proved (iii). □

Remark. If $\dim M < \infty$, the transformation formula (†) of the Christoffel symbols can also be expressed in the components $\Gamma_{ij}^k(u)$, $\Gamma'^r_{pq}(u')$, i.e.,

$$\sum_k \frac{\partial u'^r}{\partial u^k}(u) \Gamma_{ij}^k(u) = \frac{\partial^2 u'^r}{\partial u^i \partial u^j}(u) + \sum_{p,q} \frac{\partial u'^p}{\partial u^i}(u) \frac{\partial u'^q}{\partial u^j}(u) \Gamma'^r_{pq}(u).$$

Here, $u'^r(u)$ stands for the *r-th* coordinate of the transition mapping $u' \circ u^{-1}$: $u(M' \cap M'') \to u'(M'' \cap M')$

1.5.3 Lemma. *Let* ∇ *be a given a covariant derivation on* M.

(i) ∇ *determines a tensor field* T *of type* $\binom{1}{2}$ *satisfying*

$$T(X, Y) = -T(Y, X)$$

for all $(X, Y) \in \mathfrak{V}M \times \mathfrak{V}M$. *The principal part* $T(u)$ *of its local representation w.r.t.a chart* (u, M') *is given by*

$$\{X, Y \mapsto T(u)(X, Y) = \Gamma(u)(X, Y) - \Gamma(u)(Y, X)\} \in L(\mathbb{E}, \mathbb{E}; \mathbb{E}).$$

(ii) ∇ *determines a tensor field* R *of type* $\binom{1}{3}$ *satisfying*

$$R(X, Y)Z = -R(Y, X)Z$$

for all $(X, Y, Z) \in \mathfrak{V}M \times \mathfrak{V}M \times \mathfrak{V}M$.

The principal part of the local representation of $R(u)$ *w.r.t. a chart* (u, M') is given by

$$\{X, Y, Z \mapsto$$

$$R(u)(X, Y)Z = D\Gamma(u).X(Y, Z) - D\Gamma(u).Y(X, Z) +$$

$$+ \Gamma(u)\big(X, \Gamma(u)(Y, Z)\big) - \Gamma(u)\big(Y, \Gamma(u)(X, Z)\big)\} \in L(\mathbb{E}, \mathbb{E}, \mathbb{E}; \mathbb{E}).$$

Moreover, if $T \equiv 0$ *the so called Bianchi identity holds,*

$$R(X, Y)Z + R(Z, X)Y + R(Y, Z)X = 0$$

for all $(X, Y, Z) \in \mathfrak{V}M \times \mathfrak{V}M \times \mathfrak{V}M$.

T and R are called *torsion tensor* and *curvature tensor*, respectively.

Proof. We first show that the local representations $T(u)$, $T(u')$ and $R(u)$, $R(u')$ of T and R with respect to charts (u, M'), (u', M'') are related like the principal parts of sections in the bundles mentioned.

This means

$$T(u') \circ \big(D(u' \circ u^{-1}) \times D(u' \circ u^{-1})\big) = D(u' \circ u^{-1}) \,.\, T(u)$$
$$R(u') \circ \big(D(u' \circ u^{-1}) \times D(u' \circ u^{-1}) \times D(u' \circ u^{-1})\big) = D(u' \circ u^{-1}) \,.\, R(u)\,.$$

To verify these properties we use (†) in (1.5.1). For T this is immediate. To prove the tensor property for R we take the differential of (†), i.e.,

$$\begin{aligned} D^2(u' \circ u^{-1}) \,.\, \big(X, \Gamma(u)(Y, Z)\big) &+ D(u' \circ u^{-1}) \,.\, (D\Gamma(u) \,.\, X)(Y, Z) = \\ &= D^3(u' \circ u^{-1}) \,.\, (X, Y, Z) + \\ &+ \big(D\Gamma'(u') \circ D(u' \circ u^{-1}) \,.\, X\big)\big(D(u' \circ u^{-1}) \,.\, Y, D(u' \circ u^{-1}) \,.\, Z\big) + \\ &+ \Gamma'(u')\big(D^2(u' \circ u^{-1}) \,.\, (X, Y), D(u' \circ u^{-1}) \,.\, Z\big) + \\ &+ \Gamma'(u')\big(D(u' \circ u^{-1}) \,.\, Y, D^2(u' \circ u^{-1}) \,.\, (X, Z)\big). \end{aligned}$$

Moreover, from (†),

$$\begin{aligned} D^2(u' \circ u^{-1}) \,.\, \big(X, \Gamma(u)(Y, Z)\big) &= D(u' \circ u^{-1}) \,.\, \Gamma(u)\big(X, \Gamma(u)(Y, Z)\big) \\ &- \Gamma'(u')\big(D(u' \circ u^{-1}) \,.\, X, D^2(u' \circ u^{-1}) \,.\, (Y, Z)\big) \\ &- \Gamma'(u')\big(D(u' \circ u^{-1}) \,.\, X, \Gamma'(u')(D(u' \circ u^{-1}) \,.\, Y, D(u' \circ u^{-1}) \,.\, Z)\big)\,. \end{aligned}$$

Exchange X and Y and form the difference. This yields the desired result.

The antisymmetry of T is clear, and so is the antisymmetry of R in the first two arguments. As for the Bianchi identity, it follows immediately from the definition of $R(u)$, using $T(u) = 0$. □

1.5.4 Proposition. *Let* ∇ *be a covariant derivation on* M. *Then one has the identities*

$$T(X, Y) = \nabla_X Y - \nabla_Y X - [X, Y]$$
$$R(X, Y)Z = \nabla_X \nabla_Y Z - \nabla_Y \nabla_X Z - \nabla_{[X,Y]} Z,$$

for arbitrary vector fields X, Y, Z *on* M.

For $\dim M < \infty$, *these identities can serve as definitions of* T *and* R.

Proof. Once the identities have been established, the last statement is clear from our general result (1.4.14) since the mappings on the right hand side are $\mathscr{F}M$-linear in all arguments.

Now, these identities are an immediate consequence of the expressions for the principal part $[X, Y](u)$ of $[X, Y]$ in the proof of (1.4.7) and of (1.5.3). □

We generalize the concept of a covariant derivation towards a covariant derivation of a vector field along a mapping.

1.5.5 Proposition. *Let M and N be differentiable manifolds and $F: M \to N$ a morphism. N shall have a covariant derivation ∇.*

(i) *We define a* covariant derivation $\nabla_X V$ of a vector field V along F in the direction of the vector field X on M *as follows: Let (u, M') be a chart on M, (v, N') a chart on N with $F(M') \subset N'$. Put $(v \circ F \circ u^{-1})(u) = F(u)$. Let*

$$u \in u(M') \mapsto X(u) \in \mathbb{E}; \quad u \in u(M') \mapsto V(u) \in \mathbb{F}$$

be the principal parts of the local representations of X and V. Then let the principal part $\nabla_X V(u)$ of the vector field $\nabla_X V$ along F be given by

$$(*) \qquad \nabla_X V(u) = DV(u) \,.\, X(u) + \Gamma\big(F(u)\big)\big(DF(u) \,.\, X(u), V(u)\big).$$

This gives a mapping

$$(**) \qquad (X, V) \in \mathfrak{V}M \times \mathfrak{V}_F \mapsto \nabla_X V \in \mathfrak{V}_F$$

with the following properties:

(a) *It is $\mathscr{F}M$-linear in the first argument and a $\mathscr{F}M$-derivation in the second, i.e., $\nabla_X fV = f\nabla_X V + Df.XV$.*

(b) *If V is of the form $V = Y \circ F$ where Y is a vector field on N then $\nabla_X V(p)$ is given by $\big(\nabla_{T_pF\,.\,X(p)} Y\big)\big(F(p)\big)$.*

(ii) *Assume $\dim N < \infty$. Then a mapping $(**)$ satisfying* (a), (b) *necessarily has to be of the form given above. That is, it is characterized by these properties.*

Note. For $M = N$, $F = id$ we get back (1.5.2).

Proof. With the definition $(*)$ of $\nabla_X V$ by its local principal parts, we get indeed a vector field along F. This is easily verified just as we did this after (1.5.1). Also, (a) clearly holds. To verify (b) we observe, with $V(u) = Y\big(F(u)\big)$,

$$\begin{aligned}(\nabla_X V)(u) &= DY\big(F(u)\big) \,.\, DF(u) \,.\, X(u) + \Gamma\big(F(u)\big)\big(DF(u) \,.\, X(u), Y(F(u))\big) \\ &= DY\big(F(u)\big) \,.\, (TF \,.\, X)\big(F(u)\big) + \Gamma\big(F(u)\big)\big((TF \,.\, X)(F(u)), Y(F(u))\big).\end{aligned}$$

Let now $\dim N < \infty$. Writing the principal part $V(u)$ of V with respect to the basis vector fields $\{(v, f_j)\}$ on $v(N')$,

$$V(u) = \sum_j V^j(u) f_j,$$

we find for $(\nabla_X V)(u)$ using (a)

$$(\nabla_X V)(u) = \sum_j D V^j(u) \,.\, X(u) f_j + \sum_j V^j(u) (\nabla_X f_j)(u) \,.$$

The first summand is of the form $DV(u) \,.\, X(u)$, while the second one is determined by property (b) which yields

$$\sum_j V^j(u) \Gamma(F(u)) (DF(u) \,.\, X(u), f_j) = \Gamma(F(u)) (DF(u) \,.\, X(u), V(u)). \quad \square$$

As an immediate consequence we get the

1.5.6 Proposition. *Let* $F\colon M \to N$ *be a morphism and assume that* N *has a covariant derivation* ∇. *Then, with* X, Y *vector fields on* M *and* V *a vector field along* F,

$$T(TF \,.\, X, TF \,.\, Y) = \nabla_X TF \,.\, Y - \nabla_Y TF \,.\, X - TF \,.\, [X, Y] \,,$$

$$R(TF \,.\, X, TF \,.\, Y) V = \nabla_X \nabla_Y V - \nabla_Y \nabla_X V - \nabla_{[X,Y]} V \,.$$

Proof. Go back to (1.5.3) and use (1.5.4) and the definition of $\nabla_X V$ given in (1.5.5). $\square$

1.5.7 Example. Let $F\colon t \in \mathbb{R}^k \to F(t) \in M$ be a singular k-dimensional surface, cf. the example after (1.4.7). For $k = 1$ we also write t instead of t^1. Let (t, e_i) or briefly e_i be the i-th basis vector field on $\mathbb{R}^k$. Instead of $TF \,.\, (t, e_i)$ we also write $\partial F(t)/\partial t^i$. That is, $\partial F(t)/\partial t^i$ is a vector field along F. In case $k = 1$, i.e., for a curve $c : I \to M$, I an interval, we also write $\dot{c}(t)$ instead of $dc(t)/dt$.

Assume now that M possesses a covariant derivation ∇. Let V be a vector field along F. Instead of $(\nabla_{e_i} V)(t)$ we also write $\nabla V(t)/\partial t^i$. This vector field along F is called the *i-th partial covariant derivative of* V. In the case $k = 1$ we also write $\nabla V(t)/\partial t$ or simply $\nabla V(t)$.

Let (u, M') be a chart for M. Instead of $u \circ F(t)$ we again write $F(t)$. Then $DF(t) \,.\, e_i = \partial F(t)/\partial t^i$ is the principal part of $(TF \,.\, e_i)(t)$ under the representation given by the chart (u, M'). The principal part of the representation $T_{F(t)} u \,.\, V(t)$ of $V(t)$ shall also be denoted by $V(t)$. Thus, $t \mapsto V(t)$ is a differentiable mapping of some subset of $\mathbb{R}^k$ into $\mathbb{E}$, the model space of M. With this, we get from (1.5.5) $(*)$ for the principal part of the representation of $\nabla_{e_i} V(t)$ the formula

$$\frac{\nabla V}{\partial t^i}(t) = \frac{\partial V}{\partial t^i}(t) + \Gamma(F(t)) \left(\frac{\partial F}{\partial t^i}(t), V(t) \right).$$

In the case $\dim M < \infty$, we can represent the principal part $V(t)$ with the help of the basis vector fields $\{(u, f_j)\}$ on $u(M')$,

$$V(t) = \sum_j V^j(t) f_j \,.$$

Then

$$\frac{\nabla V}{\partial t^i}(t) = \sum_l \left\{ \frac{\partial V^l}{\partial t^i}(t) + \sum_{j,k} \frac{\partial F^j}{\partial t^i}(t)\, V^k(t)\, \Gamma^l_{jk}\big(F(t)\big) \right\} f_l .$$

In particular, for $k = 1$, $t^1 = t$,

$$\frac{\nabla V}{dt}(t) = \sum_k \left\{ \frac{d V^k}{dt}(t) + \sum_{i,j} \frac{dF^i}{dt}(t)\, V^j(t)\, \Gamma^k_{ij}\big(F(t)\big) \right\} f_k .$$

We write down (1.5.6) once more for the special case of a 2-dimensional surface.

1.5.8 Proposition. *Let*

$$F\colon (x, y) \in \mathbb{R}^2 \mapsto F(x, y) \in M$$

be a singular surface. On M *let there be given a covariant derivative. Then, with the abbreviations introduced in* (1.5.7),

(i) $$\frac{\nabla}{\partial y}\frac{\partial F}{\partial x} - \frac{\nabla}{\partial x}\frac{\partial F}{\partial y} = T\left(\frac{\partial F}{\partial y}, \frac{\partial F}{\partial x}\right).$$

(ii) *If* V *is a vector field along* F *then*

$$\frac{\nabla}{\partial x}\frac{\nabla}{\partial y} V - \frac{\nabla}{\partial y}\frac{\nabla}{\partial x} V = R\left(\frac{\partial F}{\partial x}, \frac{\partial F}{\partial y}\right) V . \quad \square$$

Note. In (1.5.15) we will show – in a more general setting – that there always exists a covariant derivation on a manifold M. Again in (1.8) this will also follow from the fact that there always exists a Riemannian metric on M which gives rise to the so-called Levi-Cività covariant derivation.

As pointed out already at the end of (3.1), we are lead to ask whether for a vector bundle $\pi\colon P \to M$ over a manifold M there exists a distinguished complementary 'horizontal' space to the vertical space in $T_\xi P$. The existence of such a horizontal space amounts to π having an additional structure i.e., a structure which is essentially equivalent to a covariant derivation for the sections in π.

1.5.9 Definition. Let $\pi\colon P \to M$ be a vector bundle, modelled on $\mathbb{E} \times \mathbb{F}$. *A (linear) connection* is a bundle morphism

$$\begin{array}{ccc} TP & \xrightarrow{\;K\;} & P \\ \Big\downarrow \tau_P & & \Big\downarrow \pi \\ P & \xrightarrow{\;\pi\;} & M \end{array}$$

with the following property:

In bundle charts (Pu, u, M') and (TPu, Tu, TM') for π and τ_P, this diagram has the form

$$\begin{array}{ccc}(u, \xi, X, \eta) \in \{u, \xi\} \times (\mathbb{E} \times \mathbb{F}) & \mapsto & \big(u, \eta + \Gamma_U(u)(X, \xi)\big) \in \{u\} \times \mathbb{F} \\ \Big\downarrow & & \Big\downarrow \\ (u, \xi) \in \{u\} \times \mathbb{F} & \longmapsto & \{u\}\end{array}$$

Here the upper line is the local representation $K_U = Pu \circ K \circ TPu^{-1}$ of K. The so-called *Christoffel symbols* $\Gamma_U : U \to L(\mathbb{E}, \mathbb{F}; \mathbb{F})$ of K (w.r.t. the bundle chart (Pu, u, M')) occurring here shall be differentiable.

The *horizontal space* $T_{\xi h} P$ is defined by

$$\{\ker : K | T_\xi P : T_\xi P \to P_p \equiv \mathbb{F}_p\},$$

where $p = \pi\xi$.

1.5.10 Proposition. *Assume that the vector bundle $\pi : P \to M$, modelled on $\mathbb{E} \times \mathbb{F}$, is endowed with a linear connection K.*

Then τ_P possesses the splitting

$$\tau_P = \tau_{Ph} \oplus \tau_{Pv} : T_h P \oplus T_v P \to P.$$

Here, each fibre $T_\xi P$ of τ_P is written as

$$T_\xi P = T_{\xi h} P \oplus T_{\xi v} P$$

There exist the following canonical isomorphisms, with $p = \pi\xi$,

$$T_\xi \pi | T_{\xi h} P : T_{\xi h} P \to T_p M \cong \mathbb{E},$$

$$K | T_{\xi v} P : T_{\xi v} P \to P_p \equiv \mathbb{F}_p \cong \mathbb{F}.$$

Proof. Using the bundle charts (Pu, u, M'), (TPu, Tu, TM'), we have for $T_{\xi h} P$ and $T_{\xi v} P$ the representations

$$\{u, \xi\} \times \{X, -\Gamma_u(u)(X, \xi); X \in \mathbb{E}\} \subset \{u, \xi\} \times \mathbb{E} \times \mathbb{F},$$

$$\{u, \xi\} \times \{0\} \times \mathbb{F} \subset \{u, \xi\} \times \mathbb{E} \times \mathbb{F}.$$

See the description of K_U in (1.5.9) and (1.3.14). This clearly is a splitting of the representative $\{u, \xi\} \times \mathbb{E} \times \mathbb{F}$ of $T_\xi P$.

According to (1.3.13), $T\pi | T_{\xi h} P$ is represented by

$$(u, \xi, X, -\Gamma_U(u)(X, \xi)) \mapsto (u, X)$$

and the representation of $K | T_{\xi v} P$ is given by

$$(u, \xi, 0, \eta) \mapsto (u, \eta). \quad \square$$

1.5.11 Definition. Let $\pi: P \to M$ be endowed with a connection K. Let $\xi: M \to P$ be a section in π. The *covariant derivation of* ξ *in the direction* $X \in T_p M$ is defined by

$$\nabla_X \xi(p) = K \circ T_p \xi . X .$$

In analogy to (1.5.2) we have the

1.5.12 Proposition. *Let* $X: M \to TM$ *be a vector field on* M *and* $\xi: M \to P$ *be a section in* π. *Then*

$$p \in M \mapsto \nabla_X \xi(p) \in P$$

is again a section. It is $\mathcal{F}M$*-linear in the argument* X *and a* $\mathcal{F}M$*-derivation in the argument* ξ, *i.e.,*

$$(*) \qquad \begin{cases} \nabla_{fX+gY}\xi = f\nabla_X\xi + g\nabla_Y\xi . \\ \nabla_X(f\xi + g\eta) = X(f)\xi + f\nabla_X\xi + X(g)\eta + g\nabla_X\eta , \end{cases}$$

with f, g *in* $\mathcal{F}M$ *and* X, Y *in* $\mathfrak{V}M$.

Proof. $T\xi . X$ satisfies properties analogous to those formulated under $(*)$. In particular,

$$T(f\xi) . X = X(f)\xi + fT\xi . X,$$

where on the right hand side $X(f)\xi \equiv Df . X\xi$ is being considered as an element of $T_{\xi v}P \subset T_\xi P$. Now apply $K | T_\xi P$. $\square$

Note. In the local representation given by (Pu, u, M'), (TPu, Tu, TM') we get for $\nabla_X \xi$ the formula

$$K_U \circ (u, \xi, X, D\xi . X) = \big(u, D\xi . X + \Gamma(u)(X, \xi)\big)$$

Example. A covariant derivation ∇ on a manifold M determines a linear connection K for the tangent bundle $\tau_M: TM \to M$ via the local Christoffel symbols $\Gamma = \Gamma_U: U \to L(\mathbb{E}, \mathbb{E}; \mathbb{E})$. In this way we get a 1:1 correspondence between ∇ and K.

In (1.5.5) we defined the induced covariant derivation for vector fields along a morphism $F: M \to N$, provided we have a covariant derivation for the sections (= vector fields) of the tangent bundle τ_N of N. This generalizes to the concept of an induced (linear) connection.

1.5.13 Proposition. *Let* K_π *be a linear connection for the vector bundle* $\pi: P \to N$. *Let* $F: M \to N$ *be a morphism and consider the induced bundle* $F^*\pi: F^*P \to M$ *over* M, *cf.* (1.3.10). *Then there exists an* induced connection $K_{F^*\pi}$ *determined by the condition that the following diagram is a bundle morphism:*

$$\begin{array}{ccc} TF^*P & \xrightarrow{T\pi^*F} & TP \\ \Big\downarrow K_{F^*\pi} & & \Big\downarrow K_\pi \\ F^*P & \xrightarrow{\pi^*F} & P \end{array}$$

The local representation of this diagram reads

$$\begin{array}{ccc} (u, \xi, X, \eta) & \longmapsto & (F(u), \xi, DF(u) \,.\, X, \eta) \\ \Big\downarrow & & \Big\downarrow \\ (u, \eta + \Gamma(F(u))(DF(u) \,.\, X, \xi)) & \longmapsto & (F(u), \eta + \Gamma(F(u))(DF(u) \,.\, X, \xi)) \end{array}$$

Note. Here we used a bundle chart (Pv, v, N') of N, a chart (u, M') on M with $F(M') \subset N'$ and the induced bundle chart $(F^* Pu, u, M') \equiv (Pv \circ \pi^* F, u, M')$. We have also written $F(u)$ instead of $(v \circ F \circ u^{-1})(u)$.

Proof. $K_{F^*\pi}$ is well defined, since its restriction to a tangent space can be written as

$$(\pi^* F | F^* P_p)^{-1} \circ K_\pi \circ (T\pi^* F | T_\xi F^* P) .$$

Recall that, according to the definition of the induced bundle,

$$\pi^* F | F^* P_p : F^* P_p \to P_{F(p)}$$

is a linear isomorphism.

The local representation of $K_{F^*\pi} \,.\, \eta$ with $\eta \in T_\xi F^* P$, $X \in T_p M$, $p = F^* \pi \xi$, is clear from the definition. □

Remark. Compare this with the formula (∗) for $\nabla_X V(u)$ in (1.5.5).

1.5.14 Example. Let $\pi: P \to M$ be a vector bundle and $c: I \to M$ a curve on the base manifold of π. Let π be endowed with a connection K. For a section $\xi(t)$ along $c(t)$, cf. (1.4.16), we have the covariant derivation

$$\frac{\nabla \xi}{dt}(t) = K \circ \frac{D\xi}{dt}(t) .$$

Let (Pu, u, M') be a bundle chart for π. Put $u \circ c(t) = u(t)$ and let $\xi(t)$ be represented by $(t, \xi(t))$. Then the splitting of $D\xi(t)/dt$ into its horizontal and vertical component reads – we only write down the principal part –

$$(\dot u(t), \dot\xi(t)) = (\dot u(t), -\Gamma(u(t))(\dot u(t), \xi(t))) + (0, \dot\xi(t) + \Gamma(u(t))(\dot u(t), \xi(t))) .$$

Compare this with (1.5.7).

A linear connection K on $\pi\colon P \to M$ also determines a 'curvature tensor', cf. (1.5.3), (1.5.4):

$$R(X, Y)\xi = \nabla_X \nabla_Y \xi - \nabla_Y \nabla_X \xi - \nabla_{[X,Y]}\xi = -R(Y, X)\xi\,.$$

Here, X, Y are vector fields on M and ξ is a section in π.

A torsion tensor does not make sense for a connection in a general bundle. For additional information see Flaschel und Klingenberg [1].

We will now show that a vector bundle $\pi\colon P \to M$ always has a (linear) connection. For that purpose we consider a bundle atlas $(Pu_\alpha, u_\alpha, M_\alpha)_{\alpha\in A}$ of π. From (1.0) 4 we know that the underlying atlas $(u_\alpha, M_\alpha)_{\alpha\in A}$ of M admits a partition of unity $(\phi_\beta, \tilde{M}_\beta)_{\beta\in B}$. If $\sigma\colon B \to A$ is such that $\tilde{M}_\beta \subset M_{\sigma\beta}$ then we define a bundle atlas $(Pu_\beta, u_\beta, \tilde{M}_\beta)_{\beta\in B}$ by

$$u_\beta = u_{\sigma\beta} | \tilde{M}_\beta;\ Pu_\beta = Pu_{\sigma\beta} | \pi^{-1}\tilde{M}_\beta\,.$$

The $\phi_\beta\colon M \to \mathbb{R}$ can be chosen to be differentiable, cf. (1.0) 4.

1.5.15 Theorem. *Let $\pi\colon P \to M$ be a vector bundle modelled on $\mathbb{E} \times \mathbb{F}$. π admits a bundle atlas $(Pu_\alpha, u_\alpha, M_\alpha)_{\alpha\in A}$ with an associated partition of unity $(\phi_\alpha, M_\alpha)_{\alpha\in A}$. For such a pair we construct a linear connection K for π as follows. For each $\alpha \in A$ define*

$$K_\alpha\colon TPM_\alpha \to PM_\alpha$$

by its representation $K_{U_\alpha} = Pu_\alpha \circ K_\alpha \circ TPu_\alpha^{-1}$ which shall be the canonical connection mapping

$$\{u_\alpha, \xi_\alpha\} \times (\mathbb{E} \times \mathbb{F}) \to \{u_\alpha\} \times \mathbb{F}\,;\quad (u_\alpha, \xi_\alpha, X_\alpha, \eta_\alpha) \mapsto (u_\alpha, \eta_\alpha)\,.$$

With this define

$$K | T_\xi P\colon T_\xi P \to P_p = \pi^{-1}(p);\ p = \pi\xi$$

by

$$K\eta = \sum_{\alpha\in A} \phi_\alpha(p) K_\alpha \,.\, \eta\,.$$

Proof. That K indeed is a linear connection is almost trivial. We only write down the Christoffel symbol Γ_U of K with respect to an arbitrary bundle chart (Pu, u, M') of π. Let ${}_\alpha\Gamma_U$ be the Christoffel symbol of $K_\alpha | T(PM' \cap PM_\alpha)$. Then

$$\Gamma_U(u) = \sum_\alpha \phi_\alpha(u_\alpha^{-1}(u))_\alpha\Gamma_U(u)\,. \quad \square$$

See Flaschel und Klingenberg [1] for further remarks.

Remark. In (1.2.11) we associated with r bundles $\alpha_i\colon A_i \to M$; $1 \leqslant i \leqslant r$, and a bundle $\beta\colon B \to M$ the bundle

$$L(\alpha_1, \ldots, \alpha_r; \beta): L(A_1, \ldots, A_r; B) \to M$$

of linear mappings. Assume now that on each of the α_i and on β we have given a covariant derivation ∇. Then one can define for a section ζ in $L(\alpha_1, \ldots \alpha_r; \beta)$ and a vector field X on M the covariant derivation $\nabla_X \zeta$ of ζ in the direction X by the formula – with $\xi_i \in \Gamma(\alpha_i)$ = space of sections in α_i –

$$\nabla_X \zeta(\xi_1, \ldots, \xi_r) = \nabla_X(\zeta(\xi_1, \ldots, \xi_r)) - \sum_{i=1}^{r} \zeta(\xi_1, \ldots, \nabla_X \xi_i, \ldots, \xi_r).$$

This is a $\mathscr{F}M$-linear mapping in all arguments $(X, \xi_1, \ldots, \xi_r) \in \mathfrak{V}M \times \Gamma(\alpha_1) \times \ldots \times \Gamma(\alpha_r)$. However, for infinite dimensions such a multilinear mapping does not determine a tensor in the bundle $L(\tau_M, \alpha_1, \ldots, \alpha_r; \beta)$. Only if the bundles are modelled on vector spaces of finite dimension this is possible, cf. (1.4.14) for the case of tensor bundles.

We now restrict ourselves to this case and show that a covariant derivation on M (or, more precisely, on the tangent bundle $\tau \equiv \tau_0^1: TM \equiv T_0{}^1 M \to M$ of M) extends to such a derivation on all tensor bundles τ_s^r in a canonical way.

Actually, such an extension is possible for every sort of derivation, cf. Abraham and Marsden [1] or Kobayashi and Nomizu [1] for details.

1.5.16 Lemma. *Let* dim $M < \infty$ *and assume* M *to be endowed with a covariant derivation* ∇. *For a vector field* X *on* M, ∇_X *is defined on* 1-*forms* ω *by*

$$\nabla_X \omega(X_1) = X.(\omega(X_1)) - \omega(\nabla_X X_1).$$

Thus, $\nabla_X \omega$ *is again a* 1-*form.*

For a tensor field Z *of type* $\binom{r}{s}$ *define* $\nabla_X Z$ *by the formula*

$$\begin{aligned}\nabla_X Z(X_1, \ldots, X_s, \omega_1, \ldots, \omega_r) &= X.Z(X_1, \ldots X_s, \omega_1, \ldots, \omega_r)\\ &- \sum_{j=1}^{s} Z(X_1, \ldots, \nabla_X X_j, \ldots, X_s, \omega_1, \ldots, \omega_r)\\ &+ \sum_{i=1}^{r} Z(X_1, \ldots, X_s, \omega_1, \ldots, \nabla_X \omega_i, \ldots, \omega_r),\end{aligned}$$

with X_j, $1 \leqslant j \leqslant s$, *vector fields and* ω_i, $1 \leqslant i \leqslant r$, 1-*forms.* $\nabla_X Z$ *again is a tensor field of type* $\binom{r}{s}$.

Proof. The fact that $\nabla_X X_1$ is a derivation in the second argument implies that $\nabla_X \omega$ is a $\mathscr{F}M$-linear mapping from $\mathfrak{V}M = \Gamma(\tau_0^1)$ into M, i.e., according to (1.4.14), $\nabla_X \omega \in \Gamma(\tau_1^0)$. With this one verifies that $\nabla_X Z \in \Gamma(\tau_s^r)$. □

1.6 The Exponential Mapping

In this section we consider manifolds M endowed with a covariant derivation and later on, in (1.6.13), vector bundles over M with a linear connection.

The basic construction one can carry out here is the parallel transport $\|c$ along a curve $c\colon I \to M$ of the tangent space (or, more generally, the fibre) of the initial point of c into the tangent space (or fibre) of the end point of c, cf. (1.6.2), (1.6.14).

Those curves c for which the tangent vector constitutes a parallel vector field along c are called geodesics, see (1.6.4). Given any $X_0 \in T_p M$, there always exists for small $|t|$ a geodesic $c(t)$ with $\dot{c}(0) = X_0$. This leads to the fundamental concept of the exponential mapping $\exp\colon TM \to M$, in general defined only on some open neighborhood of the zero-section $M \subset TM$ of the bundle τ_M, cf. (1.6.9).

The behaviour of geodesics on M in the neighborhood of a point resembles the behaviour of straight segments in a small ball of the model space $\mathbb{E}$ of M; this is the gist of (1.6.12)ff.

We begin with a generalization of the concept of parallelity in a Hilbert space. To make possible such a generalization to general manifolds one has to restrict oneself to vector fields along a curve. $I \subset \mathbb{R}$ denotes an interval.

1.6.1 Definition. Let $c\colon I \to M$ be a curve. A vector field V along c is called *parallel* if $\nabla V(t)/dt = 0$.

Example. Take $M = \mathbb{E}$ with the canonical derivation $\nabla = D$. Then $\nabla V(t)/dt = DV(t)/dt = 0$ means that the principal part of $V(t)$ is constant, i.e., $V(t)$ is parallel in the usual sense.

In (1.7.7), (1.7.8) we will generalize this to arbitrary Lie groups.

1.6.2 Lemma. *Let $c\colon I \to M$ be a curve; t_0, t_1 in I.*

(i) *For every $V_0 \in T_{c(t_0)}M$ there exists exactly one parallel vector field $V(t)$ along $c(t)$ such that $V(t_0) = V_0$.*

(ii) *The mapping*

$$\overset{t_1}{\underset{t_0}{\|}} c\colon T_{c(t_0)}M \to T_{c(t_1)}M$$

under which we associate with V_0 the value $V(t_1)$ of the parallel vector field $V(t)$ determined in (i) is a linear isomorphism. It is called parallel translation.

Proof. It suffices to consider the case where $c|[t_0, t_1]$ has its image in the domain of a single chart (u, M'). Indeed, there always will be a finite subdivision of the interval $[t_0, t_1]$ such that the image of each of the subdivisions lies entirely in the domain of a chart. From the validity of the lemma for each subdivision the validity of the lemma for the whole interval $[t_0, t_1]$ follows.

Put $u \circ c(t) = u(t)$. Using the notations introduced in (1.5.7), we write for the principal part of a vector field $V(t)$ along $c(t)$ again $V(t)$. $\nabla V(t)/dt$ also denotes the principal part of the covariant derivative $(\nabla_{e_1} V)(t) = \nabla V(t)/dt$ of $V(t)$.

That $V(t)$ is parallel means

$$(*) \qquad \frac{\nabla V}{dt}(t) = \frac{dV}{dt}(t) + \Gamma(u(t))\left(\frac{du}{dt}(t), V(t)\right) = 0\,.$$

The Lemma now is a consequence of a standard result on ordinary linear differential equations of the first order. The solutions $V(t)$ of $(*)$ form a vector space isomorphic to $\mathbb{E}$. The isomorphism is established by associating to a solution $V(t)$ its value $V(t^*)$ at a fixed $t^* \in I$, cf. Dieudonné [1]. □

While there always exist parallel vector fields $V(t)$ along a curve $c(t)$ with prescribed value V_0 at some $t = t_0$, in general there will not exist parallel vector fields $V(x, y)$ along a singular surface $F(x, y)$ with a prescribed value V_0 at some $(x, y) = (x_0, y_0) \in \mathbb{R}^2$.

1.6.3 Theorem. *Let M be a manifold with covariant derivation ∇. If and only if the curvature tensor R of ∇ vanishes, every singular surface in M possesses parallel vector fields with prescribed initial values.*

Note. A similar result holds for k-dimensional singular surfaces $F\colon \mathbb{R}^k \to M$ with arbitrary $k > 1$.

Proof. Let $F(x, y)$, $(x, y) \in \mathbb{R}^2$, be a singular surface in M. Fix $(x_0, y_0) \in \mathbb{R}^2$. Put $F(x_0, y_0) = q$ and choose $V_0 \in T_q M$. We want to find a vector field V along F satisfying $V(x_0, y_0) = V_0$ and $\nabla V(x, y)/\partial x = \nabla V(x, y)/\partial y = 0$.

From (1.6.2) we have a parallel vector field $V(x, y_0)$ along $\{x \to F(x, y_0)\}$ with $V(x_0, y_0) = V_0$. Again (1.6.2) yields the existence, for every x, of a parallel vector field $V(x, y)$ along $\{y \mapsto F(x, y)\}$ having at (x, y_0) the value $V(x, y_0)$.

Thus, $V(x, y)$ satisfies $V(x_0, y_0) = V_0$ and

$$\nabla V(x, y)/\partial y = 0; \ \nabla V(x, y_0)/\partial x = 0\,.$$

From (1.5.8, ii) we get: If $R \equiv 0$ then $\nabla V(x, y)/\partial x$ is parallel along $\{y \mapsto F(x, y)\}$. Since this vector field vanishes for $y = y_0$, it is zero everywhere, i.e., V is parallel along F.

Conversely, given $q \in M$ and three elements X, Y, Z in $T_q M$, here exists a surface F with $F(x_0, y_0) = q$, $\partial F(x_0, y_0)/\partial x = X$, $\partial F(x_0, y_0)/\partial y = Y$. The existence of a parallel vector field $V(x, y)$ along $F(x, y)$ with $V(x_0, y_0) = Z$ therefore implies, according to (1.5.8, ii), $R(X, Y)Z = 0$. □

We now come to the most important class of curves on a manifold with covariant derivation:

1.6.4 Definition. A curve $c\colon I \to M$ is called a *geodesic* if $\nabla \dot{c}(t)/dt = 0$.

Example. If M is a Hilbert space $\mathbb{E}$ with ∇ given by $D, D\dot{c}(t)/dt = 0$ means $c(t) = at + b$. Hence, a geodesic on $(\mathbb{E}, D)$ either is a straight line or else, a constant mapping.

1.6.5 Lemma. *For every $X \in TM$ there exists an open interval $J(X)$ containing $0 \in \mathbb{R}$ and a geodesic $c = c_X : J(X) \to M$ with $\dot{c}(0) = X$. The pair $(c_X, J(X))$ is uniquely determined if we choose $J(X)$ to be maximal with respect to these properties.*

In addition, for any real $a \neq 0$, $(c_{aX}, J(aX))$ is given by $c_{aX}(t) = c_X(at)$ and $J(aX) = a^{-1}J(X)$, for $a > 0$. For $a < 0$, $J(aX)$ is obtained from $J(|a|X)$ by reflection at the origin.

Proof. Let (u, M') be a chart containing $p = \tau X \in M$. For $c(t)$ a curve on M we put $u \circ c(t) = u(t)$. Then the principal part of the representation of $\dot{c}(t)$ is $\dot{u}(t)$. Thus, from (1.5.6) for the vanishing of the principal part of $\nabla \dot{c}(t)/dt$ we obtain the equation

$$(*) \qquad \ddot{u}(t) + \Gamma(u(t))(\dot{u}(t), \dot{u}(t)) = 0.$$

A standard result on ordinary differential equations of second order (cf. Dieudonné [1]) states: To a given $X_0 \in \mathbb{E}$, there exists an interval J containing 0 and a solution $u(t)$, $t \in J$, of $(*)$ satisfying $\dot{u}(0) = X_0$. Take now for X_0 the principal part of the representative of X. Moreover, if there is a second solution $u'(t)$, $t \in J'$, J' containing 0, with $\dot{u}'(0) = X_0$ then $u(t) = u'(t)$ for $t \in J \cap J'$.

This proves the uniqueness of the geodesic $c(t)$ and its domain of definition J on the submanifold M' of M. If $c(t_0)$ is in the domain of a chart (u', M''), consider the geodesic $c'(t)$ on M'' determined by $\dot{c}'(0) = \dot{c}(t_0)$. Then $c(t)$ can be extended into M'' by $c(t + t_0) = c'(t)$. This completes the proof of the first part.

The second part now follows from $c(at)^{\cdot} = a\dot{c}(at)$. □

1.6.6 Remark. Let ∇ and ∇^* be covariant derivations on M. We ask when the geodesics defined by ∇ and by ∇^* and having the same initial vector X coincide, for all $X \in TM$. This is equivalent to

$$\nabla_{\dot{c}}\dot{c} = \nabla^*_{\dot{c}}\dot{c}$$

for all curves $c: I \to M$.

We want to describe this property in terms of the Christoffel symbols $\Gamma(u), \Gamma^*(u)$ of ∇, ∇^* with respect to a chart (u, M') of M. First note that

$$S(u) = \Gamma^*(u) - \Gamma(u)$$

always is the principal part of a tensor field S of type $\binom{1}{2}$. This follows at once from the transformation formula (†) in (1.5.1). If the geodesics $u(t)$, $u^*(t)$, defined by $\Gamma(u)$, $\Gamma^*(u)$ with $X_0 = \dot{u}^*(0)$ are always the same, $S(u)$ must be antisymmetric in its covariant arguments,

$$\Gamma(u)(X_0, X_0) - \Gamma^*(u)(X_0, X_0) = 0.$$

In particular, if we put $S = -T/2$, where T is the torsion tensor of ∇, then we get from ∇ a torsion-free derivation ∇^* having the same geodesics as ∇. The Christoffel symbols of ∇^* are given by $\Gamma^*(u)(X, Y) = (\Gamma(u)(Y, X) - \Gamma(u)(Y, X))/2$.

We now formulate a local-global result on the existence of geodesics.

1.6.7 Lemma. *For every $p \in M$ there exists a chart (u, M') containing p with $u(p) = 0$ and open neighborhoods $0 \in U_1 \subset U_2 \subset U = u(M')$ and a $\eta > 0$ such that the following holds. Given X with $\tau X = q \in u^{-1}(U_1)$ and $|X(u(q))| < \eta$, the geodesic c_X with $\dot{c}_X(0) = X$ is defined for all $|t| < 2$ and for these t, $c_X(t) \in u^{-1}(U_2)$.*

Proof. Start with a chart (u, M') satisfying $u(p) = 0$. Rewrite the equation $(*)$ from the proof of (1.6.5) as

$$(**) \qquad \dot{u} = v;\ \dot{v} = -\Gamma(u)(v, v).$$

It is known, cf. Dieudonné [1], that there exist $\varepsilon_1 > 0$, $\varepsilon_2 > 0$ and open neighborhoods U_1, U_2 of 0, $U_1 \subset U_2 \subset U$ such that, for $u_0 \in U_1$, $v_0 \in B_{\varepsilon_1}(0) \subset \mathbb{E}$ there exists a solution

$$(***) \qquad u(t) \equiv u(t; u_0, v_0);\ v(t) \equiv v(t; u_0, v_0),\ |t| < 2\varepsilon_2$$

of $(**)$ which lies entirely in U_2 and has u_0, v_0 as initial conditions, i.e., $(u(0), v(0)) = (u_0, v_0)$.

Choose now η with $0 < \eta < \varepsilon_1 \varepsilon_2$. From (1.6.5) we have $c_X(t) = c_{X/\varepsilon_2}(t\varepsilon_2)$. This shows that we have satisfied all properties of the lemma. □

1.6.8 Corollary. *There exists an open neighborhood $\tilde{T}M$ of the submanifold M of TM such that, for every $X \in \tilde{T}M$, the geodesic $c_X(t)$ is defined for $|t| < 2$.* □

1.6.9 Definition. Let $\tilde{T}M$ be as in (1.6.8). Define the *exponential mapping*

$$\exp\colon \tilde{T}M \to M$$

by $X \mapsto c_X(1)$. The restriction $\exp|\tilde{T}M \cap T_pM$ to a fibre in $\tilde{T}M$ will also be denoted by $\exp_p$.

Remark. For an explanation of the terminology we refer to (1.7.11).

1.6.10 Proposition. *The exponential map is differentiable.*

Proof. With respect to a chart of the type considered in (1.6.7), exp possesses the local representation

$$(u_0, v_0) \in U_1 \times B_\eta(0) \subset TU \mapsto u(1; u_0, v_0) \in U_2 \subset U.$$

Observe now that the solutions $(***)$ of $(**)$ in the proof of (1.6.7) depend differentiably on the initial conditions. □

1.6.11 Definition. Denote by $C^0(I, M)$ the *space of all continuous maps*

$$c: I \to M, I = [0, 1]$$

endowed with the *compact-open topology or* – what is the same – with the *topology of uniform convergence*, where we take some metric on M – recall that M was supposed to be metrizable.

Remark. Let $\{c_n\}$ be a sequence in $C^0(I, M)$. Then $\lim c_n = c \in C^0(I, M)$ means the following: First of all, for every $t \in I$, $\lim c_n(t) = c(t)$, i.e., we have point-wise convergence. To describe the uniformity of the convergence, we observe that the limit curve c can be covered by a finite number of charts. Then for all n sufficiently large, c_n is also covered by these charts.

Consider now one of the charts, say (u, M'), and put $u \circ c_n(t) = u_n(t)$, $u \circ c(t) = u(t)$, where t in general will have to be restricted to a subinterval J of I. Then, for every $\varepsilon > 0$, there exists an integer $n(\varepsilon)$ such that $|u_n(t) - u(t)| < \varepsilon$ for all $n > n(\varepsilon)$, all $t \in J$. One easily verifies that this is independent of the choice of the finite covering of c by charts.

An alternative description of uniform convergence is given by the property: For every convergent sequence $\{t_n\}$ on I, $\lim c_n(t_n) = c(\lim t_n)$.

Indeed, if c is the uniform limit of $\{c_n\}$ and $\lim t_n = t_0$ we have in appropriate local coordinates

$$|u_n(t_n) - u(t_0)| \leqslant |u_n(t_n) - u(t_n)| + |u(t_n) - u(t_0)|$$

for n sufficiently large. Hence, $|u_n(t_n) - u(t_0)| \to 0$ for $n \to \infty$. Conversely, if $\{c_n\}$ does not uniformly converge to c, there exists an $\varepsilon > 0$ and a sequence $\{t_n\}$ on I (which we can assume to be convergent with limit t_0) such that $|u_n(t_n) - u(t_n)| \geqslant \varepsilon$, for infinitely many n. Hence,

$$|u_n(t_n) - u(t_0)| + |u(t_0) - u(t_n)| \geqslant |u_n(t_n) - u(t_n)| \geqslant \varepsilon$$

for infinitely many n. Since $\lim u(t_n) = u(t_0)$, $\{u_n(t_n)\}$ does not converge to $u(t_0)$.

We now can prove the fundamental result on the existence and local uniqueness of geodesics joining two points:

1.6.12 Theorem. *For every point p of M there exists a neighborhood W of p on M and a neighborhood $\tilde{W}$ of 0_p (= 0-vector in T_pM) in TM such that the following holds:*

(i) *For every pair q, r in W there exists exactly one geodesic*

$$c_{qr}: [0, 1] \to M$$

from q to r, i.e., $c_{pr}(0) = q$, $c_{pr}(1) = r$, such that $\dot{c}_{qr}(0) \in \tilde{W}$.

(ii) *The mapping*

$$(q, r) \in W \times W \mapsto c_{qr} \in C^0([0, 1]; M)$$

is continuous

(iii) *For every* $q \in W$, $\exp|\tilde{W} \cap T_q M$ *is a diffeomorphism onto a neighborhood* $W(q)$ *of* q, $W(q) \supset W$.

Remark. One can even choose W such that c_{qr} has its image entirely in W. Cf. Whitehead [1] for the finite dimensional case. For the Riemannian case cf. (1.9.9).

Proof. Consider the differentiable mapping

$$F\colon \tilde{T}M \to M \times M;\ X \mapsto (\tau X, \exp X).$$

We claim that

$$T_{0_p} F\colon T_{0_p} \tilde{T}M \to T_{(p,p)}(M \times M)$$

is an isomorphism.

To see this we consider the local representation of F, cf. the proof of (1.6.10),

$$(u_0, v_0) \mapsto (u_0, u(1; u_0, v_0)).$$

Since $u(1; u_0, 0) = u_0$ and $D_2 u(1; 0, 0) = id$ (D_2 denotes the differential with respect to the second set of variables v_0), the differential of the local representation of F is of the form

$$\begin{pmatrix} id & 0 \\ id & id \end{pmatrix}$$

The inverse mapping theorem (1.0.3) yields the existence of a neighborhood $\tilde{W}$ of 0_p in $\tilde{T}M$ such that

$$F|\tilde{W}\colon \tilde{W} \to W(p, p)$$

is a diffeomorphism onto a neighborhood $W(p, p)$ of (p, p) in $M \times M$.

Now choose a neighborhood W of p in M such that $W \times W \subset W(p, p)$. For $(q, r) \in W \times W$ put $(F|\tilde{W})^{-1}(q, r) = X_{qr} \in \tilde{W}$. Then $\tau X_{qr} = q$, $\exp_q X_{qr} = r$. Hence

$$c_{qr}(t) = \exp_q(t X_{qr})$$

is the geodesic satisfying the properties formulated in (i).

To prove (ii) we invoke the continuous dependence of a geodesic $c_X(t)$, $t \in [0, 1]$, on its initial value X. Finally, (iii) follows from

$$F(\tilde{W} \cap T_q M) = \{q\} \times \exp(\tilde{W} \cap T_q M)$$

where $\tilde{W} \cap T_q M$ is a submanifold of $\tilde{W}$. Since

$$(F|\tilde{W})^{-1}(\{q\} \times W) \subset \tilde{W} \cap T_q M,$$

also $W(q) \supset W$. $\square$

We conclude with the observation that the concept of parallel translation, introduced in (1.6.1), can be generalized to general vector bundles with a linear connection.

1.6.13 Definition. Let $\pi: P \to M$ be a vector bundle with connection K, c.f. (1.5.9). Let $c: I \to M$ be a curve on M and $\xi(t)$ a section along $c(t)$, ξ is called *parallel* if $\nabla \xi(t)/dt = 0$.

The generalization of (1.6.2) reads as follows.

1.6.14 Lemma. *Let $c: I \to M$ be a curve, $\pi: P \to M$ a vector bundle with connection K and fibre modelled on $\mathbb{F}$. Choose t_0, t_1 in I.*

(i) *For every $\xi_0 \in \mathbb{F}_{c(t_0)}$ (the fibre over $c(t_0)$) there exists exactly one parallel vector field $\xi(t)$ along $c(t)$ such that $\xi(t_0) = \xi_0$.*

(ii) *The mapping*

$$\overset{t_1}{\underset{t_0}{\|}} c: \mathbb{F}_{c(t_0)} \to \mathbb{F}_{c(t_1)}$$

under which we associate to ξ_0 the value $\xi(t_1)$ of the parallel vector field $\xi(t)$ determined by (i) *is a (linear) isomorphism. It is called* parallel translation.

Proof. As in (1.6.2), this is an immediated consequence of the local representation of $\nabla \xi(t)/dt = 0$, cf. (1.5.14),

$$\dot{\xi}(t) + \Gamma(u(t))(\dot{u}(t), \xi(t)) = 0. \quad \square$$

1.7 Lie Groups

We will now discuss a particularly important class of manifolds, i.e., manifolds which possess a group structure such that the composition is differentiable. As an example we have already the Hilbert space with its vector space composition.

To remain in the domain of classical examples we restrict ourselves to finite dimensions, i.e., to the so-called Lie groups G. Associated to G is its Lie algebra $\mathfrak{g}$, consisting of the left-invariant vector fields, cf. (1.7.5). We have a canonical so-called left-invariant covariant derivation on G. Its curvature tensor vanishes while its torsion tensor in general does not, see (1.7.8).

1-parameter subgroups correspond to the geodesics passing through the neutral element.

We can also define a symmetric covariant derivation on G, giving rise to the same geodesics as the left-invariant derivation, see (1.7.12).

The adjoint representation of G as automorphisms of $\mathfrak{g}$ is important. Its differential is the adjoint representation of $\mathfrak{g}$ as derivations of $\mathfrak{g}$, see (1.7.17). This leads to the Killing form on $\mathfrak{g}$ which allows us to define the concept of semi-simplicity, cf. (1.7.20).

As examples of Lie groups we study in particular the general linear group $GL(n, \mathbb{R})$ and the orthogonal group $\mathbb{O}(n)$.

1.7.1 Definition. (i) A *Lie group* is a manifold, modelled on a finite dimensional space, on which there is defined a group structure such that

$$(*) \qquad (g, h) \in G \times G \mapsto gh^{-1} \in G$$

is differentiable. The neutral element of G will be denoted by e.

(ii) For every element g of a Lie group G we define the *left translation* (*right translation*) $L_g (R_g)$ by

$$L_g: G \to G;\, h \mapsto gh,$$

$$R_g: G \to G;\, h \mapsto hg.$$

1.7.2 Proposition. *$L_g: G \to G$ ($R_g: G \to G$) is a diffeomorphism. In particular, $L_g^{-1} = L_{g^{-1}}$ ($R_g^{-1} = R_{g^{-1}}$).*

Proof. Clear from the definitions. □

1.7.3 Examples. (i) $G = \mathbb{R}^n$, with the additive composition.

(ii) $G = GL(n; \mathbb{R})$, c.f. (1.1.3, iv), with the multiplicative composition.

To see that the composition is differentiable we take the i-th row and k-th column. Then $u_{ik}(AB) = \sum_j u_{ij}(A)\, u_{jk}(B)$ and $u_{ik}(B^{-1})$ is a rational function in the coordinates $\{u_{lm}(B)\}$ of B.

(iii) The orthogonal group $\mathbb{O}(n) = \{A \in GL(n, \mathbb{R});\, A^tA = E\}$. $\mathbb{O}(n)$ is a subgroup of $GL(n, \mathbb{R})$. It also is a submanifold of $GL(n, \mathbb{R})$, cf. (1.3.9, ii). To see that the composition on $\mathbb{O}(n)$ is differentiable we consider for $g_0, h_0, g_0 h_0^{-1}$ in $\mathbb{O}(n)$ submanifold charts (u, M'), (u', M''), (u'', M'') for $G(n, \mathbb{R})$ of the type described in (1.3.5). We know that

$$(\dagger) \qquad u''(gh^{-1}) = F\big(u(g), u'(h)\big)$$

where F is a differentiable function of (u, u'). Restricting u, u', u'' to $\mathbb{O}(n)$ means putting some of the coordinates equal to zero. But then (†) still describes a differentiable dependence of the $u''(gh^{-1})$ on the $u(g), u'(h)$ with g, h in $\mathbb{O}(n)$.

1.7.4 Definition. Let G be a Lie group. A vector field $\tilde{X}$ on G is called *left-invariant* (*right-invariant*) if $TL_g \tilde{X}(h) = \tilde{X}(gh)$ $\big(TR_g \bar{X}(h) = \bar{X}(hg)\big)$.

1.7.5 Proposition. *Let G be a Lie group. The set $\mathfrak{g}$ of left-invariant vector fields on G is a sub-algebra of the Lie-algebra of all vector fields on G.*

Proof. Use (1.4.10). □

Note. As vector space, $\mathfrak{g}$ is isomorphic to any tangent space of G and in particular to T_eG, the tangent space of G at the neutral element. $\mathfrak{g}$ with its Lie-algebra structure is called *Lie algebra of* G.

1.7.6 Examples. We continue with (1.7.3). First we observe: If $X_0 \in T_e G$ we can write

$$TL_g X_0 = d(gh(s))/ds|_{s=0}$$

where $s \in]-\varepsilon, \varepsilon[\mapsto h(s) \in G$ is a curve with $h'(0) = X_0$.

(i) $G = \mathbb{R}^n$. Then the left-invariant vector fields $\tilde{X}$ on G are just the parallel vector fields. Thus, the principal part $\tilde{X}(u)$ of a left-invariant vector field with respect to the canonical coordinates is constant, $D\tilde{X}(u) = 0$. Hence, $[\tilde{X}, \tilde{Y}] = 0$ for all $\tilde{X}, \tilde{Y}$ in $\mathfrak{g}$.

(ii) $G = GL(n; \mathbb{R})$. Take the canonical coordinates. Then $\mathfrak{g} = \mathfrak{gl}(n; \mathbb{R}) \cong T_e G$ is represented by $M(n; \mathbb{R})$. Let $X \in M(n; \mathbb{R})$. According to the above formula, the principal part of the left-invariant vector field at $A \in GL(n; \mathbb{R})$, having X as value in E, is given by AX. Thus, we find for the principal part of the Lie bracket of the two left-invariant vector fields determined by $X, Y \in M(n, \mathbb{R})$:

$$[X, Y] = (D(AY) \,.\, AX - D(AX) \,.\, AY)_{A=E} = XY - YX.$$

(iii) $G = \mathbb{O}(n)$. The Lie bracket on $\mathfrak{g} = \mathfrak{o}(n) \cong T_e G$, which is represented by $\{A \in M(n; \mathbb{R}); A + {}^tA = 0\}$, is the same as for $GL(n, \mathbb{R})$.

1.7.7 Definition. Let G be a Lie group. The *left-invariant covariant derivation* on G is defined by ${}_L\nabla_X \tilde{Y} \equiv 0$, for an arbitrary left-invariant vector field $\tilde{Y}$, and all X.

Remark. Since every vector field on G can be expressed as a linear combination of a basis of left-invariant vector fields with coefficients in $\mathcal{F}G$, this determines ${}_L\nabla$ completely. Note, however, that this does not necessarily mean that the Christoffel symbols all vanish. A basis $\tilde{E}_i$, $1 \leqslant i \leqslant \dim G$, of left-invariant vector fields in general will not be a natural basis with respect to a chart. In fact, we have the

1.7.8 Proposition. *For the left-invariant covariant derivation on a Lie group the curvature tensor* ${}_LR$ *vanishes while in general the torsion tensor* ${}_LT$ *does not. In particular, for* $\tilde{X}, \tilde{Y}$ *left-invariant vector fields,* ${}_LT(\tilde{X}, \tilde{Y}) = -[\tilde{X}, \tilde{Y}]$.

Proof. It suffices to verify this for $\tilde{X}, \tilde{Y}$ elements of a basis $\{\tilde{E}_i\}$ of left-invariant vector fields. Then the claims follow at once from (1.5.4) and ${}_L\nabla \tilde{E}_i = 0$. □

1.7.9 Definition. Let G be a Lie group. A *1-parameter subgroup of* G is a Lie-group morphism

$$F\colon \mathbb{R} \to G$$

of the additive Lie group $\mathbb{R}$ into G, $TF \neq 0$.

Note. For a 1-parameter subgroup F we have

$$F(t_1 + t_2) = F(t_1)\,F(t_2), \quad F(0 + t) = F(0)\,F(t).$$

hence, $F(0) = e =$ neutral element of G.

1.7.10 Theorem. *Let* G *be a Lie group endowed with the left-invariant covariant deriv-*

ation. Then the integral curves of left-invariant vector fields on G are the geodesics on G, their maximal domain of definition is the whole real line.

Those geodesics on G which start at the neutral element e and are non-constant can also be characterized as the 1-parameter subgroups of G.

Proof. Consider the left-invariant vector field $\tilde{X}$ on G determined by $X \in T_e G$, i.e., $\tilde{X}(e) = X$. Let $c(t)$ be an integral curve of $\tilde{X}$, i.e.,

$$\dot{c}(t) = \tilde{X}(c(t)) = TL_{c(t)} \,.\, X \,.$$

Then ${}_L\nabla \dot{c}(t) = 0$, i.e., c is a geodesic.

For any $g \in G$, with $c(t)$ also $gc(t)$ is an integral curve of $\tilde{X}$, since

$$(gc(t))^{\cdot} = TL_g \,.\, \dot{c}(t) = TL_{gc(t)} \,.\, X = \tilde{X}(gc(t)) \,.$$

In particular, $c(0)^{-1} c(t)$ is the integral curve $c_X(t)$ of $\tilde{X}$ with initial condition $\dot{c}_X(0) = X$. We thus can write every integral curve of $\tilde{X}$ in the form $gc_X(t)$, some $g \in G$.

We have, for a fixed t_0,

$$c_X(t_0)\, c_X(t) = c_X(t_0 + t) \,.$$

Indeed, both curves are integral curves of the vector field $DL_g X = \tilde{X}(g)$ and they coincide for $t = 0$. The formula shows that, if $c_X(t)$ is defined for t_0 and t_1, it is defined also for $t_0 + t_1$. Thus, c_X is defined for all $t \in \mathbb{R}$ and hence it is a 1-parameter subgroup of G. □

1.7.11 Examples. We continue with (1.7.6).

(i) $G = \mathbb{R}^n$. The 1-parameter subgroups are the lines through $0 \in \mathbb{R}^n$, i.e., $c(t) = at, a \neq 0$.

(ii) $G = GL(n; \mathbb{R})$. We claim that the 1-parameter subgroup $c_X(t)$ on G is given by

$$c_X(t) = \exp(tX) \equiv \sum_{k=0}^{\infty} \frac{(tX)^k}{k!} \,.$$

Indeed, the infinite series on the right hand side converges. An elementary calculation shows

$$\exp(t_0 X) \exp(t_1 X) = \exp((t_0 + t_1) X) \,.$$

In particular, the elements $\exp(tX)$ are invertible. Thus, $t \mapsto \exp(tX)$ is the geodesic on $GL(n; \mathbb{R})$ with initial vector X.

$$X \in T_E GL(n; \mathbb{R}) = M(n; \mathbb{R}) \mapsto \exp X \in GL(n; \mathbb{R})$$

is the exponential map $\exp | T_E GL(n; \mathbb{R})$ defined in (1.6.5).

(iii) $G = \mathbb{O}(n)$.

$$X \in \mathfrak{g} = \mathfrak{o}(n) \mapsto \exp X \equiv \sum_{k=0}^{\infty} \frac{X^k}{k!} \in \mathbb{O}(n)$$

is the representation of the exponential map $\exp_e\colon T_eG \to G$. In particular, $X + {}^tX = 0$ implies

$$\exp X\,{}^t(\exp X) = \exp X \exp {}^tX = \exp(X + {}^tX) = E.$$

1.7.12 Definition. Let G be a Lie group. The *symmetric covariant derivation* ∇ is defined by

$$\nabla_X Y = {}_L\nabla_X Y - \tfrac{1}{2}\,{}_LT(X, Y).$$

1.7.13 Proposition. (i) *The* ∇*-geodesics coincide with the* ${}_L\nabla$*-geodesics, i.e., with the left translates of the* 1*-parameter subgroups.*

(ii) *The torsion tensor* T *of* ∇ *is zero, whereas in general the curvature tensor* R *is* $\neq 0$. *In particular, for left-invariant vector fields* $\tilde{X}, \tilde{Y}, \tilde{Z}$ *we have*

$$R(\tilde{X}, \tilde{Y})\tilde{Z} = \tfrac{1}{4}[\tilde{Z}, [\tilde{X}, \tilde{Y}]].$$

(iii) R *is parallel,* i.e., $\nabla_V R(X, Y)Z = 0$ *for all* V, X, Y, Z.

Proof. Clearly, ${}_L\nabla\,\dot{c}(t) = \nabla\,\dot{c}(t)$. To compute T and R we use left-invariant vector fields. Then

$$\begin{aligned} T(\tilde{X}, \tilde{Y}) &= \nabla_{\tilde{X}}\tilde{Y} - \nabla_{\tilde{Y}}\tilde{X} - [\tilde{X}, \tilde{Y}] = \\ &= -\tfrac{1}{2}\,{}_LT(\tilde{X}, \tilde{Y}) + \tfrac{1}{2}\,{}_LT(\tilde{Y}, \tilde{X}) - [\tilde{X}, \tilde{Y}] = 0, \end{aligned}$$

see (1.7.8). Moreover,

$$\begin{aligned} R(\tilde{X}, \tilde{Y})\tilde{Z} &= \nabla_{\tilde{X}}\nabla_{\tilde{Y}}\tilde{Z} - \nabla_{\tilde{Y}}\nabla_{\tilde{X}}\tilde{Z} - \nabla_{[\tilde{X}, \tilde{Y}]}\tilde{Z} = \\ &= \tfrac{1}{2}\nabla_{\tilde{X}}[\tilde{Y}, \tilde{Z}] - \tfrac{1}{2}\nabla_{\tilde{Y}}[\tilde{X}, \tilde{Z}] + \tfrac{1}{2}\,{}_LT([\tilde{X}, \tilde{Y}], \tilde{Z}) = \\ &= \tfrac{1}{4}[\tilde{X}, [\tilde{Y}, \tilde{Z}]] - \tfrac{1}{4}[\tilde{Y}, [\tilde{X}, \tilde{Z}]] - \tfrac{1}{2}[[\tilde{X}, \tilde{Y}], \tilde{Z}] = \tfrac{1}{4}[\tilde{Z}, [\tilde{X}, \tilde{Y}]]. \end{aligned}$$

In the last equation we used the Jacobi identity.

Finally, to prove (iii), we use (1.5.16) for the definition of $\nabla_{\tilde{V}}R$. We also use (1.7.12), (1.7.8) and the Jacobi identity (1.4.7, iii).

$$\begin{aligned} 8\nabla_{\tilde{V}}R(\tilde{X}, \tilde{Y})\tilde{Z} &= [\tilde{V}, [\tilde{Z}, [\tilde{X}, \tilde{Y}]]] - [[\tilde{V}, \tilde{Z}], [\tilde{X}, \tilde{Y}]] - \\ &\quad - [\tilde{Z}, [[\tilde{V}, \tilde{X}], \tilde{Y}]] - [\tilde{Z}, [\tilde{X}, [\tilde{V}, \tilde{Y}]]] = \\ &= [\tilde{V}, [\tilde{Z}, [\tilde{X}, \tilde{Y}]] - [[\tilde{V}, \tilde{Z}], [\tilde{X}, \tilde{Y}]] + [\tilde{Z}, [[\tilde{X}, \tilde{Y}], \tilde{V}]] = \\ &= [\tilde{V}, [\tilde{Z}, [\tilde{X}, \tilde{Y}]] + [[\tilde{Z}, [\tilde{X}, \tilde{Y}]], \tilde{V}] = 0. \quad \square \end{aligned}$$

For a Lie group G, the mapping

$$\operatorname{Int}(h)\colon G \to G;\ g \mapsto hgh^{-1} = L_h \circ R_{h^{-1}}g$$

is a structure preserving diffeomorphism. In particular, the neutral element $e \in G$ is transformed into itself. We can then define

1.7.14 Definition. The *adjoint representation of the Lie group* G is given by

$$\mathrm{Ad}\colon G \to GL(\mathfrak{g});\ h \to T_e \operatorname{Int}(h)$$

1.7.15. Proposition. Ad *is a differentiable group morphism.* Ad h *is a morphism of the Lie algebra, i.e.,*

$$\mathrm{Ad}\, h[X, Y] = [\mathrm{Ad}\, h\, X, \mathrm{Ad}\, h\, Y].$$

Proof. Since $\operatorname{Int}(h_1 h_2) = \operatorname{Int} h_1 \circ \operatorname{Int} h_2$ Ad is a group morphism. Define $\psi\colon G \times G \to G$ by $\psi(g, h) = \operatorname{Int}(h) g$. Then $\psi(e, h) = e$. For given $X \in T_e G = \mathfrak{g}$, $\mathrm{Ad}\, h\, X$ is given by $T_{(e,h)}\psi . (X, 0_h)$ and thus, $\mathrm{Ad}\, h$ is differentiable. Moreover, for any automorphism $\phi\colon G \to G$, $T_e\phi\colon T_e G \to T_e G \cong \mathfrak{g}$ is a Lie algebra morphism since $T\phi$ carries left invariant vector fields into left invariant vector fields and, quite generally, the tangential $T\phi$ of a diffeomorphism $\phi\colon M \to M$ is a morphism on the Lie algebra of vector fields, cf. (1.4.10). □

1.7.16 Definition. The *adjoint representation of the Lie algebra* $\mathfrak{g}$ of G is given by

$$\mathrm{ad}\colon \mathfrak{g} \to L(\mathfrak{g}; \mathfrak{g});\ X \mapsto T_e \mathrm{Ad} . X.$$

Here we have identified $T_e GL(\mathfrak{g})$ and $L(\mathfrak{g}; \mathfrak{g})$.

1.7.17 Lemma. *For the adjoint representation of the Lie algebra* $\mathfrak{g}$ *we have the formula*

$$\mathrm{ad}\, X . Y = [X, Y].$$

In addition, ad X *is a derivation for* $\mathfrak{g}$, *i.e.,*

$$\mathrm{ad}\, X . [Y, Z] = [\mathrm{ad}\, X . Y, Z] + [Y, \mathrm{ad}\, X . Z].$$

Proof. From our definitions we have

$$\begin{aligned}
\mathrm{ad}\, X . Y &= \frac{d}{dt} T_e \operatorname{Int}(\exp tX) . Y|_0 = \\
&= \frac{d}{dt} (T_{\exp(-tX)} L_{\exp tX} \circ T_e R_{\exp(-tX)} . Y)|_0 \overset{(i)}{=} \\
&= \frac{{}_L\nabla}{dt} T_e R_{\exp(-tX)} . Y|_0 = \frac{{}_L\nabla}{dt}\frac{\partial}{\partial s} \exp sY \exp(-tX)|_{0,0} \overset{(ii)}{=} \\
&= \frac{{}_L\nabla}{\partial s}\frac{\partial}{\partial t} \exp sY \exp(-tX)|_{0,0} + T(-X, Y) \overset{(iii)}{=} [X, Y].
\end{aligned}$$

Here we have used (i), the parallelity of left invariant vector fields under ${}_L\nabla$, (ii), the formula (1.5.8), (iii), again (i) and the formula (1.7.8).

The last statement now is a consequence of the Jacobi identity (1.4.7, iii). □

1.7.18 Corollary.

$$\exp \circ \operatorname{ad} = \operatorname{Ad} \circ \exp .$$

Proof. $\dfrac{d}{dt} \exp \circ \operatorname{ad} . tX|_0 = \operatorname{ad} X$; $\dfrac{d}{dt} \operatorname{Ad} \circ \exp tX|_0 = T_e \operatorname{Ad} . X = \operatorname{ad} X$. □

We add a description of the parallel translation associated to the symmetric covariant derivation.

1.7.19 Proposition. *Let* $c(t) = g . \exp(tX_0)$ *be a geodesic on* G. *Then*

$$\overset{t_0}{\underset{0}{\|}} c = T_e L_{c(t_0)} \cdot \left(\sum_0^{\infty} \frac{(-t_0/2)^k}{k!} (\operatorname{ad} X_0)^k \right) \equiv T_e L_{c(t_0)} \cdot \exp\left(-\frac{t_0}{2} \operatorname{ad} X_0 \right)$$

Proof. With the definition (1.7.12) of ∇/dt we find for the covariant derivative of the vector field

$$X(t) = T_e L_{c(t)} \cdot \exp\left(-\frac{t}{2} \operatorname{ad} X_0 \right) . X(0):$$

$$\frac{\nabla X}{dt}(t) = T_e L_{c(t)} \cdot \left(-\frac{1}{2} \operatorname{ad} X_0 . \exp\left(-\frac{t}{2} \operatorname{ad} X_0\right) . X(0) \right) - \frac{1}{2} {}_L T(\dot{c}(t), X(t))$$

$$= -\frac{1}{2} [\dot{c}(t), X(t)] + \frac{1}{2} [\dot{c}(t), X(t)] = 0 .$$

Here we have used $\dot{c}(t) = T_e L_{c(t)} . X_0$ and (1.7.8). □

1.7.20 Definition. Let G be a Lie group, $\mathfrak{g}$ its Lie algebra.

(i) The *Killing form*

$$B: \mathfrak{g} \times \mathfrak{g} \to \mathbb{R}$$

is the bilinear form given by

$$B(X, Y) = tr \operatorname{ad} X \circ \operatorname{ad} Y$$

(ii) G (or also its Lie algebra $\mathfrak{g}$) is called *semi-simple* if the Killing form is non-degenerate.

1.7.21 Proposition. *The Killing form is symmetric. It is also invariant under automorphisms of* $\mathfrak{g}$, *in particular,* $B(\operatorname{Ad} gX, \operatorname{Ad} gY) = B(X, Y)$. *Moreover,*

$$(*) \qquad B(\operatorname{ad} X . Y, Z) + B(Y, \operatorname{ad} X . Z) = 0$$

Proof. The trace of a composition of linear endomorphisms of a vector space is invariant under cyclic permutations. Thus, B is symmetric. If $\sigma: \mathfrak{g} \to \mathfrak{g}$ is a Lie algebra automorphism (e.g., $\sigma = \operatorname{Ad} g$, $g \in G$) then $\operatorname{ad} \sigma X = \sigma \circ \operatorname{ad} \circ \sigma^{-1}$. Hence, $B(\sigma X, \sigma Y)$

$= B(X, Y)$. Let $\sigma = \mathrm{Ad}(\exp tX)$. The derivation of $B(\mathrm{Ad}(\exp tX)Y, \mathrm{Ad}(\exp tX)Z)$ $= B(Y, Z)$ with respect to t at $t = 0$ yields (*). □

1.7.22 Examples. We continue with (1.7.6) and (1.7.11).

(i) $G = \mathbb{R}^n$. $\mathrm{Ad}\, g = id$, since $\mathrm{Int}(g) = id$. $B \equiv 0$.

(ii) $G = GL(n; \mathbb{R})$. If $A \in GL(n; \mathbb{R})$, $X, Y \in \mathfrak{gl}(n; \mathbb{R})$

$$= M(n; \mathbb{R}),\ \mathrm{Ad}\, A X = AXA^{-1},\ \mathrm{ad}\, X . Y = XY - YX.$$

We thus see that $\mathrm{ad}\colon \mathfrak{g} \to L(\mathfrak{g}, \mathfrak{g})$ is a Lie algebra morphism. In fact, this is the meaning of the Jacobi identy.

To determine the Killing form we introduce the elements $E_{ij} \in M(n; \mathbb{R})$, having 1 at (i, j) and 0 elsewhere. Then

$$\mathrm{ad}\, X . E_{ij} = \sum_k x_{ki} E_{kj} - \sum_k x_{jk} E_{ik}$$

$$B(X, Y) = tr\, \mathrm{ad}\, X \circ \mathrm{ad}\, Y = 2n\, tr\, X \circ Y - 2\, tr\, X\, tr\, Y.$$

If X is a multiple of the unit matrix, $\mathrm{ad}\, X = 0$, hence, $GL(n; \mathbb{R})$ is not semi-simple. However, the subgroup $SL(n, \mathbb{R})$ of matrices of determinant 1 is semi-simple. The corresponding Lie algebra consists of the $X \in M(n; \mathbb{R})$ with $tr\, X = 0$. Thus, $B(X, Y)$ $= 2n\, tr\, X \circ Y$, and hence, $B(X, {}^tX) \neq 0$ whenever $X \neq 0$.

(iii) $\mathfrak{o}(2) = \mathbb{R}$ is not semi-simple. For $n > 2$, $\mathfrak{o}(n)$ is semi-simple. Actually, we have $B(X, Y) = (n-2)\, tr\, X \circ Y$, i.e., for $n > 2$, the Killing form is negative definite.

To see this we choose for $\mathfrak{o}(n)$ the basis $\{E_{ij} - E_{ji}, i < j\}$. Then

$$\mathrm{ad}\, X . (E_{ij} - E_{ji}) = \sum_k x_{ki}(E_{kj} - E_{jk}) - \sum_k x_{jk}(E_{ik} - E_{ki}),$$

$$\mathrm{ad}\, X \circ \mathrm{ad}\, X . (E_{ij} - E_{ji}) = \sum_{k,l} x_{lk} x_{ki}(E_{lj} - E_{jl}) - \sum_{k,l} x_{jl} x_{ki}(E_{kl} - E_{lk}) -$$

$$- \sum_{k,l} x_{li} x_{jk}(E_{lk} - E_{kl}) + \sum_{k,l} x_{kl} x_{jk}(E_{il} - E_{li}).$$

The coefficient of $(E_{ij} - E_{ji})$ therefore reads

$$\sum_k x_{ik} x_{ki} - 2x_{ij} x_{ji} + \sum_k x_{kj} x_{jk}.$$

For the trace of $\mathrm{ad}\, X \circ \mathrm{ad}\, X$ we find

$$(n-1)\, tr\, X \circ X - tr\, X \circ X.$$

1.8 Riemannian Manifolds

A Riemannian manifold is a differentiable manifold in the sense of (1.1.2), endowed with a riemannian metric g. Let $\mathbb{E}$ be the model for M. Then g is a section in the bundle of symmetric bilinear forms with values in the open cone $Pos\,\mathbb{E}$ of positive definite quadratic forms, cf. (1.8.1). There always exist such sections, see (1.8.5).

On a Riemannian manifold one can define the length and the energy of a curve. Of fundamental importance is the existence of the so-called Levi-Cività (covariant) derivation. It is characterized by having torsion tensor zero and inducing a parallel transport along curves which is isometric, i.e., preserves the scalar product, cf. (1.8.11).

With the Levi-Cività derivation we define (Riemannian) normal coordinates and polar coordinates, (1.8.16). Let $B_\varrho(p)$ be the domain of the normal coordinates at p. Then the tangent spaces $T_q M, q \in B_\varrho(p)$, possess a canonical isomorphism with $T_p M$: Simply carry $T_q M$ into $T_p M$ by parallel translation along the radial geodesic from p to q.

In this way we get a bundle chart over $B_\varrho(p)$ which maps $TB_\varrho(p)$ onto $B_\varrho^{\mathbb{F}}(0) \times \mathbb{E}$. The restriction to each fibre is an isometry. This is an example for an isometric trivialization.

The construction which we just described associates to a given orthonormal basis $\{e_i(p) = e_i\}$ of $T_p M \cong \mathbb{E}$ an orthonormal basis $\{e_i(q)\}$ of $T_q M$, for each $q \in B_\varrho(p)$. Such bases for the tangent spaces over $B_\varrho(p) \subset M$ are called Cartan frames, as opposed to the natural or Gauss frames which we did introduce after (1.4.3), for any given chart (u, M') of M.

It is also possible to introduce a Riemannian metric for general vector bundles $\pi: P \to M$ with fibre modelled on a Hilbert space $\mathbb{F}$ by defining on each fibre $\mathbb{F}_p = \pi^{-1}(p), p \in M$, a scalar product, depending differentiably on p, cf. (1.8.18). There exist linear connections for π such that parallel translation becomes an isometry, see (1.8.23). In general it is not possible to define a distinguished covariant derivation, as it was the case with the Levi-Cività derivation for the tangent bundle $\tau_M: TM \to M$ over M. However, if $\dim \mathbb{E} = \dim M < \infty$ then we have for the various tensor bundles $\tau_s^r: T_s^r M \to M$ over M a Riemannian metric and the canonical extension of the Levi-Cività derivation ∇ to τ_s^r (cf. (1.5.16)) may be viewed as a distinguished derivation.

We conclude by proving the existence of isometric trivializations for an arbitrary vector bundle with Riemannian metric, see (1.8.20).

Recall from (1.2.14) that $S_2\,\mathbb{E}$ denotes the Banach space of symmetric bilinear maps from $\mathbb{E} \times \mathbb{E}$ into $\mathbb{R}$. By $Pos\,\mathbb{E}$ we denote the set of *positive definite quadratic forms on* $\mathbb{E}$. These are those $g \in S_2\,\mathbb{E}$ for which there exists a $\varepsilon = \varepsilon(g) > 0$ such that

$$g(X, X) \geqslant \varepsilon \langle X, X \rangle .$$

Here, $\langle\ ,\ \rangle$ denotes the scalar product on the Hilbert space $\mathbb{E}$.

$Pos\,\mathbb{E}$ is a so-called *positive open cone*. This means: If g and $h \in Pos\,\mathbb{E}$ and $a + b > 0, a \geqslant 0, b \geqslant 0$, then $ag + bh \in Pos\,\mathbb{E}$.

For $\dim \mathbb{E} = n < \infty$, $Pos\,\mathbb{E}$ can be described as consisting of those $g \in S_2\,\mathbb{E}$ for which $X \neq 0$ implies $g(X, X) > 0$. If $\{e_i\}$ is a basis for $\mathbb{E}$, the (n,n)-matrix defined by

$$(g_{ik}) = \big(g(e_i, e_k)\big)$$

is called *fundamental matrix of* g with respect to the basis $\{e_i\}$.

1.8.1 Definition. Let M be a manifold. *A Riemannian metric on* M is a section

$$g: M \to S_2 M$$

in the bundle $\sigma_2: S_2 M \to M$ (cf. (1.2.15)) with values $g(p) \in Pos\,T_p M$.

M, endowed with a Riemannian metric g, is called *Riemannian manifold.*

In general we simply write M instead of (M, g) to denote a Riemannian manifold.

1.8.2 Remarks. 1. $g(p) \in Pos\,T_p M$ means: If (u, M') is a chart containing p then, for the representative $g\big(u(p)\big)$ of $g(p)$,

$$g\big(u(p)\big) \in Pos\,T_{u(p)}\,U \cong Pos\,\mathbb{E}\,.$$

This condition clearly is independent of the particular choice of the chart.

2. Instead of $g(p)$ we also write g_p or $\langle\ ,\ \rangle_p$ or simply $\langle\ ,\ \rangle$, if there is no danger of confusion. $|\,|_p$ or simply $|\,|$ will denote the norm derived from $\langle\ ,\ \rangle_p$.

3. Let (u, M') be a chart on (M, g) with $u(M') = U \subset \mathbb{E}$. The principal part of the representative of $g|M'$ is a differentiable mapping

$$g: U \to Pos\,\mathbb{E}\,.$$

In the case $\mathbb{E} = \mathbb{R}^n$, $g(u), u \in U$, is determined by the n^2 differentiable functions $g_{ik}(u) = g(u)(e_i, e_k)$, i.e.,

$$g(u) = \sum_{i,k} g_{ik}(u)\, e^i \otimes e^k\,.$$

Here, e_i is the i-th basis element of $\mathbb{R}^n$. The hypothesis $g(u) \in Pos\,\mathbb{E}$ in this case means

$$\sum_{i,k} g_{ik}(u)\, \xi^i \xi^k > 0 \quad \text{for} \quad (\xi^i) \neq (0)\,.$$

If $(u, M'), (u', M'')$ are two charts containing p we have

$$g\big(u(p)\big) = g\big(u'(p)\big) \circ \big(D(u' \circ u^{-1}) \times D(u' \circ u^{-1})\big)\,.$$

This means in the finite dimensional case

$$g_{kl}(u) = \sum_{i,j} g_{ij}(u')\, \frac{\partial u'^i}{\partial u^k} \frac{\partial u'^j}{\partial u^l}\,(u)\,,$$

with $u'(u) = (u' \circ u^{-1})(u)$.

Examples. (i) $M = \mathbb{E}$ with the same scalar product g on all $T_X\,\mathbb{E}$. Here, g is equivalent to the Hilbert product $\langle\ ,\ \rangle$ on $\mathbb{E}$, i.e.,

$$a^{-1}\langle X, X\rangle \leqslant g(X, X) \leqslant a\langle X, X\rangle, \quad \text{some} \quad a \geqslant 1.$$

(ii) $M = G$, a Lie group, and $\{\tilde{\mathbb{E}}_i\}$ a basis for the left-invariant vector fields on M. Define $g(\tilde{\mathbb{E}}_i, \tilde{\mathbb{E}}_j) = a_{ij}$, elements of a positive-definite matrix.

(iii) Let (M, g), (M', g') be Riemannian manifolds. The *Riemannian product manifold* (M^*, g^*) is defined as $(M \times M', g \times g')$. I.e., on the product manifold $M^* = M \times M'$ (cf. (1.1.3, iv)) take on $T_{p*}M^* = T_{(p,p')}(M \times M') = T_pM \times T_{p'}M'$ the product $g(p) \times g'(p')$ of the metrics given on each factor.

1.8.3 Proposition. *Let $F\colon M \to N$ be an immersion and assume that N possesses a Riemannian metric h. Then*

$$g(p) = h(F(p)) \circ (TF(p) \times TF(p))$$

is a Riemannian metric on M.

Proof. That $g\colon M \to S_2M$ is a section in $\sigma_2\colon S_2M \to M$ follows also from (1.4.12). To see that $g(p) \in Pos\,T_pM$ we take charts (u, M') around p and (v, N') around $q = F(p)$. Then, with $F(u) = v \circ F \circ u^{-1}(u)$,

$$g(u)(X(u), X(u)) = h(F(u))(DF(u).X(u), DF(u).X(u)) \geqslant$$

$$\geqslant \eta\,|DF(u).X(u)|^2 \geqslant \varepsilon\,|X(u)|^2,$$

where $\varepsilon = \eta\eta'$, with $\eta' > 0$ satisfying

$$|DF(u).X|^2 \geqslant \eta'|X|^2. \quad \square$$

1.8.4 Example. A submanifold M of a Riemannian manifold N inherits a Riemannian metric from the inclusion $\iota\colon M \to N$. In particular, submanifolds of the Hilbert space $\mathbb{E}$ with its canonical metric inherit a Riemannian structure. Surfaces in $\mathbb{R}^3$ were considered in full generality for the first time by Gauss [1]. They are the original examples of Riemannian manifolds.

1.8.5 Theorem. *On every differentiable manifold there exists a Riemannian metric.*

Remark. This contrasts with the fact that the existence of an indefiinite Riemannian metric on a manifold M presupposes certain restrictions on the topological structure of M. Here we mean by an *indefinite Riemannian metric on M* a section g in the bundle $\sigma_2\colon S_2M \to M$ such that $g(p)$, for each $p \in M$, is a non-degenerate symmetric bilinear form on T_pM which is neither positive definite nor negative definite. I.e., T_pM possesses a decomposition $T_p^+M \oplus T_p^-M$ into two non-zero subspaces such that $g(p)|T_p^+M$ is positive definite and $g(p)|T_p^-M$ is negative definite. The dimension of T_p^-M is also called the *signature of* $g(p)$.

One easily sees that the existence of an indefinite Riemannian metric g of signature

k amounts to the existence of a continuous field of k-dimensional subspaces, tangent to M. A necessary condition for the existence of such a field is the vanishing of a certain Stiefel-Whitney cohomology class, for details see Steenrod [1].

Indefinite Riemannian metrics occur naturally for Lie groups, cf. (2.2.25) where the Killing form on a semi-simple symmetric space provides such a metric.

Proof. We use the existence of a differentiable partition of unity $(\phi_\beta, \tilde{M}_\beta)_{\beta \in B}$ which is subordinated to the open covering $(M_\alpha)_{\alpha \in A}$ formed by the domains M_α of an atlas $(u_\alpha, M_\alpha)_{\alpha \in A}$. See (1.0) 4 and (1.1.8) for details. In particular, we have $\sigma: B \to A$ such that $\tilde{M}_\beta \subset M_{\sigma\beta}$. Put $u_{\sigma\beta} | \tilde{M}_\beta = u_\beta$. We thus get an atlas $(u_\beta, \tilde{M}_\beta)_{\beta \in B}$. Denote by $\langle\ ,\ \rangle$ the Hilbert scalar product on $U_\beta = u_\beta \tilde{M}_\beta \subset \mathbb{E}$. With this define

$$g(p)(\ ,\ ) = \sum_\beta \phi_\beta(p) \langle Du_\beta(p)\ ,\ Du_\beta(p)\ \rangle .$$

In the case when M is compact, B can be chosen to be finite, of course. □

Remark. The existence of a Riemannian metric on a manifold M, $\dim M < \infty$, can also be deduced from the *Whitney embedding theorem* which states that every such M possesses an embedding into some Euclidean space. Apply then (1.8.4). Cf. Whitney [1].

For more on embeddings and immersions cf. Hirsch [1].

On a Riemannian manifold we can define the length and the energy integral.

1.8.6 Definition. Let $c: I \to \mathbb{R}$ be a curve on (M, g).

(i) $L(c) = \int_I \sqrt{g(c(t))(\dot{c}(t), \dot{c}(t))}\, dt$ is called *length of* c.

(ii) $E(c) = \frac{1}{2} \int_I g(c(t))(\dot{c}(t), \dot{c}(t))\, dt$ is called *energy integral of* c.

(iii) If $|\dot{c}(t)| = 1$ then we call c *parameterized by arc length.*

1.8.7 Proposition. *Let* $c: I \to M$ *be a curve on* M. *Then*

$$L(c)^2 \leqslant 2E(c)|I|$$

with $|I| =$ *length of* I *on* $\mathbb{R}$. *For* $|I| < \infty$, *equality holds if and only if* $|\dot{c}(t)| = $ *const.*

Proof. We use the Schwarz inequality:

$$\left(\int_I |\dot{c}(t)|\, dt\right)^2 \leqslant \int_I |\dot{c}(t)|^2 dt \int_I 1^2 dt = 2E(c)|I| .$$

Equality can hold only if $|\dot{c}(t)|$ is proportional to 1. □

1.8.8 Proposition. *Let* $c: t \in I \to c(t) \in M$ *be an immersion. Then there exists a parameter transformation* $s \in H \mapsto t(s) \in I$, $dt(s)/dt > 0$ *such that* $\tilde{c}(s) = c \circ t(s)$, $s \in H$, *is parameterized by arc length, i.e.,* $|\tilde{c}'(s)| = 1$.

Proof. Define $s(t)$ by

$$s(t) = \int_{t_0}^{t} |\dot{c}(t)|\,dt$$

and denote by $t(s)$ the inverse function of $s(t)$. Then $\tilde{c}'(s) = \dot{c}(t)/|\dot{c}(t)|$. □

We now consider, besides a Riemannian metric g, a covariant derivation ∇ on M. The following condition expresses a particularly close relationship between these two structures.

1.8.9 Definition. Let (M, g) be a Riemannian manifold. A covariant derivation ∇ on M is called *Riemannian* if, for all curves c on M, the parallel translation defined in (1.6.2) is an isometry.

1.8.10 Lemma. *A covariant derivation ∇ is Riemannian if and only if, for all vector fields X, Y, Z on M,*

$$(*) \qquad Xg(Y,Z) = g(\nabla_X Y, Z) + g(Y, \nabla_X Z).$$

Proof. Let X, Y, Z be vector fields on M. Fix $p \in M$. Let c be a curve on M with $\dot{c}(0) = X(p)$. $Y \circ c(t) = Y(t)$ and $Z \circ c(t) = Z(t)$ are vector fields along c. Then

$$Xg(Y,Z)(p) = \frac{d}{dt}\, g(Y(t), Z(t))|_0\,;$$

$$\nabla_X Y(p) = \frac{\nabla}{dt}\, Y(t)|_0\,;\ \nabla_X Z(p) = \frac{\nabla}{dt}\, Z(t)|_0\,.$$

Hence, for a Riemannian covariant derivation ∇, $(*)$ will follow if we show

$$(**) \qquad \frac{d}{dt}\, g(Y(t), Z(t))|_0 = g\left(\frac{\nabla Y}{dt}(t), Z(t)\right)|_0 + g\left(Y(t), \frac{\nabla Z}{dt}(t)\right)|_0\,.$$

Let now $\tilde{Y}(t), \tilde{Z}(t)$ be the parallel vector fields along $c(t)$ with $\tilde{Y}(0) = Y(0) = Y(p)$, $\tilde{Z}(0) = Z(0) = Z(p)$. Then, since ∇ is Riemannian, $dg(\tilde{Y}(t), \tilde{Z}(t))/dt = 0$ and

$$Y(t) = \tilde{Y}(t) + t\,\frac{\nabla Y}{dt}(0) + o(t);\ Z(t) = \tilde{Z}(t) + t\,\frac{\nabla Z}{dt}(0) + o(t)\,.$$

Hence,

$$\frac{d}{dt}\, g(Y(t), Z(t))|_0 = \frac{d}{dt}\, g\left(t\,\frac{\nabla Y}{dt}(0) + o(t), \tilde{Z}(t)\right)|_0 +$$

$$+\frac{d}{dt}\, g\left(\tilde{Y}(t), t\,\frac{\nabla Z}{dt}(0) + o(t)\right)|_0 + \frac{d}{dt}\, t o(t)|_0 = g\left(\frac{\nabla Y}{dt}(0), Z(0)\right) +$$

$$+ g\left(Y(0), \frac{\nabla Z}{dt}(0)\right),$$

i.e., we have $(**)$. Conversely, $(*)$ implies $(**)$. □

The next theorem is fundamental to the whole theory of Riemannian manifolds. It not only shows that, due to (1.8.5), there always exists a covariant derivation on a manifold. In addition, it states that there exists a canonical covariant derivation on a Riemannian manifold, i.e., a covariant derivation which is determined without making choices, once one has a Riemannian structure on a differentiable manifold.

The proof will show that the Theorem also holds in the case of an indefinite Riemannian metric.

This derivation was discovered by Levi-Cività after it first appeared in a geometrical context for submanifolds in Euclidean space with their induced Riemannian metric, cf. (1.10.3).

1.8.11 Theorem. *On a Riemannian manifold there exists exactly one covariant derivation which is Riemannian and has vanishing torsion. This is called the* Levi-Cività derivation. *It is characterized by the relation*

$$(*) \qquad 2\langle \nabla_X Y, Z\rangle = X\langle Y, Z\rangle + Y\langle Z, X\rangle - Z\langle X, Y\rangle + \langle Z, [X, Y]\rangle + \\ + \langle Y, [Z, X]\rangle - \langle X, [Y, Z]\rangle$$

for all vector fields X, Y, Z.

Proof. Let (u, M') be a chart. Using the expression for the Lie bracket in the proof of (1.4.7), the representation of $(*)$ in this chart becomes the relation

$$(\dagger) \qquad 2g(u)(\Gamma(u)(X, Y), Z) = Dg(u).X(Y, Z) + Dg(u).Y(Z, X) - \\ - Dg(u).Z(X, Y)$$

for arbitrary X, Y, Z in $\mathbb{E}$. Note that every $X \in T_pM$ can be extended to a vector field on M, cf. (1.1.8).

The tensor field $\Gamma(u)$ of type $\binom{1}{2}$ is uniquely determined by $(\dagger)$. Taking $\Gamma(u)$ as the Christoffel symbol of a covariant derivation on M' we see that the torsion tensor is zero, cf. (1.5.3). Moreover, the derivation is Riemannian as follows from writing down $(\dagger)$ with Y and Z exchanged and adding it to $(\dagger)$:

$$(\dagger\dagger) \qquad Dg(u).X(Y, Z) = g(u)(\Gamma(u)(X, Y), Z) + g(u)(Y, \Gamma(u)(X, Z)).$$

From $(\dagger\dagger)$ one can derive again $(\dagger)$ by adding to it the expression with (X, Y, Z) replaced by (Y, Z, X) and subtracting the one where (X, Y, Z) is replaced by (Z, X, Y). Thus, $\Gamma(u)$ is uniquely determined by the properties vanishing torsion and Riemannian, $(\dagger\dagger)$. This shows that the $\Gamma(u)$ are the Christoffel symbols of a globally defined covariant derivation. □

1.8.12 Complement. *In the case* $\dim M < \infty$, *the* $\Gamma_{ij}^k(u)$ *of the Levi-Cività derivation with respect to the cononical basis fields* $\{u, e_i\}$ *on* U *are given by*

$$\Gamma_{ij}^k(u) = \frac{1}{2}\sum_l g^{kl}(u)\left(\frac{\partial g_{jl}}{\partial u^i}(u) + \frac{\partial g_{li}}{\partial u^j}(u) - \frac{\partial g_{ij}}{\partial u^l}(u)\right).$$

Here, the $g^{kl}(u)$ *are the elements of the inverse matrix of the fundamental matrix, i.e., they are the solutions of*

$$\sum_j g_{ij}(u)\, g^{jk}(u) = \delta_i^k .$$

Proof. Put in (†) $(X, Y, Z) = (e_i, e_j, e_k)$. □

1.8.13 Lemma. *For the curvature tensor* R *of the Levi-Cività derivation we have, besides* (1.5.3, ii), *the following identities:*

$$\langle R(X, Y)Z, W\rangle = -\langle R(X, Y)W, Z\rangle = \langle R(Z, W)X, Y\rangle .$$

In particular, for every $Y \in T_pM$, *we have the so-called* curvature operator

$$R_Y\colon T_pM \to T_pM;\ X \mapsto R(X, Y)Y$$

It is a self-adjoint operator with respect to the scalar product $\langle , \rangle$ *on* T_pM, $\langle R_Y X, X'\rangle = \langle X, R_Y X'\rangle$.

Proof. We consider $X, Y, Z, W \in T_pM$ as values of vector fields on M. We can assume that for these vector fields $[X, Y] = 0$, for instance, by taking an embedding $F\colon (\mathbb{R}^2, 0) \to (M, p)$ with $\partial F(0,0)/\partial x = X(p)$, $\partial F(0,0)/\partial y = Y(p)$.

The first equation in (1.8.13) is equivalent to $\langle R(X, Y)Z, Z)\rangle = 0$. From

$$X\langle \nabla_Y Z, Z\rangle = \tfrac{1}{2}X(Y\langle Z, Z\rangle) = \tfrac{1}{2}Y(X\langle Z, Z\rangle) = Y\langle \nabla_X Z, Z\rangle$$

we get

$$\langle R(X, Y)Z, Z\rangle = \langle \nabla_X \nabla_Y Z, Z\rangle - \langle \nabla_Y \nabla_X Z, Z\rangle =$$
$$= X\langle \nabla_Y Z, Z\rangle - \langle \nabla_Y Z, \nabla_X Z\rangle - Y\langle \nabla_X Z, Z\rangle + \langle \nabla_X Z, \nabla_Y Z\rangle = 0 .$$

The second equation in (1.8.13) is a formal consequence of the first equation and (1.5.3, ii),

$$\langle R(X, Y)Z, W\rangle = -\langle R(Y, X)Z, W\rangle = \text{(Bianchi)}$$
$$\langle R(Z, Y)X, W\rangle + \langle R(X, Z)Y, W\rangle ,$$

$$\langle R(X, Y)Z, W\rangle = -\langle R(X, Y)W, Z\rangle = \text{(Bianchi)}$$
$$\langle R(W, X)Y, Z\rangle + \langle R(Y, W)X, Z\rangle .$$

Hence,

$$2\langle R(X, Y)Z, W\rangle = \langle R(Z, Y)X, W\rangle + \langle R(X, Z)Y, W\rangle$$
$$+ \langle R(W, X)Y, Z\rangle + \langle R(Y, W)X, Z\rangle .$$

The right hand side is invariant under the substitution $(X, Y) \leftrightarrow (Z, W)$. This is clear for the first and third term, and for the second and fourth term we get it from

$$\langle R(Z, X)W, Y\rangle + \langle R(W, Y)Z, X\rangle = \langle R(X, Z)Y, W\rangle + \langle R(Y, W)X, Z\rangle . \quad \square$$

For vector fields along a differentiable map as defined in (1.5.5) we have the

1.8.14 Proposition. *Let ∇ be the Levi-Cività derivation of a Riemannian manifold (N, g). Let $F: M \to N$ be differentiable. Then the covariant derivation of vector fields along F is Riemannian and* torsion-free *in the sense that*

$$Xg(V, W) = g(\nabla_X V, W) + g(V, \nabla_X W)$$

$$\nabla_X TF.Y - \nabla_Y TF.X = TF.[X, Y].$$

Here, V, W are vector fields along F and X, Y are vector fields on M.

Proof. Using charts (u, M') on M and (v, N') on N we put $v \circ F \circ u^{-1}(u) = F(u)$. According to (1.5.5), our claim amounts to the identities

$$\begin{aligned} Dg(F(u)).X(V, W) = {} & g(F(u))(\Gamma(F(u))(DF(u).X, V), W) + \\ & + g(F(u))(V, \Gamma(F(u))(DF(u).X, W)), \end{aligned}$$

$$\Gamma(F(u))(DF(u).X, DF(u).Y) - \Gamma(F(u))(DF(u).Y, DF(u).X) = 0,$$

for X and Y in $\mathbb{E}$ = model of M and V, W in $\mathbb{F}$ = model of N. Now, these identities are true since the Christoffel symbols $\Gamma(v)$ represent the Riemannian connection on N. □

We continue with a reformulation of (1.6.12) by using a chart related to the Riemannian metric for the description of a neighborhood of the 0-section of $\tau_M: TM \to M$.

Let $\varrho > 0, p \in M$. We denote by $B_\varrho(0_p)$ the ϱ-ball in T_pM around the origin 0_p of T_pM.

1.8.15 Theorem. *Let M be a Riemannian manifold. Consider the Levi-Cività derivation. Then there exists an open neighborhood $\tilde{T}M$ of M in TM such that*

$$\exp: X \in \tilde{T}M \mapsto c_X(1) \in M$$

is defined and differentiable. Here, c_X denotes the geodesic determined by $\dot{c}_X(0) = X$.

Moreover, for every $p \in M$ there exist real numbers $\varepsilon = \varepsilon(p)$, $\eta = \eta(p)$, $0 < \varepsilon < \eta$ such that

$$\exp_p | B_\varepsilon(0_p): B_\varepsilon(0_p) \to B_\varepsilon(p)$$

is a diffeomorphism onto an open neighborhood $B_\varepsilon(p)$ of p, having the following properties:

(i) *For any two points q and r in $B_\varepsilon(p)$ there exists exactly one geodesic*

$$c_{qr}: [0, 1] \to M$$

from q to r with length $< \eta$.

(ii) *The mapping $(q, r) \in B_\varepsilon(p) \times B_\varepsilon(p) \mapsto c_{qr} \in C^0([0, 1], M)$ is continuous.*

(iii) *For every $q \in B_\varepsilon(p)$, $\exp | B_\eta(0_q)$ is a diffeomorphism onto an open neighborhood $B_\eta(q)$ of q, with $B_\varepsilon(p) \subset B_\eta(q)$.*

Remark. We have used the notation $B_\eta(q)$ for the diffeomorphic image of $B_\eta(0_q)$ under $\exp_q$. In (1.9.5) we will see the $B_\eta(q)$ can indeed be viewed as η-ball around $q \in M$ with respect to a metric on M.

Proof. We refer to the proof of (1.6.12). Choose the neighborhood $\tilde{W}$ of 0_p in $\tilde{T}M$ to be of the form $\{X \in TM; |X| < \eta \text{ and } \tau X \in \exp B_{\varepsilon'}(0_p)\}$, with sufficiently small positive ε' and η. Then the W in (1.6.12) can be chosen to be of the form $\exp_p B_\varepsilon(0_p)$, with sufficiently small $\varepsilon > 0$. □

1.8.16 Corollary. *For every* $p \in M$ *there exist* (i) (Riemann) normal coordinates *and* (ii) (Riemann) polar coordinates. *That is to say:*

(i) *Let*

$$i_p\colon (T_pM, g(p)) \to (\mathbb{E}, \langle\ ,\ \rangle)$$

be an isometry from T_pM *with its scalar product* $g(p)$ *to the Hilbert space* $\mathbb{E}$. *Then*

$$\big(i_p \circ (\exp_p | B_\eta(0))^{-1}, B_\eta(p)\big)$$

is a chart with range the η*-ball* $B_\eta^{\mathbb{E}}(0)$ *in* $\mathbb{E}$. *For this chart, the Christoffel symbol vanishes in* $0 \in B_\eta^{\mathbb{E}}(0)$.

(ii) *Let*

$$j_p\colon T_pM - \{0_p\} \to \partial\bar{B}_1^{\mathbb{E}}(0) \times]0, \infty[;\ X \mapsto (i_pX/|X|, |X|)$$

be polar coordinates on the Hilbert space $(T_pM, g(p))$ *which we identify with* $(\mathbb{E}, \langle\ ,\ \rangle)$ *under an isometry* i_p. $\partial\bar{B}_1^{\mathbb{E}}(0)$ *is the unit sphere in* $\mathbb{E}$. *Then there exists* $\eta = \eta(p) > 0$ *such that*

$$\big(j_p \circ (\exp_p | B_\eta(0_p))^{-1}\colon B_\eta(p) - \{p\} \to \partial\bar{B}_1^{\mathbb{E}}(0) \times]0, \eta[$$

is a diffeomorphism.

Proof. All that remains to prove is $\Gamma(u(p)) = \Gamma(0) = 0$ for normal coordinates $u\colon B_\eta(p) \to B_\eta^{\mathbb{E}}(0)$. Now, the geodesics starting in p are represented by linear equations $u(t) = At$, $A \in \mathbb{E}$. From the differential equation (*) for $u(t)$ in the proof of (1.6.5) we have $\Gamma(0)(A, A) = 0$, for all $A \in \mathbb{E}$, hence, $\Gamma(0) = 0$. □

1.8.17 Remark. Sometimes it is convenient to have *almost normal coordinates* based on a point $p \in M$. This is a chart

$$u\colon (M', p) \to (U, 0) \subset \mathbb{E} \times \{0\}$$

such that $\Gamma(0) = 0$.

We now want to introduce a Riemannian metric on a general vector bundle $\pi\colon P \to M$, modelled on $\mathbb{E} \times \mathbb{F}$, $\mathbb{E}$ and $\mathbb{F}$ both Hilbert spaces. Recall from (1.2.17) the definition of the associated bundle

$$S_2\pi\colon S_2P \to M$$

of symmetric bilinear forms. *Pos* $\mathbb{F}_p$ denotes the subset of $S_2\pi^{-1}(p)$ of positive definite quadratic forms on the fibre $\mathbb{F}_p \equiv P_p$.

1.8.18 Definition. A *Riemannian metric on a vector bundle* $\pi: P \to M$ is a differentiable section

$$g: M \to S_2 PM$$

in $S_2\pi$ with $g(p) \in Pos\, \mathbb{F}_p$.

(1.8.5) extends immediately to general bundles.

1.8.19 Theorem. *A vector bundle* $\pi: P \to M$ *with model* $\mathbb{E} \times \mathbb{F}$, *a Hilbert space, possesses a Riemannian metric.*

Proof. The proof is the same as for (1.8.5): Construct from an atlas of bundle charts a locally finite refinement $(Pu_\beta, u_\beta, \tilde{M}_\beta)_{\beta \in B}$ with a partition of unity $(\phi_\beta, \tilde{M}_\beta)_{\beta \in B}$. Define $g(p)$ by $\sum_\beta \phi_\beta(p)\langle Pu_\beta(p), Pu_\beta(p)\rangle$. □

Let $\pi: P \to M$ be modelled on $\mathbb{E} \times \mathbb{F}$ with $\mathbb{E}$ and $\mathbb{F}$ Hilbert spaces. The scalar product in $\mathbb{F}$ will be denoted by $\langle\ ,\ \rangle$. By $\mathbb{O}(\mathbb{F})$ we denote the orthogonal group of $(\mathbb{F}, \langle\ ,\ \rangle)$, i.e., the group of linear isomorphisms which respect $\langle\ ,\ \rangle$.

A local trivialization (Pu, u, M') of π determines on each fibre $\mathbb{F}_p, p \in M'$ a scalar product $\langle\ ,\ \rangle_{u(p)}$ such that the linear isomorphism

$$Pu(p): (\mathbb{F}_p, \langle\ ,\ \rangle_{u(p)}) \to (\mathbb{F}, \langle\ ,\ \rangle)$$

becomes an isometry. For the notation see the note after (1.2.8).

Let now π be endowed with a Riemannian metric g. In general, the scalar product $\langle\ ,\ \rangle_{u(p)}$ and the scalar product $g(p)$ on $\mathbb{F}_p$ will not be the same, although these two products – by hypothesis on g – are equivalent in the sense that there exists an $a = a(p) \geqslant 1$ such that

$$a^{-1} g(p)(\xi, \xi) \leqslant \langle \xi, \xi \rangle_{u(p)} \leqslant a g(p)(\xi, \xi),$$

for all $\xi \in \mathbb{F}_p$.

With the following theorem we will show that it is always possible to find for a vector bundle π with Riemannian metric g a bundle atlas in which $g(p)$ and $\langle\ ,\ \rangle_{u(p)}$ coincide.

1.8.20 Theorem. *Let* $\pi: P \to M$ *be endowed with a Riemannian metric* g. *Then* P *possesses an atlas of bundle charts* (Pu, u, M') *such that, for every* $p \in M'$,

$$(*) \qquad Pu(p) = pr_2 \circ Pu|\, \mathbb{F}_p: (\mathbb{F}_p, g(p)) \to (\mathbb{F}, \langle , \rangle),$$

is an isometry.

Note. In this case we have for a pair (Pu, u, M'), (Pu', u', M'') of bundle charts the mappings

$$Pu'(p) \circ Pu(p)^{-1} = P(u' \circ u^{-1})(u(p)) \colon \mathbb{F} \to \mathbb{F},$$

for $p \in M' \cap M''$. These then are elements of the orthogonal group $\mathbb{O}(\mathbb{F})$ of the Hilbert space $\mathbb{F}$. We call (*) a *local isometric trivialization.*

Proof. Let (Pu, u, M') be a local trivialization of π. The principal part $g(u)$ of the local representation of the $\binom{0}{2}$-tensor is a differentiable mapping

$$U = u(M') \to Pos\,\mathbb{F} \quad \text{given by} \quad g(u(p)) = g(p) \circ (Pu(p)^{-1} \times Pu(p)^{-1}).$$

We derive from $g(u)$ a positive self-adjoint operator $A(u)$ on $\mathbb{F}$ by the relation

$$g(u)(\xi, \eta) = \langle A(u) \,.\, \xi, \eta \rangle = \langle \xi, A(u) \,.\, \eta \rangle,$$

all ξ, η in $\mathbb{F}$. $u \in U \mapsto A(u)$ is differentiable. Since it is positive definite, it possesses a positive square root $B(u)$. Define now

$$Qu \colon (u, \xi) \in U \times \mathbb{F} \mapsto (u, B(u) \,.\, \xi) \in U \times \mathbb{F}.$$

Then $(Qu \circ Pu, u, M')$ is a local isometric trivialization of π. □

1.8.21 Remarks. 1. In the case of the tangent bundle $\tau_M \colon TM \to M$ over M on can also use (1.8.11) to prove (1.8.20).

2. In the case of the tangent bundle $\tau_M \colon TM \to M$ over M, a local isometric trivialization in general cannot be obtained from the canonical trivialization (Tu, u, M') of TM, determined by a chart (u, M') of M, see (1.2.3).

Indeed, in the example after (1.4.3) we introduced the Gauss frame $\{e_i(p); p \in M'\}$ with respect to a chart (u, M') of M. (Tu, u, M') is an isometric trivialization if and only if the mapping $Du(p) \colon T_p M \to \mathbb{E}$, cf. (1.2.1, iv), which carries the $\{e_i(p)\}$ into the basis $\{e_i\}$ of $\mathbb{E}$, is an isometry, i.e., if $g_{ik}(u(p)) = \delta_{ik}$, for all $p \in M'$. But this will in general not be the case – witness (1.6.3) where we saw that then the curvature tensor on M' vanishes.

On the other hand, a local isometric trivialization of τ_M over $M' \subset M$ gives for each $T_p M, p \in M'$, an orthonormal basis which we may call a *Cartan basis*. It was Cartan [1] who demonstrated with his work the usefulness of locally defined orthonormal bases – also called *moving frames* – for the study of Riemannian geometry.

We continue with the discussion of linear connections on a bundle π with Riemannian metric which 'respect' this metric. More precisely, in analogy of (1.8.5) we define

1.8.22 Definition. Let $\pi \colon P \to M$ have a Riemannian metric. A linear connection (and its associated covariant derivation) on π is called *Riemannian* if the parallel translation along curves is an isometry.

As a refinement of (1.5.15) we show:

1.8.23 Theorem. *A bundle with Riemannian metric always possesses a Riemannian connection.*

Proof. As in the proof of (1.8.5) we start with a family $(Pu_\alpha, u_\alpha, M_\alpha)_{\alpha\in A}$ of trivializations. Due to (1.8.20) we may assume these triviliziations to be isometric. Let $(P\tilde{u}_\beta, \tilde{u}_\beta, \tilde{M}_\beta)_{\beta\in B}$ be a locally finite refinement of this family. For each $\beta \in B$ take as local connection map, with $\tilde{U}_\beta = \tilde{u}_\beta(\tilde{M}_\beta)$,

$$K_{\tilde{U}_\beta}: (\tilde{U}_\beta \times \mathbb{F}) \times \mathbb{E} \times \mathbb{F} \to \tilde{U}_\beta \times \mathbb{F}; \quad (\tilde{u}_\beta, \xi_\beta, X_\beta, \eta_\beta) \mapsto (\tilde{u}_\beta, \eta_\beta).$$

This means that we take $\Gamma_{\tilde{U}_\beta} \equiv 0$. Define now K as in (1.5.15). The parallel translation clearly is an isometry. A vector field $(\tilde{u}_\beta(t), \xi_\beta(t))$ along $\tilde{u}_\beta(t)$, $t_0 \leqslant t \leqslant t_1$, is parallel precisely if $\xi_\beta(t) = \xi_\beta(t_0) = \text{const.}$ □

1.8.24 Example. The tensor bundles $\tau_s^r: T_s^r M \to M$ over a manifold M of finite dimension with Riemannian metric naturally carry a Riemannian metric: Recall from (1.2.18) that $T_s^r \mathbb{E} \cong T_s^r \mathbb{R}^n$ possesses a canonical scalar product. If now $T_p M$ has the scalar product $g(p)$ then we get in this manner a scalar product on the fibre $(\tau_s^r)^{-1}(p) = T_s^r(T_p M)$.

1.8.25 Proposition. *Let* dim $M < \infty$. *Assume that M has a Riemannian metric g. Consider on M the associated covariant derivation ∇ of Levi-Cività. According to* (1.5.16) *this induces a covariant derivation ∇ for the sections of the various tensor bundles τ_s^r over M. This derivation is Riemannian.*

Proof. Let $\{e_i(t); 1 \leqslant i \leqslant \dim M\}$ be an orthonormal basis of $T_{c(t)}M$ which is parallel along the curve $c(t)$, $t_0 \leqslant t \leqslant t_1$. If Z is a tensor field of type $\binom{r}{s}$ along $c(t)$ denote by $Z^{i_1 \ldots i_r}_{j_1 \ldots j_s}(t)$ its components in the basis

$$\{e_{i_1}(t) \otimes \ldots \otimes e_{i_r}(t) \otimes e^{j_1}(t) \otimes \ldots \otimes e^{j_s}(t)\} \quad \text{of} \quad T_s^r(T_{c(t)}M),$$

determined by the basis $\{e_i(t)\}$. From the definition of ∇Z in (1.5.16) we see: $\nabla Z = 0$ if and only if the $Z^{i_1 \ldots i_r}_{j_1 \ldots j_s}(t)$ are constant. But since the scalar product $\langle Z(t), Z(t)\rangle$ is given by the sum of squares of the components, it clearly does not depend on t. □

1.9 Geodesics and Convex Neighborhoods

We continue to investigate the local geometric structure of a Riemannian manifold.

We begin with the Gauss Lemma (1.9.1). It is essentially a statement about the geometry in polar coordinates saying that the radial length and orthogonality to radial directions are the same as in Euclidean geometry.

Among the numerous consequences we get the existence of a distance $d(\ ,\ )$ on a Riemannian manifold, compatible with the given topology on M, see (1.9.5). $d(p, q)$ is defined as the infimum of the length of all curves which go from p to q – here we assume M to be connected.

We now can speak of the ball $B_\varrho(p)$ of radius $\varrho > 0$ around p. For sufficiently small ϱ, $B_\varrho(p)$ is convex, even strongly convex, cf. (1.9.7), (1.9.10). The convexity radius $\varkappa(p)$ of $p \in M$ is defined as the supremum of the ϱ for which $B_\varrho(p)$ is convex. $p \mapsto \varkappa(p)$ is continuous, see (1.9.11).

The existence of convex neighborhoods is but one aspect of a general phenomenon of Riemannian geometry which we term the Riemann Principle, cf. (1.9.8). Essentially it means that the Euclidean geometry of planar triangles in the ϱ-ball $B_\varrho(0_p)$ of the tangent space T_pM differs arbitrarily little from the geometry of triangles in $B_\varrho(p)$ when one takes the canonical isomorphism given by the exponential map $\exp_p | B_\varrho(0_p)$, provided ϱ is sufficiently small.

We conclude with the definition of a canonical Riemannian metric on the total tangent space TM of a Riemannian manifold M, cf. (1.9.12).

We begin with the so-called *Gauss-Lemma*. It states that the geodesic rays, starting from a point p, cut orthogonally through the 'spheres' around p. Gauss [1] proved this for surfaces.

We consider the tangent space T_pM of a point $p \in M$. Using the scalar product $g(p)$ we make T_pM into a Riemannian manifold, isometric to the Hilbert space $\mathbb{E}$ which serves as a model of M. Let $X_0 \in T_pM$, $r = |X_0| > 0$ so small that $\exp_p X_0$ is defined. Then $\tilde{c}(t) = tX_0$, $0 \leqslant t \leqslant 1$, is a geodesic in the Riemannian manifold T_pM. We also write $\tilde{p}$, $\tilde{q}$ instead of 0_p, X_0 when we consider them as points of T_pM rather than as tangent vectors.

To the geodesic $\tilde{c}(t)$ in T_pM there corresponds the geodesic $c(t) = \exp_p \tilde{c}(t)$, $0 \leqslant t \leqslant 1$, in M from p to $q = \exp \tilde{q}$. Let $Y_0 \in T_{\tilde{q}}(T_pM)$ be a vector orthogonal to $\dot{\tilde{c}}(1)$. Taking the canonical identification of $T_{\tilde{q}}T_pM$ with T_pM, under which $\tilde{c}(1)$ corresponds to X_0, this means $g(p)(X_0, Y_0) = 0$. We view Y_0 as a tangent vector to the sphere $S_r(\tilde{p})$ of radius $r = |X_0|$ at $\tilde{q}$.

1.9.1 Lemma. *With the previous notation consider*

$$T_{\tilde{q}}\exp_p \colon T_{\tilde{q}}(T_pM) \to T_qM.$$

Then the 'radial' vector $\dot{\tilde{c}}(1)$ is carried into the vector $\dot{c}(1)$, having the same length as $\dot{\tilde{c}}(1)$, and the 'spherical' vector Y_0 is carried into a vector orthogonal to $\dot{c}(1)$. In general, the length of Y_0 will not be preserved.

Remark. The Gauss Lemma really is a statement about Jacobi fields along a geodesic, and orthogonal to that geodesic. This will become evident after we have introduced this important concept in (1.12).

Proof. Consider a curve $X(s)$, $s \in]-\eta, \eta[$ in T_pM with $X(0) = X_0$, $|X(s)| = |X_0| = r$, $X'(0) = Y_0$. That is, $X(s)$ is a curve on the sphere $S_r(0_p)$. If we choose $\eta > 0$ and $\varepsilon > 0$ sufficiently small there exists the singular surface

$$F \colon (s, t) \in]-\eta, \eta[\times]-\varepsilon, 1+\varepsilon[\mapsto \exp_p(tX(s)) \in M.$$

Note: For every fixed s,

$$t \in [0,1] \mapsto F(s,t) = \exp_p\big(tX(s)\big) \in M$$

is a geodesic c_s with $|\dot{c}_s(t)| = |\partial F(s,t)/\partial t| = r$.

We have, cf. (1.5.7), (1.8.14),

$$\frac{\partial}{\partial t} g\left(\frac{\partial F}{\partial s}(s,t), \frac{\partial F}{\partial t}(s,t)\right) = g\left(\frac{\nabla}{\partial s}\frac{\partial F}{\partial t}(s,t), \frac{\partial F}{\partial t}(t,s)\right) + g\left(\frac{\partial F}{\partial s}(s,t), \frac{\nabla}{\partial t}\frac{\partial F}{\partial t}(s,t)\right) =$$

$$= \frac{1}{2}\frac{\partial}{\partial s} g\left(\frac{\partial F}{\partial t}(t,s), \frac{\partial F}{\partial t}(t,s)\right) = 0 .$$

Hence,

$$g\left(\frac{\partial F}{\partial s}(s,t), \frac{\partial F}{\partial t}(s,t)\right)$$

is independent of t. Since $F(s,0) = p$ we get

$$0 = g\left(\frac{\partial F}{\partial s}(0,0), \frac{\partial F}{\partial t}(0,0)\right) = g\left(\frac{\partial F}{\partial s}(0,1), \frac{\partial F}{\partial t}(0,1)\right) =$$

$$= g(T_{\tilde{q}}\exp_p . Y_0, T_{\tilde{q}}\exp_p . X_0) . \quad \square$$

As an immediate consequence of the Gauss Lemma we obtain a comparison between the length of a curve b from p to q and a geodesic c from p to q, provided – and this is essential – that both are images under $\exp_p$ of curves in $T_p M$.

1.9.2 Theorem. *Let $\tilde{b}(s)$, $0 \leqslant s \leqslant s_1$, be a curve in $T_p M$ from $0_p = \tilde{p}$ to $\tilde{q}$ and assume that $\tilde{b}(s) \in \tilde{T}_p M$ = domain of definition of $\exp_p$, for all s. Let $\tilde{c}(t) = t\tilde{b}(s_1)$, $0 \leqslant t \leqslant 1$, be the straight segment from $\tilde{p}$ to $\tilde{q}$. Put $\exp_p \tilde{c}(t) = c(t)$, $\exp_p \tilde{b}(s) = b(s)$. Then*

$$L(c) \leqslant L(b) .$$

Moreover, if $\tilde{b}(s) = \tilde{c}\big(t(s)\big)$, where $t(s)$ is a function satisfying $t'(s) \geqslant 0$, we have equality.

Conversely, if $L(c) = L(b)$ and if in addition $T_{t\tilde{b}(s)}\exp_p$ is of maximal rank for all s, t with $0 \leqslant t \leqslant 1$ then $\tilde{b}(s) = \tilde{c}\big(t(s)\big)$, where $t(s)$ is a differentiable function with $t'(s) \geqslant 0$.

Proof. We can assume $\tilde{b}(s) \neq \tilde{p}$ for $s > 0$. Put $|\tilde{b}(s_1)| = r > 0$. We describe $\tilde{b}(s)$, $s > 0$, by polar coordinates

$$\tilde{b}(s) = r(s)\,X(s) \quad \text{with} \quad X(s) = \tilde{b}(s)/|\tilde{b}(s)| .$$

Consider, for small $\varepsilon > 0$,

$$F\colon (s,t) \in]\varepsilon, s_1] \times [0,1] \mapsto \exp_p\big(rtX(s)\big) \in M .$$

Then we get, with $b(s) = F\big(s, r(s)/r\big)$ and $b'(s) = \dfrac{\partial F}{\partial s}\left(s, \dfrac{r(s)}{r}\right) + \dfrac{r'(s)}{r}\dfrac{\partial F}{\partial t}\left(s, \dfrac{r(s)}{r}\right)$, from (1.9.1) and $|\partial F(s,t)/\partial t| = |rX(s)| = r$,

$$|b'(s)|^2 = \left|\frac{\partial F}{\partial s}\left(s, \frac{r(s)}{r}\right)\right|^2 + \left|\frac{\partial F}{\partial t}\left(s, \frac{r(s)}{r}\right)\right|^2 \frac{r'(s)^2}{r^2} \geqslant r'(s)^2 .$$

Hence,

$$L(b) = \int_s^{s_1} |b'(s)|\, ds \geqslant \int_0^{s_1} |r'(s)|\, ds \geqslant \int_0^{s_1} r'(s)\, ds = r(s_1) = L(c).$$

Moreover, if $b(s) = c(t(s))$ with $t'(s) \geqslant 0$, $|b'(s)| = |\dot{c}(t)|\; t'(s)$, thus, $L(b) = L(c)$.

Assume now $L(b) = L(c)$. Then $r'(s) \geqslant 0$ and

$$\partial F(s, t)/\partial s = T_{rtX(s)} \exp_p . (rtX'(s)) = 0 .$$

Under our hypothesis follows $X'(s) = 0$, i.e., $X(s) = X(s_1) = \text{const}$. Thus, $\tilde{b}(s) = r(s)\, X(s_1) = (r(s)/r)\tilde{b}(s_1) = \tilde{c}(t(s))$ with $t(s) = r(s)/r$. □

Remark. For the validity of (1.9.2), the special hypotheses for the geodesic c and the curve b are important. One sometimes paraphrases this by saying that b is *embedded in a field of geodesics* starting from p.

As an example where these hypotheses are not satisfied we consider on $S_1^2(0) =$ 2-sphere of radius 1 in $\mathbb{R}^3$ the geodesic $c(t) = (0, \sin t, \cos t), 0 \leqslant t \leqslant 3\pi/2$, of length $3\pi/2$ from $p = (0, 0, 1)$ to $q = (0, -1, 0)$, and the curve (also a geodesic) $b(t) = (0, -\sin t, \cos t), 0 \leqslant t \leqslant \pi/2$, of length $\pi/2$ from p to q.

In this example c contains a conjugate point, cf. (1.12.9) for this concept. Another example where $L(c) \leqslant L(b)$ is not true is given by the surface M in $\mathbb{R}^3$ obtained from 'flattening' the 'backside' $x_0 < 0$ of the 2-sphere $S_1^2(0)$. Take for c the part $c(t) = (\cos t, \sin t, 0), -\pi/2 + \varepsilon \leqslant t \leqslant \pi/2 - \varepsilon$, some small $\varepsilon > 0$, of the equator and for b the image of the remainder of the equator under the flattening.

Both the previous examples can be subsumed under the case where we have a non-constant closed geodesic $c^*(t), 0 \leqslant t \leqslant \omega, \dot{c}^*(\omega) = \dot{c}^*(0)$. Let $c = c^* | [0, a]$ and b the inverse of $c | [a, \omega], 2a < \omega$.

The following Theorem says under what conditions a geodesic c is shortest when compared with all curves joining its end points. In this case we also speak of the *minimizing geodesic* between two points.

1.9.3 Theorem. *Let* $p \in M$ *and* $\varepsilon = \varepsilon(p)$, $\eta = \eta(p)$, $0 < \varepsilon < \eta$, *as in* (1.8.15). *Then a geodesic of length* $< \eta$, *starting from an arbitrary point* q *of* $B_\varepsilon(p) = \exp B_\varepsilon(0_p)$, *is a curve of minimal length between its end points.*

Remark. This Theorem shows that a geodesic locally always is a curve of minimal length. More precisely, given a geodesic $c(t), 0 \leqslant t \leqslant a$, then, given $t_0 \in]0, a[$, there exists an $\varepsilon = \varepsilon(c, t_0) > 0$ such that $c | [t_0 - \varepsilon, t_0 + \varepsilon]$ is a curve of minimal length among all curves going from $c(t_0 - \varepsilon)$ to $c(t_0 + \varepsilon)$.

Proof. Let $b = b(s), 0 \leqslant s \leqslant s_1$, be a curve from q to r. Put $c_{qr} = c$. If image $b \subset B_\eta(q)$, $\tilde{b}(s) = (\exp_q | B_\eta(0_q))^{-1} b(s)$ is a curve satisfying the hypotheses of (1.9.2).

Thus, $L(c) \leqslant L(b)$. Otherwise, there will be a $s^*, 0 < s^* \leqslant s_1$, where $\tilde{b}(s) = (\exp_q | B_\eta(0_q))^{-1} b(s), 0 \leqslant s \leqslant s^*$, goes from 0_q to $\tilde{b}(s^*)$ where $L(c) < |\tilde{b}(s^*)| < \eta$. But then, $L(c) < L(b|[0, s^*]) \leqslant L(b)$. □

We now can introduce a distance on M.

1.9.4 Definition. For p and q in M, the *distance* $d = d_M$ *between* p *and* q is given by $d(p, q) = \inf\{L(c);\ c$ a curve from p to $q\}$ provided p and q are in the same connected component of M. Otherwise put $d(p, q) = \infty$.

1.9.5 Theorem. *The function* $d\colon M \times M \to \mathbb{R}$ *determines a metric on* M. *The topology derived from* d *coincides with the given topology on* M.

Proof. Clearly, $d(p, q) \geqslant 0$ and $d(p, q) = 0$ if $p = q$. (1.9.3) shows that $d(p, q) > 0$ if $p \neq q$. Indeed, either $q \in B_\eta(p)$ in which case $d(p, q) =$ length of the minimizing geodesic from p to q and thus is > 0. Or else, $q \notin B_\eta(p)$ and therefore, again by (1.9.3), $d(p, q) \geqslant \eta$.

The relations $d(p, q) = d(q, p)$ and $d(p, q) + d(q, r) \geqslant d(p, r)$ are immediate consequences of the definition (1.9.4).

Finally, with $\eta > 0$ as in (1.9.3), the sets $B_\varepsilon(p), 0 < \varepsilon \leqslant \eta$, are open in the given topology on M and they also are, according to (1.9.3), the open balls of radius ε around p with respect to the distance d. □

In order to get a refinement of (1.9.3) we first show:

1.9.6 Proposition. *Let* $p \in M$ *and* $\eta = \eta(p) > 0$ *such that* $B_\eta(p)$ *is in the domain of normal coordinates around* p. *Then there exists* $\varepsilon_0, 0 < \varepsilon_0 < \eta$, *such that every sphere*

$$S_\varepsilon(p) = \exp_p(\partial \bar{B}_\varepsilon(0_p)) = \partial \bar{B}_\varepsilon(p)$$

of radius $\varepsilon, 0 < \varepsilon \leqslant \varepsilon_0$, *is convexly embedded. That is to say, if* $c(t), 0 \leqslant t \leqslant a$, *is a non-constant geodesic with* $c(t_0) \in S_\varepsilon(p)$ *and* $\dot{c}(t_0)$ *tangential to* $S_\varepsilon(p)$, *some* $t_0 \in]0, a[$, *then* $c(t)$ *is outside* $B_\varepsilon(p)$ *for all* $t \neq t_0, |t - t_0|$ *sufficiently small.*

More precisely, given λ *with* $0 < \lambda < 1$, *there exists* $\eta = \eta(\lambda, \varepsilon) > 0$ *such that*

$$d(p, c(t)) \geqslant d(p, c(t_0)) + \lambda(t - t_0)^2 = \varepsilon + \lambda(t - t_0)^2,$$

for all $|t - t_0| \leqslant \eta$.

Proof. We can assume $c(t) \in B_\eta(p)$ for all t. Let $(u, B_\eta(p))$ be normal coordinates around p. Put $u \circ c(t) = u(t)$. Denote the scalar product in $B_\eta(0)$ by a dot. Then we have for $F(t) = u(t) \cdot u(t)/2 = d(p, c(t))^2/2$

$$F(t_0) = \varepsilon^2/2;\ F(t_0) = \dot{u}(t_0) \cdot u(t_0) = 0;$$

$$\ddot{F}(t_0) = \dot{u}(t_0) \cdot \dot{u}(t_0) + \ddot{u}(t_0) \cdot u(t_0).$$

We will show that $\ddot{F}(t_0) > 0$ if only ε is sufficiently small. Since $\Gamma(0) = 0$ we get from the differentiability of the map

$$\Gamma : B_\varrho(0) \to L(\mathbb{E}, \mathbb{E}; \mathbb{E})$$

the existence of a $\varepsilon_0 > 0$ such that $\|\Gamma(u)\| \leqslant \delta < 1$, whenever $|u| \leqslant \varepsilon_0$. Thus, $\ddot{u}(t) = -\Gamma(u(t))(\dot{u}(t), \dot{u}(t))$ yields

$$\ddot{F}(t_0) \geqslant |\dot{u}(t_0)|^2 (1 - \delta) > 0. \quad \square$$

We now can prove the desired refinement of (1.9.3). It shows to which extent a small ball around $p \in M$ looks like a ball in Hilbert space.

1.9.7 Theorem. *Every point p of M possesses a* convex neighborhood. *More precisely, let $\varkappa_0^* = \varepsilon_0/3$, with $\varepsilon_0 = \varepsilon_0(p) > 0$ as in* (1.9.6). *Then, for every $\varkappa^*$, $0 < \varkappa^* \leqslant \varkappa_0^*$, the ball $B_{\varkappa^*}(p)$ of radius $\varkappa^*$ around p has the property that any two points q and r in $B_{\varkappa^*}(p)$ can be joined by a not necessarily uniquely determined geodesic c_{qr} of length $d(q, r)$ which lies entirely in $B_{\varkappa^*}(p)$.*

Proof. Let q and r in $B_{\varkappa^*}(p)$. From (1.9.3) we have the existence of a uniquely determined geodesic c_{qr}: $[0, 1] \to M$ from q to r of length $d(q, r) < 2\varkappa^*$. Thus, c_{qr} lies entirely in $B_{2\varkappa^*}(q) \subset B_{3\varkappa^*}(p) \subset B_\eta(p)$, η as in (1.9.3). Assume that c_{qr} is leaving $B_{\varkappa^*}(p)$. That is to say, there exists a $t_0 \in]0, 1[$ with $d(p, c(t_0)) \geqslant d(p, c(t))$, all t, $d(p, c(t_0)) < 3\varkappa^* \leqslant \varepsilon_0$ and $\dot{c}(t_0)$ tangential to the sphere $S_\varepsilon(p)$, $\varepsilon = d(p, c(t_0))$. But then we have from (1.9.6) that there also are t with $d(p, c(t)) > d(p, c(t_0))$ – a contradiction. $\square$

Remark. The previous discussions indicate a similarity between the local Euclidean geometry in a neighborhood of the origin 0_p of the tangent space $T_p M$ with its Hilbert metric $g(p)$ on the one hand and the local Riemannian geometry in a neighborhood of p on the other.

To make this more precise, we consider convex balls $B_\varrho(p)$ around $p \in M$ and their images $B_\varrho(0_p)$ in $T_p M$ under $(\exp_p | B_\varrho(0_p))^{-1}$. A triangle in $B_\varrho(0_p)$ is determined by its corners $\tilde{q}, \tilde{r}, \tilde{s}$. We therefore denote it by $\tilde{q}\tilde{r}\tilde{s}$. The sides of this triangle are denoted by $\tilde{q}\tilde{r}, \tilde{r}\tilde{s}, \tilde{s}\tilde{q}$.

Since $B_\varrho(p)$ is convex, three points q, r, s in $B_\varrho(p)$ also determine a triangle which we denote by qrs. Its sides, i.e., the uniquely determined minimizing geodesics between its corners, are denoted by qr, rs, sq. We can speak of the angles of a triangle qrs, provided the three points q, r, s are different.

We now formulate what we wish to call the *Riemann Principle.*

1.9.8 Scholion. *Let $B_\varrho(p)$ be a convex ball around $p \in M$ and $B_\varrho(0_p)$ the corresponding ball in the Hilbert space $T_p M$. Then the following is true. If only $\varrho > 0$ is sufficiently small then the image under $\Phi \equiv \exp_p | B_\varrho(0_p)$ of a Euclidean triangle $\tilde{q}\tilde{r}\tilde{s}$ in $B_\varrho(0_p)$ differs arbitrarily little from the Riemannian triangle qrs in $B_\varrho(p)$, determined by the images q, r, s* of *$\tilde{q}, \tilde{r}, \tilde{s}$ under Φ.*

More precisely, given $\varepsilon > 0$, then for every sufficiently small $\varrho > 0$ there exist constants $a = a(p, \varepsilon, \varrho) > 0$, $b = b(p, \varepsilon, \varrho) > 0$, $c = c(p, \varepsilon, \varrho) > 0$, not depending on $\tilde{q}\tilde{r}\tilde{s}$, *such that the images* $\Phi(\tilde{q}\tilde{r})$, $\Phi(\tilde{r}\tilde{s})$, $\Phi(\tilde{s}\tilde{q})$ *differ from the sides* qr, rs, sq *of the triangle* qrs *by less than* $a\varepsilon$ *in the metric introduced in* (1.6.11). *Moreover, the length of the sides of* $\tilde{q}\tilde{r}\tilde{s}$ *differ by less than* $b\varepsilon$ *from the length of the corresponding sides of the triangle* qrs *and the angles in* $\tilde{q}\tilde{r}\tilde{s}$ *differ by less than* $c\varepsilon$ *from the corresponding angles in the triangle* qrs.

Proof. If we choose an isometry $\iota_p\colon T_pM \to \mathbb{E}$ we can define on $B_\varrho(p)$ the normal coordinates $u = \iota_p \circ \Phi^{-1}$, cf. (1.8.16). The representation of the metric g on $B_\varrho(p)$ in these coordinates has a Taylor series which begins with

$$g(u) = g(0) + \tfrac{1}{2} D^2 g(0) \,.\, (u, u) + o(|u|^2),$$

whereas the Taylor series of the Christoffel symbol $\Gamma(u)$ reads

$$\Gamma(u) = D\Gamma(0) \,.\, u + o(|u|).$$

This shows that the differential equation for the representation $u(t)$ of a geodesic, i.e.,

$$\ddot{u}(t) + \Gamma(u(t))(\dot{u}(t), \dot{u}(t)) = 0$$

differs arbitrarily little from the differential equation $\ddot{u}(t) = 0$ of a straight segment in $B_\varrho(0_p)$, if only $|u(t)|$ is sufficiently small.

Together with (1.8.15) we get therefore the desired similarity of the image under Φ of the sides of a triangle in $B_\varrho(0_p)$ and the sides of the Riemannian triangle, having the same corners as the images. The claims on the length of the sides and the sizes of the angles follow. □

To obtain a result even sharper than (1.9.7) we introduce the

1.9.9 Definition. (i) An open set M' of M is called *strongly convex* if any two points q and r of M' can be joined by exactly one geodesic of length $d(q, r)$ which belongs entirely to M'. Moreover, every ε-ball $B_\varepsilon(p)$ in M' shall be convex.

(ii) Define the *convexity radius* of $p \in M$ by

$$\varkappa(p) = \sup\{\varkappa \in \bar{\mathbb{R}} = \mathbb{R} \cup \{\pm\infty\};\ B_\varkappa(p) \text{ is strongly convex}\}.$$

Note. If M' is strongly convex so is every convex subset of M'.

Examples. In Euclidean space, every convex set also is strongly convex. The same is true for the sphere. A maximal strongly convex set on the sphere is given by an open half sphere. Thus, whereas for the Euclidean space the convexity radius is ∞, it is $\pi\varrho/2$ on a sphere of radius ϱ.

To get an example of a convex ball $B_\varepsilon(p)$ which is not strongly convex we begin with a circular cone in $\mathbb{R}^3$ of small opening angle at its singular point p. Think of a very sharp pencil. If q has distance $\varepsilon/2$ from p and lies on one of the generators of the cone of length ε, $B_{\varepsilon/2}(q) \subset B_\varepsilon(p)$ and $B_{\varepsilon/2}(q)$ is not convex (it need not even be the diffeomorphic image of the ball of radius $\varepsilon/2$ in the tangent plane at q). If we round off the cone at its singular point p in a very small neighborhood of p we get a proper, non-singular surface. $B_\varepsilon(p)$ remains convex and not strongly convex.

For more on convex sets cf. the survey article by Walter [1]. In (2.9) we will deal again with convex sets.

1.9.10 Theorem. *Every $p \in M$ possesses a strongly convex ball $B_\varkappa(p)$. Actually for $p \in M$ there exists a $\varkappa = \varkappa(p) > 0$ such that for all $q \in B_\varkappa(p)$, $B_\varkappa(q)$ is convex.*

Proof. With $B_\varkappa(q)$ convex also $B_{\varkappa'}(q)$ is convex, $0 < \varkappa' \leqslant \varkappa$. We therefore get the first statement from the second one.

To prove the second statement we observe that the estimates in the proof of (1.9.7) can be made uniformly for all q in a certain ball around p. This is a consequence of the Riemann Principle (1.9.8).

According to (1.8.15) there exists $\varepsilon' = \varepsilon'(p) > 0$ such that

$$(\tau, \exp) \mid \bigcup_{q \in B_{3\varepsilon'}(p)} B_{3\varepsilon'}(0_q)$$

is a diffeomorphism onto an open neighborhood of the diagonal of $B_{3\varepsilon'}(p) \times B_{3\varepsilon'}(p)$. In particular, around every $q \in B_{3\varepsilon'}(p)$ there exist normal coordinates $(u_q, B_{3\varepsilon'}(q))$ on the ball of radius $3\varepsilon'$.

Consider

$$\Phi\colon (q, r) \in B_{\varepsilon'}(p) \times B_{2\varepsilon'}(p) \mapsto \Gamma(u_q(r)) \in L(\mathbb{E}, \mathbb{E}; \mathbb{E}).$$

From (1.5.1) we have the relation

$$\Gamma(u_q) = D(u_p \circ u_q^{-1})^{-1}[D^2(u_p \circ u_q^{-1}) + \Gamma(u_p) \circ (D(u_p \circ u_q^{-1}) \times D(u_p \circ u_q^{-1}))]$$

which shows that Φ is differentiable. Moreover, since $\Gamma(u_q(q)) = 0$, we have for sufficiently small ε_0, $0 < \varepsilon_0 \leqslant \varepsilon'$,

$$\|\Gamma(u_q(r))\| \leqslant \text{some } \delta < 1,$$

for all $(q, r) \in B_{\varepsilon_0}(p) \times B_{2\varepsilon_0}(p)$. That is to say, (1.9.6) is valid simultanuously for all $S_{\varepsilon'}(q)$, $0 \leqslant \varepsilon' \leqslant \varepsilon_0$, $q \in B_{\varepsilon_0}(p)$. We therefore can continue as in the proof of (1.9.7) and show the existence of a $\varkappa > 0$ such that $B_{\varkappa'}(q)$ is convex for all $q \in B_\varkappa(p)$, all $\varkappa'$ with $0 < \varkappa' \leqslant \varkappa$. □

1.9.11 Corollary. *For every $p \in M$, the* convexity radius *$\varkappa(p)$ is > 0. If $\varkappa(p_0) = \infty$ for one $p_0 \in M$, then $\varkappa(p) = \infty$ for all $p \in M$ in the connected component of p_0. Finally, $\varkappa\colon M \to \mathbb{R}$ is continuous.*

Proof. Only the last statement is not quite obvious. However, from the definition of strong convexity we have

$$\varkappa(q) \geqslant \varkappa(p) - d(p, q)$$

which shows $|\varkappa(q) - \varkappa(p)| \leqslant d(p, q)$. □

The Levi-Cività derivation on a Riemannian manifold M determines a linear connection and thus a splitting of $T_X TM$ into horizontal and vertical spaces, cf. (1.5.9). We use this to define a canonical Riemannian metric on TM:

1.9.12 Definition. Let M be a Riemannian manifold. Then we get a *Riemannian metric on the total tangent space* TM of M as follows. Let $X \in TM$ and

$$T_X TM = T_{Xh} TM \oplus T_{Xv} TM$$

be the splitting into horizontal and vertical subspace. These spaces are canonically identified with $T_p M$, $p = \tau_M X$, via $T\tau_M | T_{Xh} TM$ and $K | T_{Xv} TM$, respectively; the latter is nothing but the identification of the tangent space to the fibre $T_p M$ in X with the fibre.

Define now a scalar product on $T_X TM$ by letting the horizontal and vertical space be orthogonal and taking on each of these spaces the scalar product determined from the scalar product on $T_p M$ by the canonical identifications.

Note. This Riemannian metric on TM will be important for various subsequent applications.

1.10 Isometric Immersions

A differentiable mapping $F: M \to N$ between two Riemannian manifolds is called isometric if the tangential $T_p F: T_p M \to T_{F(p)} N$ preserves the scalar product for all $p \in M$. F then must be an immersion. We will also write M^* instead of N.

As a special case we consider a submanifold M of a Riemannian manifold M^*; take on M the Riemannian metric obtained by restricting the scalar product on $T_p M^*$, $p \in M$, to the subspace $T_p M$. For F take the inclusion.

In (1.10.3) we derive the fundamental formula of Levi-Cività by which the covariant derivation on M^* in a point $p^* = F(p)$ is related to the covariant derivation on M in p.

In (1.3.11) we had introduced for an immersion $F: M \to M^*$ the normal bundle $\nu_F: T^\perp F \to M$. If F is isometric, ν_F can be represented as a sub-bundle of the bundle $F^* \tau_{M^*}$, induced by F from the tangent bundle τ_{M^*} over M^*, see (1.10.4). This leads to the second fundamental mapping $h_p, p \in M$, associated to F – a straightforward generalization of Gauss' second fundamental form for a surface in $\mathbb{R}^3$. We get a general Gauss formula (1.10.5) as well as a general Weingarten formula (1.10.7).

We continue with a brief discussion of 1-parameter groups of isometries of a Riemannian manifold and the corresponding infinitesimal isometries, the so-called Killing vector fields. The latter form a subalgebra in the Lie algebra of all vector fields, cf. (1.10.11).

There follows the characterization of a totally geodesic submanifold $M \subset M^*$ by the vanishing of the second fundamental mapping h_p, all $p \in M$, see (1.10.14). Such submanifolds occur as fixed point sets for isometries, (1.10.15). 1-dimensional totally geodesic submanifolds are geodesics; (1.10.15) sometimes is useful to find closed geodesics. We conclude with a fixed point theorem for isometries.

We begin the definition of the structure preserving mappings (or morphisms) between Riemannian manifolds.

1.10.1 Definition. Let

$$F:(M,g)\to(N,h)$$

be a differentiable mapping of the Riemannian manifold (M,g) into a Riemannian manifold (N,h) such that

$$h(F(p))\circ(T_pF\times T_pF)=g(p)$$

all $p\in M$. Then F is called *isometric*. If in particular $(N,h)=(M,g)$ and F is an isometric diffeomorphism it is also called an *isometry*.

The set of isometries of (M,g) forms a subgroup of the group of all bijections and a subgroup of the group of diffeomorphisms. It is called the *isometry group* Iso (M,g).

1.10.2 Remarks. 1. An isometric mapping $F\colon M\to N$ is necessarily an immersion since $X\in T_pM$ and $T_pF\,.\,X=0$ implies $0=h(T_pF\,.\,X,\,T_pF\,.\,X)=g(X,X)$, i. e., $X=0$. The metric g on M which we had defined in (1.8.3) was chosen in such a way as to make the immersion F isometric.

2. Every finite dimensional Riemannian manifold possesses an isometric embedding into an Euclidean space of sufficiently high dimension – in particular, in a proper Hilbert space, see Nash [1].

3. Let M be a Riemannian manifold. With the distance d_M, cf. (1.9.4), M becomes a metric space (M,d_M). Let (M,d_M), (N,d_N) be two such metric spaces. Then any distance preserving surjective mapping $F\colon M\to N$ is an isometric diffeomorphism – at least if $\dim M<\infty$. See Myers and Steenrod [1]. A proof is also contained in Kobayashi and Nomizu [1].

In particular, $\mathrm{Iso}(M,g)$ coincides with the group of distance preserving bijections of (M,d_M).

4. In general one must expect the isometry group $\mathrm{Iso}(M,g)$ to be very small, even the identity. Let M be compact. Then one can introduce a topology on the set $\mathscr{G}M$ of Riemannian metrics of the differentiable manifold M and show that there is an open dense subset in $\mathscr{G}M$ such that, for an element g^* in this subset, $\mathrm{Iso}(M,g^*)=\{id\}$. We don't go into the details of the proof. We only indicate how $\mathscr{G}M$ can be made an open cone in a Banach space.

Recall that a Riemannian metric g on M is a section in the bundle $\sigma_2\colon S_2M\to M$ of 2-fold covariant symmetric tensors. On the vector space $\Gamma(\sigma_2)$ of sections in σ_2 we define a norm, using a fixed Riemannian metric g_0 on M, as follows: The scalar product $g_0(p)$ on T_pM induces a scalar product on the various associated tensor spaces, which we again denote by g_0, cf. (1.0) 1. In particular, for a section $g: M \to S_2 M$ in σ_2 we can define a norm by

$$|g|_0^2 = \sup_{p \in M} g_0(p)(g(p), g(p)).$$

More generally, we may take also the covariant derivative and its iterates into consideration. E. g., we can define a norm by

$$|g|_1^2 = \sup_{p \in M} \{g_0(p)(g(p), g(p)) + g_0(p)(\nabla g(p), \nabla g(p))\}.$$

Here we view $\nabla g(p)$ as a tensor field of type $\binom{0}{3}$.

The set $\mathscr{G} M$ of Riemannian metrics on M then becomes an open subset in the Banach space $\Gamma(\sigma_2)$. It even is what one calls an open cone since with g and g' in $\mathscr{G} M$ also $ag + a'g \in \mathscr{G} M$, if $a, a' \geqslant 0$ and $a + a' > 0$.

In case that M is not compact but only locally compact, one introduces on $\mathscr{G} M$ a topology (and actually a topology derivable from a metric) by using a countable covering of M by compact sets. Cf. Hirsch [1] for details on the so-called weak and strong topologies on function spaces.

5. An isometry of the Euclidean space $M = \mathbb{R}^n$ is what is classically called a *motion*: $x \mapsto Rx + x_0$ with $R \in \mathbb{O}(n)$, $x_0 \in \mathbb{R}^n$. Thus, Iso$(\mathbb{R}^n)$ is the semi-direct product $\mathbb{O}(n) \times \mathbb{R}^n$ with the composition

$$(R', x_0') \,.\, (R, x_0) = (R'R, R'x_0 + x_0')$$

6. The isometry group of the sphere $S^n = \{x \in \mathbb{R}^{n+1}; |x| = 1\}$ is the orthogonal group $\mathbb{O}(n+1)$. It operates transitively on the set of orthonormal frames. That is, given an orthogormal basis $\{X_i\}_{1 \leqslant i \leqslant n}$ in $T_p S^n$ and an orthonormal basis $\{Y_j\}_{1 \leqslant j \leqslant n}$ in $T_q S^n$ there exists (exactly one) element $g \in \mathbb{O}(n+1)$ whose tangential carries X_i into Y_i, $1 \leqslant i \leqslant n$.

The next theorem relates the image of the Levi-Cività derivation under an isometric immersion to the Levi-Cività derivation of the receiving manifold. One may view this as a certain covariance property of the Levi-Cività derivation. In the case where the receiving manifold is a euclidean space the theorem is due to Levi-Cività [1]. It shows how a covariant derivation on a submanifold defined with the help of an extrinsic operation (i. e., an orthogonal projection) can also actually be obtained by a purely intrinsic operation, i. e., the Levi-Cività derivation derived from the Riemannian metric on the submanifold.

1.10.3 Theorem. *Let*

$$F: (M, g) \to (M^*, g^*)$$

be an isometric immersion. Denote by ∇ *and* ∇^* *the Levi-Cività derivation on* M *and* M^**, respectively. Then*

$$(*) \qquad TF \,.\, \nabla_X Y = pr \circ \nabla_X^*(TF \,.\, Y)$$

where X and Y are vector fields on M and

$$pr = pr_p : T_{F(p)}M^* \to T_p F(T_p M)$$

is the orthogonal projection.

Note. (∗) can be viewed as a description of the manner in which TF and the covariant derivations on M and M^* commute.

Proof. According to (1.8.11) it suffices to show that TF^{-1} of the right hand side of (∗) is a torsion-free Riemannian covariant derivation. Here we use that $T_p F = TF|T_p M$ is an isomorphism from $T_p M$ to $pr \circ T_{F(p)}M^*$. Now, from (1.8.14) we get

$$pr \circ \nabla^*_X TF.Y - pr \circ \nabla^*_Y TF.X = pr \circ TF.[X,Y] = TF.[X,Y]$$

and $$Xg(Y,Z) = Xg^*(TF.Y, TF.Z) =$$

$$g^*(\nabla^*_X TF.Y, TF.Z) + g^*(TF.Y, \nabla^*_X TF.Z) =$$

$$= g^*(pr \circ \nabla^*_X TF.Y, TF.Z) + g^*(TF.Y, pr \circ \nabla^*_X TF.Z) =$$

$$= g(TF^{-1} \circ pr \circ \nabla^*_X TF.Y, Z) + g(Y, TF^{-1} \circ pr \circ \nabla^*_X TF.Z). \quad \square$$

In order to draw some consequences from (1.10.3) we first consider an immersion $F: M \to M^*$ and the associated normal bundle $\nu_F: T^\perp F \to M$, cf. (1.3.11). We show that in case F is isometric, ν_F possesses a particularly simple representation.

1.10.4 Lemma. *Let*

$$F: (M, g) \to (M^*, g^*)$$

be an isometric immersion. For each $p \in M$ we denote by

$$(*) \qquad T_{F(p)}M^* = T_p F \oplus T_p^\perp F$$

the orthogonal decomposition determined by the subspace $T_p F \equiv T_p F(T_p M)$ and its orthogonal complement $T_p^\perp F$. $T_p^\perp F$ is called normal space at *p*. *Then the normal bundle*

$$\nu_F : T^\perp F \to M$$

is isomorphic to the bundle which is obtained from the induced tangent bundle

$$F^*\tau_{M^*} : F^* TM^* \to M$$

by taking as fibre over $p \in M$ the subspace $T_p^\perp F$. We therefore use again $\nu_F: T^\perp F \to M$ to denote this bundle.

Remark. Here – as always – we use the canonical identification of the fibre over p, $(F^*\tau_{M^*})^{-1}(p) = \{p\} \times T_{F(p)}M^*$, with $T_{F(p)}M^*$. In the notation of (1.3.10), this identification is given by $\tau^*_{M^*}F|(F^*\tau_{M^*})^{-1}(p)$.

The tangent bundle $\tau_M : TM \to M$ is a particular sub-bundle of the induced bundle

$F^*\tau_{M^*}$. Take the restriction of the fibre over p to $T_pF(T_pM) \cong T_pM$. One thus obtains for the bundle $F^*\tau_{M^*}$ the orthogonal decomposition

$$F^*\tau_{M^*} = \tau_M \oplus \nu_F .$$

Proof. All we need to show is that $T_p^\perp F$ is linearly isomorphic to the fibre $T_{F(p)}M^*/T_pF$ of ν_F over p. But this is clear from the definitions. □

1.10.5 Lemma. *Let*

$$F: (M, g) \to (M^*, g^*)$$

be an isometric immersion. Then there exists, for every $p \in M$, a symmetric bilinear mapping

$$h_p : (X, Y) \in T_pM \times T_pM \mapsto h_p(X, Y) \in T_p^\perp F$$

which is defined as follows: Also denote by X, Y an extension of $X \in T_pM, Y \in T_pM$ towards a vector field. Then put

$$(\dagger) \qquad \nabla_X^* TF.Y(p) = TF.\nabla_X Y(p) + h_p(X, Y).$$

Note. ($\dagger$) is sometimes called *Gauss formula*. Gauss [1] had such a formula for a surface in Euclidean 3-space. The symmetric bilinear mapping h_p is called the *second fundamental mapping*. For a normal vector $N_p \in T_p^\perp F$,

$$(X, Y) \in T_pM \times T_pM \mapsto h_{N_p}(X, Y) = \langle N_p, h_p(X, Y)\rangle \in \mathbb{R}$$

is called the *second fundamental form with respect to the normal vector* N_p.

Proof. $h_p(X, Y) \in T_p^\perp F$ follows from (1.10.3). Since ∇^* and ∇ are the Levi-Cività derivations, we have

$$\nabla_X^* TF.Y - \nabla_Y^* TF.X = TF.[X, Y] = TF.(\nabla_X Y - \nabla_Y X),$$

cf. (1.8.14). This proves the symmetry of h_p. Since $\nabla_X Y(p)$ and $(\nabla_X^* TF.Y)(p)$ depend on $X(p)$ only, so does $h_p(X, Y) = h_p(Y, X)$. □

1.10.6 Example. *The distance sphere.* Let $p^* \in M^*$ and $\varrho > 0$ so small that $\bar{B}_\varrho(p^*)$ is inside the domain of normal coordinates around p^*. Let $M = \partial \bar{B}_\varrho(0_{p^*}) = S_\varrho(0_{p^*})$ be the sphere of radius ϱ in $T_{p^*}M^*$. Then

$$F: M \to M^*; p \mapsto \exp_{p^*} p$$

is an imbedding with image $= \partial \bar{B}_\varrho(p^*)$. On M we take the induced metric.

We want to determine the second fundamental form $h_{N(p)}(X, X)$ for $X \in T_pM$, $N(p)$ the 'outward' unit normal vector in $p \in M$. Here, as before, we also denote by X an extension of $X \in T_pM$ to a vector field on M.

For that purpose we consider the geodesic $c_p(t) = \exp_{p^*} tp/\varrho$, t in some interval I containing $[0, \varrho]$. $c_p(0) = p^*$, $c_p(\varrho) = \exp_{p^*} p$, $\dot{c}_p(\varrho) = N(p)$. For $X \in T_pM$, choose a curve $p(s)$, $-\eta < s < \eta$, on M with $p'(0) = X$. Consider the mapping

$$F:]-\eta, \eta[\times I \to M^*; (s, t) \mapsto \exp_{p^*} tp(s)/\varrho .$$

The vector field $X(t) = \partial F(s, t)/\partial s|_0$ is a vector field along $c_p(t) = F(0, t)$ with $X(\varrho) = X$. Here we identify $T_p M$ and $T_p F(T_p M)$. The Gauss Lemma (1.9.1) states

$$\langle \frac{\partial F}{\partial s}, \frac{\partial F}{\partial t} \rangle (s, t) = 0 .$$

Thus, using (1.5.8), we obtain

$$h_{N(p)}(X, X) = \langle \frac{\nabla^*}{\partial s} \frac{\partial}{\partial s} F, \frac{\partial F}{\partial t} \rangle |_{0,\varrho} =$$

$$-\langle \frac{\partial F}{\partial s}, \frac{\nabla^*}{\partial t} \frac{\partial}{\partial s} F \rangle |_{0,\varrho} = -\langle X, \frac{\nabla^* X}{dt} \rangle |_{\varrho}$$

In (1.12) we shall see that $X(t)$ is a so-called Jacobi field along $c_p(t)$. □

Actually, we did not make full use in this computation of the hypothesis that $\exp_{p^*}|\partial \bar{B}_\varrho(0_{p^*})$ is of maximal rank everywhere; it would have been sufficient to assume that $T_p \exp_{p^*} | T_p(T_{p^*} M^*)$ is non-degenerate and to consider a small neighborhood of p on $S_\varrho(0_{p^*}) \subset T_{p^*} M^*$, $\varrho = |p|$.

In the Gauss formula (1.10.5) we studied the covariant derivation of a tangential vector field along an immersion F. The following so-called *Weingarten formula* does the same for a normal vector field.

1.10.7 Lemma. *Let*

$$F: (M, g) \to (M^*, g^*)$$

be an isometric immersion. Let $N: M \to TM^*$ be a normal vector field along F, i. e., $N(p) \in T_p^\perp F$, all p. Then we have for $X \in T_p M$

$$\nabla_X^* N(p) = -T_p F . H_{N(p)} X + \nabla_X^{*\perp} N(p) \in T_p F \oplus T_p^\perp F .$$

Here,

$$H_{N(p)}: T_p M \to T_p M$$

is the self-adjoint operator defined by

$$\langle H_{N(p)} . X, Y \rangle = \langle X, H_{N(p)} . Y \rangle = h_{N(p)}(X, Y) .$$

Thus, the tangential component of $\nabla_X^* N(p)$ depends on $N(p)$ and $X(p)$ only.

The normal component $\nabla^{*\perp}$ of ∇^* can be viewed as a covariant derivation *on the $\mathscr{F}M$-module $\mathfrak{V}_F^\perp$* of normal vector fields *along F, cf.* (1.5.5). *That is to say,*

$$(X, N) \in \mathfrak{V}M \times \mathfrak{V}_F^\perp \mapsto \nabla_X^{*\perp} N \in \mathfrak{V}_F^\perp$$

is $\mathcal{F}M$-linear in the first argument and a derivation in the second, i. e.,

$$\nabla_X^{*\perp} fN = X(f)N + f\,\nabla_X^{*\perp} N.$$

Proof. From (1.10.5) and $\langle N, TF\,.\,Y\rangle = 0$ we get with (1.8.14)

$$\langle \nabla_X^* N, TF\,.\,Y\rangle = -\langle N, \nabla_X^* TF\,.\,Y\rangle = -h_N(X, Y).$$

The properties of $\nabla^{*\perp}$ are immediate consequences of the definitions and (1.8.14). □

In (1.11) we shall again consider the Gauss- and Weingarten formulas. We continue here with a few facts about 1-parameter groups of isometries.

1.10.8 Definition. Let M be a differentiable manifold. A *1-parameter group of diffeomorphisms* is a differentiable mapping

$$\varrho : \mathbb{R} \times M \to M;\ (s, p) \mapsto s\,.\,p \equiv \varrho(s, p)$$

such that

(i) $s\,.\,(s'\,.\,p) = (s + s')\,.\,p;\quad 0\,.\,p = p$

and (ii) $\varrho_s : M \to M;\, p \mapsto s\,.\,p$

is a diffeomorphism, for every $s \in \mathbb{R}$. ϱ determines a vector field X_ϱ on M by

$$X_\varrho(p) = \frac{d}{ds}\varrho(s, p)|_{s=0}\,.$$

If M is Riemannian, $M = (M, g)$, and if ϱ_s is an isometry for all s then ϱ is called *1-parameter group of isometries* or *isometric $\mathbb{R}$-action*. The associated vector field X_ϱ is called *Killing vector field.*

Remark. Given a vector field X on M, we have for each $p \in M$ a maximal interval $J(p)$ around $0 \in \mathbb{R}$ such that the integral curve $c_p(s)$ of X with $c_p(0) = p$ is defined, $dc_p(s)/ds = X(c_p(s))$. Denote by $J(M)$ the subset of $\mathbb{R} \times M$ formed by the union of the $J(p) \times \{p\}$. Then we have on $J(M)$ a *local 1-parameter group of diffeomorphisms* given by

$$(s, p) \in J(M) \mapsto c_p(s)\,.$$

(ii) is satisfied and (i) is satisfied as long as everything is defined.

1.10.9 Proposition. *For a Killing vector field X_ϱ on (M, g) the identity*

$$(*)\qquad g(\nabla_Y X_\varrho, Z) + g(Y, \nabla_Z X_\varrho) = 0$$

holds for all vector fields Y and Z.

Conversely, if a vector field X_ϱ satisfies $(*)$, the local group generated by X_ϱ consists of isometries.

Proof. ϱ being isometric means

$$(**)\quad g(\varrho_s(p))(T_p\varrho_s\,.\,Y, T_p\varrho_s\,.\,Z) = g(p)(Y, Z)$$

for all Y and Z in T_pM. Differentiating this with respect to s at $s=0$ and using (1.8.10) yields $(*)$.

Conversely, if X_ϱ satisfies $(*)$ then this implies that the left hand side of $(**)$ is independent of s. Thus, the mapping $p \mapsto \varrho(s,p) = c_p(s)$ is an isometry. □

Example. Consider the isometric $\mathbb{R}$-action on the euclidean (x, y, z)-space $\mathbb{R}^3$, given by the rotation around the z-axis:

$$(s,(x,y,z)) \mapsto (x\cos s - y\sin s, x\sin s + y\cos s, z).$$

Let $M \subset \mathbb{R}^3$ be a surface of revolution around the z-axis. The $\mathbb{R}$-action carries M into itself and thus defines an isometric $\mathbb{R}$-action on M. Actually, since $\varrho_{s+2\pi} = \varrho_s$, we even have a *circle* or S^1-*action* here.

We now characterize those *orbits* $\{s \in \mathbb{R} \to s\,.\,p \in M\}$ of an isometric $\mathbb{R}$-action which are geodesics:

1.10.10 Proposition. *Let ϱ be an isometric $\mathbb{R}$-action on M. Let X_ϱ be the associated Killing vector field. Consider the function*

$$\phi : M \to \mathbb{R}; p \mapsto |X_\varrho(p)|^2.$$

If and only if $D\phi(p) = 0$ and $\phi(p) \neq 0$, is the orbit $s \mapsto c(s) = s\,.\,p = \varrho(s,p)$ a non-constant geodesic.

In particular, if the $\mathbb{R}$-action induces an S^1-action, the orbit c will be a closed geodesic, *i.e., there exists $\omega > 0$ with $\dot c(\omega) = \dot c(0)$.*

Remark. In the example of a surface of revolution we thus find that those S^1-orbits (= horizontal sections) are geodesics for which the radius (= distance from the z-axis) has a stationary value.

Proof. Choose $X_0 \in T_pM$ and a curve $c^*(t)$, $t \in]-\varepsilon, \varepsilon[$, on M with $\dot c^*(0) = X_0$. Then $\phi(\varrho(s, c^*(t)))$ is independent of s. Using (1.5.7) and

$$\langle c'(s), \frac{\partial}{\partial t}\varrho(s, c^*(t))\rangle|_0 = \langle T_p\varrho_s\,.\,X_\varrho(p), T_p\varrho_s\,.\,X_0\rangle = \text{const}$$

we find

$$\frac{1}{2}D\phi(p)\,.\,X_0 = \frac{1}{2}D\phi(c(s))\,.\,T_p\varrho_s\,.\,X_0 = \frac{1}{2}\frac{d}{dt}\langle\frac{\partial}{\partial s}\varrho(s, c^*(t)), \frac{\partial}{\partial s}\varrho(s, c^*(t))\rangle|_0 =$$

$$\langle c'(s), \frac{\nabla}{\partial t}\frac{\partial}{\partial s}\varrho(s, c^*(t))\rangle|_0 = \langle c'(s), \frac{\nabla}{\partial s}\frac{\partial}{\partial t}\varrho(s, c^*(t))\rangle|_0 = -\langle\frac{\nabla}{\partial s}c'(s), T_p\varrho_s\,.\,X_0\rangle.$$

With $X_0 \in T_p M$, $T_p \varrho_s . X_0 \in T_{c(s)} M$ can also be prescribed arbitrarily. Thus, $D\phi(0) = 0$ is equivalent to $\nabla c'(s)/ds = 0$. □

The next theorem goes far towards showing that the group of isometries of a Riemannian manifold forms a Lie group – at least if the manifold has finite dimension. See Myers and Steenrod [1] for a complete proof. What we are doing here is essentially exhibiting the Lie algebra of the identity component of the group of isometries.

1.10.11 Theorem. *The set of Killing vector fields on a Riemannian manifold M is a subalgebra of the Lie algebra of all vector fields on M.*

Proof. Let X_ϱ, X_σ be Killing vector fields. According to (1.10.9) it suffices to show that

$$(*) \quad \langle \nabla_X [X_\varrho, X_\sigma], X \rangle = 0$$

for all vectors $X \in T_p M$, all $p \in M$. Here we have used pointed brackets to denote the Riemannian metric on M.

For a given $X \in T_p M$ consider the geodesic $c(t) = \exp_p tX$. Then we have to show

$$\frac{d}{dt} \langle [X_\varrho, X_\sigma](c(t)), \dot{c}(t) \rangle = 0 .$$

To see this we consider the $\mathbb{R}$-actions ϱ and σ determined by X_ϱ and X_σ. With this,

$$X_\varrho(c(t)) = \frac{\partial \varrho(s, c(t))}{\partial s}\Big|_0 ; \; X_\sigma(c(t)) = \frac{\partial \sigma(s, c(t))}{\partial s}\Big|_0 .$$

Using the fact that X_ϱ, X_σ satisfy $(*)$, (1.10.9), we obtain

$$\frac{d}{dt} \langle [X_\varrho, X_\sigma], \dot{c} \rangle (c(t)) = \frac{d}{dt} \langle \nabla_{X_\varrho} X_\sigma, \dot{c} \rangle (c(t)) - \frac{d}{dt} \langle \nabla_{X_\sigma} X_\varrho, \dot{c} \rangle (c(t))$$

$$= -\frac{d}{dt} \langle \frac{\nabla}{dt} X_\sigma, X_\varrho \rangle (c(t)) + \frac{d}{dt} \langle \frac{\nabla}{dt} X_\varrho, X_\sigma \rangle (c(t)) =$$

$$- \langle \frac{\nabla}{\partial t} \frac{\nabla}{\partial s} \frac{\partial}{\partial t} \sigma(s, c(t)), X_\varrho(c(t)) \rangle\Big|_0 + \langle \frac{\nabla}{\partial t} \frac{\nabla}{\partial s} \frac{\partial}{\partial t} \varrho(s, c(t)), X_\sigma(c(t)) \rangle\Big|_0$$

$$= -\langle R(\dot{c}(t), X_\sigma)\dot{c}(t), X_\varrho \rangle (c(t)) + \langle R(\dot{c}(t), X_\varrho)\dot{c}(t), X_\sigma \rangle (c(t)) = 0 .$$

Here we have used (1.5.8) for the exchange of $\nabla/\partial t$ and $\nabla/\partial s$ as well as the fact that, fixed s, $t \mapsto \sigma(s, c(t))$, $t \mapsto \varrho(s, c(t))$ are geodesics. □

1.10.12 Definition. An isometric immersion $F\colon (M, g) \to (M^*, g^*)$ is called *totally geodesic at* $p \in M$ if every geodesic $c_X(t)$ with initial vector $X \in T_p M$ is carried under F into a geodesic of M^*, for all t in some neighborhood of 0. F is called *totally geodesic immersion* if it is totally geodesic for every $p \in M$.

1.10.13 Examples. (i) Whereas totally geodesic immersions of dimension >1 and codimension >0 are the exception, see (v) below, there always exist (local) submanifolds which are totally geodesic in a single point $p \in M$ with an arbitrarily prescribed tangent space $T_* \subset T_p M$. Indeed, choose $\eta > 0$ so small that $\exp_p | B_\eta(0_p)$ is injective. Then $\exp_p(B_\eta(0_p) \cap T_*)$ is such a submanifold.

(ii) Any linear (closed) subspace of the Hilbert space $\mathbb{E}$ with the standard metric is totally geodesic.

(iii) The intersection of the sphere $S_\varrho^{\mathbb{E}}$ in $\mathbb{E}' = \mathbb{R} \times \mathbb{E}$ with a subspace $\mathbb{F}'$ through the origin of $\mathbb{E}'$ gives a totally geodesic submanifold of $S_\varrho^{\mathbb{E}}$. This is again a sphere. Indeed, the carriers of the geodesics on $S_\varrho^{\mathbb{E}}$ are the intersections of $S_\varrho^{\mathbb{E}}$ with 2-dimensional subspaces (planes) through the origin of $\mathbb{E}'$. Let $\{E_1, E_2\}$ be an orthonormal base for such a plane. Then $c(t) = \varrho(\cos t E_1 + \sin t E_2)$ is the geodesic on $S_\varrho^{\mathbb{E}}$ with $\dot{c}(0) = \varrho E_2 \in T_{c(0)} S_\varrho^{\mathbb{E}}$. To see this, observe $c(t)^2 = \varrho^2$. Hence, $c(t) \cdot \dot{c}(t) = 0$ and $c(t)$ is orthogonal to the orthogonal complement of the space spanned by E_1 and E_2. Therefore, $c(t) \perp T_{c(t)} S_\varrho^{\mathbb{E}}$. Since $\ddot{c}(t) = -c(t), pr \circ \ddot{c}(t) = \nabla \dot{c}(t)/dt = 0$.

(iv) Let $M^* = M \times M'$ be the Riemannian product of the Riemannian manifolds M and M', cf. the example after (1.8.2). Then the inclusion $p \in M \mapsto (p, p') \in M^*$, for any fixed $p' \in M'$, is totally geodesic.

(v) A geodesic on (M^*, g^*) is a totally geodesic submanifold. So is any open subset of (M^*, g^*). In general, a Riemannian manifold (M^*, g^*) does not possess other totally geodesic submanifolds.

In (1.10.15) we will show that non-trivial isometries give rise to totally geodesic submanifolds. However, non-trivial isometries do not exist in general, compare the remark 4 in (1.10.2).

Totally geodesic immersions allow the following characterization:

1.10.14 Proposition. *Let*

$$F: (M, g) \to (M^*, g^*)$$

be an isometric immersion.

(i) *If F is totally geodesic at $p, h_p \equiv 0$.*

(ii) *If $h_p \equiv 0$ for all $p \in M$, F is totally geodesic.*

Proof. Let $X \in T_p M$ and denote by $c_X(t)$ the geodesic with initial value X. Extend X to a vector field on a neighborhood of p on M such that $X(c_X(t)) = \dot{c}_X(t)$, $|t|$ small. Then $\nabla_X X(c_X(t)) = \nabla \dot{c}_X(t)/dt = 0$. If $F \circ c_X(t)$ is geodesic also on M^* we get: $\nabla^*_X TF.X(F \circ c_X(t)) = 0$. Thus, according to (1.10.5), $h_p(X, X) = 0$. Since h_p is symmetric and X is arbitrary, $h_p \equiv 0$.

Assume now $h_p \equiv 0$, all $p \in M$. Then our arguments show that every geodesic $c_X(t)$ under F goes into a geodesic on M^*. $\square$

1.10.15 Theorem. *Let $I: (M^*, g^*) \to (M^*, g^*)$ be an isometry. Then every connected component of the fixed point set*

$$\mathrm{Fix}(I) = \{p^* \in M^*; Ip^* = p^*\}$$

is a totally geodesic submanifold.

Proof. Let $p^* \in \mathrm{Fix}(I)$. Assume that p^* is not an isolated point of $\mathrm{Fix}(I)$ – note that $\mathrm{Fix}(I)$ is closed, since it is the counter image of the diagonal $\triangle_{M^*}$ in $M^* \times M^*$ under the differentiable mapping $p^* \mapsto (p^*, I(p^*))$.

The set H of invariant vectors

$$H = \{X \in T_{p^*}M^*; T_{p^*}I . X = X\}$$

therefore is $\neq 0$. H is a closed subspace. Since the geodesic c_X, $X \in H$, with $\dot{c}_X(0) = X$ is entirely determined by X, $c_X(t) \in \mathrm{Fix}(I)$ for all $t \in J(X)$. In an appropriatly small ball $B_\varrho(p^*)$ around p^* we therefore have that $B_\varrho(p^*) \cap \mathrm{Fix}(I)$ is carried into $B_\varrho(0_{p^*}) \cap H$ by the normal coordinates. Now apply (1.3.5). □

1.10.16 Examples. On $M = S_\varrho^{\mathbb{E}} \subset \mathbb{R} \times \mathbb{E} = \mathbb{E}'$ consider the reflection I on a subspace $\mathbb{E}_0$ through the origin of $\mathbb{E}'$, i. e.,

$$I: (x_0, x_0^\perp) \in \mathbb{E}_0 \oplus \mathbb{E}_0^\perp \mapsto (x_0, -x_0^\perp) \in \mathbb{E}_0 \oplus \mathbb{E}_0^\perp.$$

$\mathrm{Fix}(I) = S_\varrho^{\mathbb{E}} \cap \mathbb{E}_0$.

If, in particular, a connected component of $\mathrm{Fix}(I)$ is 1-dimensional, it must be a geodesic.

In this way one sees e. g. that the intersection of an *ellipsoid* in the normal form

$$\sum_0^n x_i^2/a_i = 1; 0 < a_0 \leqslant \ldots \leqslant a_n,$$

with one of the 2-dimensional coordinate planes is the carrier of a closed geodesic.

For more on fixed point sets of isometries see Kobayashi [1].

We conclude with a fixed point theorem for the action of a compact group as isometry group. The most important case is when the action occurs on a simply-connected, finite dimensional manifold of non-positive curvature, see (3.8.3) for this case. Our proof uses an idea of Eberlein.

1.10.17 Theorem. *Let G be a compact subgroup of the group of isometries of M, $\dim M < \infty$, and assume that there exists an orbit $G.p = \{g.p, g \in G\}$ of G which is contained in a $B_\sigma(p^*)$ such that $B_{2\sigma}(p^*)$ is strongly convex. Then G possesses a fixed point in $B_\sigma(p^*)$.*

Proof. Using the local compactness of M, we have the existence of a closed ball $\bar{B}_\varrho(p_0) \subset B_\sigma(p^*)$ of minimal radius ϱ containing $G.p$. Either $\bar{B}_\varrho(p_0)$ is unique (which is the case e. g. if $\varrho = 0$); and then p_0 is a fixed point. Or else, there are two different balls $\bar{B}_\varrho(p_0)$, $\bar{B}_\varrho(p_0')$ containing $G.p$. But this leads to a contradiction as follows: Let $c(t)$, $0 \leqslant t \leqslant d = d(p_0, p_0')$ be the minimizing geodesic from p_0 to p_0'. Let $r \in G.p \subset \bar{B}_\varrho(p_0) \cap \bar{B}_\varrho(p_0')$, i.e., $d(r, p_0) \leqslant \varrho$, $d(r, p_0') \leqslant \varrho$. From (1.9.6), (1.9.7) we know that $t \in [0, a] \mapsto d(r, c(t)) \in \mathbb{R}$ assumes its maximum only at a boundary point. Hence, if $q_0 = c(d/2)$ is the mid-point of the geodesic c, $d(r, q_0) < \varrho$. Since $G.p$ is compact, $\sup\{d(g.p, q_0);\ g \in G\} < \varrho$, i.e., there exist a ball of radius $< \varrho$ containing $G.p$. □

1.10.18 Remark. (1.10.17) also holds for $\dim M = \infty$, at least under the hypothesis that the balls $B_\varrho(r)$ with $r \in B_\sigma(p^*)$, $\varrho < \sigma$, are uniformly strongly convex in a sense made precise below. This hypothesis certainly is satisfied whenever σ is sufficiently small, due to the Riemann Principle (1.9.8), since it is satisfied in Euclidean (or Hilbert) spaces. More generally, the hypothesis holds for every σ in a simply connected manifold with non-positive curvature, cf. (1.11.3), (2.7.2).

Consider now the hypothesis: For every δ, $0 < \delta < \sigma$, and ϱ, $0 < \varrho < \sigma$, there exists a $\eta = \eta(\delta, \varrho) > 0$ with the following property: Whenever p and q are points in $B_\sigma(p^*)$ with $d(p, q) \geqslant 2\delta$ and r is a point with $\varrho \leqslant d(p, r) \leqslant \varrho + \varepsilon$ while $d(q, r) \leqslant d(p, r)$ then the midpoint s of the minimizing geodesic from p to q has distance $\leqslant \varrho - \eta$ from r, for all sufficiently small $\varepsilon > 0$.

If now G is a compact group of isometries which possesses an orbit $G.p$ in such a $B_\sigma(p^*)$, G has a fixed point in $\bar{B}_\sigma(p^*)$. This is seen as follows. Let ϱ be the infimum of the ϱ' such that $G.p \subset \bar{B}_{\varrho'}(p'), p' \in B_\sigma(p^*)$. If $\varrho = 0, p$ is a fixed point, since $d(p, g.p) = 0$, all $g \in G$. Thus we can assume $\varrho > 0$. Note: $\varrho < \sigma$.

Let $\{\varrho_n\}$ be a sequence with $\lim \varrho_n = \varrho$ and $\{p_n\}$ a sequence in $B_\sigma(p^*)$ such that $G.p \subset B_{\varrho_n}(p_n)$, all n. We claim that $\{p_n\}$ is a Cauchy sequence. This (or at least the existence of a convergent subsequence which is just as good) is clear if $\dim M < \infty$. Once this is shown, we put $\lim p_n = p_0$. Then $G.p \subset \bar{B}_\varrho(p_0)$, and the proof can be concluded as in the case $\dim M < \infty$.

Now, assume that $\{p_n\}$ is not a Cauchy sequence. Then there exists a $\delta > 0$ and arbitrarily large pairs (m, n) with $d(p_m, p_n) \geqslant 2\delta$. We only consider such pairs. Denote by p_{mn} the midpoint of the minimizing geodesic $c_{mn}(t)$, $0 \leqslant t \leqslant 1$, from p_m to p_n. There must exist a point $r = r_{mn} \in G.p$ with $d(p_{mn}, r) \geqslant \varrho$. Since $t \in [0,1] \mapsto d(c_{mn}(t), r) \in \mathbb{R}$ has no local maximum in the interior (see (1.9.6)), we can assume

$$\varrho \leqslant d(p_{mn}, r) \leqslant d(p_m, r);\ d(p_n, r) \leqslant d(p_m, r).$$

From $G.p \subset B_{\varrho_m}(p_m) \cap B_{\varrho_n}(p_n)$ we also have $d(p_m, r) \leqslant \varrho_m$.

But since $\varrho_m - \varrho \to 0$ for $(m, n) \to (\infty, \infty)$, we find ourselves contradicting the hypothesis that the balls $B_\varrho(r)$ are uniformly convex.

1.11 Riemannian Curvature

We continue to consider Riemannian manifolds with their Levi-Cività derivation. Let $F: M \to M^*$ be an isometric immersion. The Gauss equations (1.11.1) relate the curvature tensors R of M and R^* of M^*, bringing into play the second fundamental mapping h.

Of fundamental importance is the sectional curvature $K(\sigma)$ which is derived from the curvature tensor for every tangential 2-plane σ, see (1.11.3); $K(\sigma)$ generalizes the Gauss curvature for surfaces in $\mathbb{R}^3$. The classical euclidean and non-euclidean spaces have constant sectional curvature: $K=0$ for the Euclidean space, $K=1/\varrho^2$ for the sphere of radius ϱ, (1.11.6), and $K=-1/\varrho^2<0$ for the hyperbolic space $H_\varrho^{\mathbb{E}}$, see (1.11.7) and (1.11.8).

Next we derive the fundamental equations for isometric submersions $F\colon M^* \to M$, (1.11.9) to (1.11.13). They will be useful for the study of symmetric spaces in (2.2). A first application is given in (1.11.14) where we define the complex projective space with its canonical Riemannian metric and determine its curvature tensor.

Recall from (1.8.13) the properties of the curvature tensor R. R is a section in the bundle

$$L(\tau, \tau, \tau; \tau) \cong L(\tau^*, \tau, \tau, \tau; \varrho)$$

of tensors of type $\binom{1}{3}$. The Riemannian metric g yields a bundle isomorphism $L_{g^\cdot}$ from the cotangent bundle τ^* to the tangent bundle τ,

$$\begin{array}{ccc} T^*M & \xrightarrow{\;L_{g^\cdot}\;} & TM \\ \Big\downarrow{\scriptstyle \tau^*} & & \Big\downarrow{\scriptstyle \tau} \\ M & \xrightarrow{\;id\;} & M \end{array}$$

Here, the map $L_{g^\cdot}|T_p^*M$ is the linear isomorphism

$$\omega(p) \in T_p^*M \mapsto g(p)(\ ,\omega(\mathrm{p})) \in T_pM.$$

Thus, we also get an isomorphism

$$L(\tau^*, \tau, \tau, \tau; \varrho) \cong L(\tau, \tau, \tau, \tau; \varrho).$$

The curvature tensor R, considered as section in this latter bundle, satisfies a number of identities, see (1.8.13). We use these to derive the *Gauss equations*.

1.11.1 Theorem. *Let $F\colon M \to M^*$ be an isometric immersion. Let R and R^* be the curvature tensor on M and M^*, respectively. Then, with pointed brackets to denote the scalar product in T_pM and $T_{F(p)}M^*$,*

$$\langle R(X,Y)Z, W\rangle = \langle R^*(TF.X, TF.Y)TF.Z, TF.W\rangle +$$

$$\langle h(Y,Z), h(X,W)\rangle - \langle h(X,Z), h(Y,W)\rangle,$$

for all quadrupels X, Y, Z, W in T_pM.

Proof. From (1.10.5) we get

$$TF\,.\,\nabla_Y Z = \nabla_Y^* TF\,.\,Z - h(Y, Z), \text{ hence}$$

$$TF\,.\,\nabla_X \nabla_Y Z = pr \circ \nabla_X^* TF\,.\,\nabla_Y Z = pr \circ \big(\nabla_X^* \nabla_Y^* TF\,.\,Z - \nabla_X^* h(Y, Z)\big).$$

Write down the corresponding equation with X and Y exchanged and subtract these from each other. Using

$$\langle TF\,.\,\nabla_{[X,Y]} Z, TF\,.\,W\rangle = \langle \nabla_{[X,Y]}^* TF\,.\,Z, TF\,.\,W\rangle$$

and

$$-\langle \nabla_X^* h(Y, Z), TF\,.\,W\rangle = \langle h(Y, Z), h(X, W)\rangle$$

$$\langle \nabla_Y^* h(X, Z), TF\,.\,W\rangle = -\langle h(X, Z), h(Y, W)\rangle$$

our claim follows from (1.5.4), (1.5.6). The last two equations follow from the identities

$$\langle h(Y, Z), TF\,.\,W\rangle = 0; \langle h(X, Z), TF\,.\,W\rangle = 0$$

and (1.8.14). □

Remark. The Gauss equations (1.11.1) yield a relation between the curvature tensor R of an immersed manifold M and the curvature tensor R^* of the receiving manifold M^*. In particular, if M^* is a Hilbert space, $R^* = 0$, and (1.11.1) shows how R is determined by the second fundamental mapping h. Gauss did consider the case of a surface immersed in $\mathbb{R}^3$.

There is an important class of immersions where R and R^* coincide.

1.11.2 Corollary. *Let $F: M \to M^*$ be an isometric immersion which is totally geodesic at $p \in M$. Then*

$$\langle R(X, Y)Z, W\rangle = \langle R^*(TF\,.\,X, TF\,.\,Y)TF\,.\,Z, TF\,.\,W\rangle,$$

for all quadrupels X, Y, Z, W in $T_p M$.

Proof. Apply (1.10.12). □

Example. Assume $R^* = 0$. Let $\{e_j, e_k, e_l, e_i\}$ be elements in a Gauss frame. Then

$$\langle R(e_j, e_k)e_l, e_i\rangle = \langle h(e_i, e_j), h(e_k, e_l)\rangle - \langle h(e_i, e_k), h(e_j, e_l)\rangle.$$

1.11.3 Proposition. *Let σ be a 2-dimensional subspace of the tangent space $T_p M$ of the Riemannian manifold M. Let $\{X, Y\}$ be a basis for σ. Put*

$$K(X, Y) = \langle R(X, Y)Y, X\rangle; \; K_1(X, Y) = |X|^2 |Y|^2 - \langle X, Y\rangle^2.$$

Then $\quad K(\sigma) = K(X, Y)/K_1(X, Y)$

is independent of the choice of the basis. It is called Riemannian curvature *or* sectional curvature (of the 2-plane σ).

Proof. K as well as K_1 is symmetric. Moreover

$$K(aX+bY, Y) = a^2 K(X, Y); \; K_1(aX+bY, Y) = a^2 K_1(X, Y),$$

which implies the proposition. □

Remark. If $\dim M = 2$, then $\sigma = T_p M$. In this case, $K(\sigma)$ depends on p only. We therefore also denote it by $K(p)$. It is the *Gauss curvature*. As basis of $\sigma = T_p M$ we can take the Gauss frame $\{e_1(p), e_2(p)\}$ on the domain M' of a chart (u, M') containing p. Then, with $g_{ik} = \langle e_i, e_k \rangle$,

$$K(p) = K(\sigma) = \langle R(e_1, e_2)e_2, e_1 \rangle / (g_{11} g_{22} - g_{12}^2),$$

cf. Klingenberg [9].

As an immediate consequence of the Gauss equations (1.11.1) we find:

1.11.4 Theorem. *Let $F: M \to M^*$ be an isometric immersion. Then*

$$K(\sigma) = K^*(TF.\sigma) + \frac{\langle h(X, X), h(Y, Y) \rangle - \langle h(X, Y), h(X, Y) \rangle}{\langle X, X \rangle \langle Y, Y \rangle - \langle X, Y \rangle \langle X, Y \rangle}.$$

Here, σ is the 2-plane spanned by the two linearly independent vectors X and Y of $T_p M$. In particular, if F is totally geodesic at p, then

$$K(\sigma) = K^*(TF.\sigma). \quad \square$$

Combining the last statement with (1.10.14, (i)) we obtain the

1.11.5 Corollary. *The Riemannian curvature $K(\sigma)$ of a 2-plane $\sigma \subset T_p M$ coincides with the* Gauss curvature *of the embedded surface $\exp_p(B_\eta(0_p) \cap \sigma)$, formed by the geodesic segments of length $< \eta$ which start tangentially to σ.* □

1.11.6 Example. Let

$$F: M = S_\varrho^{\mathbb{E}} \to M^* = \mathbb{R} \times \mathbb{E} = \mathbb{E}'$$

be the canonical embedding of the sphere of radius ϱ. To determine the curvature tensor we consider a geodesic $c(t)$ on $S_\varrho^{\mathbb{E}}$ as in (1.10.13, iii). Put $\dot{c}(0) = X$. We also use X to denote an extension of X towards a vector field. Then, as in the proof of (1.10.14), $\nabla_X X = 0$ and $\nabla_X^* X = \ddot{c}(0) = -c(0)$, since on $\mathbb{E}'$, ∇^* is the ordinary differential. The Gauss formula (1.10.5) (†) reads

$$\ddot{c}(0) = \nabla_X^* X = h(X, X),$$

hence,

$$|h(X, X)| = \varrho = |X| = \frac{1}{\varrho}\langle X, X\rangle .$$

From (1.11.1) we find

$$R(X, Y)Z = \frac{1}{\varrho^2}\{\langle Y, Z\rangle X - \langle X, Z\rangle Y\}, \quad K(\sigma) = \frac{1}{\varrho^2} = \text{const.}$$

For another way of computing the curvature tensor see the proof of (1.11.8), based on the representation of the metric in stereographic coordinates, cf. (1.11.7).

1.11.7 Example. *The hyperbolic space.* In (1.1.3, ii) we had introduced the stereographic projection $u_+ : S_\varrho^{\mathbb{E}} - \{p_+\} \to \mathbb{E}$. The inverse gives a parameter representation of $S_\varrho^{\mathbb{E}} - \{p_+\}$, i. e.,

$$u \in \mathbb{E} \mapsto (x_0(u), x(u)) = \left(\varrho \frac{|u|^2/\varrho^2 - 1}{|u|^2/\varrho^2 + 1}, \frac{2u}{|u|^2/\varrho^2 + 1}\right) \in \mathbb{R} \times \mathbb{E} = \mathbb{E}'.$$

In this representation, the line element $ds^2(u) = g(u)(\ ,\ )$ of $S_\varrho^{\mathbb{E}}$ reads

$$ds(u)^2 = dx_0(u)^2 + dx(u)^2 = \frac{4\,du^2}{(|u|^2/\varrho^2 + 1)^2}.$$

We obtain the hyperbolic space $H_\varrho^{\mathbb{E}}$ as a certain dual to $S_\varrho^{\mathbb{E}}$ if we consider on $\mathbb{E}' = \mathbb{R} \times \mathbb{E}$ the *Lorentz metric*:

$$\langle (x_0, x), (y_0, y)\rangle_L = -x_0 y_0 + \langle x, y\rangle$$

For $\varrho > 0$,

$$H_\varrho^{\mathbb{E}} = \{(x_0, x) \in \mathbb{R} \times \mathbb{E} = \mathbb{E}';\ -x_0^2 + |x|^2 = -\varrho^2, x_0 > 0\}$$

is a submanifold of codimension 1 in $\mathbb{E}'$. Indeed, it is a connected component of the counter image of the regular value $-\varrho^2$ of the function $(x_0, x) \mapsto -x_0^2 + |x|^2$. For $\mathbb{E}' = \mathbb{R} \times \mathbb{R}^2$ we get one of the two shells of the hyperboloid $x_0^2 - x_1^2 - x_2^2 = \varrho^2$.

In analogy to the stereographic projection u_- of the sphere $S_\varrho^{\mathbb{E}}$ we define

$$u : (x_0, x) \in H_\varrho^{\mathbb{E}} \mapsto u(x_0, x) = x\Big/\left(\frac{x_0}{\varrho} + 1\right) \in B_\varrho^{\mathbb{E}} \subset \mathbb{E}.$$

Geometrically this means that we map $p = (x_0, x) \in H_\varrho^{\mathbb{E}}$ into the intersection of the line from $(-\varrho, 0)$ to p with the hyperplane $\{0\} \times \mathbb{E} \subset \mathbb{E}'$. The image $B_\varrho^{\mathbb{E}}$ is the ball of radius ϱ in $\mathbb{E}$.

The inverse of u reads

$$u \in B_\varrho^{\mathbb{E}} \mapsto (x_0(u), x(u)) = \left(\varrho \frac{1 + |u|^2/\varrho^2}{1 - |u|^2/\varrho^2}, \frac{2u}{1 - |u|^2/\varrho^2}\right) \in \mathbb{R} \times \mathbb{E} = \mathbb{E}'.$$

The Lorentz metric on $\mathbb{E}'$ induces on $B_\varrho^{\mathbb{E}}$ the metric

$$(*)\quad ds(u)^2 = -dx_0(u)^2 + dx(u)^2 = \frac{4du^2}{(1-|u|^2/\varrho^2)^2}.$$

Note that $(*)$ is a Riemannian metric although the Lorentz metric is not Riemannian since it is not positive definite. In other words, the restriction of the indefinite metric $-x_0^2+|x|^2$ on $T_p\mathbb{E}'$ to the tangent space $T_pH_\varrho^{\mathbb{E}}$ is positive definite.

$H_\varrho^{\mathbb{E}}$ with the metric induced from the Lorentz metric on $\mathbb{E}'$ is called *hyperbolic space of radius* ϱ (modelled on $\mathbb{E}$). We use the same name also for the representation $(B_\varrho^{\mathbb{E}}, ds^2 = (*))$ of $H_\varrho^{\mathbb{E}}$.

1.11.8 Proposition. (i) *The isometry group of the hyperbolic space* $H_\varrho^{\mathbb{E}}$ *is isomorphic to the Lorentz orthogonal group* $\mathbb{O}(1, \mathbb{E})$, *consisting of those linear isomorphisms of* $\mathbb{E}'$ *which preserve the Lorentz metric. It operates transitively on the orthonormal bases in the tangent spaces of* $H_\varrho^{\mathbb{E}}$.

(ii) *The carriers of the geodesics on* $H_\varrho^{\mathbb{E}}$ *are the intersections of* $H_\varrho^{\mathbb{E}} \subset \mathbb{E}'$ *with* 2-*planes through the origin of* $\mathbb{E}'$.

(iii) *The curvature tensor of* $H_\varrho^{\mathbb{E}}$ *reads*

$$R(X,Y)Z = -\frac{1}{\varrho^2}\{\langle Y,Z\rangle X - \langle X,Z\rangle Y\}.$$

Hence, $K(\sigma) = -1/\varrho^2 = \text{const.}$

Proof. We restrict ourselves to a few indications. First of all, since $\mathbb{O}(1,\mathbb{E})$ carries $H_\varrho^{\mathbb{E}} \subset \mathbb{E}'$ into itself, it operates as isometry group on $H_\varrho^{\mathbb{E}}$.

Consider the subgroup

$$\{id\} \times \mathbb{O}(\mathbb{E}) \subset \mathbb{O}(1,\mathbb{E}).$$

It operates transitively on the orthonormal bases of $T_{(\varrho,0)}H_\varrho^{\mathbb{E}} \cong \{0\} \times \mathbb{E} \subset \mathbb{E}'$.

Let now $q = (x_0, x) \in H_\varrho^{\mathbb{E}}, q \neq (\varrho, 0)$. Choose an orthonormal basis $\{e_i\}$ in $\mathbb{E}$ such that e_1 is proportional to x, i.e., $x = x_1 e_1$. The subgroup

$$\mathbb{O}(1, \mathbb{R}e_1) \times \{id\} \subset \mathbb{O}(1,\mathbb{E})$$

operates as identity on the space spanned by $\{e_i; i>1\}$. The element

$$\begin{pmatrix} x_0/\varrho & -x_1/\varrho \\ x_1/\varrho & -x_0/\varrho \end{pmatrix} \in \mathbb{O}(1, \mathbb{R}e_1)$$

carries q into $(\varrho, 0)$. This implies (i).

To prove (ii) we apply (1.10.15). Choose a Lorentz 2-plane $\mathbb{E}^2$ through the origin of $0 \in \mathbb{E}'$. We can find a basis $\{E_1, E_2\}$ in $\mathbb{E}^2$ with $\langle E_1, E_1\rangle_L = -1, \langle E_1, E_2\rangle_L = 0, \langle E_2, E_2\rangle_L = 1$, where $\langle\ ,\ \rangle_L$ stands for the Lorentz scalar product. Then the orthogonal space $(\mathbb{E}^2)^\perp$ to the plane $\mathbb{E}^2$ forms a complement, $\mathbb{E}' = \mathbb{E}^2 \oplus (\mathbb{E}^2)^\perp$. Let $I\colon \mathbb{E}' \to \mathbb{E}'$ be the reflection on $\mathbb{E}^2$, i.e.,

$$(x, x^\perp) \in \mathbb{E}^2 \oplus (\mathbb{E}^2)^\perp \mapsto (x, -x^\perp) \in \mathbb{E}^2 \oplus (\mathbb{E}^2)^\perp.$$

The fixed point set of $I|H_\varrho^{\mathbb{E}}$ is the curve $\mathbb{E}^2 \cap H_\varrho^{\mathbb{E}}$ which therefore is totally geodesic.

Finally, to compute the curvature tensor of $H_\varrho^{\mathbb{E}}$, we can proceed as in (1.11.6), using the formula (1.11.4). The receiving manifold M^* in this case is the space $\mathbb{E}'$ with the Lorentz metric, i. e., it is not a Riemannian manifold in the original sense. Still, one can derive the Gauss equations and the Gauss formula also for this case.

All that is needed is to show that the image $T_pF(T_pM)$ of T_pM in $T_{F(p)}M^*$ is a non-degenerate subspace of $T_{F(p)}M^*$, i. e., a subspace for which the orthogonal space forms a complement to the subspace. But this is the case if $F\colon M \to M^*$ is an isometric immersion.

Another way of computing the curvature tensor of $H_\varrho^{\mathbb{E}}$ is to use the representation (1.11.7) (*) of the metric. Since the isometry group of $H_\varrho^{\mathbb{E}}$ operates transitively on the tangential 2-planes it suffices to compute $R(u)$ at $u=0$. We use the Taylor series of $ds(u)^2$,

$$ds(u)^2 = 4(1 + 2|u|^2/\varrho^2 + \dots)\,du^2.$$

Hence, $\Gamma(0) = 0$ and therefore, according to (1.5.3),

$$R(X,Y)Z = R(0)(X,Y)Z = D\Gamma(0).X(Y,Z) - D\Gamma(0).Y(X,Z).$$

$\Gamma(u)$ is determined by (1.8.11). We find, modulo terms of order $\geqslant 2$ in $|u|$, with $\langle\ ,\ \rangle$ the scalar product $ds^2(0)$,

$$\Gamma(u)(Y,Z) = \frac{1}{2\varrho^2}\big(\langle u,Y\rangle Z + \langle u,Z\rangle Y - \langle Y,Z\rangle u\big).$$

Hence, $$D\Gamma(0).X(Y,Z) = \frac{1}{2\varrho^2}\big(\langle X,Y\rangle Z + \langle Y,Y\rangle Y - \langle Y,Z\rangle X\big).$$

This yields (iii). □

The curvature relations (1.11.2), (1.11.4) for isometric immersions possess a counterpart if we consider isometric submersions. This is due to O'Neill [1].

Recall from (1.3.2) the concept of a submersion $F\colon M^* \to M$. Then every p in the image of F is a regular value. We will restrict ourselves to the case that $F(M^*) = M$. Then, for every $p \in M$, $F^{-1}(p)$ is a submanifold N_p of M^* of codimension equal to the dimension of M.

1.11.9 Definition. Let

$$F\colon (M^*, g^*) \to (M, g)$$

be a surjective submersion between Riemannian manifolds. Then, for every $p^* \in N_p = F^{-1}(p)$, every $p \in M$, we get an orthogonal decomposition

$$T_{p^*}M^* = T_{p^*}^v M^* \oplus T_{p^*}^h M^*$$

in the so-called *vertical* and *horizontal space* (with respect to F). Here, $T_{p^*}^v M^* = T_{p^*}N_p$ and $T_{p^*}^h M^*$ is its orthogonal complement, i. e., the normal space of the submanifold

N_p in p^*. F is called *isometric* (or *Riemannian*), if $T_{p^*}F|T^h_{p^*}M^*$ is an isometry onto $T_{F(p^*)}M$, all $p^* \in M^*$.

Examples. (i) The projection $F \equiv \tau_M : TM \to M$ where TM is endowed with the Riemannian metric defined in (1.9.12).

(ii) If $M^* = M \times M'$ is a Riemannian product the projection $(p, p') \in M^* \mapsto p \in M$ is an isometric submersion.

(iii) A covering map $M^* \to M$ (cf. Steenrod [1]) is a Riemannian submersion.

We begin with a counterpart to (1.10.5).

1.11.10 Lemma. *Let*

$$F : (M^*, g^*) \to (M, g)$$

be an isometric submersion. Then a vector field X on M determines a vector field X^* on M^*, the so-called horizontal lift, *by taking in $p^* \in M^*$ the horizontal vector which projects under $T_{p^*}F$ onto $X(F(p^*))$.*

If X and Y are vector fields on M we have the following relation with their horizontal lifts X^*, Y^*:

$$(*) \qquad \nabla^*_{X^*} Y^* = (\nabla_X Y)^* + \tfrac{1}{2}[X^*, Y^*]^v.$$

Here, Z^{*v} denotes the vertical component of Z^*. $[X^*, Y^*]^v(p^*)$ depends only on $X^*(p^*)$, $Y^*(p^*)$.

Proof. Let X, Y, Z be vector fields on M and X^*, Y^*, Z^* their respective lifts. Let T be a vertical vector field on M^* which assumes a prescribed value T_0 in $p^* \in N_p$. Such a vector field exists: Simply take an arbitrary extension of T_0 to a vector field $\tilde{T}$ and then form the projection $T = \tilde{T}^v$ onto its vertical component.

With this we have the following relations:

$$\langle X^*, T\rangle = \langle Y^*, T\rangle = \langle Z^*, T\rangle = 0,$$

$$X^*\langle Y^*, Z^*\rangle = X\langle Y, Z\rangle \text{ and } T\langle X^*, Y^*\rangle = 0,$$

since $\langle X^*, Y^*\rangle = \langle X, Y\rangle$ is constant on N_p, p fixed.

Using $TF.[X^*, Y^*] = [TF.X^*, TF.Y^*] = [X, Y]$ and $TF.[X^*, T] = [X, 0] = 0$, cf. (1.4.9), we find

$$\langle [X^*, Y^*], Z^*\rangle = \langle [X, Y], Z\rangle, \langle [X^*, T], Y^*\rangle = 0.$$

With this, the formula (1.8.11) $(*)$ for the Levi-Cività derivation $\nabla^*_{X^*} Y^*$ and $\nabla_X Y$ yields

$$\langle \nabla^*_{X^*} Y^*, Z^*\rangle = \langle \nabla_X Y, Z\rangle; 2\langle \nabla^*_{X^*} Y^*, T\rangle = \langle T, [X^*, Y^*]\rangle.$$

We have thus proved $(*)$. Since

$$\langle T, [X^*, Y^*]\rangle = \langle T, \nabla^*_{X^*} Y^* - \nabla^*_{Y^*} X^*\rangle = -\langle \nabla^*_{X^*} T, Y^*\rangle + \langle \nabla^*_{Y^*} T, X^*\rangle,$$

we also see that $[X^*, Y^*]^v(p^*)$ only depends on $X^*(p^*)$ and $Y^*(p^*)$. $\square$

1.11.11 Corollary. *Let $F:(M^*,g^*)\to(M,g)$ be an isometric submersion. Let $c:I\to M$ be a curve on M and $c^*:I\to M^*$ a horizontal lift, i.e., $\dot c^*(t)$ is a horizontal lift of $\dot c(t)$. Then c is a geodesic if and only if c^* is a geodesic.*

Proof. Immediate from

$$\frac{\nabla^*\dot c^*(t)}{dt}=\left(\frac{\nabla\dot c(t)}{dt}\right)^*+\frac{1}{2}[\dot c^*(t),\dot c^*(t)]^v=\left(\frac{\nabla\dot c(t)}{dt}\right)^*. \quad\square$$

Remark. (1.11.11) may be seen more geometrically from the fact that a horizontal lift c^* of a curve c has same length as c and that a geodesic is locally minimizing and an isometric submersion does not increase distances.

We are now ready to prove the counterpart of the Gauss equations (1.11.11):

1.11.12 Theorem. *Let*

$$F:(M^*,g^*)\to(M,g)$$

be an isometric submersion. Let X, Y, Z, W be vector fields on M and X^*, Y^*, Z^*, W^* be their horizontal lifts. Then

$$\langle R(X,Y)Z,W\rangle=\langle R^*(X^*,Y^*)Z^*,W^*\rangle-\tfrac{1}{4}\langle[X^*,Z^*]^v,[Y^*,W^*]^v\rangle+$$
$$+\tfrac{1}{4}\langle[Y^*,Z^*]^v,[X^*,W^*]^v\rangle-\tfrac{1}{2}\langle[Z^*,W^*]^v,[X^*,Y^*]^v\rangle.$$

In particular, if σ is a 2-plane tangent to M and σ^* a horizontal lift of σ,

$$K(\sigma)=K^*(\sigma^*)+\tfrac{3}{4}|[X^*,Y^*]^v|^2$$

where X^*, Y^* is an orthonormal basis of σ^*.

Proof. From (1.11.10) and its proof we get, with T a vertical vector field,

$$\langle\nabla^*_T X^*,Y^*\rangle=\langle\nabla^*_{X^*}T,Y^*\rangle+\langle[T,X^*],Y^*\rangle=-\langle T,\nabla^*_{X^*}Y^*\rangle=$$
$$=-\tfrac{1}{2}\langle T,[X^*,Y^*]^v\rangle.$$

Since $\quad X^*\langle\nabla^*_{Y^*}Z^*,W^*\rangle=X\langle\nabla_Y Z,W\rangle,$

$$\langle\nabla^*_{X^*}\nabla^*_{Y^*}Z^*,W^*\rangle=X\langle\nabla_Y Z,W\rangle-\langle\nabla_Y Z,\nabla_X W\rangle-$$
$$-\tfrac{1}{4}\langle[Y^*,Z^*]^v,[X^*,W^*]^v\rangle=\langle\nabla_X\nabla_Y Z,W\rangle-\tfrac{1}{4}\langle[Y^*,Z^*]^v,[X^*,W^*]^v\rangle.$$

Moreover,

$$\langle\nabla^*_{[X^*,Y^*]}Z^*,W^*\rangle=\langle\nabla^*_{[X^*,Y^*]^h}Z^*,W^*\rangle+\langle\nabla^*_{[X^*,Y^*]^v}Z^*,W^*\rangle=$$
$$=\langle\nabla_{[X,Y]}Z,W\rangle-\tfrac{1}{2}\langle[X^*,Y^*]^v,[Z^*,W^*]^v\rangle.$$

From (1.5.4) our claim now follows. $\square$

1.11.13 Corollary. *If $F:(M^*, g^*) \to (M, g)$ is an isometric submersion then*

$$K(\sigma) \geq K^*(\sigma^*),$$

where σ^* is a horizontal lift of σ. □

Remark. As we said before, (1.11.10) should be viewed as a counterpart to the Gauss formula (1.10.5). There also exists a counterpart to the Weingarten formula (1.10.7). Just as the Gauss and Weingarten formulae together give the structural equations for an isometric immersion, their counterparts are the structural equations of an isometric submersion. That is, the submersion is determined by these equations. For details see O'Neill [1].

1.11.14 Example. *The complex projective space.*

Let $\mathbb{E}$ be a complex Hilbert space. By $\langle\ ,\ \rangle$ we denote the Hermitian scalar product on $\mathbb{E}$, i.e.,

$$\langle aX + a'X', Y\rangle = a\langle X, Y\rangle + a'\langle X', Y\rangle; \langle Y, X\rangle = \overline{\langle X, Y\rangle}$$

$$\langle X, X\rangle > 0 \text{ for } X \neq 0.$$

Note that $\langle X, Y\rangle$ in general is not real. The real part $Re\langle X, Y\rangle$ is a positive definite scalar product on $\mathbb{E}$, considered as vector space over $\mathbb{R}$. The imaginary part $Im\langle X, Y\rangle = Re\langle X, iY\rangle$ is skew symmetric, $Re\langle X, iY\rangle = -Re\langle Y, iX\rangle$. In particular, $Re(X, iX) = 0$.

Denote by $S\mathbb{E} \equiv S_1\mathbb{E}$ the unit sphere $\{\langle N, N\rangle = 1\}$ in $\mathbb{E}$. Put $\mathbb{E} - \{0\} = \mathbb{E}^*$.

The complex projective space $P\mathbb{E}$ is defined as the set of complex 1-dimensional subspaces of $\mathbb{E}$. Thus, every $N \in \mathbb{E}^*$ determines a point p in $P\mathbb{E}$, i. e., the space $\mathbb{C}N$. By

$$F: \mathbb{E}^* \to P\mathbb{E}$$

we denote the mapping which associates to $N \in \mathbb{E}^*$ the element $\mathbb{C}N \in P\mathbb{E}$.

By restricting F to the sphere $S\mathbb{E} \subset \mathbb{E}^*$ we obtain what is called the *Hopf mapping*. The counter image of $p \in P\mathbb{E}$ then is a great circle, i. e., the intersection of $S\mathbb{E}$ with a complex line, considered as real 2-plane.

We show that $P\mathbb{E}$ carries the structure of a manifold, modelled on a subspace of $\mathbb{E}$ of codimension 1. Indeed, for every $N \in S\mathbb{E}$ we can define a chart $(u_N, P\mathbb{E}_N)$ as follows: The domain $P\mathbb{E}_N$ consists of the p where $\langle N_p, N\rangle \neq 0$, $N_p \in F^{-1}(p)$. The coordinate $u_N(p)$ of $p \in P\mathbb{E}_N$ is given by

$$u_N(p) = N_p/\langle N_p, N\rangle - N.$$

Geometrically, $u_N(p) + N$ is the intersection of the complex line $\mathbb{C}N_p$ with the complex hyperplane $\{\langle X, N\rangle = 1\}$.

To see that the transition mappings $u_{N'} \circ u_N^{-1}$, N, N' in $S\mathbb{E}$, are differentiable we derive the relation

$$u_{N'} + N' = (u_N + N)/(\langle u_N, N'\rangle + \langle N, N'\rangle).$$

Note that $u_{N'} \circ u_N^{-1}$ depends on u_N only and not also on $\bar{u}_N$, when we consider $u_N, \bar{u}_N$ as independent variables in the range of the chart $(u_N, P\mathbb{E}_N)$. Thus, the transition mappings are holomorphic, $P\mathbb{E}$ is called a *complex manifold.*

To find an atlas for $P\mathbb{E}$ it would suffice to take charts $(u_N, P\mathbb{E}_N)$ where N runs through an orthonormal (Hilbert) basis of $\mathbb{E}$.

We define a Riemannian metric g^* on $\mathbb{E}^*$ by taking as scalar product $g^*(N)(\ ,\ )$ on $T_N\mathbb{E}^*$

$$g^*(N)(X^*, Y^*) = Re\langle X^*, Y^*\rangle/\langle N, N\rangle.$$

Note that the sphere $S_r\mathbb{E}$ of radius $r > 0$ in $\mathbb{E}$ thus becomes a totally geodesic submanifold of constant curvature 1. The multiplicative action of $\mathbb{C}^* = \mathbb{C} - \{0\}$ on $\mathbb{E}^*$ is isometric.

We define for each $N \in \mathbb{E}^*$ the orthogonal decomposition

$$T_N\mathbb{E}^* = T_N^v\mathbb{E}^* \oplus T_N^h\mathbb{E}^*$$

by $$T_N^v\mathbb{E}^* = \mathbb{C}N;\ T_N^h\mathbb{E}^* = \{X^*; \langle X^*, N\rangle = 0\}.$$

Then

$$T_N F: T_N^h\mathbb{E}^* \to T_{F(N)}P\mathbb{E}$$

becomes a linear isomorphism. For $N \in S\mathbb{E}$ the representation of $T_N F | T_N^h\mathbb{E}^*$ in the coordinates $(id, \mathbb{E}^*)$ of $\mathbb{E}^*$ and $(u_N, P\mathbb{E}_N)$ of $P\mathbb{E}$ is given by $X^* \mapsto X^*$.

We define on $P\mathbb{E}$ a Riemannian metric g so as to make $F: \mathbb{E}^* \to P\mathbb{E}$ a Riemannian submersion: If $p = F(N)$, $N \in S\mathbb{E}$, X, Y in $T_pP\mathbb{E}$ and X^*, Y^* their counter images in $T_N^h\mathbb{E}^*$ under $(T_N F | T_N^h\mathbb{E}^*)^{-1}$, we put

$$g(p)(X, Y) = g^*(N)(X^*, Y^*) = Re\langle X^*, Y^*\rangle.$$

Since $z \in \mathbb{C}^*$ operates on $\mathbb{E}^*$ as an isometry, the definition of g does not depend on the choice of N. Note: $g(X, iY) = -g(Y, iX)$.

We want to determine the presentation $g(u)$ of g in a chart $(u, P\mathbb{E}_N) \equiv (u_N, P\mathbb{E}_N)$. First, we compute the principal part X_u of an $X \in T_pP\mathbb{E}, p \in P\mathbb{E}_N$.

Let $X^* \in T_{N_p}^h\mathbb{E}^*$ be the counter image of X, with $N_p \in S\mathbb{E} \cap F^{-1}(p)$. $c^*(s) = N_p + sX^*$ then is a curve such that $c(s) = F \circ c^*(s)$ has initial tangent vector X. Hence,

$$\begin{aligned} X_u &= \frac{d}{ds} u \circ c(s)|_0 = \frac{d}{ds}\left(\frac{N_p + sX^*}{\langle N_p + sX^*, N\rangle} - N\right)\Big|_0 \\ &= \frac{\langle N_p, N\rangle X^* - \langle X^*, N\rangle N_p}{\langle N_p, N\rangle^2}. \end{aligned}$$

With $u(p) = N_p/\langle N_p, N\rangle - N$ one verifies, using $\langle X^*, N_p\rangle = \langle X_u, N\rangle = 0$, that

$$(*) \qquad g(u)(X_u, X_u) = \frac{\langle X_u, X_u\rangle(1+\langle u,u\rangle) - \langle X_u, u\rangle\langle u, X_u\rangle}{(1+\langle u,u\rangle)^2}$$

is $=\langle X^*, X^*\rangle = g(X,X)$. The line element $(*)$ goes back to Fubini [1] and Study [1].

We conclude by computing the curvature tensor of $(P\mathbb{E}, g)$, using the fact that $F\colon \mathbb{E}^* \to P\mathbb{E}$ is a Riemannian submersion.

Consider the vector fields $(N(N), iN(N)) = (N, iN)$ on $\mathbb{E}^*$. The Levi-Cività covariant derivation ∇^*, when restricted to a totally geodesic sphere $S_r\mathbb{E}$, coincides with the covariant derivation on the sphere of constant curvature $K=1$.

Let X^* be a horizontal vector field, tangent to $S_1\mathbb{E}$. Then

$$\nabla^*_{X^*} N = X^* \text{ and } \nabla_{X^*} iN = iX^*,$$

since $i\colon \mathbb{E}^* \to \mathbb{E}^*$ is an isometry.

Thus, for X^*, Y^* horizontal,

$$g^*([X^*, Y^*], iN) = g^*(\nabla^*_{X^*} Y^* - \nabla^*_{Y^*} X^*, iN) =$$
$$-g^*(Y^*, \nabla^*_{X^*} iN) + g^*(X^*, \nabla^*_{Y^*} iN) = 2g^*(X^*, iY^*);$$
$$g^*([X^*, Y^*], N) = 0.$$

Therefore,

$$[X^*, Y^*]^v = 2g^*(X^*, iY^*)iN.$$

We can now compute the curvature tensor R of $(P\mathbb{E}, g)$ according to (1.11.12), using the curvature tensor (1.11.6) of the sphere. Instead of $g(\ ,\ ), g^*(\ ,\ )$ we simply write $\langle\ ,\ \rangle$; this should not be confused with the Hermitian scalar product on the complex Hilbert space $\mathbb{E}$. We find

$$\langle R(X,Y)Z, W\rangle = \langle Y^*, Z^*\rangle\langle X^*, W^*\rangle - \langle X^*, Z^*\rangle\langle Y^*, W^*\rangle$$
$$-\langle X^*, iZ^*\rangle\langle Y^*, iW^*\rangle + \langle Y^*, iZ^*\rangle\langle X^*, iW^*\rangle$$
$$-2\langle Z^*, iW^*\rangle\langle X^*, iY^*\rangle.$$

In particular, if X, Y are orthonormal, we get for the sectional curvature $K(\sigma)$ of the plane σ spanned by X and Y

$$K(\sigma) = 1 + 3\cos^2\psi; \quad \cos\psi = |\langle X, iY\rangle|.$$

Here, ψ measures the angle between the complex (i.e., i-invariant) planes spanned by $\{X, iX\}$ and $\{Y, iY\}$.

We see that the range of $K(\sigma)$ is the interval [1,4]; the maximum value of $K(\sigma)$, namely 4, is assumed if σ is a complex plane, i. e., if σ is spanned by $\{X, iX\}$. Its minimum value is assumed for a plane spanned by $\{X, Y\}$ with $\langle X, Y\rangle = \langle iX, Y\rangle = 0$.

1.12 Jacobi Fields

In this section we embark upon the investigation of the geometry in the infinitesimal neighborhood of a geodesic on a Riemannian manifold M. We begin with the definition of the Fermi coordinates along a geodesic c, cf. (1.12.1). They may be viewed as the first order approximation of a neighborhood of c; in particular they represent the intrinsic product structure of the normal bundle of c.

The second order neighborhood of c we describe by the Jacobi fields $Y(t)$ along $c(t)$, (1.12.2). A Jacobi field $Y(t)$ constitutes an infinitesimal 1-parameter variation $\{c_s, -\varepsilon < s < \varepsilon\}$ of $c = c_0$ by geodesics c_s, i.e., $Y(t) = \partial c_s(t)/\partial s|_0$, (1.12.4). Of special interest are variations where $\dot{c}_s(0) \in T_{c(0)} M$, i.e., $Y(0) = 0$. This is equivalent to considering the restriction of $\exp_p, p = c(0)$, to a conelike neighborhood of the ray $\tilde{c}(t) = t\dot{c}(0), t \geqslant 0$, in $T_p M$.

Cartan's theorem (1.12.8) states that the Riemannian geometry of a ball $B_\varrho(p)$ around $p \in M$ with $\varrho > 0$ sufficiently small is completely determined by the Jacobi fields $Y(t)$ with $Y(0) = 0$ along the geodesics $c(t)$ of length $< \varrho$, emanating from p.

Consider a non-constant geodesic $c(t)$, starting from $p = c(0)$. In $T_p M$ we have the straight segment $\tilde{c}(t) = t\dot{c}(0)$ with $\exp_p \tilde{c}(t) = c(t)$. If we do not restrict ourselves to small values > 0 of t only, it may happen that the tangential $T_{\tilde{c}(t_1)} \exp_p$ is not a bijection from $T_{\tilde{c}(t_1)} T_p M$ onto $T_{c(t_1)} M$. If $\dim M$ is not finite, this may happen even when $\ker T_{\tilde{c}(t_1)} \exp_p$ is zero. Our main interest lies, however, in the case where this kernel is non-zero. In this case, we call t_1 a conjugate point of multiplicity $= \dim \ker > 0$. That t_1 is a conjugate point of multiplicity > 0 is equivalent to the existence of a non-zero Jacobi field $Y(t)$ along $c(t)$ with $Y(0) = Y(t_1) = 0$, cf. (1.12.10).

After the first conjugate point of multiplicity > 0, the infinitesimal geometry along the geodesic definitely ceases to have any resemblance with the infinitesimal geometry along a straight segment in the model space $\mathbb{E}$. The study of Jacobi fields therefore constitutes the first step towards the study of the global geometry of M.

As a first indication of the difference between the geometry in M and the geometry in $\mathbb{E}$ we show in (1.12.13) that a geodesic c ceases to be minimizing between its end points whenever it contains a conjugate point of positive multiplicity in its interior; and this, when compared with curves arbitrarily near c. On the other hand, before the first conjugate point, c is minimizing, at least when compared with nearby curves.

After determining the Jacobi fields in the spaces of constant curvature, cf. (1.12.14), we generalize the concept of conjugate points by considering singularities of $\exp_F: T^\perp F \to M^*$. Here, $T^\perp F$ is the normal bundle space of an isometric immersion $F: M \to M^*$, (1.12.15). They are called focal points.

For the special case that $F: M \to M^*$ is of codimension 1 and totally geodesic at $p \in M$, the property to be focal point on the normal geodesic $c^\perp(t)$ which starts at p only depends on the Jacobi fields along $c^\perp$.

The first focal point of multiplicity > 0 on a normal geodesic $c^\perp(t)$ to $F(M)$ is characterized by the fact that, after this point, $c^\perp$ ceases to be minimizing, when compared with nearby curves, starting from $F(M)$ near $c^{\perp}(0)$. This is in analogy with the characterization of the first conjugate point.

We begin by introducing Fermi coordinates along a geodesic c. The existence of these coordinates is based on two special properties of a geodesic. On the one hand, c is a totally geodesic immersion and on the other hand, the $\mathscr{F}M$-module $\mathfrak{V}_c^T$ of tangential vector fields along c (see (1.4.8)) has a basis consisting of parallel vector fields – in fact, the vector field $\dot{c}$.

Whenever an immersion F has both of these properties one can introduce so-called Fermi coordinates for F. Note, however, that for an immersion it is the exception rather than the rule, that even one of these two properties holds, cf. (1.6.3) and (1.10.13, v).

Let $c: I \to M$ be a geodesic with $|\dot{c}| = 1$. The interval $I \subset \mathbb{R}$ shall contain 0. For I to be a manifold, I should be open. But we will also allow I to be closed in which case c should be viewed as the restriction of a map of an open interval containing I. Since c is an immersion it has a normal bundle ν_c. ν_c forms part of the commutative diagram

$$\begin{array}{ccc} T_c^{\perp} & \xrightarrow{\tau^* c} & TM \\ {\scriptstyle \nu_c}\downarrow & & \downarrow{\scriptstyle \tau} \\ I & \xrightarrow{c} & M \end{array}$$

cf. (1.10.4). Here we have briefly written $\tau^* c$ instead of $\tau^* c | F^{\perp} c$. The fibre $\nu_c^{-1}(t)$ over $t \in I$ is via $\tau^* c$ isomorphic to the subspace $T_t^{\perp} c$ of $T_{c(t)} M$ which is the orthogonal complement in $T_{c(t)} M$ to the tangent space $T_t c$ of c, generated by $\dot{c}(t)$.

Parallel translation along c preserves the scalar product and $\dot{c}(t)$ is a parallel vector field along $c(t)$. Thus, if we choose an orthogonal decomposition $\mathbb{E} = \mathbb{R} \times \tilde{\mathbb{E}}$ of the model space $\mathbb{E}$ of M and an isometric identification of $\mathbb{E}$ with $T_{c(0)} M$ such that $\mathbb{R} \subset \mathbb{E}$ corresponds to $T_0 c$ and $\tilde{\mathbb{E}} \subset \mathbb{E}$ corresponds to its orthogonal space $T_0^{\perp} c$, we obtain a trivialization ϕ_c of the normal bundle ν_c

$$\phi_c \colon T_c^{\perp} \to I \times \tilde{\mathbb{E}};\ \tilde{X} \mapsto \Big(\nu_c \tilde{X}, \big(\overset{0}{\underset{\nu_c \tilde{X}}{\|}} c\big)\tau^* c \tilde{X}\Big).$$

Denote by $\tilde{B}_{\varrho}(0)$ the ball of radius ϱ around $0 \in \tilde{\mathbb{E}}$.

1.12.1 Proposition. *Let $c: I \to M$ be a geodesic with $|\dot{c}| = 1$. For every bounded interval $I' \subset I$ there exists a $\varrho = \varrho(I') > 0$ such that*

$$(*) \qquad \exp \circ \tau^* c \circ \phi_c^{-1} | I' \times \tilde{B}_{\varrho}(0) \to M$$

is defined and is a local diffeomorphism. In particular, every $t_0 \in I'$ is contained in an interval $J = J(t_0) \subset I'$ such that $(*)$, restricted to $J \times \tilde{B}_{\varrho}(0)$, is a diffeomorphism onto its image $M_{c,\varrho,J} \subset M$. Thus,

$$u_{c,\varrho,J} \equiv \left(\exp \circ \tau^* c \circ \phi_c^{-1} | J \times \tilde{B}_\varrho(0)\right)^{-1} : M_{c,\varrho,J} \to J \times \tilde{B}_\varrho(0)$$

is a chart, called a Fermi chart (determined by c). *In the* Fermi coordinates *so defined, the principal part of the covariant derivative of a vector field along c is the ordinary derivative of the principal part.*

Proof. Since $c(I')$ is contained in a compact subset of M there exists $\eta > 0$ such that exp is defined for all $X \in TM$ with $|X| < \eta$, $X \in T_{c(t)}M$, $t \in I'$. See also (1.9.10). Observe that

$$T_{(t,0)}(\tau^* c \circ \phi_c^{-1}) : T_{(t,0)} T_c^{\perp} \to T_{0_{c(t)}} T_{c(t)} M \cong T_{c(t)} M$$

is a linear isomorphism. Thus, the same is true for the tangential of (*) at $(t, 0)$. Now apply the inverse mapping theorem.

It remains to show that the Christoffel symbols for the Fermi coordinates vanish for $(t_0, 0) \in I \times \tilde{B}_\varrho(0) \subset I \times \mathbb{E}$. To see this we note that the geodesic $t \mapsto c(t_0 + t)$ through $c(t_0)$ in Fermi coordinates is represented by

$$u : t \mapsto (t_0 + t, 0).$$

The geodesic c_X with initial direction $X \in T_{t_0}^{\perp} c$ also has a linear representation,

$$u : t \mapsto (t_0, t\tilde{X}),$$

where $\tau^* c \circ \phi_c^{-1}(t_0, \tilde{X}) = X$.

Since $u(t)$ is geodesic if and only if $\ddot{u}(t) + \Gamma(u(t))(\dot{u}(t), \dot{u}(t)) = 0$ we see

$$\Gamma((t, 0)) | (\mathbb{R} \times \{0\}) \times (\mathbb{R} \times \{0\}) = 0,$$

$$\Gamma((t, 0)) | (\{0\} \times \tilde{\mathbb{E}}) \times (\{0\} \times \tilde{\mathbb{E}}) = 0.$$

Moreover, a normal parallel vector field along c is represented with constant principal part. Thus,

$$\Gamma((t, 0)) | (\mathbb{R} \times \{0\}) \times (\{0\} \times \tilde{\mathbb{E}}) = 0,$$

which proves our claim. □

We now define Jacobi fields along a geodesic. A deeper understanding of these vector fields can be obtained only later when we identify them with the flow-invariant vector fields along the orbits of the geodesic flow on the tangent bundle, cf. (3.1.17).

1.12.2 Definition. Let $c(t)$, $t \in [0, a]$, be a geodesic, possibly constant. *A Jacobi field along c* is a vector field $Y(t)$ satisfying the so-called *Jacobi equation*

$$(*) \qquad \frac{\nabla^2 Y}{dt^2}(t) + R(Y(t), \dot{c}(t))\dot{c}(t) = 0.$$

Or, more briefly, with the operator $R_{\dot{c}(t)}$ introduced in (1.8.13),

$$(*) \qquad \nabla^2 Y(t) + R_{\dot{c}(t)} Y(t) = 0.$$

1.12.3 Proposition. (i) *The set* $\mathscr{I}_c$ *of Jacobi fields along* c *is a vector space naturally isomorphic to* $T_{c(t_0)}M \times T_{c(t_0)}M \cong \mathbb{E} \times \mathbb{E}$ *under the mapping*

$$Y \mapsto (Y(t_0), \nabla Y(t_0)).$$

(ii) *On* $\mathscr{I}_c$ *we have a canonical* symplectic structure, *given by the non-degenerate skew symmetric 2-form*

$$\alpha(Y, Y') = \langle Y(t), \nabla Y'(t)\rangle - \langle Y'(t), \nabla Y(t)\rangle .$$

This means in particular that the right hand side is independent of t.

(iii) *Assume* $|\dot c| \neq 0$. *Then* $\mathscr{I}_c$ *splits into a 2-dimensional* α*-non-degenerate subspace* $\mathscr{I}_c^{\top}$ *of tangential Jacobi fields and a* α*-orthogonal complementary space* $\mathscr{I}_c^{\perp}$ *of Jacobi fields orthogonal to* $\dot c$. *A tangential Jacobi field is of the form*

$$Y^{\top}(t) = (a + bt)\dot c(t).$$

Note. Denote by $\mathscr{I}_c^0$ the subspace of $\mathscr{I}_c$ consisting of the Jacobi fields which vanish for $t = 0$. Then α, restricted to $\mathscr{I}_c^0$, is identically zero. $\mathscr{I}_c^0$ therefore is also called a *Lagrangian subspace,* cf. (3.1.1).

Proof. If $\dot c = 0$, i. e., $c(t) = c(0)$, the Jacobi fields are of the form $Y(t) = A + tB$, where $A, B \in T_{c(0)}M$. So we can assume $\dot c \neq 0$. Let $t \mapsto s(t)$ be the arc length parameter on c, cf. (1.8.8). I. e., $\dot s(t) = |\dot c(t)|$, $s(0) = 0$. Denote by $s \mapsto t(s)$ the inverse of $t \mapsto s(t)$. Then the Jacobi equation (*) remains invariant. That is, both $Y(t)$ and $Y(t(s))$ are Jacobi fields. We therefore will assume $|\dot c| = 1$.

If we introduce Fermi coordinates along c the Jacobi equation (*) gives for the principal part $Y(t)$ of the representation of $Y(t)$ the equation

$$(**) \qquad \ddot Y(t) + R(Y(t), e_1)e_1 = 0$$

$$\text{with} \qquad R(Y(t), e_1)e_1 = D\Gamma((t, 0)) \,.\, Y(t)(e_1, e_1).$$

Here, e_1 is the basis vector field for the first factor in $\mathbb{R} \times \tilde{\mathbb{E}} = \mathbb{E}$ and we have written e_1 instead of $(t, e_1) \in T_{(t,0)}\mathbb{E}$.

(**) is a homogeneous differential equation of second order for $Y(t)$. (i) now is nothing but the fact that the solutions of such an equation are in 1 : 1 correspondence with the values $Y(t_0)$, $\dot Y(t_0)$ at some $t_0 \in I$.

To prove (ii) we differentiate the expression for $\alpha(Y, Y')$ with respect to t. From (*) we get zero for the derivative. Thus, α is a $\mathbb{R}$-valued skew symmetric 2-form on $\mathscr{I}_c$. Actually, using the identification of $\mathscr{I}_c$ with $T_{c(t_0)}M \times T_{c(t_0)}M$ given in (i) we see that α is nothing but the canonical symplectic 2-form on $\mathbb{E} \times \mathbb{E}$ where $\mathbb{E}$ has the Hilbert scalar product $\langle , \rangle$.

Finally, if $Y(t) = f(t)\dot c(t)$ is a tangential Jacobi field, (**) becomes $\ddot f = 0$, thus, $f(t) = a + bt$. $\mathscr{I}_c^{\top}$ is generated by $Y_1^{\top}(t) = \dot c(t)$ and $Y_2^{\top}(t) = t\dot c(t)$. Thus, $\alpha(Y_1^{\top}, Y_2^{\top}) = 1$, i.e., $\mathscr{I}_c^{\top}$ is α-non-degenerate.

If $Y \in \mathscr{I}_c$ is in the α-orthogonal complement of $\mathscr{I}_c^{\top}$ this means

$$0 = \alpha(Y_1^{\top}, Y) = \langle \dot{c}(t), \nabla Y(t) \rangle,$$

$$0 = \alpha(Y_2^{\top}, Y) = t\langle \dot{c}\ (t), \nabla Y(t) \rangle - \langle Y(t), \dot{c}(t) \rangle,$$

i. e., $\langle Y(t), \dot{c}(t) \rangle = 0$. □

We now characterize Jacobi fields as infinitesimal variations of c by geodesics.

1.12.4 Lemma. *Let $c: I \to M$ be a geodesic, $\dot{c} \neq 0, 0 \in I$. If*

$$F:]-\varepsilon, \varepsilon[\times I \to M$$

is a differentiable map such that, for each $s \in]-\varepsilon, \varepsilon[$, $c_s = F|\{s\} \times I$ is a geodesic and $c_0 = c$, then the vector field

$$Y(t) = \frac{\partial F}{\partial s}(s, t)|_{s=0}$$

is a Jacobi field. Every Jacobi field can be obtained in this manner.

Proof. From (1.5.8) we get

$$\frac{\nabla^2}{\partial t^2} \frac{\partial F}{\partial s}(s, t) = \frac{\nabla}{\partial t} \frac{\nabla}{\partial s} \frac{\partial F}{\partial t}(s, t) = \frac{\nabla}{\partial s} \frac{\nabla^2 F}{\partial t^2}(s, t) + R\left(\frac{\partial F}{\partial t}, \frac{\partial F}{\partial s}\right)\frac{\partial F}{\partial t}(s, t).$$

On the right hand side the first term vanishes since c_s is a geodesic. Observe now that $\partial F(0, t)/\partial t = \dot{c}(t)$.

Conversely, let $Y(t)$ be a Jacobi field along $c(t)$, $c(0) = p$. Consider the curve $b(s) = \exp_p s\, Y(0)$ (or any curve $b(s)$ with $b'(0) = Y(0)$) in M and in TM a curve $X(s)$ in the domain of definition of exp with $\tau \circ X(s) = b(s)$, $X(0) = \dot{c}\,(0)$, and initial direction

$$X'(0) = Y(0) + \nabla Y(0) \in T_{X(0)h} TM \oplus T_{X(0)v} TM.$$

Then $F(s, t) = \exp_{b(s)} t X(s)$ is a 1-parameter familiy of geodesics. We claim that the Jacobi field $\partial F(0, t)/\partial s$ coincides with $Y(t)$.

Using the decomposition

$$\frac{dX(s)}{ds} = b'(s) + \frac{\nabla X}{ds}(s) \in T_{X(s)h} TM \oplus T_{X(s)v} TM,$$

where we have identified the horizontal as well as the vertical part of $T_{X(s)} TM$ with $T_{b(s)} M$, we get

$$\frac{\partial F}{\partial s}(0, t) = T_{t\dot{c}(0)} \exp_p . \big(Y(0) + t \nabla Y(0)\big).$$

Hence, $\partial F(0, 0)/\partial s = Y(0)$ and

$$\frac{\nabla}{\partial t} \frac{\partial F}{\partial s}(0,0) = \frac{\nabla}{\partial s} \frac{\partial F}{\partial t}(0,0) = \frac{\nabla X}{ds}(0) = \nabla Y(0). \quad \square$$

1.12.5 Corollary. *Let $c:[0,a]\to M$ be a non-constant geodesic with $c(0)=p$. Then a Jacobi field $Y(t)$ along $c(t)$ with $Y(0)=0$ can be written as*

$$Y(t) = T_{t\dot c(0)}\exp_p . t\nabla Y(0). \quad \square$$

Remark. Let $c(t)$ be a geodesic, $c(0)=p$. Then $\tilde c(t)=t\dot c(0)$ can be viewed as a geodesic on $\tilde T_p M = T_p M \cap \tilde T M$. For any $A\in T_p M$,

$$t\mapsto \tilde Y(t) = tA \in T_{\dot c(t)} T_p M$$

is a Jacobi field along the geodesic $\tilde c$, vanishing for $t=0$. The corollary states that $T\exp_p$, restricted to the tangent spaces $T_{\tilde c(t)} T_p M$, carries $\mathscr{J}^0_{\tilde c}$ into $\mathscr{J}^0_c$.

Jacobi fields carry a large amount of information on the structure of a Riemannian manifold. As a preliminary step towards making this clear we introduce the following

1.12.6 Definition. Let M, M^* be Riemannian manifolds modelled on the same Hilbert space $\mathbb{E}$. Let $I=[0,a]$ and

$$c: I\to M;\ c^*: I\to M^*$$

be geodesics of equal length. Choose an isometry

$$\imath_0 : (T_{c(0)}M, \dot c(0)) \to (T_{c^*(0)}M^*, \dot c^*(0)).$$

With this define

$$\imath_t \equiv \overset{t}{\underset{0}{\|}} c^* \circ \imath_0 \circ \overset{0}{\underset{t}{\|}} c : (T_{c(t)}M, c(t)) \to (T_{c^*(t)}M^*, \dot c^*(t)).$$

1.12.7 Proposition. (i) *Each $\imath_t$, $t\in I$, is an isometry.*

(ii) *Let $Y(t)$ be a vector field along $c(t)$, $t\in I$. Then $\imath_t Y(t)$ is a vector field along $c^*(t)$, $t\in I$. This mapping commutes with the covariant derivations ∇ in M and ∇^* in M^*, i.e., $\nabla^*\circ\imath_t = \imath_t\circ\nabla$.*

Proof. (i) is clear from the definition. As for (ii) we get from (1.5.8) with $F(s,t) = c(t+s)$, $Y(t,s) = (\overset{t+s}{\underset{t}{\|}} c)\, Y(t)$,

$$\frac{\nabla}{\partial s}\frac{\nabla}{\partial t} Y(t,s) = \frac{\nabla}{\partial t}\frac{\nabla}{\partial s} Y(t,s) + R(\dot c(t+s), \dot c(t+s))\, Y(t,s) = 0.$$

That is, $\nabla Y(t,s)/\partial t$ is a parallel field in s which coincides with $(\overset{t+s}{\underset{t}{\|}} c)\nabla Y(t)/dt$. $\square$

We are now able to prove *Cartan's Theorem* in which the existence of a local isometry is characterized by a certain property of the curvature operator, cf. Cartan [1].

1.12.8 Theorem. *Let M, M^* be Riemannian manifolds modelled on the same Hilbert space $\mathbb{E}$. Let $p\in M, p^*\in M^*$ and $\eta>0$ be such that $B_\eta(p), B_\eta(p^*)$ are domains of* *normal coordinates on* M *and* M^*, *respectively. Choose an isometry*

$$\iota_p : T_p M \to T_{p^*} M^*.$$

(i) *The mapping*

$$(*) \qquad \Phi \equiv \exp_{p^*} \circ \iota_p \circ (\exp_p | B_\eta(0_p))^{-1} : B_\eta(p) \to B_\eta(p^*)$$

is a diffeomorphism carrying radial geodesics $c(t)$, *i. e., geodesics in* $B_\eta(p)$ *with* $c(0) = p$ *into radial ones. The tangential of* Φ *carries Jacobi fields* $Y(t)$ *along radial geodesics* $c(t)$ *satisfiying* $Y(0) = 0$ *into Jacobi fields* $Y^*(t)$ *along* $c^*(t)$ *with* $Y^*(0) = 0$.

(ii) *Assume now that for all radial geodesics* $c(t)$ *and their images* $c^*(t) = \Phi \circ c(t)$ *we have*

$$(\dagger) \qquad \iota_t \circ R_{\dot c(t)} = R^*_{\dot c^*(t)} \circ \iota_t ,$$

where $\iota_t : T_{c(t)} M \to T_{c^*(t)} M^*$ *is defined as in* (1.12.6) *and* $R_{\dot c(t)}, R^*_{\dot c^*(t)}$ *are the self-adjoint operators defined by the curvature tensors* R, R^* *of* M *and* M^*, *respectively, cf.* (1.8.13).

Then Φ *is an isometry.*

Note. Conversely, if Φ is an isometry, the condition $(\dagger)$ is satisfied.

There exists an extension of Cartan's Theorem concerning the existence of a global isometry – at least if $\dim \mathbb{E} < \infty$. See Ambrose [1].

Proof. Clearly, Φ carries radial geodesics into radial ones. Since a Jacobi field $Y(t)$ with $Y(0) = 0$ along a radial geodesic $c(t)$ can be written as $\partial F(s, t)/\partial s|_0$ where each $t \to F(s, t)$ is a radial geodesic and $F(0, t) = c(t)$, all of (i) is true.

As for (ii), we recall from (1.12.7) that ι_t and the covariant derivation commute. Thus, if $Y(t)$ with $Y(0) = 0$ satisfies the Jacobi equations $(*)$ in (1.12.2), $(\dagger)$ implies that in addition its image $Y^*(t) = \iota_t Y(t)$ is a Jacobi field. Since $\iota_t : T_{c(t)} M \to T_{c^*(t)} M^*$ is an isometry, it only remains to check whether $T_{c(t)} M$ is spanned by the value at t of the Jacobi fields Y along c with $Y(0) = 0$. But this is indeed the case, see (1.12.5). □

We now come to the important concept of a conjugate point.

1.12.9 Definition. Let $c : [0, a] = J \to M$ be a geodesic starting from $p = c(0)$. Put $t\dot c(0) = \tilde c(t)$. $t_1 \in [0, a]$ is called *conjugate point* (of $c(0)$ along $c|[0, t_1]$) if

$$T_{\tilde c(t_1)} \exp_p : T_{\tilde c(t_1)} T_p M \to T_{c(t_1)} M$$

is not an isomorphism. If $\dim \ker T_{\tilde c(t_1)} \exp_p = k$, k is called the *multiplicity* of the conjugate point.

Remarks. 1. For proper Hilbert spaces $\mathbb{E}, \mathbb{E}'$, a linear continuous mapping $L : \mathbb{E} \to \mathbb{E}'$ with $\ker L = 0$ need not be an isomorphism. Therefore, it is possible that t_1 is a conjugate point even if $k = 0$. Cf. Flaschel und Klingenberg [1] for further remarks. We do not know, however, whether this can occur for the Riemannian Hilbert manifolds $H^1(I, M)$ and their various subspaces which we are going to consider in chapter 2.

In any case, to simplify the subsequent results, we frequently will make the assumption $\dim M < \infty$. In that case, a conjugate point t_1 of multiplicity 0 actually is a non-conjugate point, i. e., $T_{\tilde{c}(t_1)} \exp_p$ is an isomorphism.

2. The kernel of $T_{\tilde{c}(t_1)} \exp_p$ always is orthogonal to the radial vector $\dot{\tilde{c}}(t_1)$, as follows from (1.9.1).

Example. Consider $p = (\varrho, 0) \in S_\varrho^{\mathbb{E}} \subset \mathbb{R} \times \mathbb{E}$, cf. (1.1.3, ii). $\exp_p$ carries the sphere $\partial \bar{B}_{\pi\varrho}(0_p) \subset T_p S_\varrho^{\mathbb{E}}$ of radius $\pi\varrho$ into $(-\varrho, 0)$. Hence, every geodesic c on $S_\varrho^{\mathbb{E}}$, $|\dot{c}| = 1$, has a conjugate point at $t = \pi\varrho$.

1.12.10 Lemma. *Let $c: [0, a] \to M$ be a geodesic, $\dot{c} \neq 0$. Then t_1 is a conjugate point of multiplicity $k > 0$ if and only if the space $\mathcal{J}_c^{\perp 0} = \mathcal{J}_c^0 \cap \mathcal{J}_c^\perp$ of Jacobi fields $Y(t)$ along $c(t)$ with $Y(0) = Y(t_1) = 0, \langle Y(t), \dot{c}(t) \rangle = 0$, has dimension k.*

Proof. Let $t_1 > 0$. From (1.12.5) we know that, for a vector $A \in T_{\tilde{c}(t_1)} T_p M$, orthogonal to $\tilde{c}(t_1)$, the image under $T_{\tilde{c}(t_1)} \exp_p$ is the value $Y(t_1)$ of the Jacobi field $Y(t)$ along $c(t)$ with $Y(0) = 0, A = t_1 \nabla Y(0)$. Thus, $\ker T_{c(t_1)} \exp_p$ corresponds to the subspace of $\mathcal{J}_c^{\perp 0}$ of Jacobi fields Y which vanish at t_1.

Remark. This Lemma can be interpreted as saying that at a conjugate point of c of multiplicity $k > 0$ geodesics starting from $p = c(0)$ nearly meet again infinitesimally at c. In (2.1.12) we will make a more precise statement.

We show that a geodesic is a locally minimizing curve between its end points whenever it has no conjugate points in its interior. This ceases to be true for a geodesic which has a conjugate point of multiplicity > 0 in its interior, cf. (1.12.13).

We begin with the

1.12.11 Definition. Let $c: J = [0, a] \to M$ be a geodesic from $p = c(0)$ to $q = c(a)$.

(i) A *(piecewise differentiable) variation* of c (with fixed end points) is a continuous mapping

$$F:]-\varepsilon, \varepsilon[\times J \to M, \text{ for some } \varepsilon > 0$$

with $$F|\{0\} \times J = c,\ F(s, 0) = p,\ F(s, a) = q, \text{ for all } s \in]-\varepsilon, \varepsilon[,$$

and such that $F|]-\varepsilon, \varepsilon[\times [t_{j-1}, t_j]$, $1 \leqslant j \leqslant l$, is differentiable, for a certain subdivision $0 = t_0 < \ldots < t_l = a$ of J.

The *variation vector field associated with* F is the piecewise differentiable vector field $Y(t) = \partial F(s, t)/\partial s|_0$.

(ii) A *(piecewise differentiable) variation vector field* along c is a continuous piecewise differentiable vector field $Y(t)$ along $c(t)$ with $Y(0) = Y(a) = 0$.

Any such vector field can be derived from a variation F, take e.g., $F(s,t) = \exp_{c(t)} s\,Y(t)$.

(iii) On the vector space of piecewise differentiable variation vector fields along c we define a *symmetric bilinear form* $D^2E(c)$ by

$$D^2E(c)(X,Y) = \int_J \langle \nabla X, \nabla Y\rangle(t)\,dt - \int_J \langle R_{\dot c}X, Y\rangle(t)\,dt$$

Here, $R_{\dot c}X(t) = R(X(t),\dot c(t))\dot c(t)$ is the self-adjoint curvature operator defined in (1.8.13).

Remarks. 1. Note that for a piecewise differentiable variation $F(s,t)$, we have continuous partial derivatives $\partial^k F(s,t)/\partial s^k$ for all k.

2. In the notation $D^2E(c)$ for the symmetric bilinear form we have alluded to the possibility that it can be interpreted as the Hessian of a differentiable function E on a certain manifold of curves from p to q. (1.12.12) points towards this possibility, but only in (2.5) we will be able to make this precise, at least when $\dim M < \infty$.

3. Let $0 = t_0 < \ldots < t_l = a$ be a subdivision of J such that the two variation vector fields $X(t)$ and $Y(t)$ are differentiable, when restricted to $[t_{j-1}, t_j], j = 1, \ldots, l$. Then we get, using partial integration on the first term of $D^2E(c)$,

$$D^2E(c)(X,Y) = -\int_J \langle \nabla^2 X + R_{\dot c}X, Y\rangle(t)\,dt + \sum_1^l \langle \nabla X, Y\rangle(t)\Big|_{t_{j-1}}^{t_j}$$

$$= -\int_J \langle \nabla^2 X + R_{\dot c}X, Y\rangle(t)\,dt + \sum_1^{l-1} \langle \nabla X(t_j-) - \nabla X(t_j+), Y(t_j)\rangle.$$

1.12.12 Lemma. *Let $c: J = [0,a] \to M$ be a geodesic on M and $F(s,t)$, $(s,t) \in]-\varepsilon, \varepsilon[\times J$, be a variation of c with associated variation vector field $Y(t)$. Put $F|\{s\} \times J = c_s$ and*

$$\tfrac{1}{2}\int_J \langle \dot c_s(t), \dot c_s(t)\rangle\,dt = E(c_s) = E(s),$$

cf. (1.8.7.) *Then*

$$E'(0) = 0\,;\ E''(0) = D^2E(c)(Y,Y).$$

Proof. With (1.5.8) we find

$$E'(s) = \int_J \langle \frac{\nabla}{\partial t}\frac{\partial F}{\partial s}, \frac{\partial F}{\partial t}\rangle\ (s,t)\,dt;\ E''(0) = \int_J \langle \nabla Y, \nabla Y\rangle(t)\,dt +$$

$$+ \int_J \langle \frac{\nabla}{\partial t}\frac{\nabla^2 F}{\partial s^2}, \frac{\partial F}{\partial t}\rangle\ (0,t)\,dt + \int_J \langle R(Y,\dot c)Y, \dot c\rangle(t)\,dt.$$

Using partial integration and $\partial F(0,t)/\partial t = \dot c(t)$, $\partial F(0,t)/\partial s = Y(t)$, $\nabla \dot c(t) = 0$, we get

$$E'(0) = \int_J \langle \nabla Y, \dot{c} \rangle (t) \, dt = \int_J \frac{d}{dt} \langle Y, \dot{c} \rangle (t) \, dt = 0 .$$

A similar argument using $F(s, 0) = p$, $F(s, a) = q$ shows that the middle term on the right hand side of the expression for $E''(0)$ vanishes. □

The following is the geometric characterization of the first conjugate point, alluded to above, at least for $\dim M < \infty$.

1.12.13 Theorem. *Let $c: J = [0, a] \to M$ be a geodesic from $p = c(0)$ to $q = c(a)$. Put $t\dot{c}(0) = \tilde{c}(t)$.*

(i) Assume that $T_{\tilde{c}(t)} \exp_p : T_{\tilde{c}(t)} T_p M \to T_{c(t)} M$ is an isomorphism for all $t \in J$. Then every curve b from p to q, sufficiently near c, has length $L(b) \geqslant L(c) =$ length c. It follows that the form $D^2 E(c)$ is positive semi-definite.

(ii) If c has a conjugate point of multiplicity > 0 in its interior, then there exists a variation vector field X along c with $D^2 E(c)(X, X) < 0$. Hence, for a variation F of c having X as variation vector field, the curve $c_s = \{t \to F(s, t)\}$ from p to q is shorter than c, for all sufficiently small $|s| \neq 0$.

Proof. Under the assumption made under (i), we have for every $t \in [0, a]$ an open neighborhood $\mathscr{U}(\tilde{c}(t))$ of $\tilde{c}(t)$ *in* $T_p M$ such that $\exp_p|\mathscr{U}(\tilde{c}(t))$ is a diffeomorphism onto an open neighborhood $\mathscr{U}(c(t))$ of $c(t)$. By compactness of $[0, a]$ we can assume that there exists a $\varrho > 0$ such that $\mathscr{U}(\tilde{c}(t))$ contains the ϱ-ball $B_\varrho(\tilde{c}(t))$, for all $t \in [0, a]$.

We now consider curves $b(s)$, $s \in [0, \tilde{a}]$ from p to q which are near c in the following sense: There exists a subdivision $s_0 = 0 < \ldots < s_l = \tilde{a}$ of $[0, \tilde{a}]$ and a subdivision $t_0 = 0 < \ldots < t_l = a$ of $J = [0, a]$ such im $b|[s_{j-1}, s_j] \subset \operatorname{im} \exp_p|B_\varrho(\tilde{c}(t_j))$, $j = 1, \ldots, l$. Define now $\tilde{b}|[s_0, s_1]$ by applying $(\exp_p|B_\varrho(\tilde{c}(t_1)))^{-1}$ to $b|[s_0, s_1]$. Similarly, define $\tilde{b}|[s_1, s_2]$ by applying $(\exp_p|B_\varrho(\tilde{c}(t_2)))^{-1}$ to $b|[s_1, s_2]$ such that $\tilde{b}(s_1)$ has the given value, and so on. In this way we construct a curve $\tilde{b}(s)$, $s \in [0, \tilde{a}]$, in $T_p M$ from $\tilde{c}(0)$ to $\tilde{c}(a)$ with $b(s) = \exp_p \tilde{b}(s)$. From (1.9.2) then follows $L(b) \geqslant L(c)$.

Let now Y be a variation vector field along c. Let F be a variation having Y as variation vector field. Put $F|\{s\} \times J = c_s$. Then, with $|J| = a$,

$$2E(c_s)|J| \geqslant L(c_s)^2 \geqslant L(c)^2 = 2E(c)|J|,$$

for all sufficiently small $s \geqslant 0$. Hence, with (1.12.12), $D^2 E(c)(Y, Y) \geqslant 0$. This completes the proof of (i).

For the proof of (ii) we take a non-zero Jacobi field $Y(t)$ along $c(t)$ with $Y(0) = Y(t_1) = 0$. Choose t_0, t_2 with $0 < t_0 < t_1 < t_2 \leqslant a$. We even can assume $Y(t_0) \neq 0$, since the zeros of a non-zero Jacobi field are isolated, such a field being the solution of a second order differential equation.

Let $Z(t)$ be a differentiable vector field along $c(t)$ with $Z(t) = 0$ for $0 \leqslant t \leqslant t_0$ and $t_2 \leqslant t \leqslant a$, and $Z(t_1) = -\nabla Y(t_1) \neq 0$. We define $Y^*(t) = Y(t)$, for $0 \leqslant t \leqslant t_1$ and $= 0$, for $t_1 \leqslant t \leqslant a$. Note that Y^* is a once broken Jacobi field. Put $Y^*(t) + \eta Z(t) = X_\eta(t)$, for some $\eta \geqslant 0$. Then

$$D^2E(c)(X_\eta, X_\eta) = D^2E(c)(Y^*, Y^*) + 2\eta D^2E(c)(Y^*, Z) + \eta^2 D^2E(c)(Z, Z).$$

The formula for $D^2E(c)(X, Y)$ obtained by partial integration shows that the first term on the right hand side vanishes while the second term becomes

$$2\eta\langle\nabla Y^*(t_1^-) - \nabla Y^*(t_1^+), Z(t_1)\rangle = -2\eta|\nabla Y(t_1)|^2.$$

Thus, for $\eta > 0$ sufficiently small, $D^2E(c)(X_\eta, X_\eta) < 0$. □

Remarks. 1. For the proof of (ii) we had constructed a variation vector field $X_\eta(t)$ which is not differentiable at $t = t_1$. However, it is clear from an approximation argument that there also exist differentiable X with $D^2E(c)(X, X) < 0$. Similarly, we also get differentiable curves b near c from p to q with $L(b) < L(c)$.

2. While a geodesic $c: J \to M$ with no conjugate point $t_1 \in J$ is shortest curve between its end points p and q, when compared with nearby curves from p to q, this need not be the case when we compare $L(c)$ with the length of 'far away' curves from p to q, cf. the example after (1.9.2).

3. In case the end point a of a geodesic $c: [0, a] \to M$ is a conjugate point of multiplicity > 0, while there is no conjugate point in the interior of $[0, a]$, one cannot make a general statement as to whether c is locally shortest curve from $p = c(0)$ to $q = c(a)$ or not. There exist examples for each of these possibilities. Actually, $D^2E(c)$ is positive semidefinite in this case, but $D^2E(c)$ possesses a non-trivial nullspace. More precisely, we have a non-zero Jacobi field $Y(t)$ along $c(t)$, with $Y(0) = Y(a) = 0$, and for this Y, $D^2E(c)(Y, Z) = 0$, for all variation vector fields Z of c.

In (2.5) we will see more clearly that this corresponds to the well-known phenomenon occuring for a differentiable function $E: (\mathbb{R}^n, 0) \to (\mathbb{R}, 0)$ with $DE(0) = 0$: If the Hessian $D^2E(0)$ is positive semidefinite but not positive definite, then E may or may not have a local minimum at 0.

1.12.14 Jacobi fields on spaces of constant curvature. Let M have constant sectional curvature $K = K_0$. Let $c: \mathbb{R} \to M$ be a geodesic $\neq$ const. Using (1.11.6), (1.11.8), the equation (1.12.2) for a Jacobi field $Y(t) \perp \dot{c}(t)$ reads

$$\frac{\nabla^2 Y(t)}{dt^2} + K_0|\dot{c}(t)|^2 Y(t) = 0.$$

Hence, the general solution is of the form $Y(t) = y(t)A(t)$, where $\nabla A(t) = 0$ and $\langle A(t), \dot{c}(t)\rangle = 0$ with

$$y(t) = \begin{cases} a\cos(\sqrt{K_0}|\dot{c}|t) + b\sin(\sqrt{K_0}|\dot{c}|t), & \text{for } K_0 > 0 \\ a + bt, & \text{for } K_0 = 0 \\ a\cosh(\sqrt{-K_0}|\dot{c}|t) + b\sinh(\sqrt{-K_0}|\dot{c}|t) & \text{for } K_0 < 0. \end{cases}$$

This shows that a geodesic c on a space of constant curvature $K_0 \leqslant 0$ has no conjugate points. If $K_0 > 0$ then there are conjugate points at precisely $t = m\pi/\sqrt{K_0}|\dot{c}|$, $m = 1, \ldots$ Hence, there are no conjugate points if $L(c|[0,1]) = |\dot{c}| < \pi/\sqrt{K_0}$, whereas for $L(c|[0,1]) = |\dot{c}| \geqslant \pi/\sqrt{K_0}$ there are conjugate points.

We shall generalize the concept of a conjugate point. Recall that a conjugate point on a non-constant geodesic $c(t)$, starting at $c(0) = p \in M$, is a point t_1 on the ray $\tilde{c}(t) = t\dot{c}(0)$ in T_pM where $T_{\tilde{c}(t_1)} \exp_p$ is not bijective.

T_pM may be viewed as the total space of the normal bundle of the 0-dimensional submanifold $\{p\} \mapsto M$ of M.

With the concepts of (1.10.4) we consider an isometric immersion $F: M \to M^*$ and the associated normal bundle

$$\nu_F : T^\perp F \to M .$$

We have the following bundle morphism

$$\begin{array}{ccc} T^\perp F & \xrightarrow{\ \tau_{M^*}^* \cdot F|T^\perp F\ } & TM^* \\ \Big\downarrow \nu_F & & \Big\downarrow \tau_{M^*} \\ M & \xrightarrow{\quad F \quad} & M^* \end{array}$$

The restriction $\tau_{M^*}^* \cdot F|T_p^\perp F$ to the fibre $T_p^\perp F$ over p gives the canonical identification with the subspace $T_p^\perp F$ in $T_{F(p)}M^*$ which is the orthogonal complement of $T_pF(T_pM)$. In particular, $\tau_{M^*}^* \cdot F|T^\perp F$ is an immersion

Recall from (1.9.12) the splitting

$$T_{X^*}TM^* = T_{X^*h}TM^* \oplus T_{X^*v}TM^*$$

of the tangent space $T_{X^*}TM^*$ at X^* of the total tangent space TM^* of a Riemannian manifold M^*. We also defined l. c. a scalar product on $T_{X^*}TM^*$, using this splitting. We use this to define a Riemannian metric on $T^\perp F$.

1.12.15 Definition. Let $F: M \to M^*$ be an isometric immersion and denote by $\nu_F : T^\perp F \to M$ the associated normal bundle, cf. (1.10.4).

Let $X \in T^\perp F$ and put $\nu_F(X) = p$, $\tau_{M^*}^* \cdot FX = X^* \in T_{F(p)}M^*$.

(i) Using the identification under $T_X\tau_{M^*}^* \cdot F$ of $T_XT^\perp F$ with a subspace of $T_{X^*}TM^*$, define a splitting

$$T_XT^\perp F = T_{Xh}T^\perp F \oplus T_{Xv}T^\perp F$$

into a *horizontal* and a *vertical subspace* as follows: $T_{Xv}T^\perp F$ is the subspace tangent to the fibre $T_p^\perp F$ over p whereas $T_{Xh}T^\perp F$ corresponds to $(T_X\tau_{M^*}^* \cdot F)^{-1}T_{X^*h}TM^*$.

(ii) Define a Riemannian metric on $T^\perp F$ by taking on $T_X T^\perp F$ the metric induced from the metric on $T_{X^*} TM^*$ by this identification. Then $\nu_F : T^\perp F \to M$ becomes an isometric submersion, cf. (1.11.9).

(iii) Define the *exponential mapping*

$$\exp_F : T^\perp F \to M^*$$

by $\exp_F X = \exp \circ \tau^*_{M^*} FX$. Put $\exp_F | T^\perp_p F = \exp_{F,p}$.

(iv) Let $\tilde{c}^\perp(t) = tX$ be a non-constant normal ray in $T^\perp F$. Then $c^\perp(t) = \exp_F \tilde{c}^\perp(t)$ is a geodesic in M^*, starting at $p^* = F(p)$ orthogonally to $T_p F(T_p M) \subset T_{p^*} M^*$. We call t_1 or also $c^\perp(t_1)$ a *focal point* (of F along $c^\perp|[0, t_1]$) if

$$(*) \qquad T_{\tilde{c}^\perp(t_1)} \exp_{F,p} : T_{c^\perp(t_1)} T^\perp F \to T_{c^\perp(t_1)} M^*$$

is not an isomorphism. The *multiplicity* k of a focal point is the dimension of the kernel of $(*)$. Note: $k = 0$ does not imply that $(*)$ is an isomorphism, if $\dim M = \infty$.

1.12.16 Example. Take an orthogonal decomposition $\mathbb{E} = \mathbb{E}_1 \times \mathbb{E}_2$ intor two non-zero subspaces. Consider for the sphere $S^{\mathbb{E}}_\varrho \subset \mathbb{R} \times \mathbb{E}$ the mapping

$$F : S^{\mathbb{E}_1}_\varrho \to S^{\mathbb{E}}_\varrho,$$

obtained from restricting the inclusion

$$\tilde{F} : \mathbb{R} \times \mathbb{E}_1 \to \mathbb{R} \times \mathbb{E}_1 \times \{0\} \subset \mathbb{R} \times \mathbb{E}_1 \times \mathbb{E}_2 = \mathbb{R} \times \mathbb{E}$$

to the sphere $S^{\mathbb{E}_1}_\varrho \subset \mathbb{R} \times \mathbb{E}_1$.

Since $T^\perp \tilde{F} = (\mathbb{R} \times \mathbb{E}_1) \times \mathbb{E}_2$ we have

$$T^\perp F = S^{\mathbb{E}_1}_\varrho \times \mathbb{E}_2 .$$

The normal geodesics $c^\perp$ are parameterized great circles on $S^{\mathbb{E}}_\varrho$ which pass through the sphere $S^{\mathbb{E}}_\varrho \cap \mathbb{E}_2$ at distance $\varrho\pi/2$. That is,

$$\exp_F | S^{\mathbb{E}_1}_\varrho \times \partial \bar{B}^{\mathbb{E}_2}_{\pi\varrho/2}$$

is singular in all points; every normal geodesic $c^\perp$ to the submanifold $S^{\mathbb{E}_1}_\varrho \subset S^{\mathbb{E}}_\varrho$ has a focal point of multiplicity $= \dim \mathbb{E}_2 > 0$ at distance $\pi\varrho/2$.

We can prove the following generalization of the Gauss Lemma (1.9.1) which corresponds to the special case $M = p^* \in M^*$, $F =$ inclusion. For the case where $F : M \to M^*$ is a regular curve, (i. e., a curve with non-zero tangent vector) on a surface it goes back to Gauss [1]. Here it yields to the existence of geodesic parallel coordinates.

1.12.17 Lemma. *Consider the restriction of the exponential mapping*

$$\exp_F : T^\perp F \to M^*$$

to a normal ray $\tilde{c}^\perp(t) = tX_0$, *which starts at* $p = \nu_F(X_0)$. *Put* $\exp_{F,p} \tilde{c}^\perp(t) = c^\perp(t)$. *Here,* $|X_0| > 0$ *shall be so small that* $\exp_{F,p} X_0$ *is defined. Then* $T_{X_0} \exp_{F,p}$ *maps the 'radial'* *vector* $\dot{\tilde{c}}^{\perp}(1)$ *isometrically into* $\dot{c}^{\perp}(1) \in T_{c^{\perp}(1)}M^*$. *A vector* Y_0 *orthogonal to* $\dot{\tilde{c}}^{\perp}(1)$ *(also called a 'spherical' vector) is mapped into a vector orthogonal to* $\dot{c}^{\perp}(1)$.

Proof. The first statement follows from the fact that $|\dot{\tilde{c}}^{\perp}(t)| = |\dot{\tilde{c}}(0)|, |\dot{c}^{\perp}(t)| = |\dot{c}^{\perp}(0)|$.

If Y_0 is orthogonal to $\dot{\tilde{c}}^{\perp}(1)$ in $T_{\tilde{c}^{\perp}(1)}T^{\perp}F$, we can describe it with the help of a normal section $X(s)$ along a curve $b(s)$ on $M: X(s) \in T^{\perp}_{b(s)}F$, $X(0) = X_0$, $|X(s)| = |X_0|$ and $X'(0) = Y_0$. Then $F(s, t) = \exp_{F,b(s)} tX(s)$ is a variation of $F(0, t) = c^{\perp}(t)$ by normal geodesics. We have to show that $\partial F(s, 1)/\partial s|_0$ is orthogonal to $\partial F(0, t)/\partial t|_1$.

Just as in the proof of (1.9.1) we see that $\langle \frac{\partial F}{\partial s}, \frac{\partial F}{\partial t} \rangle (s, t)$ is independent of t.

Hence,

$$0 = \langle b'(0), X_0 \rangle = \langle \frac{\partial F}{\partial s}, \frac{\partial F}{\partial t} \rangle |_{0,0} = \langle \frac{\partial F}{\partial s}, \frac{\partial F}{\partial t} \rangle |_{0,1}. \quad \square$$

In (1.12.10) we had characterized a conjugate point of multiplicity $k > 0$ by the vanishing of a non-zero Jacobi field $Y(t) \perp \dot{c}(t)$ with initial condition $Y(0) = 0$. We can generalize this, provided we replace the initial condition $Y(0) = 0$ by one adapted to the immersion $F: M \to M^*$, cf. also Bishop-Crittenden [1].

1.12.18 Lemma. *Let* $c^{\perp}(t) = \exp_{F,p} \tilde{c}^{\perp}(t)$ *be a normal geodesic of the isometric immersion* $F: M \to M^*, p = c^{\perp}(0)$. *Put* $\dot{c}^{\perp}(0) = X_0 \in T_p F$.

Denote by $\mathcal{J}_{c^{\perp}}^{F\perp}$ *the vector space of those Jacobi fields* $Y(t)$ *along* $c^{\perp}(t)$ *which satisfy* $\langle Y(t), \dot{c}^{\perp}(t) \rangle = 0$ *and*

$$(*) \qquad Y(0) \in T_p M; \ \nabla^* Y(0) + H_{X_0} . Y(0) \in T_p^{\perp} F.$$

Then $c^{\perp}(t_1)$ *is a focal point of* $p = c^{\perp}(0)$ *along* $c^{\perp}|[0, t_1]$ *of multiplicity* > 0 *if and only if there exists a non-zero* $Y \in \mathcal{J}_{c^{\perp}}^{F\perp}$ *with* $Y(t_1) = 0$.

Note. For simplicity, we have identified $T_p M$ with $T_p F(T_p M) \subset T_{F(p)} M^*$. $H_{X_0}: T_p M \to T_p M$ is the operator defined by the second fundamental form h_{X_0} of F for the normal vector X_0, cf. (1.10.7).

Using (1.10.7) one sees that $\mathcal{J}_{c^{\perp}}^{F\perp}$ is a Lagrangian subspace of $\mathcal{J}_{c^{\perp}}^{\perp}$ in the sense of (3.1.1), i. e., the symplectic form α vanishes identically $\mathcal{J}_{c^{\perp}}^{F\perp}$.

Proof. kernel $T_{\tilde{c}^{\perp}(t_1)} \exp_{F,p}$ is orthogonal to $\tilde{c}^{\perp}(t_1)$, cf. (1.12.17).

Let $t_1 > 0$. To describe the image under $T_{\tilde{c}^{\perp}(t_1)} \exp_{F,p}$ of a tangent vector $A \in T_{\tilde{c}^{\perp}(t_1)} T^{\perp} F$, orthogonal to the radial vector $\dot{\tilde{c}}^{\perp}(t_1)$, we consider a curve $X(s)$ in $T^{\perp}F$ with $X(0) = \tilde{c}^{\perp}(t_1) = t_1 \dot{c}^{\perp}(0)$, $X'(0) = A$, $|X(s)| = \text{const.}$ Put $v_F X(s) = b(s)$. Then $Y(t) = \partial \exp_{F,b(s)}(tX(s)/t_1)/\partial s|_0$ is a Jacobi field $Y(t)$ along $\exp_{F,p}(tX(0)/t_1) = c^{\perp}(t)$ with $Y(t_1) = T_{\tilde{c}^{\perp}(t_1)} \exp_{F,p} . A$, cf. (1.12.4).

We only have to check that $Y \in \mathcal{J}_{c^{\perp}}^{F\perp}$. From (1.12.4) we know $Y(0) = b'(0) \in T_p M$, $\nabla^* Y(0) = \nabla^* X(s)/ds|_0$. From the Weingarten formula (1.10.7) (with $X(s)$ a normal vector field which is differentiated in the direction $b'(0)$

$= Y(0))$ we get

$$\nabla^* Y(0) + H_{X_0} \cdot Y(0) \in T_p^\perp F.$$

$\langle Y(t), \dot{c}^\perp(t) \rangle = 0$ follows from $\langle Y(0), \dot{c}^\perp(0) \rangle = 0$ and $\langle X(s), X(s) \rangle = \text{const}$, i. e., $\langle \nabla^* Y(0), \dot{c}^\perp(0) \rangle = 0$. □

1.12.19 Corollary. *That $c^\perp(t_1)$ is a focal point of multiplicity > 0 of the immersion $F: M \to M^*$ along $c^\perp|[0, t_1]$ depends only on the second fundamental form of F with respect to the normal vector $\dot{c}^\perp(0)$ at $p \in M$. In particular, if F is totally geodesic at $p \in M$ and has codimension 1, that $c^\perp(t_1)$ is a focal point of multiplicity > 0 depends only on $c = c^\perp$. It means that there exists a non-zero Jacobi field $Y(t)$ along $c(t)$, with $\langle Y(0), \dot{c}(0) \rangle = 0$, $\nabla^* Y(0) = 0$, which vanishes for $t = t_1 > 0$.*

Proof. Immediate from (1.12.18). In particular F being totally geodesic in p means that the second fundamental form $h_{\dot{c}(0)}$ vanishes identically. If, moreover, F has codimension 1 then the condition $(*)$ in (1.12.18) reduces to

$$\langle Y(0), \dot{c}(0) \rangle = 0; \ \nabla^* Y(0) \text{ proportional to } \dot{c}(0).$$

Since also $\langle \nabla^* Y(0), \dot{c}(0) \rangle = 0$ we get $\nabla^* Y(0) = 0$. □

We are thus lead to the

1.12.20 Definition. Let $c: [0, a] \to M$ be a non-constant geodesic on M. $c(t_1)$ is called *focal point of* $p = c(0)$ (along $c|[0, t_1]$) *of multiplicity* $k > 0$ if the space of non-zero Jacobi fields Y along c with

$$\langle Y(0), \dot{c}(0) \rangle = 0, \ \nabla^* Y(0) = 0, \ Y(t_1) = 0.$$

has dimension k.

Note. In (1.12.13) we have characterized – at least for $\dim M < \infty$ – the first conjugate point on a geodesic as the point where the geodesic ceases to be minimizing between its end points when compared with nearby curves. Similarly, it is possible to characterize the first non-trivial focal point on a geodesic c starting at $F(p)$ as the point after which there exist shorter curves from $F(p'), p'$ near p, to the end point of $c^\perp$. See (2.5.15) for details.

1.12.21 Example. Let M^* be a Hilbert space with the canonical Riemannian metric and $F: M \to M^*$ a submanifold. Then the Jacobi fields $Y \in \mathcal{J}_{c^\perp}^{F\perp}$ are of the form $Y(t) = A + Bt$, with $A \perp X_0, B \perp X_0, A \in T_p M, B + H_{X_0} \cdot A \in T_p^\perp M$.

Thus, $Y(t_0) = 0$ for $t_0 > 0$ is equivalent to $B = -A/t_0 \in T_p M$, i. e., $B = -H_{X_0} \cdot A = -A/t_0$. The focal points t_0 on $c^\perp$, $\dot{c}^\perp(0) = X_0$, therefore correspond to the inverse of the negative eigenvalues of the operator H_{X_0}. The value t_0 possesses the following geometric interpretation: Let b be a curve on M which near $p = c(0)$ coincides with the intersection of M with the 2-plane through $X_0 \in T_p^\perp M$ and an eigenvector to the eigenvalue t_0. Then b has in p the curvature radius t_0.

Chapter 2: Curvature and Topology

In this chapter we restrict ourselves to finite dimensional Riemannian manifolds. The concept of completeness and the cut locus are our first topics in the study of the topology of such manifolds M. There follows a rather detailed account of the symmetric spaces from the point of view of Riemannian geometry. After this, in sections (2.3) to (2.5) we develop the theory of the Hilbert manifold of paths on M with its various important submanifolds. The Lusternik-Schnirelmann Theory is presented in great detail and it finds its first applications in existence proofs for closed geodesics as well as in the proof of a fundamental estimate on the injectivity radius. There follows in (2.7) the proof of the Comparison Theorem of Alexandrov-Topogonov. Section (2.8) is dedicated to the Sphere Theorem and its complements. We conclude with an exposition of the structure theory of noncompact manifolds of curvature $\geqslant 0$.

2.1 Completeness and Cut Locus

We restrict ourselves to finite dimensional manifolds. From (1.9.4), (1.9.5) we know that a Riemannian manifold M carries a natural distance d. We call M (metrically) complete if it is complete as a metric space. M is called geodesically complete if every geodesic can be extended to all $t \in \mathbb{R}$. The fundamental theorem of Hopf und Rinow (2.1.3) states that these two concepts of completeness are equivalent.

We now assume M to be complete and connected. Then there always exists a minimizing geodesic between two points on M, i.e., the length of this geodesic is equal to the distance between these points, cf. (2.1.3).

Fix now $p \in M$. For $q \in M$, there possibly exists more than one minimizing geodesic between p and q. The cut locus $C(p)$ of p is the complement on M of the maximal open set formed by those q where the minimizing geodesic to q is unique; this is the characterization of $C(p)$ given in (2.1.14). The definition of $C(p)$ in (2.1.4) was different. In (2.1.6) till (2.1.11) we derive various properties of $C(p)$ and of the distance from p to $C(p)$. This latter is called the injectivity radius $\iota(p)$ of p; it is the supremum of those $\varrho \in \mathbb{R}^+$ where $\exp_p | B_\varrho(0_p)$ is a diffeomorphism onto its image. In a preparatory step we give an important characterization of a conjugate point, see (2.1.12), (2.1.13).

The infimum $\iota(p), p \in M$, is called the injectivity radius $\iota(M)$ of M. $\iota(M) > 0$ if M is compact, (2.1.10). If p is such that $\iota(p) = \iota(M) < \infty$ and if there is no conjugate point on a minimizing geodesic from p to $q \in C(p)$ with $d(p, q) = d(p, C(p))$ then there passes a unique closed geodesic of length $2d(p, q)$ through p and q, (2.1.11). This will have important consequences later on.

For a survey see the article by Kobayashi [3].

In the appendix we study the concept of orientation for vector bundles over a manifold.

2.1.1 Definition. Let M be a Riemannian manifold.

(i) M is called *(metrically) complete* if it is complete as a metric space, where the distance $d(\, , \,)$ is derived from the Riemannian structure, cf. (1.9.4).

(ii) Let $p \in M$. M is called *geodesically complete at* p if $\exp_p$ is defined on all of $T_p M$. If this is true for all $p \in M$ then M is called *geodesically complete.*

Remarks. 1. The geodesic completeness of M simply means, that, for every vector $X \in TM$, the geodesic $c_X(t)$ with initial vector $\dot{c}_X(0) = X$ is defined for all $t \in \mathbb{R}$.

2. If the underlying topological manifold of M is compact, M is complete. For non-compact M, however, completeness depends on the choice of the metric. Take e.g. the open unit ball $B_1^n(0)$ in $\mathbb{R}^n$. $B_1^n(0)$, endowed with the Euclidean metric of $\mathbb{R}^n$, is not complete. However, if we take on $B_1^n(0)$ the metric which yields the hyperbolic space H_1^n, cf. (1.11.7), then we obtain a complete manifold.

3. Actually, one can always find on every non-compact differentiable manifold a Riemannian metric which yields a complete Riemannian manifold. According to Whitney, any differentiable manifold can be embedded as a closed submanifold in some Euclidean space. Take now the induced Riemannian metric. Cf. Whitney [1].

For the existence of complete Riemannian metrics with special properties cf. Nomizu and Ozeki [1], Greene [1].

4. By removing a point or a larger closed subset from a complete Riemannian manifold we can construct examples where not always two points can be joined by a minimizing geodesic. Take e.g. $\mathbb{R}^2 - \{x \leqslant 0; y \leqslant 0\}$. Such a manifold is not complete, cf. (2.1.3). Cf. also Note 3 after (2.1.3).

2.1.2 Lemma. *Let* $p \in M$ *and* $\varrho > 0$ *be such that* $\exp_p$ *is defined on the ball* $B_\varrho(0_p) \subset T_p M$. *Then every* $q \in M$ *with* $d(p, q) < \varrho$ *can be joined to* p *by a minimizing geodesic* c_{pq}.

Proof. Let $\varepsilon > 0$ be such that $B_{2\varepsilon}(p)$ is the domain of normal coordinates, cf. (1.8.16). From (1.8.15) we have our claim for $\varrho \leqslant \varepsilon$. We therefore assume $\varepsilon < d(p, q) < \varrho$. Since $S_\varepsilon(p) = \exp(\partial \bar{B}_\varepsilon(0_p))$ is compact, there exists an $r \in S_\varepsilon(p)$ satisfying

$$d(p, q) = d(p, r) + d(r, q) = \varepsilon + d(r, q).$$

Also, there exists a well determined geodesic $c(t), |\dot{c}| = 1$ with $c(0) = p, c(\varepsilon) = r$.

The set T of $t \in [0, d(p, q)]$ satisfying

$$(*) \qquad d(p, c(t)) + d(c(t), q) = d(p, q)$$

is closed and contains ε. Let t' be the supremum of T. In particular, $\varepsilon \leqslant t' \leqslant d(p, q)$. If $t' = d(p, q)$ we have with $c|[0, d(p, q)]$ a minimizing geodesic c_{pq} from p to q. We will derive a contradiction from $t' < d(p, q)$.

Put $c(t') = p'$. Choose ε' with $0 < \varepsilon' < d(p, q) - t'$ and such that $B_{2\varepsilon'}$ is the domain of normal coordinates around p'. Just as before there exists a $r' \in S_{\varepsilon'}(p')$ with

$$d(p', q) = d(p', r') + d(r', q) = \varepsilon' + d(r', q).$$

Let c' be the geodesic determined by $|\dot{c}'| = 1$, $c'(0) = p'$, $c'(\varepsilon') = r'$. From

$$(\dagger) \qquad d(p, p') + d(p', r') + d(r', q) = d(p, q)$$

and the triangle inequality for p, r', q we get

$$d(p, p') + d(p', r') = d(p, r').$$

This means that the curve from p to r', formed by the geodesic $c|[0, t']$ followed by the geodesic $c'|[0, \varepsilon']$ has length $d(p, r')$ and hence is minimizing. The local geometry of the normal ball $B_{\varepsilon'}(p')$ shows that these curves cannot have a corner at p', i.e., $c'(\varepsilon') = c(t' + \varepsilon') = r'$. Thus, $(\dagger)$ implies that $(*)$ does hold also for $t' + \varepsilon' > t'$. □

We now can prove the fundamental Theorem of Hopf und Rinow [1]. The basic ideas of the proof go back to Hilbert [1]. Our proof follows de Rham [1].

2.1.3 Theorem. *Let M be a Riemannian manifold. Then the following properties are equivalent.*

(i) *M is metrically complete.*
(ii) *M is geodesically complete at one $p \in M$.*
(iii) *M is geodesically complete.*

Any of these properties implies
(iv) *For any pair (p, q) of points there exists a minimizing geodesic c_{pq} from p to q.*

Notes. 1. Property (iv) can hold also if M is not complete. Take e.g. the interior of a convex domain in $\mathbb{R}^n$ or, more generally, a strongly convex ball in the sense of (1.9.9).

2. The minimizing geodesic c_{pq} from p to q need not be unique. Take for instance on the sphere for p and q a pair of antipodal points. For more on this phenomenon see (2.1.4)ff.

3. Since there exist non-complete Riemannian manifolds which cannot be extended to a complete Riemannian manifold (e.g., the universal covering of $\mathbb{R}^2 - \{0\}$, cf. Hopf und Rinow [1]), there arises the question for the possible obstructions to extend a non-complete Riemannian manifold to a complete one. Besides Hopf und Rinow cf. also Dubois [1], Alexander and Bishop [1].

4. In case M has infinite dimension, (2.1.3) (iv) needs not to be true for a geodesically complete manifold, i.e., for a manifold where exp is defined on all of TM.

As an example consider on the proper Hilbert space $\mathbb{E}$ with coordinates $(x_0, x_1, x_2, \dots)$ the differentiable function $F(x) = x_0^2 + \sum_1^\infty x_i^2/a_i$. Here $a_1 > a_2 > \dots > 1$ and $\lim a_i = 1$. 1 is a regular value of F and thus, $M = \{x_0^2 + \sum_1^\infty x_i^2/a_i = 1\}$ is a submanifold of codimension 1 of $\mathbb{E}$ – an ellipsoid. M is complete with the induced Riemannian metric. The points $p = (1, 0, 0, \dots)$ and $q = (-1, 0, 0, \dots)$ have distance π, since the length of the half ellipse from p to q through the point $x_i = \sqrt{a_i}$ goes to π, as i goes to ∞. But there is no curve of length π on M from p to q.

Proof. (i) implies (iii). To see this consider the geodesic $c(t) = c_X(t)$ determined by $\dot{c}(0) = X$. We can assume $X \neq 0$ and, since $c_{aX}(t) = c_X(at)$, even $|X| = 1$.

Let $J = J(X)$ be the maximal domain of definition of $c(t)$, cf. (1.6.5). It suffices to derive a contradiction from the assumption that the upper bound t_+ of J in $\bar{\mathbb{R}} = \mathbb{R} \cup \{\pm\infty\}$ is finite. In that case let $\{t_i\}$ be a Cauchy sequence in J with t_+ as limit. Since for $t \leqslant t'$, t, t' in J,

$$d(c(t), c(t')) \leqslant L(c|[t, t']) = |t - t'|,$$

$\{c(t_i)\}$ is a Cauchy sequence on M. Let $q \in M$ be its limit. If $B_\varepsilon(q)$ denotes a domain of normal coordinates around q we choose t' with $0 < t_+ - t' < \varepsilon/2$. Let $c'(t)$ be the geodesic with initial value $\dot{c}(t')$, i.e., $c'(t) = c(t' + t)$. Now, c' is defined for $t = \varepsilon/2$, i.e., c is defined for $t' + \varepsilon/2 > t_+$ – a contradiction.

From (2.1.2) we know that (ii) implies (iv). Clearly, (iii) implies (ii). So it only remains to show that (ii) implies (i).

To see this we consider a Cauchy sequence $\{q_n\}$ on M. Let $c_n = c_{pq_n}$ be a minimizing geodesic from p to q_n, $|\dot{c}_n| = 1$. That is, if we put $d(p, q_n) = L(c_n) = L_n$, $\{L_n\}$ is a Cauchy sequence. Let L be its limit. $\{\dot{c}_n(0)\}$ is on the unit sphere of T_pM, hence, by going to a subsequence if necessary, we can assume that $\{\dot{c}_n(0)\}$ has a limit, say X. Let c be the geodesic determined by $\dot{c}(0) = X$. The continuous dependence of the value $c^*(L^*)$ of a geodesic c^*, $|\dot{c}^*| = 1$, on its initial value $\dot{c}^*(0)$ and L^* implies that $\{q_n = c_n(L_n)\}$ has $q = c(L)$ as limit. □

From now on we will always assume that our manifolds are complete.

In general, the differentiable map $\exp_p\colon T_pM \to M$ will not be injective. This is true in particular if M is compact. More precisely, one can show the following: Whenever M possesses a nontrivial homotopy group $\pi_i M$ for some $i \geqslant 1$ then there exist infinitely many geodesics joining p with a given q. I.e., $(\exp|T_pM)^{-1}(q)$ is infinite. See Serre [1].

On the other hand, we know that $\exp_p$, restricted to a sufficiently small ball $B_\varrho(0_p)$ around the origin 0_p of T_pM, is injective. We will now describe a maximal open domain in T_pM, which contains 0_p, which is star shaped with respect to 0_p and on which $\exp_p$ is injective. In this way we will find a description of M as a cell in T_pM

which in general has a non-empty boundary. M then can be obtained from this cell by certain identifications of its boundary.

2.1.4 Definition. (i) For a Riemannian manifold M we denote by $T_1 M$ the *total space of the unit tangent bundle,* i.e.,

$$T_1 M = \{X \in TM; |X| = 1\}.$$

(ii) Let $X \in T_1 M, \tau X = p$. Then $\sigma(X)$ shall be the supremum in $\bar{\mathbb{R}}^+ = \mathbb{R} \cup \{+\infty\}$ of the t with $d(\tau X, \exp_p tX) = t$. For $p \in M$ we denote by $\tilde{C}(p)$ the set of elements $\sigma(X) X \in T_p M$ with $X \in T_p M \cap T_1 M = T_{p1} M$ for which $\sigma(X)$ is finite. $\tilde{C}(p)$ is called *tangential cut locus of* p. Its image $C(p)$ under $\exp_p$ is called *cut locus of* p.

Note. For convex surfaces the cut locus has been introduced by Poincaré [2] under the name 'ligne de partage'. Every compact surface admits a Riemannian metric of constant curvature. For such a metric, the cut locus $C(p)$ consists of finitely many smooth arcs, if we exclude the case of the sphere S^2 where $C(p)$ consists of a single point. 'Cutting up' the surface M along $C(p)$ yields a particularly simple representation of the surface as a domain bounded by a polygon on which the edges are pairwise identified, cf. Wolf [1].

We will see that a similar fact also holds in general.

2.1.5 Lemma. *The mapping*

$$\sigma: T_1 M \to \bar{\mathbb{R}}^+; X \mapsto \sigma(X)$$

is continuous.

Proof. Let $\{X_n\}$ be a sequence in $T_1 M$ with limit X. We can assume that the sequence $\{\sigma_n = \sigma(X_n)\}$ possesses a limit in the compact space $\bar{\mathbb{R}}^+$, say σ^*. We want to derive a contradiction from $\sigma^* \neq \sigma(X) = \sigma$.

Consider first the case $\sigma^* < \sigma$. With $\tau X_n = p_n$, $\tau X = p = \lim p_n$ we choose $\eta > 0$ such that $\sigma^* + \eta < \sigma$. For large n, the geodesic $c_{X_n}|[0, \sigma^* + \eta]$ from p_n to q_n will not be a minimizing curve between its end points. Hence, minimizing geodesics from p_n to q_n will have initial directions $\tilde{X}_n \neq X_n$, $|\tilde{X}_n| = 1$. We can assume that $\lim \tilde{X}_n = \tilde{X}$ and $\lim q_n = q$ exist. We claim $\tilde{X} \neq X$.

Indeed, since $c_X|[0, \sigma]$ contains no conjugate point in its interior, the mapping

$$\tau_M \times \exp: TM \to M \times M,$$

considered already in the proof of (1.6.12), is of maximal rank on the segment in TM

$$[0, \sigma^* + \eta] X = \{tX; 0 \leqslant t \leqslant \sigma^* + \eta < \sigma\}.$$

Moreover, it is injective. The same therefore is true for an appropriate neighhborhood of $[0, \sigma^* + \eta] X$ in TM. If now $\lim \tilde{X}_n = X$, the sets $[0, \sigma^* + \eta] \tilde{X}_n \subset TM$ belong to this neighborhood, for large n. But the elements $d(p_n, q_n) \tilde{X}_n$ and $(\sigma^* + \eta) X_n$ have the same image (p_n, q_n) under $\tau_M \times \exp$, a contradiction.

Thus, $\tilde{X} \neq X$, hence, $c_{\tilde{X}}|[0, d(p,q)]$ is a minimizing geodesic from p to $q = c_X(\sigma^* + \eta)$, different from $c_X|[0, \sigma^* + \eta]$, a contradiction to the definition of $\sigma = \sigma(X) > \sigma^* + \eta$.

Assume now $\sigma < \sigma^*$. Choose $\eta > 0$ with $\sigma + \eta < \sigma^*$. Then, for large n, $c_{X_n}|[0, \sigma + \eta]$ will be minimizing between its end points p_n and q_n. From $\lim p_n = p$, $\lim q_n = q$ and $d(p_n, q_n) = \sigma + \eta$ follows $d(p,q) = \sigma + \eta = L(c|[0, c + \eta])$. That is, $c|[0, \sigma + \eta]$ is a minimizing geodesic from p to q, in contradition to the definition of σ. □

2.1.6 Examples. 1. For the sphere S^n_ϱ is $\tilde{C}(p) = \partial B_{\pi\varrho}(0_p)$, $C(p)$ is the antipodal point of p.

2. Conversely, if $C(p)$ consists of a single point q, M is homeomorphic to the sphere. Indeed, whenever $\sigma(X)$ for $X \in T_{p1} M$ is finite, its value is $d(p,q) = \pi r_0$. The continuity of σ implies that $\sigma(X) = \pi r_0$ for all $X \in T_{p1} M$. Hence, if M_0 is the sphere of radius r_0 and $p_0 \in M_0$, q_0 its antipodal point, the mapping

$$\phi \equiv \exp_{p_0} \circ (\exp_p | B_{\pi r_0}(0_p))^{-1} : M - \{q\} \to M_0 - \{q_0\}$$

possesses a continuous extension by putting $\phi(q) = q_0$.

Note that M need not be isometric to M_0; take e.g., for M any surface of revolution of the type of S^2 and p on the axis of revolution.

3. Let $P^n_\varrho = S^n_\varrho / \mathbb{Z}_2$ be the *real projective space*, obtained from the sphere S^n_ϱ by taking the quotient of the isometric $\mathbb{Z}_2$-action $p \mapsto$ antipodal point of p. If $p \in P^n_\varrho$ is represented by $(\pm\varrho, 0) \in S^n_\varrho \subset \mathbb{R} \times \mathbb{R}^n$ then $C(p)$ is $S^{*n-1}_\varrho / \mathbb{Z}_2$ where S^{*n-1}_ϱ is the sphere of radius ϱ in the subspace $\{0\} \times \mathbb{R}^n$ of $\mathbb{R} \times \mathbb{R}^n$.

4. Let M be a circular cylinder in $\mathbb{R}^3$. Then $C(p)$ is the straight line on M, opposite to p.

5. Let $M = \{x^2 + y^2 = 2z\}$ be a paraboloid of revolution in $\mathbb{R}^3$. For $p = p_0 = (0,0,0) =$ vertex of M, $C(p)$ is empty. If we let move p upwards along one branch of a generating parabola, $C(p)$ will form the upper part of the other branch of that parabola. As p moves on, the lower part of $C(p)$ moves down until it comes arbitrarily close to the vertex p_0. For details see (2.9.2) and (2.9.12).

6. Construct on $S^2 = \{x^2 + y^2 + z^2 = 1\} \subset \mathbb{R}^3$ a new metric g as follows: On the equator $z = 0$ take the sequence $\{p_n = (\cos 2\pi/(n+1), \sin 2\pi/(n+1), 0\}$, with $p_0 = (1,0,0)$ as limit. For each n attach to S^2 a small tangential cone with vertex p_n and a base of radius $< 2\pi/(n+1)(n+2)$. Round off the vertices. The cut locus $C(p)$ of the south pole $p = (0,0,-1)$ then will consist of a sequence $\{c_n\}$ of geodesic segments, emanating from the north pole and pointing towards p_n, getting shorter and shorter as n approaches infinity with limit the north pole.

2.1.7 Proposition. *Let $c(t)$ be a geodesic starting at $p = c(0)$ with $\dot{c}(0) = X$, $|X| = 1$. Then $c|[0, \sigma(X)[$ contains no conjugate points.*

Note. We say briefly that the cut point on a geodesic c appears before, or coincides with the first conjugate point.

Proof. This is an immediate consequence of (1.12.13, ii). □

2.1.8 Theorem. *Let M be a complete Riemannian manifold of dimension n. Choose a $p \in M$.*

(i) *The complement $M - C(p)$ of the cut locus $C(p)$ of M can be retracted homeomorphically onto a ball around p.*

(ii) *Let M be compact. Then $M - \{p\}$ possesses the cut locus $C(p)$ as strong deformation retract. Hence, the inclusion $i\colon C(p) \to M$ induces an isomorphism i_* in homotopy and homology up to dimension $n-2$. In dimension $n-1$, it still is an isomorphism in homology whereas in homotopy i_* will in general only be surjective.*

Note. For homology and homotopy we refer to Spanier [1]. That $M - C(p)$ is actually diffeomorphic to a ball around p – at least for M compact – is proved in Ozols [1]. In the case of dimension $\neq 4$, any differentiable manifold which is homeomorphic to the Euclidean space, is actually diffeomorphic to the Euclidean space, cf. Stallings [1].

Proof. Let $B_\varrho(p)$ be a domain of normal coordinates based at p. For $q \in M - C(p)$, $q \neq p$, define $d(q) > 0$ and $X(q) \in T_pM$ by $q = \exp_p d(q)\, X(q)$; $|X(q)| = 1$. I.e., $d(q) = d(p, q)$. With this we define a family of homeomorphisms

$$H\colon [0,1] \times (M - C(p)) \to M - C(p)$$

by

$$H(t, p) = p \quad \text{and, for } q \neq p,$$

$$H(t,q) = \exp_p \left\{ \left[(1-t)\, d(q) + t\varrho \frac{d(q)^2}{d(q)^2 + \varrho^2} \cdot \frac{\sigma(X(q))^2 + \varrho^2}{\sigma(X(q))^2} \right] X(q) \right\}.$$

Thus, $H(1, \,)$ is a homeomorphism of $M - C(p)$ onto $B_\varrho(p)$.

Now let M be compact. We define a retraction

$$K\colon [0,1] \times (M - \{p\}) \to M - \{p\}$$

of $M - \{p\}$ onto $C(p)$ by $K(t,q) = q$, for $q \in C(p)$ and

$$K(t,q) = \exp_p \{ [(1-t)\, d(q) + t\sigma(X(q))]\, X(q) \}$$

for $q \notin C(p)$. That is to say, we push $q \notin C(p)$ along the half geodesic emanating from p until we hit $C(p)$.

To prove the statements about homotopy and homology groups we look at the exact sequences of the pair $(M, M - \{p\}) \sim (M, C(p))$: Since $(M, M - \{p\})$ is homotopically equivalent to $(S^n, *)$ we can write these sequences in the form

$$\to \pi_{k+1} M \xrightarrow{j_*} \pi_{k+1} S^n \xrightarrow{\partial} \pi_k C(p) \xrightarrow{i_*} \pi_k M \xrightarrow{j_*} \pi_k S^n \to,$$

$$\to H_{k+1} M \xrightarrow{j_*} H_{k+1} S^n \xrightarrow{\partial} H_k C(p) \xrightarrow{i_*} H_k M \xrightarrow{j_*} H_k S^n \to.$$

Thus, for $1 \leqslant k \leqslant n-2$, i_* determines isomorphisms whereas for $k = n-1$ we get surjective morphisms. In homology, $j_*|H_n M$ is injective. Hence

$$i_*: H_{n-1} C(p) \to H_{n-1} M$$

is an isomorphism. □

Remark. Instead of referring to the exact sequence for homotopy, one gets the properties of i_* also directly by observing that a mapping $f: S^k \to M$ always is homotopic to one having its image in $M - \{p\}$, provided $k \leqslant n-1$. And a homotopy between two such maps can also be realized in $M - \{p\}$, provided $k \leqslant n-2$. Similar arguments apply to homology.

2.1.9 Definition. Let $p \in M$. The *injectivity radius* $\iota(p)$ is the supremum in $\bar{\mathbb{R}}^+$ of the numbers ϱ such that $\exp_p|B_\varrho(0_p)$ is injective. The *injectivity radius* $\iota(M)$ *of* M is defined as the infimum of $\iota(p), p \in M$.

2.1.10 Proposition. *The injectivity radius $\iota(p)$ depends continuously on p. If $K \subset M$ is a compact set then $\iota(K) = \inf\{\iota(p); p \in K\}$ is positive. In particular, if M is compact, the injectivity radius $\iota(M)$ of M is positive.*

Remark. If M is non-compact, $\iota(p), p \in M$, need not have a positive lower bound. Take for instance a surface of revolution for which the generating curve is asymptotic to one end of the z-axis, like $z = 1/x, x > 0$.

Proof. Let $\{p_n\}$ be a sequence with limit p. Choose $X_n \in T_{p_n 1} M = T_{p_n} M \cap T_1 M$ with $\sigma(X_n) = \iota(p_n)$. We can assume $\lim X_n = X \in T_{p1} M$. Then (2.1.5) yields $\iota(p) \leqslant \sigma(X) = \lim \sigma(X_n)$. If $\iota(p) < \sigma(X)$ we choose $X^* \in T_{p1} M$ with $\sigma(X^*) = \iota(p)$. There will be a sequence $\{X_n^*\}$ with $\tau X_n^* = p_n$ such that $\lim X_n^* = X^*$. Since $\lim \sigma(X_n^*) = \sigma(X^*)$ it follows $\sigma(X_n^*) < \sigma(X_n)$, for large n — a contradiction. □

We continue with three geometric properties of the cut locus.

2.1.11 Lemma. (i) *If $q \in C(p)$ then $p \in C(q)$. For M compact, actually*

$$p = \bigcap_{q \in C(p)} C(q)$$

(ii) *Let $q \in C(p)$. If among the minimizing geodesics from p to q there is one, say c, such that q is not conjugate to p along c, then there exists a second minimizing geodesic c' from p to q.*

(iii) *Let $q \in C(p)$ have minimal distance from p, $d(p, C(p)) = d(p, q)$. Assume that none of the minimizing geodesics from p to q possesses q as conjugate point. Then there exist exactly two minimizing geodesics $c: [0, 1] \to M$ and $c': [0, 1] \to M$ from p to q and they meet at q with opposite directions, $\dot{c}(1) = -\dot{c}'(1)$. If, moreover, $d(p, q) = d(p, C(p))$ is minimal for all $p \in M$, i.e., $d(p, q) = \iota(M)$, c and c' together actually form a closed geodesic, i.e., $\dot{c}(0) = -\dot{c}'(0)$.*

Proof. Put $d(p, q) = a$. Let $c(t), t \in \mathbb{R}$, be a minimizing geodesic from p to q, $|\dot{c}|$

$=1$, $c(0)=p$, $c(a)=q$. Then $\sigma(-\dot c(a+\varepsilon))<a+\varepsilon$ for $\varepsilon>0$. In the limit, for ε going to zero, we get $\sigma(-\dot c(a))\leqslant a$. If $\sigma(-\dot c(a))<a$, $c|[0,a]$ would not be minimizing. Thus, $p\in C(q)$.

Assume now M compact. Then, for every $X\in T_{p1}M$, $\sigma(X)<\infty$. If $p'\in C(q)$ for all $q\in C(p)$ we know $p'\notin C(p)$. Actually, $p'=p$ for such a p'. Indeed, consider a minimizing geodesic $c(t)$, $|\dot c|=1$, from $p=c(0)$ to $p'=c(d(p,p'))$. Then $\sigma(\dot c(0))>d(p,p')$, i.e., c meets $C(p)$ only at a point q after p'. But then $p'\in C(q)$ can hold only for $p'=p$.

To prove (ii) let c be a minimizing geodesic from p to q, $d(p,q)=a$, q not conjugate to p along c. For a sequence $\{\varepsilon_n\}$ of positive numbers with $\lim\varepsilon_n=0$ put $c(a+\varepsilon_n)=q_n$. A minimizing geodesic c_n from p to q_n will be different from $c|[0,a+\varepsilon_n]$. We can assume that the sequence $\{c_n\}$ has a limit geodesic c' of length $d(p,q)=a$ from p to q. Since, for some small $\eta>0$, $c|[0,a+\eta]$, contains no conjugate points the arguments used in the proof of (2.1.6) show $c'\neq c$.

For the proof of (iii) we observe that under our hypothesis there exists neighborhoods U and U' of $\dot c(0)$ and $\dot c'(0)$ in T_pM such that $\exp_p|U$ and $\exp_p|U'$ are diffeomorphisms onto neighborhoods V and V' of q on M.

On V and V' we consider the distance functions from p,

$$d(r)=|(\exp_p|U)^{-1}(r)| \quad\text{and}\quad d'(r')=|(\exp_p|U')^{-1}(r')|.$$

Obviously,

$$T_q d\,.\,X=\left\langle\frac{\dot c(1)}{|\dot c(1)|},X\right\rangle;\; T_q d'\,.\,X=\left\langle\frac{\dot c'(1)}{|\dot c'(1)|},X\right\rangle,$$

cf. the Gauss Lemma (1.9.1).

If now $\dot c(1)\neq-\dot c'(1)$, we can pick an $X\in T_qM$ such that $T_qd\,.\,X=T_qd'\,.\,X<0$. Thus, for small $\delta>0$, the point $q^*=\exp_q\delta X$ will belong to $V\cap V'$. Put $(\exp_p|U)^{-1}(q^*)=\tilde q^*$, $(\exp_p|U')^{-1}(q^*)=\tilde q'^*$. Then $\tilde q^*\neq\tilde q'^*$ and we have two geodesics

$$c^*(t)=\exp_p(t\tilde q^*);\; c'^*(t)=\exp_p(t\tilde q'^*),\quad 0\leqslant t\leqslant 1,$$

from p to q^*, of length $d(q^*)$ and $d'(q^*)$, respectively. But

$$d(q^*)=d(q)+T_qd\,.\,\delta X+\ldots<d(q)=d(p,q)$$

$$d'(q^*)=d'(q)+T_qd'\,.\,\delta X+\ldots<d'(q)=d(p,q).$$

That is, there are points on $C(p)$ having distance $<d(p,q)=d(p,C(p))$ from p – a contradiction.

The last statement now simply follows by exchanging the rôles of p and q. □

Before giving a final characterization of $C(p)$, we insert a result on conjugate points. Recall from (1.12.10) and the subsequent remark that a conjugate point $c(t_1)$ of $c(0)$ along the geodesic $c|[0,t_1]$ is characterized by the fact that there exists a non-zero Jacobi field $Y(t)$ along $c(t)$ with $Y(0) = Y(t_1) = 0$. Thus, there exists a geodesic $c' \neq c$, starting from $c(0)$, which meets c again infinitesimally at t_1.

The next result shows that this is true not only infinitesimally.

2.1.12 Theorem. *Let $c(t)$, $0 \leqslant t \leqslant a$, be a geodesic on M from $p = c(0)$ to $q = c(a)$. Assume that t_1, $0 < t_1 < a$, is a conjugate point. Put $t\dot{c}(0) = \tilde{c}(t)$, $\tilde{c}(t_1) = \tilde{r}$, $c(t_1) = \exp_p \tilde{c}(t_1) = r$. Then $\exp_p$ is not locally bijective at $\tilde{r}$.*

That is, in every neighborhood of r there exists a point r' where two different geodesics c', c'' starting from p meet again. c', c'' can be chosen to belong to any prescribed neighborhood of c.

Remarks. 1. We thus see that $\ker T_r \exp_p \neq 0$ implies that $\exp_p$ is not 1:1 near $\tilde{r}$. For a general differentiable mapping ϕ between manifolds of equal dimension this need not be true. Take e.g. the bijective mapping $\phi\colon t \in \mathbb{R} \mapsto t^3 \in \mathbb{R}$, where $D\phi(0) = 0$.

2. In case $\dim M = 2$, (2.1.12) can be sharpened, cf. (2.1.13).

Proof. Assume that there were an open neighborhood $\tilde{U}$ of $\tilde{r}$ in $T_p M$ such that $\exp_p | \tilde{U}$ is injective and hence – due to the invariance of domain theorem – a bijective mapping onto a neighborhood U of r.

To derive a contradiction from this assumption we first observe that all $t \neq t_1$, $|t - t_1|$ sufficiently small, are non-conjugate. This simply follows from the fact that a conjugate point $t' > 0$ can be characterized by the vanishing of the determinant $\det\big(Y_1(t), \ldots, Y_{n-1}(t)\big)$ of $(n-1)$ linearly independent Jacobi fields $Y_j(t)$ along $c(t)$ with $Y_j(0) = 0$, $\langle Y_j(t), \dot{c}(t) \rangle = 0$. If t' is a conjugate point of multiplicity k, the k-th derivative of this determinant at t' is non-zero, cf. also (2.5.8).

Thus, we can choose t_0, t_2, $0 < t_0 < t_1 < t_2 < a$, such that $\operatorname{im} \tilde{c} | [t_0, t_2] \subset U$ and t_1 is the only conjugate point in $[t_0, t_2]$.

We proceed as in the proof of (1.12.13, ii) by taking a non-zero Jacobi field $Y(t)$ along $c(t)$ with $Y(0) = Y(t_1) = 0$. Define the field $Y^*(t)$ by $Y(t)$, for $t \leqslant t_1$, and by 0, for $t \geqslant t_1$. Let $Z(t)$ be a differentiable vector field with $Z(t) = 0$ for $t \leqslant t_0$ and $t \geqslant t_2$, while $Z(t_1) = -\nabla Y(t_1)$. Put $Y^*(t) + \eta Z(t) = X(t)$. Then $D^2 E(c)(X, X) < 0$ for sufficiently small $\eta > 0$. Such an η shall be fixed.

Since $T_{\tilde{c}(t_0)} \exp_p$ is bijective, there exists a curve $\tilde{e}(s)$ in $T_p M$ with $\exp_p \tilde{e}(s) = \exp_{c(t_0)} s Y(t_0)$, $|s|$ sufficiently small. We now define a variation $F(s, t)$ of $c(t)$ by

$$F(s,t) = \begin{cases} \exp_p\big(t\tilde{e}(s)/t_0\big), & \text{for } 0 \leqslant t \leqslant t_0 \\ \exp_{c(t)}\big(sX(t)\big), & \text{for } t_0 \leqslant t \leqslant a. \end{cases}$$

This variation may not be differentiable in t at $t = t_1$. Its variation vector field is $X(t)$. Under our assumption, for all sufficiently small $|s| \neq 0$, the length $L(c_s)$ of the curve $c_s(t) = F(s, t)$ from p to $q = c(a)$ is $< L(c)$, cf. the proof of (1.12.13, ii).

But on the other hand, for a small $s = s_0 \neq 0$, the part $c_{s_0} | [t_0, t_2]$ will be in U. Put $c_{s_0} = b$. Then there exists a curve $\tilde{b}(t)$, $0 \leqslant t \leqslant a$, with $b(t) = \exp_p \tilde{b}(t)$. Indeed, for

$0 \leqslant t \leqslant t_0$ take $\tilde{b}(t) = t\tilde{e}(s_0)/t_0$. For $t_0 \leqslant t \leqslant t_2$ take the counter image of $b(t)$, $t_0 \leqslant t \leqslant t_2$, under $(\exp_p | \tilde{U})^{-1}$, and for $t_2 \leqslant t \leqslant a$ take $\tilde{b}(t) = \tilde{c}(t)$.

Now, for such a curve b we have from (1.9.2) $L(b) \geq L(c)$ – the desired contradiction. □

2.1.13 Complement. *Let* $\dim M = 2$. *Let* $c(t)$, $0 \leqslant t \leqslant a$, *be a geodesic on* M *with* $|\dot{c}| = 1$, $c(0) = p$. *Assume that* t_1, $0 < t_1 < a$, *is a conjugate point of* c. *Denote by* c_s *the geodesic which starts from* p *with initial direction* $\dot{c}_s(0) = \cos s\,\dot{c}(0) + \sin s\, e_2$, *where* $\{\dot{c}(0), e_2\}$ *is an orthonormal basis for* $T_p M$. *Then, given* $\varepsilon > 0$, *every geodesic* c_s, *with* $|s| \neq 0$ *sufficiently small, intersects* c *transversally in a point* $c(t)$ *with* $|t - t_1| < \varepsilon$.

Proof. Let $e_2(t)$ be the parallel vector field along $c(t)$ determined by $e_2(0) = e_2$. Put $\dot{c}(t) = e_1(t)$. We then can define Fermi coordinates $u = (u^1, u^2)$ on a neighborhood M' of $c(t_1)$, cf. (1.12.1). For an appropriate open interval I of $\mathbb{R}$ containing t_1, $c|I$ has coordinates $u^1(t) = t$, $u^2(t) = 0$.

More generally, for every sufficiently small $|s|$, $c_s|I$ has coordinates $u^1(s, t)$, $u^2(s, t)$ with $u^1(0, t) = t$, $u^2(0, t) = 0$ and $\partial u^2(s, t)/\partial s|_0 = y(t)$, i.e., the $e_2(t)$-component of the Jacobi field $Y(t) = \partial c_s(t)/\partial s|_0$ determined by the family $\{c_s\}$ of geodesics.

It follows $y(t_1) = 0$, $\dot{y}(t_1) \neq 0$. Hence,

$$u^2(s, t) = s\big((t - t_1)\dot{y}(t_1) + o(|t - t_1|)\big) + o(|s|)h(t).$$

This shows: Given $\varepsilon > 0$, for every sufficiently small $|s| \neq 0$ there is a t_1' with $|t_1' - t_1| < \varepsilon$, $u^2(s, t_1') = 0$ and $\partial u^2(s, t_1')/\partial s \neq 0$. That is to say, for such an s, c_s has a transversal intersection with c in $c(t_1')$, where $d\big(c(t_1'), c(t_1)\big) < \varepsilon$. □

We can now give the characterization of the cut locus $C(p)$ of a point p, which helps explain the name. See in particular the second part which may be thought of as saying that M can be obtained from an open, starlike set $\tilde{M}(p)$ in $T_p M$ by identifying points on a dense set of its boundary $\tilde{C}(p)$. For this and related results cf. also Warner [1], Karcher [1], Bishop [1], Wolter [1].

2.1.14 Theorem. *Choose* $p \in M$. *The complement* $M(p) = M - C(p)$ *of the cut locus* $C(p)$ *of* p *is the maximal open set in* M *with the property that each of its points can be joined to* p *by exactly one minimizing geodesic.*

Let $\tilde{M}(p) \subset T_p M$ *be the range of the polar coordinates on* $M(p)$ *based at* p. *Thus,* $\tilde{M}(p)$ *is formed by the* $\tilde{q} \in T_p M$ *such that* $q = \exp_p \tilde{q} \in M(p)$ *and the straight segment from* 0_p *to* $\tilde{q}$ *is carried into the unique minimizing geodesic from* p *to* q. *Then the boundary of* $\tilde{M}(p)$ *is formed by the tangential cut locus* $\tilde{C}(p)$, *cf.* (2.1.4). *Now, the set of those* $\tilde{q}' \in \tilde{C}(p)$ *for which there exists* $\tilde{q}'' \neq \tilde{q}'$ *in* $\tilde{C}(p)$ *with* $\exp_p \tilde{q}' = \exp_p \tilde{q}''$ *is dense in* $\tilde{C}(p)$.

The cut locus $C(p)$ *contains no interior points with respect to* M.

Proof. Let $\{q_n\}$ be a convergent sequence on $C(p)$. We want to show that the limit point q also belongs to $C(p)$. Now, we know from (2.1.11, ii) that, for each n, either there exist two different minimizing geodesics from p to q_n or else, q_n is conjugate to p

along the minimizing geodesic from p to q_n. It follows that the limit point q then must have also one or both of these properties, i.e., $q \in C(p)$. Thus, $M - C(p)$ is open.

To show that $M(p) = M - C(p)$ is maximal open with the property that for each $r \in M(p)$ there exists exactly one minimizing geodesic c_r from p to r we assume that there were a $q \in C(p)$ and an open neighborhood U of q on M such that for each $q' \in U$, there is a unique minimizing geodesic $c_{q'}, |\dot{c}_{q'}| = 1$, from p to q'. The mapping

$$\phi: q' \in U \mapsto d(p, q')\dot{c}_{q'}(0) \in T_pM$$

is continuous injective and hence, the invariance of domain theorem yields a homeomorphism with an open neighborhood $\tilde{U}$ of the point $\tilde{q} = d(p, q)\dot{c}_q(0)$. But this contradicts (2.1.12), since $\phi^{-1} = \exp_p|\tilde{U}$ and $d(p, q)$ is a conjugate point of p on c_q.

Let $\tilde{q} \in \tilde{C}(p)$ such that there is no $\tilde{r} \in \tilde{C}(p)$, $\tilde{r} \neq \tilde{q}$, with $q = \exp_p \tilde{q} = \exp_p \tilde{r} \in C(p)$. Thus, there is but one minimizing geodesic c_q from p to q. From (2.1.12) we know that $\tilde{q}$ is not isolated on $\tilde{C}(p)$ and the last argument shows that every neighborhood $\tilde{U}$ of $\tilde{q}$ in T_pM contains points $\tilde{q}'$ such that $q' = \exp_p \tilde{q}'$ can be joined by two minimizing geodesics. That is to say, there exists $\tilde{q}'' \neq \tilde{q}'$ on $\tilde{C}(p)$ with $\exp_p \tilde{q}'' = \exp_p \tilde{q}' = q'$.

To see that $C(p)$ contains no interior points observe that $C(p) = \exp_p \tilde{C}(p)$ with $\tilde{C}(p) = \partial \tilde{M}(p)$. Or else, if $q \in C(p)$ and $c(t), 0 \leqslant t \leqslant a = d(p, q)$ is a minimizing geodesic from p to q then, for some $\varepsilon > 0$, the minimizing geodesic $c|[0, a - \varepsilon]$ has none of the properties mentioned in (2.1.11, ii). Thus, $c(a - \varepsilon) \notin C(p)$. □

Remarks. 1. It may very well happen that there is a $q \in C(p)$ with exactly one minimizing geodesic going from p to q. A simple example is an ellipsoid of revolution, $x^2 + y^2/b + z^2/b = 1$, $b > 1$: For $p = (0, 0, \sqrt{b})$, $C(p)$ is an arc on the half ellipse $x^2 + z^2/b = 1, z > 0$. Its two end points are conjugate points of p along the ellipse $y = 0$. Cf. Alkier [1] and (3.5.13).

2. For every compact differentiable manifold except the 2-sphere, there exist Riemannian metrics g such that the resulting Riemannian manifold M possesses a point p where $C(p)$ contains no conjugate point, cf. Weinstein [1].

3. For the case of a real compact analytic Riemannian manifold (M, g) the structure of the cut locus $C(p)$ is relatively well understood. For surfaces, $C(p)$ consists of finitely many arcs. In the case that M is the 2-sphere, $C(p)$ is a tree, i.e., there are no cycles on $C(p)$. Always, a free end point q of $C(p)$ is a point which can be joined by exactly one minimizing geodesic. Also, the free end points of $C(p)$ are those points where the distance in T_pM to the singular set of the exponential map $\exp|T_pM$ has a local minimum. See Myers [1].

4. For differentiable Riemannian manifolds the cut locus can be quite complicated. We saw this already in (2.1.6, 5). Actually, for every compact differentiable manifold M one can find a metric such that $C(p)$ is not triangulable, at least for one point $p \in M$. In the case $M = S^2$, there even exists an example of a metric such that we get a

surface of revolution and $C(p)$ is not triangulable for all p in a certain open subset of M. See Gluck and Singer [1].

Nevertheless, for any compact manifold M and fixed point $p \in M$ there exists a dense open set $\mathcal{U}$ in the space $\mathcal{G}$ of C^∞-Riemannian metrics such that for any $g \in \mathcal{U}$, the cut locus $C(p)$ is triangulable and C^0-stable under small perturbations. It is an open question whether the same holds for the cut locus of all points on M, at least for a residual set of metrics on M; see Wall [1].

2.1 Appendix – Orientation

We recall the concept of an orientation for a finite dimensional vector space $\mathbb{F}$. Call two bases $\{f_i\}$, $\{f'_i\}$ of $\mathbb{F}$ *equally oriented* if the linear isomorphism $F: \mathbb{F} \to \mathbb{F}$ determined by $F(f_i) = f'_i$ has determinant $\det F > 0$. In this way, the set of bases of $\mathbb{F}$ is divided into two classes. To provide $\mathbb{F}$ with an orientation $\mathfrak{o}$ means distinguishing one of these two classes. The elements of this distinguished class usually are called the *positively oriented bases,* whereas the elements of the other class are called *negatively oriented.*

Let $(\mathbb{F}, \mathfrak{o})$, $(\mathbb{F}', \mathfrak{o}')$ be oriented vector spaces, $\dim \mathbb{F} = \dim \mathbb{F}'$. An isomorphism $F: \mathbb{F} \to \mathbb{F}'$ is called *orientation preserving* or *positive* if it carries a positively oriented base into a positively oriented base. Also note that an orientation $\mathfrak{o}$ of $\mathbb{F}$ and an isomorphism $F: \mathbb{F} \to \mathbb{F}'$ determines an orientation $\mathfrak{o}'$ on $\mathbb{F}'$ by calling the image under F of a positively oriented base positively oriented.

Before extending the concept of an orientation to general vector bundles we first consider the case of a product bundle.

2.1.A.1 Definition. Let $\pi: P \to M$ be a vector bundle modelled on $\mathbb{E} \times \mathbb{F}$, with $\dim \mathbb{F} < \infty$. Assume that π is equivalent to the product bundle $pr_1: U \times \mathbb{F} \to U$:

$$\begin{array}{ccc} P & \xrightarrow{\quad Pu \quad} & U \times \mathbb{F} \\ \Big\downarrow \pi & & \Big\downarrow pr_1 \\ M & \xrightarrow{\quad u \quad} & U \end{array}$$

Then π is called *oriented* if each fibre $\mathbb{F}_p = \pi^{-1}(p)$ is provided with an orientation $\mathfrak{o}_p$ and $\mathbb{F}$ is provided with an orientation $\mathfrak{o}_0$ such that, for all $p \in M$, $Pu(p): \mathbb{F}_p \to \mathbb{F}$ is orientation preserving. In this case the bundle isomorphism Pu is called *orientation preserving*.

We use this to give the

2.1.A.2 Definition. Let $\pi: P \to M$ be a bundle modelled on $\mathbb{E} \times \mathbb{F}$ with $\dim \mathbb{F} < \infty$. Let there be given an orientation $\mathfrak{o}_0$ of $\mathbb{F}$. π is called *oriented* with the orientation $\mathfrak{o}$ if, for each $p \in M$, $\mathbb{F}_p = \pi^{-1}(p)$ is provided with an orientation $\mathfrak{o}_p$ subject to the following condition:

There exists an atlas for M and for each chart (u, M') of this atlas a bundle chart (Pu, u, M') such that the isomorphism Pu of PM' with the oriented product bundle $U \times \mathbb{F}$ is orientation preserving.

Remark. The charts (Pu, u, M') of such an atlas are called *positively oriented.* For the special case of the tangent bundle $\tau_M: TM \to M$ of a manifold we also call M *oriented* if τ_M is oriented. Note that here M must have finite dimension. For M modelled on a proper Hilbert space, it does not make sense to speak of an orientation, cf. Flaschel und Klingenberg [1] for more details.

There remains the question of the existence of an orientation for a bundle π. Clearly, this is equivalent to the existence of a bundle atlas $(Pu_\alpha, u_\alpha, M_\alpha)_{\alpha \in A}$ such that the transition mappings

$$Pu_\beta^{-1} \circ (Pu_\alpha | \mathbb{F}_p): \mathbb{F}_p \to \mathbb{F}_p;\ p \in M_\alpha \cap M_\beta$$

all are orientation preserving. One way to decide whether there exists an orientation for π is the following:

First assume M to be connected – clearly the question of orientability of π only depends on the orientability of π, restricted to each of the connected components of M.

Choose a base point $p \in M$. Moreover, assume given a linear connection on π. This allows us to associate to every closed curve $c: (I, \partial I) \to (M, p)$, $I = [0, 1]$, the linear isomorphism

$$(*) \qquad \overset{1}{\underset{0}{\|}} c: \mathbb{F}_p \to \mathbb{F}_p$$

The sign of the determinant of $(*)$ does not depend on the choice of the linear connection employed in the definition of the parallel translation. In fact, if we have two linear connections,

$$K_0: TP \to P \quad \text{and} \quad K_1: TP \to P,$$

we can form the homotopy $K_t, 0 \leqslant t \leqslant 1$,

$$K_t = (1 - t) K_0 + t K_1: TP \to P$$

of linear connections, cf. the definition (1.5.9). The associated covariant derivations $\nabla_t, 0 \leqslant t \leqslant 1$, each give rise to a linear mapping of the type $(*)$ with non-zero determinant.

We start with the following observation.

2.1.A.3 Proposition. *The sign of the determinant of the mapping* (∗), *associated with the closed curve* $c\colon (I, \partial I) \to (M, p)$, *only depends on the homotopy class of* c.

We obtain a group morphism

$$\delta\colon \pi_1 M \to \mathbb{Z}_2 \cong \mathbb{R}^*/\mathbb{R}^{+*}$$

by associating to a representative $c\colon (I, \partial I) \to (M, p)$ *of an element* γ *of the fundamental group* $\pi_1 M$ *of* M *the determinant of* (∗), *divided by its absolute value.*

Proof. Since (∗) depends continuously on c and is never zero, the determinant of (∗), divided by its absolute value, is constant for a continuous family of closed curves from p to p. The composition $c_1 * c_2$ of two such closed curves c_1 and c_2 determines a mapping (∗) which is the composition of the mappings (∗) determined by c_1 and c_2. □

We now give a useful characterization for the existence of an orientation.

2.1.A.4 Lemma. *Let* $\pi\colon P \to M$ *be a vector bundle with fibre modelled on the finite dimensional vector space* $\mathbb{F}$. *Assume* M *to be (arcwise) connected. Then* π *permits an orientation if and only if* $\pi_1 M$ *is mapped by* δ, (2.1.A.3), *into the identity* $+1$. *That is, for each closed curve* c, (∗) *has positive determinant.*

Proof. Assume π to be oriented. For $c\colon (I, \partial I) \to (M, p)$ and $t \in I$ consider

$$(*)_0^t \quad \overset{t}{\underset{0}{\|}}\, c\colon \mathbb{F}_p = \mathbb{F}_{c(0)} \to \mathbb{F}_{c(t)}$$

We claim that these isomorphisms are orientation preserving and hence, since $(*)_0^0$ is the identity, $(*)_0^1 \equiv (*)$, has positive determinant.

To see this we first observe that, for any interval $I' = [t', t''] \subset I$ where $c(I')$ lies entirely in the domain M' of a bundle chart (Pu, u, M') the mapping

$$(\dagger)_{t'}^t \quad \overset{t}{\underset{t'}{\|}}\, c\colon \mathbb{F}_{c(t')} \to \mathbb{F}_{c(t)}, \quad t \in I'$$

possesses a representation of the form

$$_u(\dagger)_{t'}^t \quad \overset{t}{\underset{t'}{\|}}\, u \circ c\colon \mathbb{F} \to \mathbb{F}, \quad t \in I'.$$

The determinant of the latter mapping depends continuously on t and never vanishes. Therefore, it does not change its sign. Thus, the mappings $(\dagger)_{t'}^{t''}$ are orientation preserving.

Since I is compact, we have a subdivision $0 = t_0 < t_1 < \ldots t_k = 1$ of $I = [0, 1]$ such that the intervals $I_j = [t_{j-1}, t_j]$, $1 \leqslant j \leqslant k$, belong entirely to the domain M_j of a positively oriented bundle chart (Pu_j, u_j, M_j). For each j, $(\dagger)_{t_{j-1}}^{t_j}$ is orientation preserving; so we find by induction that (∗) has positive determinant.

Conversely, assume that (∗) always has positive determinant. Choose an orientation $\mathfrak{o}_p$ for $\mathbb{F}_p$. For an arbitrary $q \in M$ define the orientation $\mathfrak{o}_q$ of $\mathbb{F}_q$ such that

$$\overset{1}{\underset{0}{\|}} c_0 : \mathbb{F}_p \to \mathbb{F}_q$$

is orientation preserving, where $c_0 : I \to M$ is a curve from p to q.

It must be shown that this does not depend on the choice of the curve c_0. Indeed, if $c_1 : I \to M$ is another such curve, c_0, followed by the inverse of c_1, gives a closed curve c of which we know that $(*)$ has positive determinant.

Finally, our assignment of an orientation, when restricted to the $\mathbb{F}_q$ with q in the domain M' of a bundle chart (Pu, u, M'), is determined by the sign of a mapping of the type ${}_u(\dagger)^t_{t'}$ where $u \circ c$ lies entirely in $U = u(M')$. Therefore it is such as to make the mapping $Pu: PM' \to U \times \mathbb{F}$ orientation preserving, for an appropriate choice of the orientation of $\mathbb{F}$. □

2.1.A.5 Corollary. *If M is simply connected then every $\pi: P \to M$ possesses an orientation.* □

In general, if $\pi: P \to M$ does not possess an orientation, one needs not to go over to the universal covering of M, as we will show now.

2.1.A.6 Proposition. *Assume that the bundle $\pi: P \to M$ does not permit an orientation, with M connected. Then there exists a 2-fold connected covering M_0 of M with covering map $f_0: M_0 \to M$ such that the bundle*

$$\pi_0 \equiv f_0^* \pi : P_0 \equiv f_0^* P \to M_0,$$

induced from π by f_0, permits an orientation.

Note. The precise construction of M_0 will be given below. In case π does permit an orientation, this construction still applies. However, M_0 will then consist of two copies of M and thus will not be connected.

Proof. For each $q \in M$, we define $f_0^{-1}(q)$ to consist of two elements, q_+ and q_-, represented by $\mathbb{F}_q$ together with one of the two possible orientations. Fix, moreover, an orientation $\mathfrak{o}_0$ for the model $\mathbb{F}$ of the fibres of π.

$$\text{Put} \bigcup_{q \in M} f_0^{-1}(q) = M_0 \quad \text{and let} \quad f_0: M_0 \to M$$

be the canonical projection. We construct an atlas for M_0 by starting with a bundle atlas for π. For a bundle chart (Pu, u, M') of π denote by $M'_+ \subset M_0$ the set of those oriented $\mathbb{F}_q$, $q \in M'$ for which

$$Pu: M'_+ \cong PM' \to U \times \mathbb{F}, \; U = u(M'),$$

is orientation preserving. Define now

$$u_+ = u \circ f_0 : M'_+ \to U.$$

Let $\iota\colon \mathbb{F} \to \mathbb{F}$ be an orientation reversing map, e.g., the reflection on a subspace of odd codimension. Then let $M'_- \subset M_0$ be the set of those $\mathbb{F}_q, q \in M'$, for which

$$(id \times \iota) \circ Pu\colon M'_- \cong PM' \to U \times \mathbb{F}$$

is orientation preserving. Define

$$u_- = u \circ f_0\colon M'_- \to U.$$

M_0 now is a manifold, with the family of charts $\{(u_+, M'_+), (u_-, M'_-)\}$ as an atlas. The mapping f_0 is represented by $u \circ f_0 \circ u_-^{-1}$ and $u \circ f_0 \circ u_+^{-1}$, each of which is the identity.

In addition, M_0 is connected. By hypothesis there exists a closed curve $c\colon (I, \partial I) \to (M, p)$ for which the mapping (∗) has negative determinant. This means that its 'lift' into M_0 joins the points p_+ and p_- in $f_0^{-1}(p)$.

Finally, our construction shows that the induced bundle $f_0^* \pi \equiv \pi_0$ can be oriented. A positively oriented bundle atlas is given by the family of charts $\{(Pu, u_+, M'_+), ((id \times \iota) \circ Pu, u_-, M'_-)\}$. □

2.1.A.7 Examples. (i) Every simply connected manifold can be oriented. This is true therefore in particular for the sphere S^n, the Euclidean space $\mathbb{R}^n$ and the hyperbolic space H^n.

(ii) Let $P^n = S^n/\mathbb{Z}_2$ be the real projective space. It is obtained from S^n by identifying antipodal points, cf. (2.1.6,3). An atlas for P^n can be obtained from an atlas of S^n in which each chart (u, M') has as domain M' an open half sphere (or a part of it). If

$$\varrho\colon S^n \to P^n$$

denotes the quotient map, $(u \circ (\varrho | M')^{-1}, \varrho M')$ is a chart for P^n.

We claim that P^n is orientable if and only if n is odd.

Since the fundamental group of P^n is $\mathbb{Z}_2$, we have to investigate the sign of the determinant of (∗) where c is a closed curve, not homotopic to 0. We consider $c = \varrho \circ c'$ where c' is a half great circle joining p to its antipodal point p'.

Choose an orthonormal base $\{e_i\}$ in $T_p S^n$ where e_1 is tangential to c' in $c'(0) = p$. Under parallel translation along c' we get an orthonormal base $\{e'_i\}$ for $T_{p'} S^n$,

$$T_{p'} \varrho \,.\, e'_1 = T_p \varrho \,.\, e_1 \in T_{\varrho p} P^n, \quad T_{p'} \varrho \,.\, e'_i = - T_p \varrho \,.\, e_i \in T_{\varrho p} P^n, \quad i > 1.$$

Hence, the sign of the determinant of the linear mapping (∗) is given by $(-1)^{n-1}$.

(iii) Let $\pi\colon P \to S$ be a vector bundle over the circle $S = I/\partial I$. π is orientable if and only if π is equivalent to the product bundle $pr_1\colon S \times \mathbb{F} \to S$.

To see this we can assume P to be covered by finitely many bundle charts (Pu_i, u_i, S'_i), $1 \leqslant i \leqslant k$. Each S'_i is an open interval on S. Moreover, there is subdivision $0 = t_0 < \ldots < t_k = 1$ of $I = [0, 1]$ such that $[t_{i-1}, t_i] \subset S'_i$.

As usual, we denote by $P_t u_i$ the restriction of Pu_i to the fibre $\mathbb{F}_t$ over $t \in S'_i$. Starting with Pu_1, we modify Pu_2 by the linear isomorphism $P_{t_1} u_1 \circ P_{t_1} u_2^{-1}$:

$\{t_1\} \times \mathbb{F} \to \{t_1\} \times \mathbb{F}$. Write again Pu_2 for this modified trivialization and modify Pu_3 by $P_{t_2}u_2 \circ P_{t_2}u_3^{-1}$, and so on. In this way, we eventually get a continuous mapping from $\pi^{-1}[0,1[$ into $[0,1[\times \mathbb{F}$, linear on each fibre. For the $\mathbb{F}_0 = \mathbb{F}_1$ over $0 = 1 \in S$ we get the two linear isomorphisms $P_0 u_1$ and $P_1 u_k$. π is orientable if and only if the mapping

$$P_1 u_k \circ P_0 u_1^{-1} \colon \{0\} \times \mathbb{F} \to \{1\} \times \mathbb{F}$$

is orientation preserving, i.e., if this mapping, which we also denote by $A(1)$, belongs to the identity component of $GL(\mathbb{F})$.

If this is the case, choose a path $A(t)$, $0 \leqslant t \leqslant 1$, in $GL(\mathbb{F})$ from $A(0) = id$ to $A(1)$. Then

$$(id, A(t)^{-1}) \circ P_t u \colon P_t \to (t, \mathbb{F}),\ 0 \leqslant t \leqslant 1,$$

gives a trivialization of the full bundle π.

Remark. For manifolds M, the existence of an orientation also can be charaterized by the existence of a n-form on M, $n = \dim M$, which never vanishes. Cf. Dieudonné [1] for details.

2.2 Symmetric Spaces

One owes to E. Cartan [2]the discovery of an important class of Riemannian manifolds which contains as particular cases the spheres, the Euclidean spaces and the hyperbolic spaces. These so-called symmetric spaces M are defined by the property that, for every $p \in M$, there exists an isometry $\sigma_p \colon M \to M$ which near p is the reflection on p along the geodesics through p.

It can be shown that the connected identity component G of the group of isometries of a symmetric space M is a Lie group. G operates transitively on M. If one fixes a $p \in M$ and considers the reflection $\sigma_p \colon (M, p) \to (M, p)$ on p, then this induces an involution σ on G. The fixed point set G_σ of σ is a subgroup of G which has the same identity component as the isotropy group H of G at p, i.e., the group formed by those $g \in G$ which leave p invariant. H is compact.

Conversely, let (G, H, σ) be a tripel consisting of a connected Lie group G with an involution σ and a compact subgroup H which is contained in G_σ = fixed point set of σ, such that H and G_σ have the same identity component. An Ad(H)-invariant and $T\sigma$-invariant metric on the negative eigen space of $T\sigma \,|\, T_e G$ then determines a symmetric space $M = G/H$, where σ induces the reflection on $H \in M$.

The equivalence between symmetric spaces and tripels (G, H, σ) of the type just described has the effect that the structure of the symmetric spaces is usually developed as part of the theory of Lie groups. Our interest in symmetric spaces is different. We want to study them as particular examples of Riemannian manifolds. To investigate their structure we will employ the methods of Riemannian geometry rather than Lie group theory.

As a consequence, we will not present here everything that is known about symmetric spaces. In particular, no attempt is made to classify the symmetric spaces. For this and further developments we must refer to the books of Helgason [1], Kobayashi and Nomizu [1] and Wolf [1].

We begin by showing that the isometry group of a symmetric space M contains, for every geodesic c, a 1-parameter group which operates on the tangent spaces $T_{c(t)}M$, $t \in \mathbb{R}$, by parallel translations. This implies that the connected component of the isometry group of M operates transitively, cf. (2.2.5), (2.2.6).

A symmetric space is characterized – at least locally – by the vanishing of the covariant derivative of its curvature tensor, (2.2.8) and (2.2.12). This fact allows an explicit description of the Jacobi fields, (2.2.9) and (2.2.10). Among the numerous consequences of this fact we get that a compact symmetric space has sectional curvature $\geqslant 0$, (2.2.15).

Next we enter into a rather detailed description of the Lie algebra $\mathfrak{g}$ of Killing vector fields on M. As we stated already above, the connected component G of the isometry group of M is a Lie group and the structure of $\mathfrak{g}$ therefore reflects the infinitesimal structure of G. More precisely, we fix $p \in M$ and decompose $\mathfrak{g}$ into the subalgebra $\mathfrak{h}$, formed by those Killing vector fields which vanish at p, and the vector space complement $\mathfrak{m}$ of $\mathfrak{h}$ in $\mathfrak{g}$, formed by the infinitesimal translations along geodesics passing through p, cf. (2.2.19). The reflection $\sigma_p: M \to M$ on p induces an involution σ on $\mathfrak{g}$ with $\sigma | \mathfrak{h} = id$, $\sigma | \mathfrak{m} = -id$.

The negative of the Killing form restricted to $\mathfrak{h}$ is a positive definite scalar product. On $\mathfrak{m}$ we get a positive definite scalar product from the canonical identification of $\mathfrak{m}$ with T_pM, cf. (2.2.22). We thus have derived the "orthogonal symmetric Lie algebra" $(\mathfrak{g}, \mathfrak{h}, \sigma)$, associated to the symmetric space M. The classification of the symmetric spaces is based mainly on the classification of these $(\mathfrak{g}, \mathfrak{h}, \sigma)$. We give only a first step towards this classification, (2.2.23), (2.2.24).

In (2.2.25) we define the various types of symmetric spaces. In (2.2.26) we give an estimate for the injectivity radius of spaces of compact type (which have curvature $K \geqslant 0$) and show that a space M of non-compact type is simply connected and diffeomorphic to T_pM.

In (2.2.27) we study the complex projective space as symmetric space. Another example is given by the orthogonal group (2.2.28). In (2.2.29) we give an outline of the theory of symmetric homogeneous spaces G/H.

In preparing this section, we found notes of H. Karcher from a course at Bonn in 1975/76 very helpful.

2.2.1 Definition. A connected Riemannian manifold M of finite dimension is called *symmetric* if for every point $p \in M$ there exists an isometry

$$\sigma_p: (M, p) \to (M, p)$$

with $T_p\sigma_p: T_pM \to T_pM$ equal $-id$. In particular, $\sigma_p^2 = id$.

A symmetric Riemannian manifold is also called a *symmetric space*.

2.2.2 Examples. (i) The simply connected manifolds M of constant curvature are symmetric. Since the group of isometries operates transitively on these spaces, it suffices to exhibit the existence of a symmetry σ_p for a single point $p \in M$. For $M = \mathbb{E}, p = 0$, this is clear. For the sphere $S_\varrho^{\mathbb{E}}$, (1.1.3), or the hyperbolic space $H_\varrho^{\mathbb{E}}$, (1.11.7), take $p = (1, 0)$, $\sigma_p(x_0, x) = (x_0, -x)$.

(ii) The complex projective space $P\mathbb{E}$, (1.11.14), is symmetric. Take the representation $\mathbb{C}N_p \subset \mathbb{E}$ of a point $p = F(N_p) \in P\mathbb{E}$. Let $\sigma_p^*: \mathbb{E} \to \mathbb{E}$ be the reflection on the complex line $\mathbb{C}N_p$ i.e., $\sigma_p^* | \mathbb{C}N_p = id$, $\sigma_p^* | (\mathbb{C}N_p^\perp =$ orthogonal complement of $\mathbb{C}N_p) = -id$. Then $\sigma_p = F \circ \sigma_p^* \circ F^{-1}$ is a reflection on $p \in P\mathbb{E}$.

2.2.3 Proposition. *A symmetric space M is complete.*

Proof. According to (2.1.3) it suffices to show that M is geodesically complete. Assume now that a geodesic $c(t)$ has been defined on the interval $[0, t_0]$. Put $c(t_0) = p$. $T_{c(t_0)}\sigma_p$ carries $-\dot{c}(t_0)$ into $\dot{c}(t_0)$. Since σ_p is an isometry, it carries $c(t), t \in [0, t_0]$ into $c(2t_0 - t)$, i.e., c is defined on $[0, 2t_0]$. $\square$

2.2.4 Definition. Let $c: \mathbb{R} \to M$ be a geodesic in the symmetric space M. Then we define the *c-translation* (or briefly *translation*) *by* $t \in \mathbb{R}$ to be the isometry

$$ {}_c\tau_t \equiv \tau_t = \sigma_{c(t/2)} \circ \sigma_{c(0)} . $$

2.2.5 Lemma. *Let τ_{t_0} be the c-translation by t_0. Then $\tau_{t_0} c(t) = c(t + t_0)$ and*

$$ T_{c(t)}\tau_{t_0}: T_{c(t)} M \to T_{c(t+t_0)} M $$

coincides with the parallel translation $\overset{t+t_0}{\underset{t}{\|}} c$.

The family $\{\tau_t, t \in \mathbb{R}\}$ of translations along c forms a 1-parameter group of isometries.

Proof. $\tau_{t_0} c(t) = \sigma_{c(t_0/2)} \circ \sigma_{c(0)} c(t) = \sigma_{c(t_0/2)} c(-t) = c(t + t_0)$. Moreover,

$$ T_{c(-t)}\sigma_{c(t_0/2)} = - \overset{t+t_0}{\underset{t_0/2}{\|}} c \circ \overset{t_0/2}{\underset{-t}{\|}} c , $$

since an isometry carries parallel vector fields into parallel vector fields and $T_{c(t_0/2)}\sigma_{c(t_0/2)} = -id$. Hence,

$$ T_{c(t)}\tau_{t_0} = -\left(\overset{t+t_0}{\underset{t_0/2}{\|}} c\right) \circ \left(\overset{t_0/2}{\underset{-t}{\|}} c\right) \circ -\left(\overset{-t}{\underset{0}{\|}} c\right) \circ \left(\overset{0}{\underset{t}{\|}} c\right) = \overset{t+t_0}{\underset{t}{\|}} c. $$

The family of parallel translations along c forms a 1-parameter group of isometries on the Riemannian vector bundle $c^*\tau_M: c^*TM \to \mathbb{R}$, induced from $\tau_M: TM \to M$ by $c: \mathbb{R} \to M$. For a complete connected manifold, an isometry $F: M \to M$ is determined by its tangential $T_{p_0}F$ at a single point $p_0 \in M$, because $T_{p_0}F$ will determine the images of the points on geodesics starting on p_0 and every $p \in M$ occurs on one of these geodesics. In particular, in our case where $F = \tau_{t_0}$, we see that $\tau_{t_0} \circ \tau_{t_1} = \tau_{t_1 + t_0}$. □

2.2.6 Corollary 1. *The identity component of the isometry group of a symmetric space M operates transitively on M.*

Remark. We did not prove that the isometry group of a Riemannian manifold is a Lie group and, in particular, has a topology. See Myers and Steenrod [1] for this result. Therefore, we define the identity component to consist of those elements which belong to the 1-parameter subgroups of the isometry group.

Proof. Given p and q on M, there exists a geodesic $c: (\mathbb{R}, 0, 1) \to (M, p, q)$. Then ${}_c\tau_1 p = q$. □

2.2.7 Corollary 2. *Let M be a symmetric space. Then every geodesic loop $c: (I, \partial I) \to (M, p)$, $I = [0, 1]$, is a closed geodesic, $\dot{c}(1) = \dot{c}(0)$.*

Consequently, the fundamental group $\pi_1 M$ of M is abelian.

Proof. Denote by X the Killing field derived from the 1-parameter group of isometries $\tau_t = {}_c\tau_t$, $t \in \mathbb{R}$. That is, cf. (1.10.8),

$$X(q) = d\tau_t q/dt|_0.$$

In particular, $X(c(t)) = \dot{c}(t)$. Hence, $X(p) = \dot{c}(0) = \dot{c}(1)$.

The claim about the fundamental group is equivalent to saying that the mapping $\gamma \in \pi_1 M \to \gamma^{-1} \in \pi_1 M$ is a group morphism.

Now choose $p \in M$. An element $\gamma \in \pi_1 M$ is represented by a class of mutually homotopic curves $c': (I, \partial I) \to (M, p)$. In each class there is a geodesic loop, see (2.4.19) or employ the methods from (3.7). This will even be a closed geodesic as we just saw.

$\sigma_p: (M, p) \to (M, p)$ induces an involution on $\pi_1 M$. Let $c_\gamma: (I, \partial I) \to (M, p)$ be a closed geodesic representing $\gamma \in \pi_1 M$. Since $\sigma_p c_\gamma(t) = c_\gamma(-t)$, $\sigma_p c_\gamma$ represents $-\gamma$. □

We now come to the most important property of a symmetric space.

2.2.8 Theorem. *For a symmetric space M, the curvature tensor is parallel, i.e.,*

$$\nabla_U R(X, Y)Z = 0,$$

for all U, X, Y, Z.

Proof. We determine the effect of $T_p\sigma_p$ on the vector $\nabla_U R(X, Y)Z(p)$ in two ways:

$$T_p\sigma_p \,.\, \nabla_U R(X, Y)Z(p) = -\nabla_U R(X, Y)Z(p);$$

$$T_p\sigma_p \,.\, \nabla_U R(X, Y)Z(p) = \nabla_{T_p\sigma_p . U} R(T_p\sigma_p \,.\, X, T_p\sigma_p \,.\, Y)\, T_p\sigma_p \,.\, Z(p) =$$

$$= \nabla_{-U} R(-X, -Y)(-Z)(p) = \nabla_U R(X, Y)Z(p). \quad \square$$

For a geodesic $c(t)$ on a Riemannian manifold M we defined in (1.8.13) the self-adjoint operator

$$R_{\dot c(t)}\colon T_{c(t)}M \to T_{c(t)}M;\ X \mapsto R(X, \dot c(t))\dot c(t).$$

The subspace $T_t^{\perp}c$ of vectors orthogonal to $\dot c(t)$ is thereby carried into itself. We now consider this operator for a symmetric space M.

2.2.9 Theorem. *Let M be a symmetric space, $c\colon \mathbb{R} \to M$ a geodesic. Put $c(0) = p, \dot c(0) = X_0$. Assume $X_0 \neq 0$. Then the operators $R_{\dot c(t)}, t \in \mathbb{R}$ are constant on the parallel vector fields $X(t)$ along c,*

$$\overset{t_1}{\underset{t_0}{\|}} c\, R_{\dot c(t_0)} X(t_0) = R_{\dot c(t_1)} \overset{t_1}{\underset{t_0}{\|}} c\, X(t_0).$$

It follows that a Jacobi field $Y(t)$ along $c(t)$ with initial values $Y(0) = A_0$, $\nabla Y(0) = A_1$ can be written as

$$Y(t) = \Big(\overset{t}{\underset{0}{\|}} c\Big) A(t),$$

with $$A(t) = \sum_0^\infty \frac{(-1)^k t^{2k+1}}{(2k+1)!} R_{X_0}^k A_0 + \sum_0^\infty \frac{(-1)^k t^{2k}}{(2k)!} R_{X_0}^k A_1.$$

2.2.10 Complement. *Let M be a symmetric space.* (i) *The range of the sectional curvature K on M is determined by the range of the eigenvalues of the self-adjoint operators R_{X_0}, where X_0 runs through the unit vectors in some T_pM.*

(ii) *Let $c(t)$ be a geodesic on M, $|\dot c| \neq 0$. Then the Jacobi fields $Y(t)$ along $c(t)$ with $\langle Y(t), \dot c(t)\rangle = 0$ possess a basis consisting of vectors ${}_AY_0(t), {}_AY_1(t)$ of the following form:*

Choose an orthonormal basis of the space $T_0^{\perp}c$, consisting of eigenvectors of the self-adjoint operator R_{X_0}, $X_0 = \dot c(0)$. If now A is such a basis vector and λ its eigenvalue, put

$$Y_0(t) \equiv {}_AY_0(t) = \Big(\overset{t}{\underset{0}{\|}} c\Big) A_0(t);\ Y_1(t) \equiv {}_AY_1(t) = \Big(\overset{t}{\underset{0}{\|}} c\Big) A_1(t),$$

where

$$A_0(t) = \begin{cases} \dfrac{\sin\sqrt{\lambda}\, t}{\sqrt{\lambda}} A, & \text{if } \lambda > 0 \\[2ex] tA, & \text{if } \lambda = 0 \\[2ex] \dfrac{\sinh\sqrt{-\lambda}\, t}{\sqrt{-\lambda}} A & \text{if } \lambda < 0 \end{cases}$$

$$A_1(t) = \begin{cases} \cos\sqrt{\lambda}\,t A, & \text{if } \lambda > 0 \\ A & \text{if } \lambda = 0 \\ \cosh\sqrt{-\lambda}\,t A, & \text{if } \lambda < 0. \end{cases}$$

Proof. In (1.8.13) we showed that the operator $R_{\dot{c}(t)}$ is self-adjoint and that $R_{\dot{c}(t)}$ carries the subspace $T_t^{\perp} c$ into itself.

Let now M be a symmetric space. Then (2.2.8) states that $R_{\dot{c}(t)}$ commutes with parallel translation along $c(t)$. It is a matter of simple verification to show that the $A(t)$, (2.2.9), yield Jacobi fields.

Let now A be an eigenvector of R_{X_0} for the eigenvalue λ, $|X_0| = |A| = 1$, $\langle X_0, A\rangle = 0$. Then

$$\langle R(A, X_0)X_0, A\rangle = \langle R_{X_0}A, A\rangle = \lambda$$

is the value of the sectional curvature of the plane spanned by X_0 and A.

The last statement in (2.2.10) is an immediate consequence of the fact that $T_0^{\perp} c$ possesses an orthonormal basis formed by eigenvectors of the self-adjoint operator R_{X_0}. □

2.2.11 Definition. A Riemannian manifold is called *locally symmetric* if $\nabla R \equiv 0$.

This definition is motivated by the following

2.2.12 Lemma. *Let M be locally symmetric. Then, for every $p \in M$, every $\varrho > 0, \varrho \leqslant \iota(p) =$ injectivity radius of p, there exists an isometry*

$$\sigma_p : (B_\varrho(p), p) \to (B_\varrho(p), p)$$

with $T_p\sigma_p = -id$.

Proof. We have defined σ_p in such a way that $\sigma_p c(t) = c(-t)$, where $c(t)$, $-\varrho < t < \varrho$, is a geodesic with $c(0) = p$. The local symmetry implies

$$T\sigma_p \,.\, R(X, Y)Z(q) = R(T\sigma_p \,.\, X, T\sigma_p \,.\, Y)\,T\sigma_p \,.\, Z(\sigma_p q).$$

Thus, the hypotheses for Cartan's theorem (1.12.8) are satisfied. □

2.2.13 Remark. Whereas it can be shown that a simply connected locally symmetric manifold is a symmetric space, there do exist examples of compact locally symmetric spaces which are not (globally) symmetric.

One way of obtaining such examples is to consider manifolds which have as their universal covering a symmetric space. Take e. g. the 3-sphere

$$S^3 = \{(z_0, z_1) \in \mathbb{C}^2, z_0\bar{z}_0 + z_1\bar{z}_1 = 1\}.$$

On S^3 we have the $S^1 \times S^1$-action

$$((e^{i\varphi_0}, e^{i\varphi_1}), (z_0, z_1)) \in (S^1 \times S^1) \times S^3 \mapsto (e^{i\varphi_0} z_0, e^{i\varphi_1} z_1) \in S^3.$$

Consider now in $S^1 \times S^1$ the cyclic group $\mathbb{Z}_m$ of order m given by

$$\{(e^{2\pi ir/m}, e^{2\pi ilr/m}); r = 1, \ldots, m\}$$

Then $S^3/\mathbb{Z}_m$ is a manifold, the so-called *lens space* $L(m, l)$. Here, l is an integer, $1 \leqslant l < m$, (l, m) relatively prime.

$L(2,1)$ is the real projective space. If $m > 2$, $L(m, l)$ is not symmetric. Take e. g. the reflection $\sigma_p(z_0, z_1) = (\bar{z}_0, -z_1)$ on $p = (1, 0)$. Then σ_p does not commute with the $\mathbb{Z}_m$-action.

That $L(m, l)$ for $m > 2$ cannot be symmetric also will follow from (2.2.26) where we show that the distance to the cut locus on a symmetric space of constant curvature $K = 1$ must be at least $\pi/2$.

The following Lemma will be quite useful.

2.2.14 Lemma. *Let M be a symmetric space, $c: \mathbb{R} \to M$ a non-constant geodesic. Assume that $Y(t)$ is a Jacobi field along $c(t)$ orthogonal to $c(t)$, such that $\nabla Y(0) = 0$. Then $Y(t)$ can be generated by a Killing vector field,*

$$Y(t) = \partial \tau_s c(t)/\partial s|_{s=0}.$$

Here, $\{\tau_s \equiv {}_b\tau_s\}$ is the 1-parameter group of translations along the geodesic $b(s) = \exp sY(0)$.

Proof. We only need to verify that the Jacobi field $\tilde{Y}(t) = \partial \tau_s c(t)/\partial s|_{s=0}$ satisfies $\tilde{Y}(0) = Y(0)$, $\nabla \tilde{Y}(0) = 0$. The first is clear from the definition. As for the second, we know from (2.2.5) that $\partial \tau_s c(t)/\partial t|_0$ is parallel along $b(s)$. Hence,

$$0 = \frac{\nabla}{\partial s}\frac{\partial}{\partial t}\tau_s c(t)|_{0,0} = \frac{\nabla}{\partial t}\frac{\partial}{\partial s}\tau_s c(t)|_{0,0} = \nabla \tilde{Y}(0). \quad \square$$

As a first application we mention the

2.2.15 Theorem. *If a symmetric space M is compact then its sectional curvature K is $\geqslant 0$.*

Proof. The range of K is given by the range of the eigenvalues of R_{X_0}, $|X_0| = 1$, cf. (2.2.10). Assume now that there were an eigenvalue $\lambda < 0$. Let A be a corresponding eigenvector. Then we get a Jacobi field $Y_1(t)$ along the geodesic $c(t) = \exp_p tX_0$ of the form given in (2.2.10). According to (2.2.14), $Y_1(t)$ can be generated by a 1-parameter group. Actually, $Y_1(t)$ is the value of the Killing field of this group at $c(t)$. Since M is compact, $|Y_1(t)|$ must be bounded, which contradicts the form of $Y_1(t)$. $\square$

With essentially the same argument we can show:

2.2.16 Lemma. *Let c be a closed geodesic on a symmetric space M, i.e., $\dot{c}(0) = \dot{c}(a) \neq 0$, for some $a > 0$. Then the sectional curvature $K(\sigma)$ of a 2-plane σ tangent to c is $\geqslant 0$.*

Proof. Assume the contrary. Then there would exist a Jacobi field $Y_1(t)$ along $c(t)$ of the form (2.2.10) with $\lambda < 0$. But this would mean that $|Y_1(t)|$ is unbounded, which is impossible. □

The conjugate points on a symmetric space are of a very special type.

2.2.17 Lemma. *Let $c: \mathbb{R} \to M$ be a geodesic on a symmetric space M. Assume that $c(t_1)$ is conjugate to $c(0)$ along $c|[0, t_1]$. Then there exists a 1-parameter family $c_s = \tau_s c$ of geodesics, $c_0 = c$, all of which pass through $c(0)$ and $c(t_1)$. $L(c_s|[0, t_1]) = \text{const}$.*

Proof. We can assume that there exists a Jacobi field of the form $Y_0(t)$, (2.2.10), with $\lambda > 0$, $Y_0(0) = Y_0(t_1) = 0$. Then $\nabla Y_0(t_1/2) = 0$. Let τ_s be the 1-parameter group generated by $Y_0(t_1/2)$. Since $Y_0(t)$ is the Killing field of τ_s in $c(t)$ we find that the geodesics $\tau_s c$ pass through $c(0)$ and $c(t_1)$. □

We want to determine the structure of the Lie algebra of Killing fields for a symmetric space. For this reason we first prove the following

2.2.18 Lemma. *A Killing vector field X on a symmetric space M satisfies the second order differential equation*

$$\nabla_A \nabla_B X - \frac{1}{2} \nabla_{[A,B]} X + R(X, A) B = 0,$$

for arbitrary vector fields A and B on M.

Proof. We first show

$$\nabla_A \nabla_A X + R(X, A) A = 0.$$

This is immediate from the fact that along a geodesic $c(t)$, $\dot{c}(0) = A$, $X(t) = \partial \tau_s c(t)/\partial s|_0$ is a Jacobi field, where τ_s is the 1-parameter group associated to X. Hence,

$$\nabla_A \nabla_B X + \nabla_B \nabla_A X + R(X, A) B + R(X, B) A = 0.$$

Now adding the formula (1.5.4) for the curvature tensor $R(A, B) X$ and using the Bianchi identity (1.5.3) we prove our claim. □

2.2.19 Definition. Let M be a symmetric space. Fix a $p \in M$. Let $\mathfrak{g}$ be the Lie algebra of Killing vector fields on M. Then define

$$\mathfrak{h} = \text{set of those } X \in \mathfrak{g} \text{ with } X(p) = 0,$$

$$\mathfrak{m} = \text{set of those } Y \in \mathfrak{g} \text{ with } \nabla Y(p) = 0.$$

2.2.20 Theorem. *Let M be a symmetric space and define, for a fixed $p \in M$, $\mathfrak{g}, \mathfrak{h}, \mathfrak{m}$ as in (2.2.19). Then*

$$\mathfrak{g} = \mathfrak{h} + \mathfrak{m},$$

$$[\mathfrak{h}, \mathfrak{h}] \subset \mathfrak{h},\ [\mathfrak{h}, \mathfrak{m}] \subset \mathfrak{m},\ [\mathfrak{m}, \mathfrak{m}] \subset \mathfrak{h}.$$

For each $Y \in \mathfrak{m}$, *the orbit* $\tau_t p$ *through* p *of the* 1*-parameter group of isometries generated by* Y *is a geodesic* c_Y. *Thus, this* 1*-parameter group is the group formed by the* c_Y*-translations.*

Finally, if $\sigma : \mathfrak{g} \to \mathfrak{g}$ *denotes the involution induced on* $\mathfrak{g}$ *by the reflection* $\sigma_p : (M, p) \to (M, p)$ *on* p, $\sigma | \mathfrak{h} = id$, $\sigma | \mathfrak{m} = -id$.

Remark. This is the Lie-algebra version of the structure of the identity component G of the group of isometries of M, together with its isotropy subgroup H at the point $p \in M$. See the remarks at the beginning of this section.

Proof. If X and X' are in $\mathfrak{h}$, $[X, X'](p) = \nabla_X X'(p) - \nabla_{X'} X(p) = 0$.

If Y and Y' are in $\mathfrak{m}$ then the definition yields $[Y, Y'](p) = \nabla_Y Y'(p) - \nabla_{Y'} Y(p) = 0$.

Before proving that $X \in \mathfrak{h}$, $Y \in \mathfrak{m}$ implies $[X, Y] \in \mathfrak{m}$ we characterize the $Y \in \mathfrak{m}$. Assume $Y \neq 0$. Then $\nabla Y(p) = 0$ implies $Y(p) \neq 0$, cf. (2.2.18). Let $c(t) = \exp_p t\, Y(p)$ be the geodesic with $\dot c(0) = Y(p)$. To show that the 1-parameter group $\{\tau_t\}$ of translations along c gives rise to the Killing field Y it suffices, according to (2.2.18), to show that $\nabla \tilde Y(p) = 0$, where $\tilde Y(q) = \frac{d}{dt} \tau_t q|_0$. But we know from (2.2.5) that $T_p \tau_t$ is the parallel translation along c from 0 to t. If $A \in T_p M$ is the initial vector $b'(0)$ of a curve $b(s)$ then

$$0 = \frac{\nabla}{dt} T\tau_t \,.\, A|_0 = \frac{\nabla}{dt}\frac{\partial}{\partial s} \tau_t b(s)|_{0,0} = \frac{\nabla}{\partial s}\frac{\partial}{\partial t} \tau_t b(s)|_{0,0} = \nabla_A \tilde Y(p).$$

Thus, if $X \in \mathfrak{h}$, $Y \in \mathfrak{m}$, $[X, Y](p) = \nabla_X Y(p) - \nabla_Y X(p) = -\nabla_Y X(p)$, and it suffices to show that $\nabla_A \nabla_Y X(p) = 0$, for all $A \in T_p M$.

To see this let $b(u)$ be a curve with $b'(0) = A$. Let τ_s, τ_t be the 1-parameter groups associated with X and Y, respectively. Then, with $X(p) = 0$ and (1.5.8),

$$\nabla_A \nabla_Y X(p) = \frac{\nabla}{\partial u}\frac{\nabla}{\partial t}\frac{\partial}{\partial s} \tau_s \tau_t b(u)|_{0,0,0} = \frac{\nabla}{\partial t}\frac{\nabla}{\partial u}\frac{\partial}{\partial s} \tau_s \tau_t b(u)|_{0,0,0} =$$

$$\frac{\nabla}{\partial t} T\tau_s \,.\, \tau_t A|_{0,0} = \frac{\nabla}{\partial t} T\tau_t \,.\, A|_0 = 0.$$

Finally, since $T\sigma_p | T_p M = -id$, $T\sigma_p|\mathfrak{m} = -id$. To determine $T\sigma_p|\mathfrak{h}$ we note that, for $X \in \mathfrak{h}$, $T\sigma_p | \nabla X(p) = id$, since $\nabla X(p)$ is a linear transformation on $T_p M$. Hence, $T\sigma_p|\mathfrak{h} = id$. □

2.2.21 Corollary. *Let* M *be a symmetric space,* $\mathfrak{g}, \mathfrak{h}, \mathfrak{m}$ *as in* (2.2.20). *Then the curva-*

ture tensor can be written as

$$R(X, Y)Z(p) = -[[X, Y], Z](p),$$

with $X, Y, Z \in \mathfrak{m}$.

In particular, the sectional curvature $K(\sigma)$ *of the 2-plane spanned by a pair of orthonormal vectors* $Y(p), Y'(p)$ *in* $T_p M$ *is given by*

$$K(\sigma) = -\langle[[Y, Y'], Y'], Y\rangle(p),$$

with Y, Y' *in* $\mathfrak{m}$ *having the values* $Y(p), Y'(p)$ *at* $p \in M$.

Proof. We know from (2.2.18), using $[Y, Y'](p) = 0$ for $Y, Y' \in \mathfrak{m}$,

$$\nabla_Y \nabla_Z X(p) + R(X, Y)Z(p) = 0, \quad \nabla_Z \nabla_X Y(p) + R(Y, Z)X(p) = 0.$$

Taking the difference and inserting the formula (1.5.4) for the curvature and torsion tensor we get

$$R(X, Y)Z(p) = \nabla_Z(\nabla_X Y - \nabla_Y X)(p) = \nabla_Z[X, Y] = [Z, [X, Y]]. \quad \square$$

2.2.22 Lemma. *Let M be a symmetric space,* $\mathfrak{g} = \mathfrak{h} + \mathfrak{m}$ *the decomposition of the Lie algebra* $\mathfrak{g}$ *of its Killing vector fields according to* (2.2.20). *Then the scalar product on* $\mathfrak{m}$,

$$\langle Y, Y'\rangle_{\mathfrak{m}} = g(p)(Y(p), Y'(p)),$$

can be complemented by a (positive definite) scalar product on $\mathfrak{h}$ *by defining*

$$\langle X, X'\rangle_{\mathfrak{h}} = -\mathit{tr}\, \mathrm{ad}\, X \circ \mathrm{ad}\, X'.$$

$\langle\ ,\ \rangle_{\mathfrak{h}}$ *is nothing but the negative of the Killing form of* $\mathfrak{g}$, *restricted to* $\mathfrak{h}$, *cf.* (1.7.20).

Construct from $\langle , \rangle_{\mathfrak{h}}$ *and* $\langle , \rangle_{\mathfrak{m}}$ *a scalar product on* $\mathfrak{g}$ *by letting* $\mathfrak{h}$ *and* $\mathfrak{m}$ *be orthogonal.*

The thus defined scalar product $\langle , \rangle_{\mathfrak{g}}$ *on* $\mathfrak{g}$ *is* ad $\mathfrak{h}$*-invariant, i.e., for* $X \in \mathfrak{h}$,

$$\langle \mathrm{ad}\, X . Z, Z'\rangle_{\mathfrak{g}} + \langle Z, \mathrm{ad}\, X . Z'\rangle_{\mathfrak{g}} = 0.$$

Proof. The ad $\mathfrak{h}$-invariance of $\langle , \rangle_{\mathfrak{m}}$ is proved in (1.10.9), since it means, with Y, Y' in $\mathfrak{m}$, X in $\mathfrak{h}$,

$$g(p)(-\nabla_Y X(p), Y'(p)) + g(p)(Y(p), -\nabla_{Y'} X(p)) = 0.$$

The ad $\mathfrak{h}$-invariance of $\langle\ ,\ \rangle_{\mathfrak{h}}$ will follows from (1.7.21). To see that it is positive definite, we only need observe that the mapping

$$\mathrm{ad}\, X \circ \mathrm{ad}\, X' : \mathfrak{m} \to \mathfrak{m}$$

is given by

$$U \mapsto [X, [X', U]](p) = [X, -\nabla_U X'](p) = \nabla_{\nabla_U X'} X(p).$$

Since $\nabla X(p): T_p M \to T_p M$, $X \in \mathfrak{h}$, is skew symmetric with respect to $g(p)$, $\nabla X(p)$ is represented by a skew symmetric matrix (x_{ik}) with respect to an orthonormal basis in T_pM. Hence, $\langle X, X\rangle_{\mathfrak{h}} \geq \sum_{i,k} (x_{ik})^2 > 0$, whenever $X \neq 0$ – note that $-\, tr\, \mathrm{ad}\, X|\mathfrak{h} \circ \mathrm{ad}\, X'|\mathfrak{h}$ is positive semi-definite. □

We now prove a large part of the decomposition theorem for the so-called "orthogonal involutive" (in the terminology of Wolf [1]) or "orthogonal symmetric" (in the terminology of Kobayashi and Nomizu [1]) Lie algebras $(\mathfrak{g}, \mathfrak{h}, \sigma)$.

2.2.23 Lemma. *Let M be a symmetric space, $\mathfrak{g} = \mathfrak{h} + \mathfrak{m}$ the decomposition of the Lie algebra $\mathfrak{g}$ of its Killing vector fields, together with its scalar product, see* (2.2.20), (2.2.22).

Let $B_{\mathfrak{m}}$ be the restriction of the Killing form B of $\mathfrak{g}$ to $\mathfrak{m}$. Let

$$\beta : \mathfrak{m} \to \mathfrak{m}$$

be the self-adjoint operator determined by $B_{\mathfrak{m}}$ with respect to the scalar product $\langle\, ,\rangle_{\mathfrak{m}}$. Denote by

$$\mathfrak{m} = \mathfrak{m}_1 \oplus \ldots \oplus \mathfrak{m}_r$$

the decomposition of $\mathfrak{m}$ into the eigenspaces of β, belonging to the different eigenvalues of β. Then this is an $\langle\, ,\rangle_{\mathfrak{m}}$-orthogonal decomposition. Moreover, $[\mathfrak{m}_i, \mathfrak{m}_j] = 0$ for $i \neq j$.

2.2.24 Complement. *Assume, in addition, that $B_{\mathfrak{m}}$ is nondegenerate. This is equivalent to saying that $\mathfrak{g}$ is semi-simple. Then the scalar product $\langle\, ,\rangle_{\mathfrak{m}}$ on $\mathfrak{m}$ can be written as*

$$\frac{1}{\lambda_1} B_{\mathfrak{m}_1} \oplus \ldots \oplus \frac{1}{\lambda_r} B_{\mathfrak{m}_r}.$$

Here, $B_{\mathfrak{m}_j}$ is the restriction of $B_{\mathfrak{m}}$ to the subspace $\mathfrak{m}_j$, and λ_j is the eigenvalue of $\beta|\mathfrak{m}_j$.

Proof. The orthogonality of $\mathfrak{m}_i$ and $\mathfrak{m}_j$ with respect to $\langle\, ,\rangle_{\mathfrak{m}}$ and $B_{\mathfrak{m}}$ is evident. Let now $Y_i \in \mathfrak{m}_i$, $Y_j \in \mathfrak{m}_j$, $i \neq j$, λ_i, λ_j their different eigenvalues. Then

$$B([Y_i, Y_j], [Y_i, Y_j]) = -B(Y_j, \mathrm{ad}\, Y_i\, [Y_i, Y_j]) =$$

$$-\lambda_j \langle Y_j, \mathrm{ad}\, Y_i [Y_i, Y_j]\rangle = \lambda_j \langle [Y_i, Y_j], [Y_i, Y_j]\rangle.$$

The same formula holds with i and j interchanged.

As for the complement, we simply observe that $B(Y_i, Y_i') = \lambda_i \langle Y_i, Y_i' \rangle$ in $\mathfrak{m}_i$. □

2.2.25 Definition. Let M be a symmetric space. Let $\mathfrak{g} = \mathfrak{h} + \mathfrak{m}$ be the decomposition of the Lie algebra of its Killing fields.

(i) M is said to be of *Euclidean type* if $[\mathfrak{m}, \mathfrak{m}] = 0$, i. e., $B_{\mathfrak{m}} \equiv B|\mathfrak{m} \equiv 0$.

(ii) M is called *semi-simple* if $\mathfrak{g}$ is semi-simple, i. e., if $B_{\mathfrak{m}} = B|\mathfrak{m}$ is non-degenerate, cf. (2.2.22).

(iii) M is said to be of *compact type (non-compact type)* if it is semi-simple and $K \geq 0$ $(K \leq 0)$.

Remark. One usually employs the terms Euclidean, compact (non-compact) type only for the Lie algebras $\mathfrak{g}$ with their decomposition $\mathfrak{h}+\mathfrak{m}$. It is easy to prove, using (2.2.23), that each such Lie algebra can be written as a direct Lie algebra sum of Lie algebras of Euclidean, compact and non-compact type. This then corresponds to a decomposition of the universal covering of the symmetric space M, see Kobayashi and Nomizu [1], Wolf [1].

2.2.26 Theorem. *Let M be a symmetric space.*

(i) *If M is of compact type, M is compact, and so is its universal covering. Hence, the fundamental group $\pi_1 M$ is finite.*

The injectivity radius $\iota(M)$ of M satisfies $\pi/2\sqrt{K_1} \leqslant \iota(M) \leqslant \pi/\sqrt{K_1}$ where $K_1 = \sup K =$ the greatest eigenvalue of the operator $R_{X_0}, |X_0| = 1$.

(ii) *If M is of non-compact type, M is simply connected. For every $p \in M$, the exponential mapping $\exp_p\colon T_p M \to M$ is a diffeomorphism.*

Remark. The estimates on the injectivity radius in (2.2.26, i) are optimal in general as is seen by taking for M the real projective space P^n and the sphere S^n. If $\pi_1 M = 0$, one can prove that $\iota(M) = \pi/\sqrt{K_1}$, see Cheeger and Ebin [1].

Proof. M being semi-simple implies that for every $Y \in \mathfrak{m}$, $Y \neq 0$, $\operatorname{ad} Y\colon \mathfrak{m} \to \mathfrak{h}$ is non-zero. On account of (2.2.23), (2.2.24) we can assume that given Y in $\mathfrak{m}$, $|Y|_{\mathfrak{m}} = 1$, Y has a non-zero component in some eigenspace $\mathfrak{m}'$ of the operator $\beta\colon \mathfrak{m} \to \mathfrak{m}$ with an eigenvalue $\lambda \neq 0$. There exists $Y' \in \mathfrak{m}'$ with $\langle Y, Y' \rangle_{\mathfrak{m}} = 0$, $|Y'|_{\mathfrak{m}} = 1$ such that $[Y, Y'] \neq 0$. If σ is the plane spanned by Y and Y', then we find according to (2.2.21)

$$K(\sigma) = -\langle [[Y, Y'], Y'], Y \rangle_{\mathfrak{m}} = -\frac{1}{\lambda} B([[Y, Y'], Y'], Y) =$$

$$\frac{1}{\lambda} B([Y, Y'], [Y, Y']) = -\frac{1}{\lambda} \langle [Y, Y'], [Y, Y'] \rangle_{\mathfrak{h}} \neq 0.$$

Let now M be of compact type. On every geodesic $c(t) = \exp tY$, $|Y| = 1$, there exist conjugate points. Indeed, as we just saw, there exist 2-planes σ tangent to c with $K(\sigma) \neq 0$. Actually, under our hypothesis, $K(\sigma) > 0$. Thus, M is compact, and so is its universal covering $\tilde{M}$, cf. (2.1.5).

Since there exist conjugate points on M on geodesics of length $\pi/\sqrt{K_1}$, $\iota(M) \leqslant \pi/\sqrt{K_1}$. If $\iota(M) < \pi/\sqrt{K_1}$, then there exists a closed geodesic c of length $\leqslant 2\iota(M) < 2\pi/\sqrt{K_1}$ on M, see (2.1.11, iii). The point $c(0)$ possesses a conjugate point $c(t_1)$ along $c|[0, t_1]$. If we assume $|\dot{c}| = 1$, then $t_1 \geqslant \pi/\sqrt{K_1}$.

From (2.2.17) we know that a whole family $c_s = \tau_s c$ of geodesics passes through $c(0), c(t_1)$, where $\{\tau_s\}$ is a 1-parameter group of isometries. Thus, all c_s are closed geodesics of length $= L(c)$. But then, since $L(c)$ is the minimum of all possible (non-trivial) geodesic biangles on M, there are only two possibilities for $c(t_1)$. Either, $c(t_1)$

is the point opposite to $c(0)$ on c — but then $c(t_1)$ would be conjugate to $c(0)$ along c at distance $\iota(M) = L(c)/2 < \pi/\sqrt{K_1}$ which is impossible.

The other possibility is that $c(t_1)$ coincides with $c(0)$. Hence, $L(c) \geqslant \pi/\sqrt{K_1}$, i. e., $\iota(M) \geqslant \pi/2\sqrt{K_1}$.

Now assume M to be of non-compact type. According to (2.2.16), there can be no closed geodesic c on M, since not all sectional curvatures $K(\sigma)$ with σ tangent to c are $= 0$. Also, no geodesic on M carries pairs of conjugate points, as is shown by the form of the Jacobi equations (2.2.10) for $\lambda \leqslant 0$. Thus, for every $p \in M$, $C(p)$ must be empty for otherwise there would exist a geodesic loop through p, cf. (2.1.11, iii), which is even a closed geodesic, see (2.2.7). (2.1.14) implies that $\exp_p: T_p M \to M$ is a diffeomorphism. □

2.2.27 Example. *The complex projective space.*

We continue with (1.11.14). Note that there we did not assume that the complex projective space $P\mathbb{E}$ has finite dimension. For the subsequent considerations this hypothesis also is not necessary, although we will use a few facts about symmetric spaces which we defined for finite dimension only.

Recall that we had the isometric submersion

$$F: \mathbb{E}^* \to P\mathbb{E}$$

where $\mathbb{E}^* = \mathbb{E} - \{0\}$ is isometric to $S\mathbb{E} \times \mathbb{R}$, $S\mathbb{E}$ = sphere of constant curvature = 1 in $\mathbb{E}$.

Choose $p \in P\mathbb{E}$, $N \in F^{-1}(p) \cap S\mathbb{E}$. A unit vector $X_0 \in T_p P\mathbb{E}$ then has a unique horizontal lift X_0^* in the space $T_N^h \mathbb{E}^*$. Let

$$c^*(t) = \cos t N + \sin t X_0^*$$

be the geodesic in $S\mathbb{E} \subset \mathbb{E}^*$ with initial direction X_0^*. Since $\langle \dot{c}^*(t), c^*(t)\rangle = \langle \dot{c}^*(t), ic^*(t)\rangle = 0$, this is a horizontal curve, i. e., $\dot{c}^*(t) \in T^h_{c^*(t)} \mathbb{E}^*$, all t.

From (1.11.11) we therefore know that $c(t) = F \circ c^*(t)$ is a geodesic on $P\mathbb{E}$ with initial direction X_0. Since $T_N F . \dot{c}^*(0) = T_{-N} F . \dot{c}^*(\pi)$, $c(t)$ is a closed geodesic of prime period π. We thus have shown:

Every geodesic on $P\mathbb{E}$ is closed of length π.

Let $\mathbb{E} = \mathbb{E}' \oplus \mathbb{E}''$ be an orthogonal decomposition of $\mathbb{E}$ into complex (i. e., i-invariant) subspaces. Let $\sigma_{\mathbb{E}'}: \mathbb{E} \to \mathbb{E}$ be the reflection on $\mathbb{E}'$, i. e., $\sigma_{\mathbb{E}'}|\mathbb{E}' = id$, $\sigma_{\mathbb{E}'}|\mathbb{E}'' = -id$. This induces an isometry

$$\sigma_{P\mathbb{E}'}: P\mathbb{E} \to P\mathbb{E}$$

on $P\mathbb{E}$ having as fixed point set

$$\mathrm{Fix}(\sigma_{P\mathbb{E}'}) = P\mathbb{E}' \cup P\mathbb{E}''.$$

According to (1.10.15), each of these projective subspaces of $P\mathbb{E}$ is totally geodesic embedded.

In particular, if $\mathbb{E}' = \{N, iN\}$ is complex 1-dimensional, i.e., $\mathbb{E}' \cong \mathbb{C}$, then $P\mathbb{E}' \cong P\mathbb{C}$ is a point $p \in P\mathbb{E}$. We will see below that in this case $P\mathbb{E}''$ is the cut locus $C(p)$ of p.

If $\mathbb{E}'$ is complex 2-dimensional, $P\mathbb{E}' \cong P\mathbb{C}^2$ is isometric to the 2-sphere $S^2_{1/2}$ of curvature $K = 4$.

Consider $p \in P\mathbb{E}$. Let $\mathbb{E}^1_p = \mathbb{C}N$ be the representative space of p, i.e., $F(N) = p$, $N \in S\mathbb{E}$. The tangent space $T_p P\mathbb{E}$ carries a complex structure stemming from the isomorphism $T_N F\colon T^h_N \mathbb{E}^* \to T_p P\mathbb{E}$.

Let $\mathbb{E}^2_p$ be a 2-space containing $\mathbb{E}^1_p$. Then $F(\mathbb{E}^2_p)$ is a 2-sphere $S^2_{1/2}$. $T_p S^2_{1/2}$ is given as the image under $T_N F$ of $\mathbb{E}^2 \cap T^h_N \mathbb{E}$. Thus, $T_p S^2_{1/2}$ also carries a complex structure. The geodesics with initial direction $X_0 \in T_p S^2_{1/2}$, $|X_0| = 1$, are the parameterized great circles on $S^2_{1/2}$, starting in p. Every such geodesic passes for $t = \pi/2$ through the antipodal point $\tilde{p} = F(\tilde{\mathbb{E}}^1_p)$ of p on $S^2_{1/2} = F(\mathbb{E}^2)$, where $\tilde{\mathbb{E}}^1_p$ is the complex line in $\mathbb{E}^2$ orthogonal to $\mathbb{E}^1_p$.

We paraphrase this by saying that through each $p \in P\mathbb{E}$ there passes a 'bouquet' of totally geodesic embedded 2-spheres of constant curvature $K = 4$; the tangent planes of these 2-spheres in p are precisely the i-invariant 2-planes in $T_p P\mathbb{E}$.

It follows that the cut locus $C(p)$ of $p \in P\mathbb{E}$ is the projective subspace $P\mathbb{E}^{1\perp}_p$ where $\mathbb{E}^{1\perp}_p$ is the orthogonal complement of $\mathbb{E}^1_p$ in $\mathbb{E}$. Indeed, $P\mathbb{E}^{1\perp}_p$ is the union of the antipodal points $\tilde{p}$ of p on the bouquet of 2-spheres $S^2_{1/2}$ through p. To see this we only have to exclude the possibility that, for some geodesic $c, |\dot{c}| = 1$, starting from p, the cut locus is reached before $c(\pi/2)$. Since the sectional curvature of $P\mathbb{E}$ is $\leqslant 4$, we then would have from (2.1.11, iii) the existence of a closed geodesic lf length $< \pi$. But every such geodesic lies on some sphere $S^2_{1/2}$ of curvature 4, and there, a closed geodesic has length $\geqslant \pi$.

Let $X_0 \in T_p P\mathbb{E}$ be a unit vector. Then we have from (1.11.14) for the curvature operator R_{X_0} the expression

$$R_{X_0} Y = Y + 3\langle Y, iX_0\rangle iX_0; \quad \langle Y, X_0\rangle = 0.$$

Thus, R_{X_0} has a 1-dimensional eigenspace spanned by iX_0 for the eigenvalue $\lambda = 4$ whereas its orthogonal complement is the eigenspace for $\lambda = 1$.

We therefore have according to (2.2.10) the following two types of Jacobi fields along $c(t) = \exp_p tX_0$, vanishing for $t = 0$:

$$Y(t) = \frac{1}{2}\sin 2t A(t); \quad A(t) = \overset{t}{\underset{0}{\|}} c\, iX_0$$

$$Z(t) = \sin t B(t); \quad B(t) = \overset{t}{\underset{0}{\|}} c\, Y_0 .$$

Here, $|Y_0| = 1, \langle Y_0, X_0\rangle = \langle Y_0, iX_0\rangle = 0$.

2.2.28 Example. *The orthogonal group.* We consider the identity component $M = S\mathbb{O}(n)$ of the orthogonal group $\mathbb{O}(n)$, i.e., those $A \in GL(n, \mathbb{R})$ with $A\,{}^tA$

$= I, \det A = 1$, cf. (1.7.3), (1.7.6), (1.7.11), (1.7.22). Let $n > 2$. A Riemannian metric on M we take

$$g(X, Y) \equiv \langle X, Y \rangle = \frac{1}{2-n} B(X, Y) = \sum_{i,k} x_{ik} y_{ik},$$

where B is the Killing form. Here we use the fact that every $X \in T_p M$ is the value at p of a well determined element of the Lie algebra $\mathfrak{o}(n)$ of $S\mathbb{O}(n)$.

We claim that the symmetric covariant derivation ∇ introduced in (1.7.12) coincides with the Levi-Cività derivation associated to the Riemannian metric g. Since ∇ has vanishing torsion tensor, we only need to show that the parallel translation defined by ∇ respects the scalar product g. The formula (1.7.19) shows that parallel translation along geodesics is an isometry if and only if $\operatorname{ad} X$ is skew symmetric w.r.t. g. But this is the case for the Killing form, see (1.7.21).

2.2.29 Example. *Symmetric homogeneous spaces.*

We consider a pair (G, H) where G is a connected Lie group and H a compact connected subgroup. We will assume that H does not contain a normal (= invariant) subgroup of G – otherwise we take the quotient of (G, H) with respect to such a subgroup.

Let $(\mathfrak{g}, \mathfrak{h})$ be the Lie algebra of (G, H). We assume the existence of a vector space complement $\mathfrak{m}$ of $\mathfrak{h}$ in such that

$$[\mathfrak{h}, \mathfrak{h}] \subset \mathfrak{h};\ [\mathfrak{h}, \mathfrak{m}] \subset \mathfrak{m};\ [\mathfrak{m}, \mathfrak{m}] \subset \mathfrak{h},$$

where the first relation is always true, of course. Note that this allows us to define a Lie algebra automorphism $\sigma: \mathfrak{g} \to \mathfrak{g}$ by $\sigma|\mathfrak{h} = id, \sigma|\mathfrak{m} = -id$.

We begin by showing that the set $\bar{G} = G/H$ of left cosets of H carries the structure of a manifold such that the quotient map

$$F: G \to \bar{G};\ g \mapsto gH$$

becomes a submersion, $\dim \bar{G} = \dim \mathfrak{m}$, cf. Warner [2]. For this we do not need all the hypotheses made for (G, H) and $(\mathfrak{g}, \mathfrak{h})$.

First observe that G operates transitively on $\bar{G}$ by

$$g \in G \mapsto \{\bar{L}_g : \bar{G} \mapsto \bar{G};\ g'H \mapsto gg'H\}.$$

For the topology on $\bar{G}$ we take as open sets $\bar{U} \subset \bar{G}$ those for which $F^{-1}\bar{U}$ is open. F is an open mapping since for $U \subset G$ open

$$F^{-1}(F(U)) = \bigcup_{h \in H} U.h$$

is open.

$\bar{G}$ also is Hausdorff: If $\bar{g} \neq \bar{g}'$ then we can construct disjoint neighborhoods $\bar{U}, \bar{U}'$ of $\bar{g}, \bar{g}'$ as follows. The counter image of H in $G \times G$ under the mapping

$$(g, g') \in G \times G \to g^{-1}g' \in G$$

is closed. Thus, if $g^{-1}g' \notin H$, there exists a neighborhood of the form $U \times U'$ of (g, g') in $G \times G$ which does not meet this counter image. Put now $\bar{U} = FU$, $\bar{U}' = FU'$.

To construct a chart for a neighborhood of $\bar{e} = Fe \in \bar{G}$ we choose a neighborhood $U + V \subset \mathfrak{h} + \mathfrak{m} = \mathfrak{g}$ of $0_e \in \mathfrak{g}$ so small that $\exp_e: U + V \to G$ (the exponential map defined by the left-invariant covariant derivation, see (1.7.7)) is a diffeomorphism onto its image G_e. Consider the mapping

$$\Phi: V \times V \to G; (Y, Y') \mapsto (\exp_e Y)^{-1} . \exp_e Y'.$$

H (or at least $H \cap G_e$) is a submanifold of G and we see that Φ is transversal to H at e, $\Phi \pitchfork_e H$ in the notation used in (1.3.6). $T_{(0_e, 0_e)}\Phi$ carries the subspace $\{0_e\} \times \mathfrak{m}$ into $\mathfrak{m}$. It follows – at least by replacing $U \times V$ and hence G_e by suitably smaller neighborhoods for which we employ the same notation – that $\Phi \pitchfork (H \cap G_e)$.

According to (1.3.8), $\Phi^{-1}(H \cap G_e)$ is therefore a submanifold of $\mathfrak{m} \times \mathfrak{m}$ of codimension $\operatorname{codim} \mathfrak{h} = \dim \mathfrak{m}$. But clearly, the diagonal of $V \times V$ is such a manifold, having as its image under Φ the element $e \in H$. That is, $\Phi(Y, Y') \in H$ if and only if $Y = Y'$.

We define an atlas for $\bar{G} = FG$ by first defining a chart around Fe. Put $FG_e = G_e H = \bar{G}_e$ and consider

$$u_e: \bar{G}_e \to V; \exp_e Y . H \mapsto Y.$$

$(u_e, \bar{G}_e)$ is then a chart covering $\bar{e} = Fe$.

To get an atlas for $\bar{G}$ we translate this chart. For every $g \in G$ we define $(u_g, \bar{G}_g)$ by

$$u_g = u_e \circ \bar{L}_g : \bar{G}_g = \bar{L}_g^{-1} \bar{G}_e \to V.$$

The quotient map $F: G \to \bar{G}$ thus becomes a submersion; the representation of $F|G_e$ in the charts $((\exp_e|G_e)^{-1}, G_e)$ and $(u_e, \bar{G}_e)$ is the projection

$$pr_2 : \mathfrak{g} = \mathfrak{h} + \mathfrak{m} \to \mathfrak{m}$$

onto the second summand, restricted to $U \times V$.

The G-action

$$G \times \bar{G} \to \bar{G}; (g, \bar{g}') \mapsto \bar{L}_g \bar{g}' = gg'H$$

is an action by diffeomorphisms since $\bar{L}_g \bar{g}' = Fgg'$, $g' \in \bar{g}'$. $H \subset G$ is the isotropy group of $\bar{e} = eH$, $T_{\bar{e}}\bar{L}_h: T_{\bar{e}}\bar{G} \to T_{\bar{e}}\bar{G}$ is given by $\operatorname{Ad} h: \mathfrak{m} \to \mathfrak{m}$ if we identify $T_{\bar{e}}\bar{G}$ with $\mathfrak{m}$. Or else, $T\bar{L}_h = F \circ \operatorname{Ad} h$.

We now use the remaining hypotheses, in particular that H is compact connected, to define Riemannian metrics on G and $\bar{G}$ such as to make the G-action on these spaces an action by isometries and $F: G \to \bar{G}$ an isometric submersion.

From our assumptions we have that the adjoint representation $\operatorname{Ad} H$ of H in $GL(\mathfrak{g})$ leaves the subspaces $\mathfrak{h}$ and $\mathfrak{m}$ of $\mathfrak{g}$ invariant. Denote the restriction to these subspaces by $\operatorname{Ad}_{\mathfrak{h}} H$ and $\operatorname{Ad}_{\mathfrak{m}} H$, respectively. Both are compact subgroups of $GL(\mathfrak{h})$ and $GL(\mathfrak{m})$, respectively. We assume it to be known that it then is possible to define scalar products $\langle\ ,\ \rangle_{\mathfrak{h}}$ and $\langle\ ,\ \rangle_{\mathfrak{m}}$ on $\mathfrak{h}$ and $\mathfrak{m}$ in such a manner that $\mathrm{Ad}_{\mathfrak{h}} H$ and $\mathrm{Ad}_{\mathfrak{m}} H$ become subgroups of the such defined orthogonal groups $\mathbb{O}(\mathfrak{h})$ and $\mathbb{O}(\mathfrak{m})$, cf. Chevalley [1]. Indeed, any real representation of a compact lie group is equivalent to representation to orthogonal matrices.

We take as scalar product $\langle,\rangle_{\mathfrak{g}}$ on $\mathfrak{g} \cong T_e G$ the orthogonal sum of $\langle,\rangle_{\mathfrak{h}}$ and $\langle,\rangle_{\mathfrak{m}}$. Translating with TL_g the tangent space $T_e G$ into $T_g G$ we make G into a Riemannian manifold.

In $T_{\bar{e}}\bar{G}$ we define the scalar product $\langle\ ,\ \rangle_{\bar{e}}$ such as to make $T_e F|\mathfrak{m} : \mathfrak{m} \to T_{\bar{e}}\bar{G}$ an isometry. The action of $T\bar{L}_h, h \in H$, on $T_{\bar{e}}G$ then is by isometries since it is given by $T_e F \circ \mathrm{Ad}\, h$. Define the scalar product $\langle\ ,\ \rangle_{\bar{g}}$ on $T_{\bar{g}}G$ by letting $T_{\bar{e}}\bar{L}g : T_{\bar{e}}\bar{G} \to T_{\bar{g}}\bar{G}$ be an isometriy, with some $g \in \bar{g}$. If also $g' \in \bar{g}, g^{-1}g' = h \in H$ and, as we just saw, $T\bar{L}_h = T\bar{L}_g^{-1} \circ T\bar{L}_{g'}$ is an isometry.

We can use (1.11.12) to determine the sectional curvature $K(\bar{\sigma})$ of a 2-plane $\bar{\sigma} \subset T_{\bar{e}}\bar{G}$. We denote by $\sigma \subset \mathfrak{m}$ its horizontal lift – note the change in notations as compared to (1.11.12). Then, if $\sigma = \mathrm{span}\{X, Y\}$, X, Y orthonormal,

$$K(\bar{\sigma}) = K(\sigma) + \tfrac{3}{4}|[X, Y]_{\mathfrak{h}}|^2.$$

The expression for $K(\sigma)$ can also be given in terms of Lie brackets and ad, cf. Cheeger and Ebin [1].

We defined above the involution $\sigma : \mathfrak{g} \to \mathfrak{g}$ with $\sigma|\mathfrak{h} = id, \sigma|\mathfrak{m} = -id$. If we knew that there exists an isometric symmetry $\sigma_e : G \to G$ with $T_e\sigma_e = \sigma$ then G as well as $\bar{G}$ would be a symmetric space. Such an extension σ_e of σ exists locally, but not necessarily globally, i.e., $\bar{G} = G/H$ is a locally symmetric space. However, if e.g. G is simply connected, $\bar{G}$ is (globally) symmetric.

We now consider the case when the scalar product $\langle,\rangle_{\mathfrak{g}}$ on $\mathfrak{g}$ is not only $\mathrm{Ad}H$-invariant but also $\mathrm{Ad}G$-invariant. This is equivalent to saying that

$$\langle \mathrm{ad}X \,.\, Y, Z\rangle_{\mathfrak{g}} + \langle Y, \mathrm{ad}X \,.\, Z\rangle_{\mathfrak{g}} = 0,$$

for all X, Y, Z in $\mathfrak{g}$. From (1.7.19) it follows that the symmetric covariant derivation ∇ on G, see (1.7.12), coincides with the Levi-Cività derivation. We therefore have with (1.7.13) for $\sigma = \mathrm{span}\{X, Y\}$:

$$K(\sigma) = \tfrac{1}{4}\langle \mathrm{ad}\, Y \,.\, [X, Y], X\rangle = \tfrac{1}{4}|[X, Y]|^2.$$

Hence, if $\bar{\sigma} = \mathrm{span}\{T_e F \,.\, X, T_e F \,.\, Y\}$,

$$K(\bar{\sigma}) = \tfrac{1}{4}|[X, Y]_{\mathfrak{m}}|^2 + |[X, Y]_{\mathfrak{h}}|^2,$$

i.e., on G and $\bar{G}$ the sectional curvature is $\geqslant 0$.

From (1.11.11) we know that the geodesics $\bar{c}(t)$ in $\bar{G}$ with $\dot{\bar{c}}(0) = \bar{X} \in T_{\bar{e}}\bar{G}$ can be written as $\bar{c}(t) = F \circ c(t)$ where $c(t)$ is a horizontal lift of $\bar{c}(t)$. We can assume $\dot{c}(0) \in \mathfrak{m}$. Thus, the geodesics $\bar{c}$ on $\bar{G}$, starting at $\bar{e}$, are the projections $F \circ c$ of 1-parameter subgroups in G with initial direction in $\mathfrak{m}$.

Here are a few examples:

$$S^n = S\mathbb{O}(n+1)/S\mathbb{O}(n), \quad \text{cf. } (1.10.2, 6)$$

$$H^n = S\mathbb{O}(1, n)/S\mathbb{O}(n), \quad \text{cf. } (1.11.8)$$

$$P\mathbb{C}^n = S\mathbb{U}(n+1)/S(\mathbb{U}(n) \times \mathbb{U}(1)).$$

Here, $S\mathbb{U}(k)$ is the special unitary group, formed by the (k, k)-matrices $A \in M(k; \mathbb{C})$ with complex elements satisfying $A^t\bar{A} = id$, $\det A = 1$.

$G(k, l) = S\mathbb{O}(k+l)/S\mathbb{O}(k) \times S\mathbb{O}(l)$, the *Grassmann manifold of the (real) k-dimensional subspaces of the (real) space* $\mathbb{R}^{k+l}$. In particular, for $l = 1$, we get the real projective space P^{k+1}.

$G_{\mathbb{C}}(k, l) = S\mathbb{U}(k+l)/S(\mathbb{U}(k) \times \mathbb{U}(l))$, the *Grassmann manifold of the (complex) k-dimensional subspaces in the complex space* $\mathbb{C}^{k+l}$. In particular, for $l = 1$, we get the complex projective space $P\mathbb{C}^{k+1}$.

$S\mathbb{U}(n)/S\mathbb{O}(n)$, the space of Lagrangian (= totally isotropic) n-dimensional spaces in $\mathbb{R}^{2n}$ with the standard symplectic structure – also called *Lagrangian Grassmann manifold.*

$S\mathbb{O}(2n)/\mathbb{U}(n)$, the *space of Lagrangian subspaces of* $\mathbb{C}^{2n}$, endowed with the bilinear form $(z, w) \in \mathbb{C}^{2n} \times \mathbb{C}^{2n} \mapsto \sum_i z_i w_i \in \mathbb{C}$.

$S\mathbb{U}(2n)/Sp(n)$, the *space of Lagrangian subspaces of* $\mathbb{H}^{2n}$, $\mathbb{H}$ the quaternions, endowed with the bilinear form $(z, w) \in \mathbb{H}^{2n} \times \mathbb{H}^{2n} \mapsto \sum_i z_i w_i \in \mathbb{H}$.

For the geometry of symmetric spaces, in particular, for the structure of their cut locus, see Sakai [1].

2.3 The Hilbert Manifold of H^1-curves

We begin in this section the construction of a proper Hilbert manifold, associated canonically to a finite dimensional Riemannian manifold M. The Hilbert manifold $H^1(I, M)$ in question is formed by the maps $c: I \to M$ of class H^1, $I = [0, 1]$, cf. (2.3.1). Of greater importance than the space of all H^1-maps $c: I \to M$ are certain submanifolds of finite codimension which we will introduce in (2.4).

Here we describe in full detail the natural atlas for $H^1(I, M)$. The charts $(\exp_c^{-1}, \mathcal{U}(c))$, $c \in H^1(I, M)$, are defined with the help of the inverse of the exponential map, restricted to 'short' vector fields along c, cf. (2.3.10), (2.3.12).

Next we describe the two canonical bundles α^0 and α^1 over $H^1(I, M)$. The fibre $(\alpha^0)^{-1}(c)$ over c consists of the H^0-vector fields along c while $(\alpha^1)^{-1}(c)$, consists of the H^1-vector fields along c, is identified with the tangent space $T_c H^1(I, M)$ at c, cf. (2.3.13). This is used then to show that there exists a canonical Riemannian metric on $H^1(I, M)$, cf. (2.3.19), and that the energy integral $E(c)$ is differentiable, cf. (2.3.20). The only critical points of E are the constant maps.

It is well known that the set $C^0(I, \mathbb{R}^n)$ of continuous curves $c: I = [0, 1] \to \mathbb{R}^n$ in euclidean n-space $\mathbb{R}^n$ is a Banach space. $C^0(I, \mathbb{R}^n)$ can be viewed as the completion of the space $C'^{\infty}(I, \mathbb{R}^n)$ of piecewise differentiable curves with respect to the maximum norm

$$\|c\|_\infty = \sup_{t \in I} |c(t)|.$$

The associated distance is

$$d_\infty(c, c') = \sup_{t \in I} |c(t) - c'(t)|.$$

For an element $c \in C^0(I, \mathbb{R}^n)$ in general there exists neither the length $L(c)$ nor the energy integral $E(c)$. Therefore, in differential geometry one considers a different norm on $C'^{\infty}(I, \mathbb{R}^n)$, i.e., the norm $\|c\|_1$ derived from the scalar product

$$\langle c, c' \rangle_1 = \langle c, c' \rangle_0 + \langle \partial c, \partial c' \rangle_0 .$$

Here, $\langle e, e' \rangle_0 = \int_I e(t) \cdot e'(t)\, dt$ and $\partial c(t) = \dot{c}(t)$. The completion of $C'^{\infty}(I, \mathbb{R}^n)$ with respect to the norm $\|c\|_1$ is denoted by $H^1(I, \mathbb{R}^n)$. According to a classical result of Lebesgue, an H^1-curve $c: I \to \mathbb{R}^n$ can be described as an absolutely continuous curve for which $\partial c(t) = \dot{c}(t)$ exists for almost all t and ∂c is square integrable, i.e., $\langle \partial c, \partial c \rangle_0 < \infty$.

We recall that $c: I \to \mathbb{R}^n$ is called *absolutely continuous*, if, for every $\varepsilon > 0$, there exists a $\delta > 0$ such that

$$0 \leqslant t_0 < \ldots < t_{2k+1} \leqslant 1 \quad \text{and} \quad \sum_{i=0}^{k} |t_{2i+1} - t_{2i}| < \delta$$

imply $\displaystyle\sum_0^k |c(t_{2i+1}) - c(t_{2i})| < \varepsilon .$

Note that in particular $H^1(I, \mathbb{R}^n)$ is a subset of $C^0(I, \mathbb{R}^n)$. Actually, as one can see easily, the inclusion is continuous, cf. (2.3.5). The functional $E(c) = \langle \partial c, \partial c \rangle_0 / 2$ on $C'^{\infty}(I, \mathbb{R}^n)$ possesses an extension to the Hilbert space $H^1(I, \mathbb{R}^n)$ and there it defines an $\mathbb{R}$-valued C^∞-function. Its differential is given by $DE(c) . \eta = \langle \partial c, \eta \rangle_0$.

We now extend these constructions to the case of curves on a n-dimensional Riemannian manifold M. We assume M to be complete, $I = [0, 1]$.

2.3.1 Definition (i) Denote by $C'^{\infty}(I, M)$ the *set of piecewise differentiable curves* $c: I \to M$.

(ii) By $C^0(I, M)$ denote the *space of continuous curves* $c: I \to M$, endowed with the metric

$$d_\infty(c, c') = \sup_{t \in I} d(c(t), c'(t))$$

Here, d is the metric derived from the Riemannian metric on M, cf. (1.6.11), (1.9.4).

(iii) A curve $c: I \to M$ is called of *class* H^1 if, for a chart (u, M') of M and $I' = c^{-1}(M')$ the mapping $t \in I' \mapsto u \circ c(t) \in \mathbb{R}^n$ is in $H^1(I', \mathbb{R}^n)$. By $H^1(I, M)$ we denote *the set of* H^1*-maps* $c: I \to M$.

Note. In the definition of a mapping $c: I \to M$ to be of class H^1, the particular choice of the chart does not play a rôle. Indeed, if $\phi: U \subset \mathbb{R}^n \to U' \subset \mathbb{R}^n$ is a diffeomorphism and $u: J \to U$ is of class H^1 then also $\phi \circ u: J \to U'$ is of class H^1.

We have the following canonical inclusions:

$$C'^{\infty}(I, M) \hookrightarrow H^1(I, M) \hookrightarrow C^0(I, M).$$

2.3.2 Proposition. *$C'^{\infty}(I, M)$ is a dense subspace of the complete metric space $\{C^0(I, M), d_{\infty}\}$.*

Proof. A curve $c \in C^0(I, M)$ can be covered by finitely many charts. Thus, the proposition is reduced to the well-known fact that $C'^{\infty}(I', \mathbb{R}^n)$ is dense in $C^0(I', \mathbb{R}^n)$, I' an interval in I. □

On $H^1(I, M)$, we can consider the *energy integral* E.

2.3.3 Proposition. *For* $c \in H^1(I, M)$,

$$E(c) = \tfrac{1}{2} \int_I \langle \dot{c}(t), \dot{c}(t) \rangle \, dt$$

is well defined.

Proof. Let $u \circ c: I' \to \mathbb{R}^n$ be a local representation of $c|I'$ with respect to some chart (u, M'). Then we can define $2E(c|I')$ to be the integral over I' of the function $g(u \circ c(t))((u \circ c)^{\cdot}(t), (u \circ c)^{\cdot}(t))$ – note that $u \circ c \in H^1(I', \mathbb{R}^n)$ possesses a square integrable derivative. In the same way as for curves of class C^{∞}, one shows that this is independent of the particular choice of the chart. □

As a preparation for the introduction of the structure of a differentiable manifold on $H^1(I, M)$, modelled on a Hilbert space, we first consider the infintesimal approximation of $C'^{\infty}(I, M)$ at $c \in C'^{\infty}(I, M)$, given by the vector space of piecewiese differentiable vector fields along c. This vector space can be viewed as the 'tangent space' $T_c C'^{\infty}(I, M)$ of the 'manifold' $C'^{\infty}(I, M)$ at the point c. Our main interest lies in its completion with respect to a H^1-norm. To get a precise formulation, we take $T_c C'^{\infty}(I, M)$ to be the space of sections in the induced bundle $c^*\tau$.

2.3.4 Definition. Let $c \in C'^{\infty}(I, M)$.

(i) Define $c^*\tau$ to be *the induced bundle over* I,

$$\begin{array}{ccc} c^*TM & \xrightarrow{\tau^* c} & TM \\ \big\downarrow{\scriptstyle c^*\tau} & & \big\downarrow{\scriptstyle \tau} \\ I & \xrightarrow{c} & M \end{array}$$

For each closed subinterval $I_j \subset I$ with $c|I_j$ differentiable, the restriction of $c^*\tau$ to I_j is the (differentiable) induced bundle in the sense of (1.3.10). The fibre $c^*\tau^{-1}(t)$ over t is also denoted by $T_{c,t}$.

(ii) By $C'^{\infty}(c^*TM)$ we denote the vector space of piecewise differentiable sections of $c^*\tau$. We also write $T_c C'^{\infty}(I, M)$ instead and call this the *tangent space to* $C'^{\infty}(I, M)$ *at* c.

(iii) Using the scalar product $\langle , \rangle$ on the fibres $T_{c,t}$ of $c^*\tau$, stemming from the Riemannian metric on the corresponding $T_{c(t)}M$, we define, for ξ, η in $C'^{\infty}(c^*TM)$,

(a) $$\|\xi\|_{\infty} = \sup_{t \in I} |\xi(t)|,$$

(b) $$\langle \xi, \eta \rangle_0 = \int_I \langle \xi(t), \eta(t) \rangle dt,$$

(c) $$\langle \xi, \eta \rangle_1 = \langle \xi, \eta \rangle_0 + \langle \nabla \xi, \nabla \eta \rangle_0 .$$

The norm derived from the scalar product $\langle , \rangle_r$ is denoted by $\| \, \|_r$, $r = 0,1$.

(iv) The completion of $C'^{\infty}(c^*TM)$ with respect to the norms $\| \, \|_{\infty}$, $\| \, \|_r$ is denoted by $C^0(c^*TM)$ and $H^r(c^*TM)$, $r = 0,1$, respectively.

Remark. A piecewise differentiable section of $c^*\tau$ is a continuous map $\xi: I \to c^*TM$ with $c^*\tau \circ \xi = id$ and such that there exists a subdivision of I into closed intervals J with $c|J$ differentiable and $\xi|J$ differentiable. For $\xi|J$ we have the covariant derivative $\nabla\xi|J$ with respect to the induced connection on $c^*\tau|J$, cf. (1.5.14). Alternatively, ξ or rather $\tau^*c \circ \xi$, can be viewed as a piecewise differentiable vector field along c.

Note that $C^0(c^*TM)$ is a Banach space whereas $H^r(c^*TM)$, $r = 0,1$ are Hilbert spaces.

2.3.5 Proposition. *The inclusions*

$$H^1(c^*TM) \hookrightarrow C^0(c^*TM) \hookrightarrow H^0(c^*TM)$$

are continuous. More precisely,

(i) *if* $\xi \in C^0(c^*TM)$, *then* $\|\xi\|_0 \leqslant \|\xi\|_{\infty}$; *and*

(ii) *if* $\xi \in H^1(c^*TM)$, *then* $\|\xi\|_{\infty} \leqslant \sqrt{2}\,\|\xi\|_1$.

Proof. (i)

$$\|\xi\|_0^2 = \int_I \langle \xi(t), \xi(t) \rangle \, dt \leqslant \int_I \max_t |\xi(t)|^2 dt = \|\xi\|_{\infty}^2$$

(ii) Choose

$t_1 \in I$ with $\|\xi\|_{\infty} = |\xi(t_1)|$. Then

$$\|\xi\|_{\infty}^2 = |\xi(t)|^2 + \int_t^{t_1} \frac{d}{ds} |\xi(s)|^2 ds \leqslant |\xi(t)|^2 + 2\int_I |\xi(s)| \, |\nabla\xi(s)| \, ds \leqslant$$

$$\leqslant \|\xi\|_0^2 + \|\xi\|_0^2 + \|\nabla\xi\|_0^2 \leqslant 2\|\xi\|_1^2 . \quad \square$$

Note. Besides vector bundles of the type $c^*\tau: c^*TM \to I$, we will also be led to consider associated vector bundles like product bundles, bundles of multilinear mappings etc., cf. section (1.2) on tensor bundles. In each case, such a bundle

$$\pi: E \to I$$

carries a Riemannian metric and a Riemannian connection, cf. (1.8.18). This allows us to form, just as in (2.3.4) for the bundle $c^*\tau$, the completions

$$C^0(E);\ H^0(E);\ H^1(E)$$

of the bundle $C'^\infty(E)$ of piecewise differentiable sections in π. (2.3.5) holds also in this case. The proof is exactly the same.

2.3.6 Definition. Let

$$\pi: E \to I;\ \phi: F \to I$$

be vector bundles with Riemannian metric and Riemannian connection. Let $\mathcal{O} \subset E$ be such that $\mathcal{O}_t = \mathcal{O} \cap \pi^{-1}(t) \neq \emptyset$, all $t \in I$.

(i) Denote by $H^1(\mathcal{O})$ *the set of* $\xi \in H^1(E)$ *with* $\xi(t) \in \mathcal{O}_t$, all $t \in I$.

(ii) Let

$$f: \mathcal{O} \to F$$

be a differentiable fibre mapping. That is, for each $t \in I, f|\mathcal{O}_t$ is differentiable with image in $F_t = \phi^{-1}(t)$. Then define

$$\tilde{f}: H^1(\mathcal{O}) \to H^1(F);\ (\xi(t)) \mapsto (f \circ \xi(t)).$$

Our next goal is show that $H^1(\mathcal{O}) \subset H^1(E)$ is open and $\tilde{f}$ is differentiable. We begin with the

2.3.7 Proposition. *With the notions of* (2.3.6), $\tilde{f}$ *is a continuous mapping defined on an open subset of* $H^1(E)$.

Proof. Denote by $C^0(\mathcal{O})$ the set of continuous sections ξ of π with $\xi(t) \in \mathcal{O}_t$. Then there exists a $\varrho > 0$ such that $\xi \in C^0(\mathcal{O})$, $\eta \in C^0(E)$, $\|\xi - \eta\|_\infty < \varrho$ implies $\eta(t) \in \mathcal{O}_t$, all $t \in I$. Hence, $C^0(\mathcal{O})$ is open in $C^0(E)$. $H^1(\mathcal{O})$ is the counter image of $C^0(\mathcal{O})$ under the continuous inclusion $H^1(E) \hookrightarrow C^0(E)$. Therefore, $H^1(\mathcal{O})$ is open.

To see that $\tilde{f}$ is continuous we note that with $\|\eta - \xi\|_1$ small also $\|\eta - \xi\|_\infty$ and $\|\nabla\eta - \nabla\xi\|_0$ become small. Moreover, $\|\tilde{f}(\eta) - \tilde{f}(\xi)\|_0 \leqslant \|\tilde{f}(\eta) - \tilde{f}(\xi)\|_\infty$.

Let now $\dot{\eta}(t) = \dot{\eta}(t)_h + \dot{\eta}(t)_v$ be the decomposition into the horizontal and the vertical part respectively. $\dot{\eta}(t)_h$ has the local representation

$$(t, \eta(t), \partial t, -\Gamma_t(\partial t, \eta(t))),$$

cf. (1.5.14), and $\dot{\eta}(t)_v$ is canonically identified with $\nabla\eta(t)$. Therefore, if

$$T_{\eta(t)}f = T_{\eta(t),1}f + T_{\eta(t),2}f$$

denotes the decomposition into the restrictions upon the horizontal and vertical subspaces, respectively, we have

$$\nabla(f\circ\eta)(t)-\nabla(f\circ\xi)(t)=T_{\eta(t),2}f.\nabla\eta(t)-T_{\xi(t),2}f.\nabla\xi(t)=$$
$$=T_{\eta(t),2}f.\left(\nabla\eta(t)-\nabla\xi(t)\right)+\left(T_{\eta(t),2}f-T_{\xi(t),2}f\right).\nabla\xi(t).$$

Hence, with $\|\xi-\eta\|_1\to 0$, also $\|\nabla\tilde{f}(\eta)-\nabla\tilde{f}(\xi)\|_0\to 0$. □

2.3.8 Proposition. Let

$$\pi: E\to I;\ \phi: F\to I$$

be vector bundles as in (2.3.7). *Let*

$$L(\pi;\phi): L(E;F)\to I$$

be the associated bundle of linear mappings with the induced Riemannian metric and Riemannian connection. Then the canonical inclusions

$$H^1\left(L(E;F)\right)\hookrightarrow L\left(H^0(E);H^0(F)\right)\quad \textit{and}$$
$$H^1\left(L(E;F)\right)\hookrightarrow L\left(H^1(E);H^1(F)\right),$$

given by

$$A=\left(A(t)\right)\mapsto\{\tilde{A}:\xi=\left(\xi(t)\right)\mapsto\left(A(t).\xi(t)\right)\},$$

are continuous. More precisely,

(*) $\qquad \|\tilde{A}(\xi)\|_0^2\leqslant\|A\|_\infty^2\|\xi\|_0^2\leqslant 2\|A\|_1^2\|\xi\|_0^2$ and

(**) $\qquad \|\tilde{A}(\xi)\|_1^2\leqslant 8\|A\|_1^2\|\xi\|_1^2.$

Proof. We use (2.3.5). Then (*) is obvious. To prove (**) we note

$$\nabla A(\xi)(t)=\nabla A(t).\xi(t)+A(t).\nabla\xi(t).$$

Hence,

$$\|\nabla\tilde{A}(\xi)\|_0^2=\|\nabla A.\xi+A.\nabla\xi\|_0^2\leqslant 2\|\nabla A.\xi\|_0^2+2\|A.\nabla\xi\|_0^2\leqslant$$
$$\leqslant 4\|\nabla A\|_0^2\|\xi\|_1^2+4\|A\|_1^2\|\xi\|_1^2\leqslant 8\|A\|_1^2\|\xi\|_1^2.\quad\square$$

Remark. A similar result holds for the canonical inclusions

$$H^1\left(L(E_1,E_2,\dots,E_k;F)\right)\hookrightarrow L\left(H^0(E_1),H^1(E_2),\dots,H^1(E_k);H^0(F)\right)\quad\text{and}$$
$$H^1\left(L(E_1,E_2,\dots,E_k;F)\right)\hookrightarrow L\left(H^1(E_1),H^1(E_2),\dots,H^1(E_k);H^1(F)\right).$$

2.3.9 Lemma. *Let* $f:\mathcal{O}\to F$ *be a fibre map. Then* $\tilde{f}: H^1(\mathcal{O})\to H^1(F)$ *is once continuously differentiable with the tangential given by* $T_2f^{\sim}$.

Note: The differentiability of arbitrarily high order is proved along the same lines, starting with the differentiable fibre map

$$T_2 f\colon \mathcal{O} \to L(E; F)$$

See Klingenberg [10], Flaschel und Klingenberg [1] for details. (2.3.9) is a slight generalization of a result contained in Palais [1] and Eliasson [2].

Proof. From (2.3.7) we know that $\tilde{f}$ is continuous. The Taylor formula gives

$$f(\eta(t)) - f(\xi(t)) - T_{\xi(t),2} f \,.\, (\eta(t) - \xi(t)) = r(\xi(t), \eta(t)) \,.\, (\eta(t) - \xi(t)).$$

Here,

$$r(\xi(t), \eta(t)) = \int_I T_{\xi(t) + s(\eta(t) - \xi(t)),2} f ds - T_{\xi(t),2} f$$

is a fibre map of some $\mathcal{O}' \times \mathcal{O}' \subset \mathcal{O} \times \mathcal{O}$, $\mathcal{O}'$ open, into the bundle $L(\pi; \phi)\colon L(E; F) \to I$.

Consider the associated continuous mapping

$$r^{\sim}\colon H^1(\mathcal{O}' \times \mathcal{O}') \to H^1(L(E; F)).$$

Then, with $T_2 f^{\sim}\colon H^1(\mathcal{O}') \to H^1(L(E; F))$,

$$\|\tilde{f}(\eta) - \tilde{f}(\xi) - T_2 f^{\sim}(\xi) \,.\, (\eta - \xi)\|_1 =$$

$$\|r^{\sim}(\xi, \eta) \,.\, (\eta - \xi)\|_1 \leqslant \text{const}\, \|r^{\sim}(\xi, \eta)\|_1 \|\eta - \xi\|_1 .$$

Since $r^{\sim}(\xi, \xi) = 0$ we get $\|r^{\sim}(\xi, \eta)\|_1 \to 0$ with $\|\xi - \eta\|_1 \to 0$, i.e., $T_2 f^{\sim} \in H^1(L(H^1(E); H^1(F))$ has the properties of a differential, $T\tilde{f} = T_2 f^{\sim}$. □

From (2.1.10) we have the existence of an open neighborhood $\mathcal{O}$ of the 0-section of $\tau\colon TM \to M$ with the property that $\exp|(\mathcal{O}_p = T_p M \cap \mathcal{O})$ is a diffeomorphism onto its image = open neighborhood of $p \in M$. For M compact, we can choose $\mathcal{O}$ to be the ε-ball bundle of τ, some small $\varepsilon > 0$, i.e., $\mathcal{O}_p = B_\varepsilon(0_p)$, all $p \in M$.

2.3.10. Definition. Let $\mathcal{O}$ be as above. For $c \in C'^{\infty}(I, M)$ denote by $\mathcal{O}_c$ the subset of $c^* TM$ formed by the $\mathcal{O}_{c,t} = \mathcal{O}_c \cap T_{c,t}$ which corresponds under $\tau^* c$ to $T_{c(t)} M \cap \mathcal{O}$.

Define

(†) $\qquad \exp_c\colon H^1(\mathcal{O}_c) \to H^1(I, M)$

by

$$(\xi(t)) \mapsto (\exp_{c(t)} \tau^* c \, \xi(t))$$

and denote the image by $\mathcal{U}(c)$.

2.3.11 Proposition. (†) *is bijective. Let* $c, d \in C'^{\infty}(I, M)$.

$$\exp_d^{-1} \circ \exp_c\colon \exp_c^{-1}(\mathcal{U}(c) \cap \mathcal{U}(d)) \to \exp_d^{-1}(\mathcal{U}(d) \cap \mathcal{U}(c))$$

is a diffeomorphism between open sets in the Hilbert spaces $H^1(c^*TM)$ *and* $H^1(d^*TM)$.

Proof. $\mathscr{U}(c)$ consists precisely of those $e \in H^1(I, M)$ with $e(t) \in \exp_{c(t)}(\mathcal{O}_{c(t)})$. This shows that (†) is a bijection.

For each $t \in I$ we form

$$\mathcal{O}_{c,d,t} = \mathcal{O}_{c,t} \cap (\exp \circ \tau^* c)^{-1} \circ (\exp \circ \tau^* d) \mathcal{O}_{d,t}$$

and put $\bigcup_{0 \leqslant t \leqslant 1} \mathcal{O}_{c,d,t} = \mathcal{O}_{c,d}$, if $\mathcal{O}_{c,d,t} \neq \emptyset$ for all $t \in I$. Otherwise put $\mathcal{O}_{c,d} = \emptyset$. $\mathcal{O}_{c,d}$ is an open subset of $\mathcal{O}_c$ and

$$H^1(\mathcal{O}_{c,d}) = \exp_c^{-1}\big(\mathscr{U}(c) \cap \mathscr{U}(d)\big).$$

The map

$$f_{d,c}\colon (\exp \circ \tau^* d)^{-1} \circ (\exp \circ \tau^* c)\colon \mathcal{O}_{c,d} \to d^* TM$$

is a fibre map, $\exp_d^{-1} \circ \exp_c = \tilde{f}_{d,c}$. Hence, (2.3.9) applies. □

2.3.12 Theorem. *The set* $H^1(I, M)$ *of* H^1*-mappings* $c\colon I \to M$ *is a Hilbert manifold; its differentiable structure is given by the* natural atlas

$$\{\exp_c^{-1}, \mathscr{U}(c); c \in C'^{\infty}(I, M)\}.$$

Proof. The charts are modelled on a proper separable Hilbert space, with typical representative $H^1(c^*TM) \cong H^1(I, \mathbb{R}^n)$.

The family $\mathscr{U}(c)$, $c \in C'^{\infty}(I, M)$, is an open covering of $H^1(I, M)$ since $C'^{\infty}(I, M)$ is dense in $H^1(I, M) \subset C^0(I, M)$, see (2.3.2). (2.3.11) shows that the natural atlas is differentiable.

To see that $H^1(I, M)$ has a countable base it suffices to show that the natural atlas has a countable subatlas. Actually, it suffices to show that, for a sequence $\{M^k\}$ of relatively compact open subsets M^k of M with $\bigcup_k M^k = M$, $M^k \subset M^{k+1}$, there exists a countable subatlas of the natural atlas covering $H^1(I, M^k)$.

To see this we show that for fixed k and any integer $l > 0$ the set

$$H^1(I, M^k)^l = \{c \in H^1(I, M^k); E(c) \leqslant l\}$$

can be covered by a finite subset of the natural atlas.

To prove this we choose $\iota = \iota(M^k) > 0$ to be the injectivity radius on M^k. Let $m = m(\iota, l)$ be an integer satisfying $18l < m\iota^2$. Then $e \in H^1(I, M^k)^l$ implies, for $t \in [(j-1)/m, j/k]$,

$$d\big(e(j-1/m), e(t)\big)^2 \leqslant \Big(\int_{(j-1)/m}^{j/m} |\dot{e}(t)|\, dt\Big)^2 \leqslant \frac{1}{m} \int_I \langle \dot{e}(t), \dot{e}(t)\rangle\, dt \leqslant 2\, l/m \leqslant \iota^2/9.$$

Hence, $e \mid [(j-1)/m, j/m]$ lies entirely in a $\iota/3$-ball.

There exists a finite set P of points on M^k such that the $\iota/3$-balls around these points will cover M^k. Given $e \in H^1(I, M^k)^l$, we can find a sequence $\{p_0, \ldots, p_m\}$ in P such that $e(j/m) \in B_{\iota/3}(p_j)$. For each of these finitely many sequences we choose a $c \in C'^{\infty}(I, M)$ such that $c|[(j-1)/m, j/m]$ is the minimizing geodesic from p_{j-1} to p_j. Then $e \in H^1(I, M^k)^l$ implies $e \in \mathscr{U}(c)$, for one of these c's. □

Remark. Given a differentiable mapping $f: M \to N$ from a manifold M into a manifold N, we obtain a mapping

$$H^1(I, f): H^1(I, M) \to H^1(I, N); (c(t)) \mapsto (f \circ c(t)).$$

This mapping is differentiable. And if $g: N \to L$ is another differentiable mapping we get $H^1(I, g \circ f) = H^1(I, g) \circ H^1(I, f)$. Thus, $H^1(I, \ )$ constitutes a covariant functor from the category {finite dimensional manifolds and differentiable mappings} into the category {Hilbert manifolds and differentiable mappings}.

Associated with the manifolds $H^1(I, M)$ there are in a natural way two vector bundles

$$\alpha^r: H^r(H^1(I, M)^* TM) \to H^1(I, M), \ r = 0, 1.$$

The total space is the union of the spaces $H^r(c^* TM)$, $c \in H^1(I, M)$. α^1 is canonically isomorphic to the tangent bundle $\tau_{H^1(I,M)}$ of $H^1(I, M)$.

To make this precise we define for $\xi \in \mathcal{O}$, $\mathcal{O}$ an open neighborhood of M in TM as above,

$$T_{\xi,1}\exp = T_\xi \exp \circ (T\tau | T_{\xi h} TM)^{-1}: T_{\tau\xi} M \to T_{\exp \xi} M,$$

$$T_{\xi,2}\exp = T_\xi \exp \circ (K | T_{\xi v} TM)^{-1}: T_{\tau\xi} M \to T_{\exp \xi} M.$$

Here, $T\tau | T_{\xi h} TM: T_{\xi h} TM \to T_{\tau\xi} M$ and $K | T_{\xi v} TM: T_{\xi v} TM \to T_{\tau\xi} M$ are the canonical isomorphisms. Under our assumptions, $T_{\xi,1}\exp$ and $T_{\xi,2}\exp$ are linear isomorphisms.

Note. In the above definitions, we have written exp instead of $\exp_p = \exp | T_p M$, where $p = \tau\xi$. There can be no misunderstanding as to where the base point of the vector is. A similar simplification will also be used further down.

Define, for $c \in C'^{\infty}(I, M)$ and $r = 0, 1$,

$$\phi_{r,c}^{-1}: H^1(\mathcal{O}_c) \times H^r(c^* TM) \to (\alpha^r)^{-1} \mathscr{U}(c)$$

by

$$(\xi(t), \eta_c(t)) \mapsto (T_{\tau^* c\xi(t), 2}\exp) \,.\, \tau^* c\eta_c(t)).$$

Here, the right hand side is viewed as a H^r-mapping $I \to TM$ which under τ goes into the base H^1-curve $(\exp \circ \tau^* c\xi(t))$ belonging to $\mathscr{U}(c)$.

2.3.13 Lemma. *The family*

$$\{(\phi_{r,c}, \exp_c^{-1}, \mathscr{U}(c)); c \in C'^{\infty}(I, M)\}$$

constitutes a bundle atlas for a bundle α^r *over* $H^1(I, M)$, *associated with the natural atlas of* $H^1(I, M)$. *The typical fibre of the bundle is the separable Hilbert space* $H^r(I, \mathbb{R}^n)$.

The bundle α^1 *is canonically isomorphic to the tangent bundle* $\tau_{H^1(I,M)}$.

Proof. First consider the case $r = 1$. Then we see that, for c, d in $C'^{\infty}(I, M)$,

$$\phi_{1,d} \circ \phi_{1,c}^{-1}: H^1(\mathcal{O}_{c,d}) \times H^1(c^* TM) \to H^1(\mathcal{O}_{d,c}) \times H^1(d^* TM)$$

is of the form

$$(\exp_d^{-1} \circ \exp_c, T(\exp_d^{-1} \circ \exp_c)) \equiv (f_{c,d}^{\sim}, Tf_{c,d}^{\sim}),$$

with $f_{c,d}$ as in the proof of (2.3.11). This shows that the above atlas is precisely the tangent atlas associated with the natural atlas of $H^1(I, M)$.

In the case $r = 0$ we observe that the composition maps $\phi_{0,d} \circ \phi_{0,c}^{-1}$ are again of the form $(f_{d,c}^{\sim}, Tf_{d,c}^{\sim})$ and the composition mapping

$$H^1(\mathcal{O}_{c,d}) \xrightarrow{Tf_{d,c}^{\sim}} H^1(L(c^* TM; d^* TM)) \hookrightarrow L(H^0(c^* TM); H^0(d^* TM))$$

is differentiable, see (2.3.8). □

Note. The previous result shows that we obtain an intrinsic description of the tangent space $T_e H^1(I, M)$ of $H^1(I, M)$ at an arbitrary element $e \in H^1(I, M)$ by considering the vector space of H^1-maps $\eta: I \to TM$ satisfying $\tau \circ \eta = e$. That is, η is an H^1-vector field along the H^1-curve e.

Before we prove that we also have a natural scalar product on $T_e H^1(I, M)$, we show that natural charts exist for every $e \in H^1(I, M)$. This will follow from the next Lemma.

To formulate our result we put, with $\mathcal{O} \subset TM$ as before,

$$\tilde{\mathcal{O}} = \{\eta \in TH^1(I, M); \eta(t) \in \mathcal{O}\}.$$

2.3.14 Lemma. *The mapping*

$$\tilde{F} = \tau_{H^1(I,M)} \times \exp^{\sim}: \tilde{\mathcal{O}} \subset TH^1(I, M) \to H^1(I, M) \times H^1(I, M);$$

$$\eta(t) \mapsto (\tau \circ \eta(t), \exp \eta(t))$$

is differentiable. It maps a sufficiently small open neighborhood $\tilde{\mathcal{O}}' \subset \tilde{\mathcal{O}}$ *of the zero section of* $\tau_{H^1(I,M)}$ *onto an open neighborhood of the diagonal of* $H^1(I, M) \times H^1(I, M)$.

Note. This resembles the property of the map $F = \tau \times \exp$ in the proof of (1.6.12). Note, however, that in the present case $\exp^{\sim}$ is not the exponential map of a covariant derivation on $H^1(I, M)$.

Proof. Using the local representation of $T\mathcal{U}(c)$, see (2.3.13), we obtain for $\exp^{\sim}$ the representation

$$(\xi(t), \eta_c(t)) \mapsto ((\exp \circ \tau^* c)^{-1} \circ \exp \circ T_{\tau^* c\, \xi(t), 2} \exp . \tau^* c \eta_c(t)).$$

This is a differentiable fibre map from $\mathcal{O}_c \times c^* TM$ into $c^* TM$. Apply now (2.3.9).

$\tilde{F}$ maps the zero section of $\tau_{H^1(I,M)}$ bijectively onto the diagonal of $H^1(I, M) \times H^1(I, M)$. It only remains to be shown that $T\tilde{F}$ at $0_c \in T_c H^1(I, M)$ is a bijection. But this follows by looking at the local representation: It carries $(\xi(t), 0)$ into $(\xi(t), \xi(t))$, and $(0, \eta(t))$ into $(0, \eta(t))$. □

2.3.15 Corollary. *For every* $e \in H^1(I, M)$, *there exists a natural chart*

$$(\exp_e^{-1}, \mathcal{U}(e))$$

with

$$\exp_e \equiv \exp^{\sim} | \tilde{\mathcal{O}}' \cap T_e H^1(I, M) : \tilde{\mathcal{O}}' \cap T_e H^1(I, M) \to \mathcal{U}(e). \quad \square$$

2.3.16 Proposition. *The mapping*

$$\partial : H^1(I, M) \to H^0(H^1(I, M)^* TM); \quad (e(t)) \mapsto (\partial e(t) \equiv \dot{e}(t))$$

is a differentiable section in the bundle α^0.

For $e \in \mathcal{U}(c)$, $\xi = \exp_c^{-1} e$, *the representation of* ∂e *in the bundle chart over* $\mathcal{U}(c)$ *is given by*

$$\partial_c \xi(t) = \nabla \xi(t) + \theta_c \xi(t)$$

with

$$\theta_c \xi(t) = \tau^* c^{-1} \circ (T_{\tau^* c\xi(t), 2} \exp)^{-1} \circ (T_{\tau^* c\xi(t), 1} \exp) \circ \tau^* c \,.\, \partial t.$$

Proof. We have

$$\partial e(t) = T_{\tau^* c\xi(t)} \exp \,.\, (\tau^* c\xi(t)_h + \tau^* c\xi(t)_v) =$$

$$= (T_{\tau^* c\xi(t), 1} \exp) \circ \tau^* c \,.\, \partial t + (T_{\tau^* c\xi(t), 2} \exp) \circ \tau^* c \,.\, \nabla \xi(t).$$

This gives the expressions for $\partial_c \xi(t)$ and $\theta_c \xi(t)$.

In particular,

$$\theta_c : \mathcal{O}_c \to c^* TM$$

is a fibre mapping. Hence, the mapping which associates to $\xi = \exp_c^{-1} e \in H^1(\mathcal{O}_c)$ the principal part $\nabla \xi + \theta_c \tilde{\xi} \in H^0(c^* TM)$ of the representation of ∂e is differentiable. □

2.3.17 Theorem. *The bundle* α^0 *of* H^0*-vector fields along* H^1*-curves on* M *has a Riemannian metric which is characterized by the property that on* $(\alpha^0)^{-1}(c) = H^0(c^* TM)$, $c \in C'^{\infty}(I, M)$, *it is given by* $\langle , \rangle_0$. *We therefore denote this metric by* $\langle , \rangle_0$ *in general.*

Proof. With $\mathcal{O}$ as above define

$$G : \mathcal{O} \subset TM \to L(TM, TM)$$

by

$$\langle G(\xi), \rangle_{\tau(\xi)} = \langle T_{\xi, 2} \exp \,, T_{\xi, 2} \exp \rangle_{\exp \xi}.$$

From the properties of $\mathcal{O}$ it follows that $G(\xi)$ is a positive self-adjoint operator of class C^∞.

$$G_c\colon (\tau^* c)^{-1} \circ G \circ (\tau^* c)\colon \mathcal{O}_c \to L(c^* TM, c^* TM)$$

is a fibre map. Thus, the composite mapping

$$H^1(\mathcal{O}_c) \xrightarrow{G_c^\sim} H^1(L(c^* TM; c^* TM)) \hookrightarrow L(H^0(c^* TM); H^0(c^* TM))$$

is a positive self-adjoint operator of class C^∞; it therefore defines a Riemannian metric on the representation $H^1(\mathcal{O}_c) \times H^0(c^* TM)$ of $(\alpha^0)^{-1}\mathcal{U}(c)$. Clearly, this metric does not depend on the particular representation. □

As a preparation for defining a Riemannian metric also on $\alpha^1 = \tau_{H^1(I,M)}$ we prove an analogue of (2.3.10):

2.3.18 Proposition. *The Levi-Cività covariant derivation ∇ on M determines a riemannian covariant derivation ∇_{α^0} on the bundle α^0. In particular, if η is a vector field on $H^1(I, M)$, the mapping $\eta \mapsto \nabla_{\alpha^0}(\partial) \,.\, \eta$ is a section in α^0. Since, for $\alpha^1(\eta) \in C'^\infty(I, M)$, it coincides with the H^0-vector field $\nabla\eta(t)$ along $c(t)$, we also write $\nabla\eta$ instead of $\nabla_{\alpha^0}(\partial) \,.\, \eta$.*

Proof. Denote by $\Gamma(\xi)$, $\xi \in \mathcal{O}$, the Christoffel symbol of the Levi-Cività connection in the normal coordinates based at $\tau\xi \in M$. Let $c \in C'^\infty(I, M)$. Then we get a bundle mapping

$$\Gamma_c\colon \mathcal{O}_c \to L(c^* TM, c^* TM; c^* TM)$$

by

$$\Gamma_c = (\tau^* c)^{-1} \circ (\Gamma \circ \tau^* c)\,(\tau^* c \times \tau^* c)\,.$$

The associated mapping, cf. (2.3.9),

$$\begin{aligned} \Gamma_c^{\sim}\colon H^1(\mathcal{O}_c) &\to H^1(L(c^* TM, c^* TM; c^* TM)) \\ &\hookrightarrow L(H^1(c^* TM), H^0(c^* TM); H^0(c^* TM)) \end{aligned}$$

represents the desired riemannian connection ∇_{α^0}.

To compute $\nabla_{\alpha^0}(\partial) \,.\, \eta$ we consider the bundle trivializations of α^0 and α^1 over $(\exp_c^{-1}, \mathcal{U}(c))$. According to (2.3.13), $\eta \in (\alpha^1)^{-1}(e)$ is represented by $(\xi(t), \eta_c(t))$, with $\xi = \exp_c^{-1} e$. (2.3.16) gives the representation $\partial_c \xi(t)$ of ∂e. Thus, the principal part of the representation of $\nabla_{\alpha^0}(\partial) \,.\, \eta$ reads

$$D_2(\partial_c \xi(t)) \,.\, \eta_c(t) + \Gamma_c(t)\,(\eta_c(t), \partial_c \xi(t))\,.$$

Substituting here the formula (2.3.16) for $\partial_c \xi(t)$ we find

$$\nabla\eta_c(t) + D_2(\theta_c \xi(t)) \,.\, \eta_c(t) + \Gamma_c(t)\,(\eta_c(t), \partial_c \xi(t))\,.$$

Thus, if $e = c$, i.e., if $\xi = 0$, this simplifies to $\nabla\eta_c(t)$. □

2.3.19 Theorem. *The Hilbert manifold $H^1(I, M)$ has a Riemannian metric; for each* *element* $c \in C'^{\infty}(I, M) \subset H^1(I, M)$, *the scalar product on* $T_c H^1(I, M) \cong H^1(c^* TM)$ *coincides with the product* $\langle\ ,\ \rangle_1$. *We therefore denote this Riemannian metric on* $H^1(I, M)$ *by* $\langle\ ,\ \rangle_1$.

Proof. Define the Riemannian scalar product on $T_e H^1(I, M)$ by

$$(\xi, \eta) \mapsto \langle \xi, \eta \rangle_0 + \langle \nabla \xi, \nabla \eta \rangle_0 .$$

That this is a differentiable section in $L_s^2(\alpha^1)$ follows from (2.3.16) and (2.3.18). □

We conclude this section with the

2.3.20 Theorem. *The energy integral*

$$E\colon H^1(I, M) \to \mathbb{R}$$

is differentiable with

$$DE(c) \, . \, \eta = \langle \partial c, \nabla \eta \rangle_0 .$$

Proof. The differentiability of E follows from (2.3.16) and (2.3.17). To determine its differential we only need to recall from the proof of (2.3.18) that the local representation of $\nabla \eta$ for $\eta \in T_c H^1(I, M)$ yields

$$\nabla_{\alpha^0} \partial c \, . \, \eta = \nabla \eta . \text{ That is, } D(\tfrac{1}{2}\langle \partial c, \partial c \rangle_0) \, . \, \eta = \langle \partial c, \nabla \eta \rangle_0 . \quad □$$

2.3.21 Corollary. *The only critical points of* E *on* $H^1(I, M)$ *are the constant maps.*

Proof. Clearly, $c = \text{const}$, i.e., $\partial c = 0$ implies $DE(c) = 0$. Conversely, $DE(c) \, . \, \eta = \langle \partial c, \nabla \eta \rangle_0 = 0$ for all $\nabla \eta \in H^0(c^* TM)$ implies $\partial c = 0$. □

Note. This reflects the fact that $H^1(I, M)$ possesses a canonical retraction onto M = the space of constant maps,

$$H^1(I, M) \times I \to H^1(I, M); \; (\{c(t)\}, s) \mapsto \{c(t(1 - s))\} .$$

Only for certain submanifolds of $H^1(I, M)$ do there exist nontrivial critical points of E. Regarding this, see the next section.

2.4 The Loop Space and the Space of Closed Curves

In this section we consider certain geometrically important submanifolds of the Hilbert manifold $H^1(I, M)$, introduced in (2.3). The most important ones are the loop space $\Omega_{pq} M = \Omega M$ of H^1-curves going from a fixed point $p \in M$ to a fixed point $q \in M$ and the space of closed H^1-curves. Putting $[0, 1]/\{0, 1\} = S$ we also write for this space $H^1(S, M)$ or ΛM.

The critical points of $E|\Omega_{pq} M$ are the geodesics from p to q and the critical points of $E|\Lambda M$ are, besides the constant maps, the closed geodesics, cf. (2.4.3). Ω and Λ are complete with respect to the distance derived from the Riemannian metric, cf. (2.4.7). It also can be shown that these spaces are geodesically complete. Of great importance is the fact that the gradient vector field grad E on ΩM and on ΛM satisfies the condition (C) of Palais and Smale – for ΛM one must assume here that M is compact, cf. (2.4.9). This condition is a substitute for the failure of a proper Hilbert manifold to be locally compact.

We now can construct critical points with the help of the minimax method on appropriate families of subsets of ΩM or ΛM – called ϕ-families, cf. (2.4.17), (2.4.18). As a first application we show that E assumes its infimum on every connected component of ΩM or ΛM. This implies e.g. that in every class of freely homotopic closed curves, not homotopic to a constant curve, there exists a shortest closed curve which is a closed geodesic, (2.4.19). In (2.4.20) we show that also on a simply connected compact manifold there exists at least one closed geodesic.

We continue with the the concepts and notations introduced in (2.3).
We start with the following general observation:

2.4.1 Proposition. *The mapping*

$$P: H^1(I, M) \to M \times M;\; c \mapsto (c(0), c(1))$$

is a submersion. As a consequence, if $N \subset M \times M$ *is a submanifold of codimension* k, $H^1_N M = P^{-1}(N)$ *is a submanifold of* $H^1(I, M)$ *of codimension* k.

Proof. The differentiability of P is clear from the local representation. The tangential $T_c P$ at c is given by

$$\eta \in T_c H^1(I, M) \cong H^1(c^*\tau) \mapsto (\eta(0), \eta(1)) \in T_{c(0)} M \times T_{c(1)} M,$$

which shows that P is a submersion. The remainder follows from (1.3.8). □

2.4.2 Definition. The submanifold $P^{-1}(N)$ of $H^1(I, M)$ is denoted by $H^1_N M$. By $C^0_N M$ we denote its completion in $C^0(I, M)$.

For the special case $N = \{p, q\}$ we get the *H^1-loop space* $\Omega_{pq} M$ or ΩM. For $N = \Delta =$ diagonal of $M \times M$ we get the *space ΛM of (parameterized) closed curves.*

Remarks. 1. The tangent space $T_c \Omega M$ of $\Omega M = \Omega_{pq} M$ at c can be identified with the H^1-vector fields $\xi(t)$ along $c(t)$ satisfying $\xi(0) = \xi(1) = 0$. The tangent space $T_c \Lambda M$ consists of the periodic H^1-vector fields $\xi(t)$ along $c(t)$, i.e., $\xi(0) = \xi(1)$.

2. Among the various other submanifolds $H^1_N M = P^{-1}(N)$ of $H^1(I, M)$ we only mention the case $N = M_0 \times M_1 \subset M \times M$, where M_0, M_1 are submanifolds of M. In this case, $H^1_N M$ is formed by the H^1-curves c with $c(0) \in M_0, c(1) \in M_1$. $T_c H^1_N M$ consists of the $\xi(t)$ along $c(t)$ with $\xi(0) \in T_{c(0)} M_0, \xi(1) \in T_{c(1)} M_1$.

We can now characterize the critical points of E in $H^1_{M_0 \times M_1} M$ and ΛM.

2.4.3 Lemma. (i) *The critical points* c *of* $E | H^1_{M_0 \times M_1} M$ *are the geodesics with*

$c(0) \in M_0, \dot{c}(0) \in T^{\perp}_{c(0)} M_0;$ $c(1) \in M_1, \dot{c}(1) \in T^{\perp}_{c(1)} M_1$. *In particular, if* $M_0 \times M_1 = \{p\} \times \{q\}$, *the critical points are the geodesics from* p *to* q.

(ii) *The critical points of* $E|\Lambda M$ *are the constant maps or the closed geodesics, i.e.,* $\dot{c}(0) = \dot{c}(1)$, $\nabla \dot{c} = 0, |\dot{c}| \neq 0$.

Proof. First assume $\partial c \in H^1(c^* TM)$. Then we get by partial integration

$$DE(c) \,.\, \eta = \langle \partial c, \nabla \eta \rangle_0 = - \langle \nabla \partial c, \eta \rangle_0 + \langle \dot{c}(1), \eta(1) \rangle - \langle \dot{c}(0), \eta(0) \rangle .$$

Thus, $\nabla \partial c = 0$ and $\dot{c}(0) \perp T_{c(0)} M_0$, $\dot{c}(1) \perp T_{c(1)} M_1$ implies $DE(c) \,.\, \eta = 0$, whenever $\eta \in T_c H^1_{M_0 \times M_1} M$.

Similarly one sees that a closed geodesic or a constant map is a critical point for $E|\Lambda M$.

Conversely, let $DE(c)| T_c H^1_{M_0 \times M_1} M = 0$. Determine H^1-vector fields $\zeta(t)$ and $\xi(t)$ along $c(t)$ by

$$\nabla \zeta(t) = \partial c(t) = \dot{c}(t), \zeta(0) = 0; \quad \nabla \xi(t) = 0, \ \xi(1) = \zeta(1).$$

Put $\zeta(t) - t\xi(t) = \eta(t)$. Then

$$\eta(0) = \eta(1) = 0; \ \nabla \eta(t) = \partial c(t) - \xi(t).$$

Moreover,

$$\langle \xi, \partial c - \xi \rangle_0 = \langle \xi, \nabla \eta \rangle_0 = \int_0^1 \frac{d}{dt} \langle \xi(t), \eta(t) \rangle \, dt = 0$$

and

$$0 = DE(c) \,.\, \eta = \langle \partial c, \nabla \eta \rangle_0 = \langle \partial c, \partial c - \xi \rangle_0 .$$

Hence, $\|\partial c - \xi\|_0^2 = 0$, i.e., $\partial c = \xi$ is of class H^1 and $\nabla \partial c = \nabla \zeta = 0$; thus, c is a geodesic.

From the formula

$$DE(c) \,.\, \eta = \langle \dot{c}(1), \eta(1) \rangle - \langle \dot{c}(0), \eta(0) \rangle = 0$$

we get $\dot{c}(1) \perp T_{c(1)} M_1$, $\dot{c}(0) \perp T_{c(0)} M_0$.

In the case $c \in \Lambda M$, $c \neq$ const, we need to show $\dot{c}(0) = \dot{c}(1)$. But this follows from the same argument, applied to $c(t + \frac{1}{2})$ instead of $c(t)$. □

We continue with some preliminary estimates.

2.4.4 Proposition. *Let* $d = d_M$ *be the distance on* M, *derived from the Riemannian metric on* M. *Let* $c \in H^1(I, M)$. *Then*

$$d(c(t_0), c(t_1)) \leq |t_1 - t_0|^{1/2} \sqrt{2E(c)} .$$

Proof. We apply Cauchy-Schwarz:

$$d(c(t_0), c(t_1))^2 \leq \Big(\int_{t_0}^{t_1} |\dot{c}(t)| \, dt\Big)^2 \leq |t_1 - t_0| 2E(c) . \quad \square$$

2.4.5 Proposition. *Let* c, c' *in* $H^1(I, M)$. *Then*

$$|\sqrt{2E(c)} - \sqrt{2E(c')}| \leqslant d_{H^1}(c, c'),$$

where d_{H^1} *denotes the distance derived from the Riemannian metric on* $H^1(I, M)$. *This relation is true a fortiori if* c *and* c' *belong to some submanifold* $H^1_N M$ *and we replace* d_{H^1} *by the distance on* $H^1_N M$.

Remark. In the case $M = \mathbb{R}^n$, (2.4.5) becomes the trivial estimate

$$|\|\partial c\|_0 - \|\partial c'\|_0| \leqslant \|c - c'\|_1.$$

Proof. We can assume that c, c' belong to the same connected component of $H^1_N M$. Then $d_{H^1}(c, c')$ is the infimum of the length $L(F)$ of curves $F\colon [0,1] \to H^1(I, M)$ from $c = F(0)$ to $c' = F(1)$. An approximation argument shows that it suffices to consider the case $E(F(s)) > 0$, for all $s \in [0,1]$. Putting $F(s)(t) = F(t, s)$ we find

$$\frac{d}{ds}\|\partial F\|_0 = \frac{1}{\|\partial F\|_0}\int_I \langle \frac{\partial F}{\partial t}, \nabla \frac{\partial F}{\partial s} \rangle (t, s)\, dt \leqslant$$

$$\leqslant \Big(\int_I |\nabla \frac{\partial F}{\partial s}(t, s)|^2 dt\Big)^{1/2} \leqslant \|\frac{dF}{ds}(s)\|_1.$$

Integration yields the claim. □

2.4.6 Lemma. *The inclusion*

$$H^1_N M \hookrightarrow C^0_N M$$

is continuous and is also compact whenever one of the projections $pr_1 N \subset M$, $pr_2 N \subset M$ *of* $N \subset M \times M$ *is compact. This is true in particular for* ΩM *and it is also true for* ΛM, *if* M *is compact.*

Proof. The continuity is clear from the definition, cf. also (2.3.2). As for the compactness, consider a bounded sequence $\{c_m\}$ on $H^1_N M$. (2.4.5) then implies that $\{E(c_m)\}$ is bounded and (2.4.4) shows that $\{c_m\}$ is equicontinuous. The claim now follows if the evaluation set $\{c_m(t_0)\}$, for every $t_0 \in I$, is relatively compact. To see this, assume that $pr_1 N$ is compact, i.e., $\{c_m(0)\}$ is relatively compact. (2.4.4) then shows that this is true also for $\{c_m(t_0)\}$. □

2.4.7 Theorem. $H^1_N M$, *with its distance deduced from the Riemannian metric, is a complete metric space, provided* $pr_1 N \subset M$ *or* $pr_2 N \subset M$ *is compact. In particular,* $\Omega M = \Omega_{pq} M$ *and* ΛM *(for* M *compact) are complete metric spaces.*

Proof. Let $\{c_m\}$ be a Cauchy sequence. From (2.4.6) we then known that there exists a limit $c_0 \in C^0_N M$ for this sequence. Since c_0 can be approximated by an H^1-curve c, we may assume that, for all large m, c_m is in a closed set contained in the domain $\mathscr{U}(c)$ of the natural chart based at c. Appealing to the Riemann Principle (1.9.8) we only need to show that $\{\xi_m = \exp_c^{-1} c_m\} \in T_c \equiv T_c H_N^1 M$ is convergent. But this is clear, since T_c is a closed linear subspace of the Hilbert space $T_c H^1(I, M) = H^1(c^* TM)$. $\square$

2.4.8 Definition. Let $H_N^1 M$ be one of the submanifolds of $H^1(I, M)$ introduced in (2.4.2). Then we define the gradient vector $\operatorname{grad} E(c)$ in $T_c H_N^1 M$ as a representation of the 1-form $DE(c)$, i.e.,

$$\langle \operatorname{grad} E(c), \eta \rangle_1 = DE(c) \,.\, \eta, \text{ all } \eta \in T_c H_N^1 M.$$

Note. If c is differentiable and $H_N^1 M = \Omega M$ or $= \Lambda M$ we can apply partial integration and find that $\operatorname{grad} E$ is the solution ξ of

$$\nabla^2 \xi(t) - \xi(t) = \nabla \partial c(t)$$

satisfying the boundary conditions $\xi(0) = \xi(1) = 0$ or $\xi(1) - \xi(0) = 0$, $\nabla \xi(1) - \nabla \xi(0) = \partial c(1) - \partial c(0)$.

We can now formulate the *condition* (C) of Palais and Smale [1]:

"Let $\{c_m\}$ be a sequence such that $\{E(c_m)\}$ is bounded and $\{\|\operatorname{grad} E(c_m)\|_1\}$ is a null sequence. Then $\{c_m\}$ possesses accumulation points and each such point c is a critical point, i.e., $\operatorname{grad} E(c) = 0$."

2.4.9 Theorem. *Condition* (C) *holds on* ΛM, *provided* M *is compact, and it always holds on* ΩM.

Note. Our proof will show that, more generally, condition (C) holds on $H_N^1 M$, whenever either $pr_1 N$ or $pr_2 N$ is compact. See Grove [1].

Proof. We know from the proof of (2.4.7) that we can assume – by going to a subsequence, if necessary, which shall again be denoted by $\{c_m\}$ – that all c_m belong to the domain $\mathscr{U}(c)$ of a natural chart and form a Cauchy sequence in the d_∞-metric. Put $\exp_c^{-1} c_m = \xi_m$.

Invoking the Riemann Principle (1.9.8), we know that it suffices to show that $\{\xi_m\}$ is a Cauchy sequence in $T_c \equiv T_c H_N^1 M$.

We know already that $\{\xi_m\}$ is a Cauchy sequence in the norm $\| \ \|_\infty$ and hence also in the norm $\| \ \|_0$, cf. (2.3.5). The local representative on $\partial_c \xi_m$ of ∂c_m, see (2.3.16), tells us that $\{\partial_c \xi_m\}$ is a Cauchy sequence in the $\| \ \|_0$-norm if and only if this is the case for the sequence $\{\nabla \xi_m\}$.

We first show that $\{\|\xi_m\|_1\}$ is bounded. Indeed, from the Riemann Principle we find that $\|\partial_c \xi_m\|_0^2$ is of a size comparable to $\langle \partial c_m, \partial c_m \rangle_0 = 2E(c_m)$ with a factor independent of m.

By looking at the formula (2.3.10) for $DE(c)$ we see by the same argument that

$$\langle \partial_c \xi_l, \partial_l - \partial_c \xi_m \rangle_0 = DE(c_l) \,.\, (\xi_l - \xi_m),$$

modulo terms which to go 0, as $l, m \to \infty$. Here we have interpreted ξ_l, ξ_m as elements in $T_{c_l} H_N^1 M$, which has in our chart the representative $T_c H_N^1 M$.

We now write, modulo such vanishing terms,

$$\|\partial_c\xi_l - \partial_c\xi_m\|_0^2 = \langle\partial_c\xi_l, \partial_c\xi_l - \partial_c\xi_m\rangle_0 + \\ + \langle\partial_c\xi_m, \partial_c\xi_m - \partial_c\xi_l\rangle_0 = DE(c_l)\,.\,(\xi_l - \xi_m) + \\ + DE(c_m)\,.\,(\xi_m - \xi_l)\,.$$

But

$$|DE(c_m)\,.\,(\xi_l - \xi_m)| \leqslant \|\operatorname{grad} E(c_m)\|_1 \, \|\xi_l - \xi_m\|_1$$

and $\|\operatorname{grad} E(c_m)\|_1 \to 0$ for $m \to \infty$ while $\|\xi_l - \xi_m\|_1$ remains bounded.Therefore, $\|\partial_c\xi_l - \partial_c\xi_m\|_0 \to 0$, for $l, m \to \infty$. □

There are a few immediate consequences of the validity of condition (C). First we introduce the following notation:

2.4.10 Definition. (i) As joint notation for the spaces $\Omega = \Omega M = \Omega_{pq} M$ and ΛM (for M compact) we use P.

(ii) For any real number $\varkappa$ we define

$$P^\varkappa = \{c \in P;\, E(c) \leqslant \varkappa\} \quad \text{and} \quad P^{\varkappa -} = \{c \in P;\, E(c) < \varkappa\}\,.$$

Note. For $P = \Omega_{pq} M$, $\varkappa_0 = d(p,q)^2/2$ is the smallest value of $E|P$. Hence, for $\varkappa < \varkappa_0$, $P^\varkappa$ is empty. For $P = \Lambda M$, $\inf E|\Lambda M = 0$.

2.4.11 Proposition. *Denote by $Cr P$ the set of critical points on P. Choose a real number $\varkappa$. Put $Cr P \cap E^{-1}(\varkappa) = Cr\varkappa$. Then $Cr\varkappa$ is compact. The same is true for $Cr P^\varkappa = Cr P \cap P^\varkappa$.*

Proof. Immediate from (2.4.9). □

2.4.12 Lemma. *Choose an open neighborhood $\mathscr{U}$ of $Cr\varkappa$ in P. Then there exist $\varepsilon > 0, \eta > 0$ such that*

$$c \in \complement\mathscr{U} \cap (P^{\varkappa + \varepsilon} - P^{\varkappa - \varepsilon}) \quad \textit{implies} \quad \|\operatorname{grad} E(c)\|_1 \geqslant \eta$$

Proof. Let $\{c_m\}$ be a sequence with $\{E(c_m)\}$ converging towards $\varkappa$ and $\{\|\operatorname{grad} E(c_m)\|_1\}$ going to zero. Condition (C) implies that the elements c_m, for all sufficiently large m, will lie in the prescribed neighborhood $\mathscr{U}$. □

2.4.13 Definition. The integral curve of the vector field $-\operatorname{grad} E$ on P which starts for $s = 0$ at c is denoted by $\phi_s c$.

Remarks. One knows from the theory of differential equations that, for every $c \in P$, there exists a maximal interval $J = J(c)$, containing $0 \in \mathbb{R}$, on which $\phi_s c$, $s \in J$ is defined, cf. Dieudonné [1]. If $\operatorname{grad} E(c) = 0$, $\phi_s c = c$, for all $s \in \mathbb{R}$, i.e., in this case $J(c) = \mathbb{R}$. As we will show in (2.4.15), $J(c) = \mathbb{R}$ also for all other $c \in P$.

2.4.14 Lemma.

(i) $$\frac{d}{ds} E(\phi_s c) = DE(\phi_s c) \,.\, (-\operatorname{grad} E(\phi_s c)) = -\,\|\operatorname{grad} E(\phi_s c)\|_1^2 \leqslant 0 .$$

Hence, for $s_0 \leqslant s_1$,

$$E(\phi_{s_1} c) - E(\phi_{s_0} c) \leqslant 0 .$$

(ii) *Denote by* $d_P(\ ,\ )$ *the distance on* P, *derived from the Riemannian metric. Then, if* $s_0 \leqslant s_1$,

$$d_P(\phi_{s_1} c, \phi_{s_0} c)^2 \leqslant \left(\int_{s_0}^{s_1} \left\| \frac{d\phi_s c}{ds} \right\|_1 ds \right)^2 \leqslant$$

$$\leqslant \left| \int_{s_0}^{s_1} \|\operatorname{grad} E(\phi_s c)\|_1^2 \, ds \right| |s_1 - s_0| \leqslant$$

$$|E(\phi_{s_1} c) - E(\phi_{s_0} c)| \, |s_1 - s_0| \leqslant E(\phi_{s_0} c) |s_1 - s_0| .$$

(iii) $$\|\operatorname{grad} E(c)\|_1^2 \leqslant 2E(c) .$$

(iv) $$E(\phi_s c) \leqslant E(c) + E(c) e^{-2s}, \quad \textit{for all } s \in J(c) .$$

Proof. (i) is an immediate consequence of the definitions. In (ii) we use the Schwarz inequality. As for (iii), we have

$$\|\operatorname{grad} E(c)\|_1^2 = DE(c) \,.\, \operatorname{grad} E(c) = \langle \partial c, \nabla \operatorname{grad} E(c) \rangle_0 \leqslant$$

$$\leqslant \|\partial c\|_0 \, \| \nabla \operatorname{grad} E(c)\|_0 \leqslant \sqrt{2E(c)} \, \|\operatorname{grad} E(c)\|_1 .$$

To prove (iv) we have from (i) and (iii), for $E(c) > 0$,

$$\frac{d}{ds} E(\phi_s c) \geqslant -2E(\phi_s c); \quad \frac{d}{ds} \ln E(\phi_s c) \geqslant -2 .$$

Therefore, if $s \leqslant 0$,

$$E(\phi_s c) \leqslant E(c) e^{-2s} .$$

From this we get, for all $s \in J(c)$,

$$E(\phi_s c) \leqslant \max\big(E(c), E(c) e^{-2s}\big) \leqslant E(c) + E(c) e^{-2s} . \quad \square$$

2.4.15 Theorem. *The integral curve* $\phi_s c$ *is defined for all* $s \in \mathbb{R}$.

Note. The fact that $\phi_s c$ is defined not only for $s \geqslant 0$ (when it is a simple consequence of $E(\phi_s c) \geqslant 0$) was observed by Solà-Morales [1].

Proof. Put $J(c) =]\, s_-, s_+ [$. If $s_- > -\infty$ or $s_+ < +\infty$, there exists a Cauchy sequence $\{s_m\}$, $s_- < s_m < s_+$, with limit a finite boundary point of $J(c)$. From (2.4.14) (iv) and (ii) it then follows that $\{\phi_{s_m} c\}$ is a Cauchy sequence. Let $\tilde{c}$ be its limit.

There exists a neighborhood $\mathscr{U}$ of $\tilde{c}$ and a $\varepsilon > 0$ such that $\phi_s c^*$ is defined for all $c^* \in \mathscr{U}$, all $|s| < 2\varepsilon$. In particular, $\phi_s \phi_{s_m} c = \phi_{s+s_m} c$ is defined for $|s| = \varepsilon$, m large. But $\varepsilon + s_m$ or $-\varepsilon + s_m$ lies outside $J(c)$ for large m – a contradiction. □

2.4.16 Lemma. *Let $\varkappa$ be a non-critical value of E. Then there exists $\varepsilon > 0$, $s_0 > 0$ such that*

$$\phi_s P^{\varkappa+\varepsilon} \subset P^{\varkappa-\varepsilon}, \quad \text{all } s \geqslant s_0 .$$

Remark. This Lemma will be fundamental for the subsequent existence proofs of critical points.

Proof. From (2.4.12) we have, with $\mathscr{U} = \emptyset$, an $\eta > 0$ such that $\|\operatorname{grad} E(c)\| \geqslant \eta$ for c with $\varkappa - \varepsilon \leqslant E(c) \leqslant \varkappa + \varepsilon$. Put $2\varepsilon/\eta^2 = s_0$.

Whenever $c \in P^{\varkappa-\varepsilon}$ so $\phi_s c \in P^{\varkappa-\varepsilon}$ for $s \geqslant 0$. Therefore it remain to consider c's with $\varkappa - \varepsilon < E(c) \leqslant \varkappa + \varepsilon$. We derive a contraction from the assumption that $E(\phi_s c) > \varkappa - \varepsilon$ for $0 \leqslant s \leqslant s_0$. Indeed, if this were the case we would get from (2.4.14)

$$E(\phi_{s_0} c) = E(c) - \int_0^{s_0} \|\operatorname{grad} E(\phi_s c)\|_1^2 \, ds \leqslant \varkappa + \varepsilon - \eta^2 s_0 \leqslant \varkappa - \varepsilon . \quad \square$$

2.4.17 Definition. Let $\varkappa_0$ be such that there are no critical values in $]\varkappa_0, \varkappa_0 + \varepsilon]$, for some small $\varepsilon > 0$.

A ϕ-family of P mod $P^{\varkappa_0}$ is a family $\mathscr{A}$ of non-empty subsets A of P such that

(i) $E|A$ is bounded,
(ii) $A \in \mathscr{A}$ implies $\phi_s A \in \mathscr{A}$, for $s \geqslant 0$,
(iii) $\sup E|A \geqslant \varkappa_0 + \varepsilon$.

The *critical value* of such a family $\mathscr{A}$ is defined by

$$\varkappa_{\mathscr{A}} = \inf_{A \in \mathscr{A}} \sup E|A .$$

In the case $\varkappa_0 < \inf E|P$, we simply speak of a ϕ-family of P.

Examples. (i) Consider the elements $\{\phi_s c\}$ of a ϕ-trajectory, all $s \geqslant 0$. Let $\lim_{s \to \infty} E(\phi_s c) = \varkappa^*$. Then this is a ϕ-family of P mod $P^{\varkappa_0}$ for any $\varkappa_0 < \varkappa^*$. In the same way one gets from A with $E|A$ bounded the ϕ-family $\{\phi_s A;\ \text{all } s \geqslant 0\}$.

(ii) Let P' be a connected component of P. Then the $c \in P'$ form a ϕ-family.

(iii) Consider $P = \Lambda M$. $\varkappa_0 = 0$ is a critical value such that there are no critical values in $]\varkappa_0, \varkappa_0 + \varepsilon] =]0, \varepsilon]$, $\varepsilon > 0$ sufficiently small. Indeed, take some ε, $0 < \varepsilon < 2\iota(M)^2$ where $\iota(M)$ is the injectivity radius of the compact manifold M, cf. (2.1.10). Then $E(c) \leqslant \varepsilon$ implies $L(c) \leqslant \sqrt{2E(c)} < 2\iota(M)$. Thus, for $c \in \Lambda^\varepsilon M$, $c(t)$ belongs to the ball $B_{\iota(M)}(c(0))$ around $c(0)$. c cannot be a closed geodesic since then $c(\frac{1}{2}) = \exp \frac{1}{2}\dot{c}(0)$ must be equal to $\exp(-\frac{1}{2}\dot{c}(0))$. Thus, $0 = \inf E|\Lambda M$ is always an isolated critical value.

While $\varkappa_0 = \inf E|\Omega M$ certainly is a critical value, it need not be isolated, c.f. the example (2.1.6; 5).

(iv) Let $F: (\bar{B}^k, \partial \bar{B}^k) \to (\Lambda M, \Lambda^0 M)$ be homotopically non-trivial. Let $\varepsilon > 0$ be as in (iii). Then there can be no F' homotopic to F with im $F' \subset \Lambda^\varepsilon M$. Indeed, we could then deform each $F'(x)$ into $F'(x)(0) \in \Lambda^0 M$, and this simultaneously for all $x \in \bar{B}^k$. Therefore, the set $\{F'(\bar{B}^k); F'$ homotopic to $F\}$ forms a ϕ-family of $\Lambda M \bmod \Lambda^0 M$.

(v) Let $w \in H_*(P, P^{\varkappa_0})$ be a non-trivial homology class. Assume that $P^{\varkappa_0+\varepsilon'}$ can be deformed into some subset of $P^{\varkappa_0}$, for all sufficiently small ε'. We obtain a ϕ-family of $P \bmod P^{\varkappa_0}$ from w by taking as its sets the images of the singular simplices, belonging to the cycles u representing w.

The importance of the concept of a ϕ-family lies in the fact that its critical value is a critical value of E.

2.4.18 Theorem. *The critical value $\varkappa = \varkappa_{\mathscr{A}}$ of a ϕ-family $\mathscr{A}$ of $P \bmod P^{\varkappa_0}$ is $> \varkappa_0$ and it is a critical value of E.*

Assume that the members A of the family $\mathscr{A}$ are compact. Let $\mathscr{U}$ be an open neighborhood of the set $Cr\varkappa$ of critical points with E-value $\varkappa = \varkappa_{\mathscr{A}}$. Then there exists a $A' \in \mathscr{A}$ such that $\phi_s A' \subset \mathscr{U} \cap P^{\varkappa-}$, for all $s \geq 0$.

Note. One occasionally describes this by saying 'A remains hanging at $Cr\varkappa$'.

Proof. The definition of $\varkappa = \varkappa_{\mathscr{A}}$ implies that for every $\varepsilon > 0$ there is a $A \in \mathscr{A}$ with $\sup E|A < \varkappa + \varepsilon$. Thus, $\varkappa > \varkappa_0$. If $\varkappa$ were non-critical, (2.4.16) would yield an $\varepsilon > 0$ and $s_0 > 0$ with $\sup E|\phi_{s_0} A \leq \varkappa - \varepsilon$ with $A \in \mathscr{A}$ as above. This proves the first statement.

Fix now an open neighborhood $\mathscr{U}$ of $Cr\varkappa$. Let $\mathscr{V}, \mathscr{W}$ be open neighborhoods of $Cr\varkappa$ satisfying

$$\mathscr{U} \supset \mathscr{V} \supset \mathscr{W}$$

and such that there exists a $\varrho > 0$ with

$$d_P(c', c'') > \varrho, \quad \text{for } c' \in \mathscr{V}, c'' \notin \mathscr{U};$$

$$d_P(c, c') \geq \varrho, \quad \text{for } c \in \mathscr{W}, c' \notin \mathscr{V}.$$

That there exist such neighborhoods follows from the fact that $Cr\varkappa$ is compact and d_P is a distance on P.

According to (2.4.12) there exist $\varepsilon > 0$ and $\eta > 0$ such that $\|\operatorname{grad} E(c)\|_1 \geq \eta$ for $c \notin \mathscr{W}$ and $\varkappa - \varepsilon \leq E(c) \leq \varkappa + \varepsilon$. Let $\phi_s c, 0 \leq s \leq s_1$, be a trajectory from a point $c \in \mathscr{W}$ with $\varkappa - \varepsilon \leq E(c) \leq \varkappa + \varepsilon$, to a point $\phi_{s_1} c = c' \notin \mathscr{V}$ with $\varkappa - \varepsilon \leq E(\phi_s c) \leq \varkappa + \varepsilon$. Since this trajectory contains an arc of length $> \varrho$ in the set

$$(*) \qquad \complement \mathscr{W} \cap (P^{\varkappa+\varepsilon} - P^{(\varkappa-\varepsilon)-}),$$

we can estimate the decrease of E along this trajectory by

$$E(c') - E(c) = -\int_0^{s_1} \|\operatorname{grad} E(\phi_s c)\|_1^2\, ds \leqslant -\eta^2 s_1 .$$

From (2.4.14, ii) we have $\varrho^2 \leqslant E(c)s_1 \leqslant (\varkappa+\varepsilon)s_1$, i.e.,

$$E(c') \leqslant E(c) - \eta^2 \varrho^2/(\varkappa+\varepsilon) < E(c) - \eta^2\varrho^2/\varkappa .$$

Hence, by taking $\varepsilon > 0$ sufficiently small, we can assume that the trajectory $\phi_s c$, $s \geqslant 0$, of an element $c \in \mathscr{W} \cap P^{(\varkappa+\varepsilon)-}$ remains in the set $\mathscr{V} \cup P^{(\varkappa-\varepsilon)-}$ because when it leaves $\mathscr{V}$ it has lost an amount $> 2\varepsilon$ of its E-value and therefore has entered $P^{(\varkappa-\varepsilon)-}$.

For the same reason we can assume that the trajectory $\phi_s c'$, $s \geqslant 0$, of an element $c' \in \mathscr{V} \cup P^{\varkappa+\varepsilon}$ remains in $\mathscr{U} \cup P^{(\varkappa-\varepsilon)-}$.

Now choose $A \in \mathscr{A}$, $A \subset P^{\varkappa+\varepsilon}$. For every $c \in A$ there is an $s_0 = s(c) \geqslant 0$ such that $\phi_s c$, $s \geqslant s_0$, lies in $\mathscr{V} \cup P^{(\varkappa-\varepsilon)-}$. Indeed, since $\|\operatorname{grad} E\|_1$ is bounded away from zero on the set (*), there will be a $s_1 \geqslant 0$ with $\phi_{s_1} c \in \mathscr{W} \cap P^{(\varkappa+\varepsilon)-}$ and we can apply our previous result.

With such a $s_0 = s(c)$ we can find an open neighborhood $\mathscr{U}(c)$ of c such that

$$\phi_{s_0} \mathscr{U}(c) \subset \mathscr{V} \cup P^{(\varkappa-\varepsilon)-}$$

As we have seen above, we then have

$$\phi_s \mathscr{U}(c) \subset \mathscr{U} \cup P^{(\varkappa-\varepsilon)-}, \quad \text{for all } s \geqslant s_0 .$$

Since A is compact, a finite number of such $\mathscr{U}(c)$ cover A. Therefore there will be a $s_0 \geqslant 0$ such that $\phi_s A \subset \mathscr{U} \cup P^{\varkappa-}$, all $s \geqslant s_0$. Put $\phi_{s_0} A = A'$. □

As a first consequence we mention the

2.4.19 Theorem. *On every connected component P' of P, E assumes its infimum. If c is an element in P' with minimal E-value then c is critical. That is, in the case $P = \Omega_{pq} M$, c is a geodesic from p to q. In the case $P = \Lambda M$, c is a constant map, if $\Lambda' M$ consists of the nullhomotopic curves; otherwise c is a closed geodesic.* □

Remarks with examples. The connected components of $\Omega M = \Omega_{pq} M$ are in 1:1 correspondence with the elements of the fundamental group $\pi_1 M$ of M. The connected components of ΛM correspond to the conjugacy classes of $\pi_1 M$.

Thus, if we take e.g. for M a torus of revolution in Euclidean 3-space, there are at least as many geodesics from a point p to a point q as there elements in $\pi_1 M = \mathbb{Z} \times \mathbb{Z}$. Since $\pi_1 M$ is abelian, the same is true for the number of closed geodesics. Actually, in each class of freely homotopic, not null homotopic closed curves on a torus, there are at least two closed geodesics, not just differing by their parameterizations. One is of minimal length in its class, the other of minimax type, i.e., with index plus nullity equal to one. This is due to the fact that a connected component of $\Lambda(S^1 \times S^1)$, when we divide out by the parameterizations, has the homotopy type of S^1.

In the case of a manifold of negative sectional curvature, each connected component of ΛM not containing the trivial curves has, modulo parameterization, the homotopy type of a point. This is the reason for the uniqueness of a curve of minimal length (up to parameterization) in such a homotopy class. For a proof see (3.8.13), (3.8.14).

The previous theorem does not yield the existence of a closed geodesic in case the manifold is simply connected. Appealing to some well known facts in the topology of manifolds we can show:

2.4.20 Theorem. *On any compact Riemannian manifold there exists a closed geodesic.*

Remark. This theorem was proved by Lyusternik and Fet [1]. Our proof is somewhat different from the original one. For a more elementary proof, avoiding the full theory of the Hilbert manifold ΛM, cf. (3.7.7).

Proof. If $\pi_1 M \neq 0$, the theorem is contained in (2.4.19). So assume $\pi_1 M = 0$. Since M is compact, there exists k, $0 < k < \dim M$, such that the homotopy group $\pi_{k+1} M$ of M is non-trivial. That is, there exists a

(*) $\quad f: S^{k+1} \to M$

which is not homotopic to a constant mapping, cf. Spanier [1].

We associate with such an f a mapping

(**) $\quad F = F(f): (B^k, \partial B^k) \to (\Lambda M, \Lambda^0 M)$

as follows: First, identify the closed k-ball $\bar{B}^k = \{x \in \mathbb{R}^k; |x| \leqslant 1\}$ with the half equator on $S^{k+1} \subset \mathbb{R}^{k+2}$ given by

$$\{x = (x_0, \ldots, x_{k+1}) \in S^{k+1};\ x_0 \geqslant 0 \quad \text{and} \quad x_1 = 0\}.$$

Denote by $c_p(t)$, $0 \leqslant t \leqslant 1$, the circle which starts out from $p \in \bar{B}^k$ orthogonally to the hyper-plane $\{x_1 = 0\}$ and enters the half sphere $\{x_1 \geqslant 0\}$. If $p \in \partial \bar{B}^k$, c_p is in the trivial circle $c_p(t) = p$, of course. With this we put

$$F(p) = \{f \circ c_p(t); 0 \leqslant t \leqslant 1\}.$$

Note that, conversely, a map F of $(\bar{B}^k, \partial \bar{B}^k)$ into $(\Lambda M, \Lambda^0 M)$ determines a map $f = f(F): S^{k+1} \to M$ such that $F(f(F)) = F$ and $f(F(f)) = f$. Indeed, describe $q \in S^{k+1}$ by $c_p(t)$. t is uniquely determined, except when $q \in \partial \bar{B}^k \subset S^{k+1}$ in which case $q = c_q(t) = \text{const}$. Define now $f(q)$ by $F(p)(t)$.

Even more is true. If f_s, $0 \leqslant s \leqslant 1$, is a homotopy of $f = f_0$, (*), then the associated family $F_s = F(f_s)$, (**), is a homotopy of $F = F_0$. And a homotopy F_s of $F = F_0$ determines a homotopy $f_s = f(F_s)$ of $f(F) = f(F_0)$.

Recall from example (iii) after (2.4.17) that the family of sets $F'(\bar{B}^k)$, F' homotopic to F, is a ϕ-family of ΛM mod $\Lambda^0 M$. □

One can ask whether there exists more than one closed geodesic on a compact Riemannian manifold M. Of course, two closed geodesics which only differ by the choice of the initial point or the orientation should, in this connections, not be counted as different. Also, closed geodesics which are just different coverings of an underlying prime closed geodesic should not be counted as geometrically different. See (2.5.A) and in particular the remark there at the end for a precise formulation of what it means that two geodesics are geometrically different.

In our present exposition of Riemannian geometry we will give only a few results concerning the existence of geometrically different geodesics. For a full account we must refer the reader to Klingenberg [10]. One of the main results there is the existence of infinitely many geometrically distinct closed geodesics, provided M has finite fundamental group.

The corresponding question for the loop space $\Omega = \Omega_{pq}$ of curves from p to q is much easier to answer. Serre [1] has shown that there exist infinitely many geodesics from p to q if and only if some homotopy group $\pi_k M, k \geqslant 1$, of M is non-zero. For M compact, this is known always to be true. What Serre shows is that $\pi_k M \neq 0$, for some $k \geqslant 1$, implies that the sequence $\{b_i \Omega M\}$ of Betti numbers of ΩM contains infinetely many non-zero elements. This then yields the existence of infinitely many critical points in ΩM.

A note of caution must be added. As shown by the sphere with the standard metric, it can happen that the infinitely many geodesics from p to q all lie on the same closed geodesic. Thus one might ask whether it is really justified to speak of infinitely many geometrically different geodesics from p to q.

However, such a phenomenon presumably can happen only for very special Riemannian metrics: In general, for almost all p and q on M, different geodesics from p to q have no arcs in common

2.5 The Second Order Neighborhood of a Critical Point

We continue to consider the functional E on one of the Hilbert submanifolds of $H^1(I, M)$ which satisfy condition (C). Our main interest is again directed towards the loop space $\Omega M = \Omega_{pq} M$ and the space ΛM of closed curves – sometimes also called a free loop space.

We study the second order neighborhood of a critical point c of E, i. e., the Hessian $D^2 E(c)$. For the cases mentioned above, $D^2 E(c)$ has a particularly simple expression, involving the curvature tensor only, cf. (2.5.1). The associated self-adjoint operator A_c is of the form identity plus compact operator, cf. (2.5.2). In particular, the negative eigenspace of A_c has finite dimension; it is called the index of c, cf. (2.5.3), (2.5.6). The null space of A_c consists of the Jacobi fields which belong to the tangent space at c to the submanifold under consideration, cf. (2.5.6). In (2.5.7) we determine the eigenspaces of A_c for c a critical point of ΛS^n.

The Morse Index Theorem (2.5.9) for a critical point c of $\Omega_{pq} M$ relates the index of c to the number of conjugate points on c. (2.5.14) is the analogue for a closed geodesic and in (2.5.15) the Focal Index Theorem is formulated.

We conclude with an important generalization (2.5.16) of (2.4.16) where the index of the critical points comes into play. Usually, this result is proved by first considering the case where all critical points are non-degenerate (this is the hypothesis for the so-called Morse theory proper) and then employing an approximation argument. Our direct proof uses a novel technique. The result may be viewed as to belong to the common realm of Lusternik-Schnirelman and Morse theory.

In the appendix we describe the natural S^1-action and $\mathbb{Z}_2$-action on the manifold ΛM of closed H^1-curves.

We continue to consider the functional E on one of the Hilbert manifolds $H^1_{M_0 \times M_1} M$ or ΛM of $H^1(I, M)$. In the first case we assume M_0 or M_1 to be compact. In the second case, M shall be compact. Thus, condition (C) holds in all cases.

We begin by defining a linear operator with the help of the curvature tensor.

Let c be a geodesic. We start with the linear bundle mapping

$$R_c : c^* TM \to c^* TM\,;\ \xi_t \in (c^*\tau)^{-1}(t) \mapsto R_{\dot c(t)}\xi_t \in (c^*\tau)^{-1}(t).$$

Here, $R_{\dot c(t)}\xi_t = R(\xi_t, \dot c(t))\dot c(t)$ is the curvature operator defined in (1.8.13). We consider the induced mapping $\tilde{R_c}$, cf. (2.3.6),

$$\tilde{R_c} : H^1(c^* TM) \cong T_c H^1(I, M) \to T_c H^1(I, \mathrm{M}).$$

Note that $\xi_0 = \xi_1 = 0$ implies $R_c\xi_0 = R_c\xi_1 = 0$ and $\xi_0 = \xi_1$ implies $R_c\xi_0 = R_c\xi_1$, if $\dot c(0) = \dot c(1)$. Thus, $\tilde{R_c}$ transforms the subspace $T_c\Omega_{c(0)c(1)}M$ into itself and, if c is closed geodesic, then $T_c\Lambda M$ is also transformed into itself.

We compute $D^2E(c)$:

2.5.1 Lemma. (i) *Let c be a critical point of $E|H^1_{M_0\times M_1}M$. Let ξ, ξ' be elements of $T_c H^1_{M_0\times M_1}M$. Then*

$$D^2E(c)(\xi, \xi') = \langle \nabla\xi, \nabla\xi'\rangle_0 - \langle \tilde{R_c}\xi, \xi'\rangle_0 + h_{\dot c(1)}(\xi(1), \xi'(1)) - h_{\dot c(0)}(\xi(0), \xi'(0)).$$

Here, $h_{\dot c(i)}(\,,\,)$ is the second fundamental form of M_i at $c(i)$ in the direction of the normal vector $\dot c(i)$, $i = 0,1$.

(ii) *If M_i is totally geodesic at $c(i)$, $i = 0,1$, the formula $(*)$ reduces to*

$$(**)\qquad D^2E(c)(\xi, \xi') = \langle \nabla\xi, \nabla\xi'\rangle_0 - \langle \tilde{R_c}\xi, \xi'\rangle_0.$$

This holds in particular for $\Omega M = \Omega_{c(0)c(1)}M$.

(iii) *The formula $(**)$ also holds for c a critical point of $E|\Lambda M$.*

Proof. We proceed as in the proof of (1.12.12). It suffices to take $\xi = \xi'$ and then apply polarization. Let $c_s, s \in\,]-\varepsilon, \varepsilon[$ be a curve on $H^1_{M_0\times M_1}M$ with $c_0 = c$ and $dc_s/ds|_0 = \xi$. Put $c_s(t) = F(s, t)$. Then

$$\frac{d}{ds}E(c_s) = \int_0^1 \langle \frac{\nabla}{\partial s}\frac{\partial}{\partial t}F, \frac{\partial F}{\partial t}\rangle(s,t)\,dt = \int_0^1 \langle \frac{\nabla}{\partial s}\frac{\partial}{\partial t}F, \frac{\partial F}{\partial t}\rangle(s,t)\,dt,$$

$$D^2E(c)(\xi,\xi) = \int_0^1 \left\langle \frac{\nabla}{\partial t}\frac{\partial}{\partial s}F, \frac{\nabla}{\partial t}\frac{\partial}{\partial s}F \right\rangle (0,t)\,dt +$$

(§)

$$+ \int_0^1 \left\langle R\left(\frac{\partial F}{\partial s}, \frac{\partial F}{\partial t}\right)\frac{\partial F}{\partial s}, \frac{\partial F}{\partial t} \right\rangle (0,t)\,dt + \int_0^1 \frac{d}{dt}\left\langle \frac{\nabla}{\partial s}\frac{\partial}{\partial s}F, \frac{\partial F}{\partial t} \right\rangle (0,t)\,dt.$$

The last term in (§), when integrated, is of the form

$$h_{\dot c(1)}(\xi(1), \xi(1)) - h_{\dot c(0)}(\xi(0), \xi(0)),$$

cf. the Gauss formula (1.10.5). It vanishes for M_i totally geodesic at $c(i)$, $i = 0, 1$, see (1.10.14).

The same arguments apply to a critical point c of ΛM, the only difference being that this time the last term in (§) vanishes due to

$$F(s,1) = F(s,0); \quad \frac{\partial F}{\partial t}(0,1) = \frac{\partial F}{\partial t}(0,0). \quad \square$$

We analyse the structure of the operator associated to $D^2E(c)$ in the case ΩM and ΛM and $H^1_{M_0 \times \{p_1\}}$ with M_0 of codimension 1 and totally geodesic at $p_0 = c(0)$. As common notation for these three cases we use P.

2.5.2 Lemma. *Let c be a critical point of $E|P$. Then the self-adjoint operator A_c defined by*

$$\langle A_c\xi, \xi'\rangle_1 = \langle \xi, A_c\xi'\rangle_1 = D^2E(c)(\xi, \xi')$$

is of the form

$$A_c = 1 + k_c.$$

Here, 1 is the identity and k_c is the compact operator given by

$$k_c = -(1-\nabla^2)^{-1} \circ (\tilde{R}_c + 1)$$

Proof. Partial integration and the boundary conditions for differentiable $\eta, \eta' \in T_cP$ give the relation

$$\langle \eta, \eta'\rangle_1 = \langle \nabla\eta, \nabla\eta'\rangle_0 + \langle \eta, \eta'\rangle_0 = \langle -\nabla^2\eta, \eta'\rangle_0 + \langle \eta, \eta'\rangle_0 =$$
$$\langle (1-\nabla^2)\eta, \eta'\rangle_0.$$

By continuity, this relation extends to all of T_cP. Thus, for ξ, ξ' in T_cP,

$$\langle A_c\xi, \xi'\rangle_1 = \langle \nabla\xi, \nabla\xi'\rangle_0 - \langle \tilde{R}_c\xi, \xi'\rangle_0 = \langle \xi, \xi'\rangle_1 - \langle (\tilde{R}_c + 1)\xi, \xi'\rangle_0 =$$
$$= \langle \xi, \xi'\rangle_1 - \langle (1-\nabla^2)^{-1} \circ (\tilde{R}_c + 1)\xi, \xi'\rangle_1.$$

As for the compactness of k_c we can either refer to general facts about the compactness of the inverse of an elliptic differential operator such as $-(1-\nabla^2)$, or we can derive it this directly as follows: From

$$\langle k_c\xi, k_c\xi\rangle_1 = -\langle(\tilde{R_c}+1)\xi, k_c\xi\rangle_0$$

we get $\|k_c\xi\|_1^2 \leqslant \|\tilde{R_c}+1\|_\infty \cdot \|k_c\xi\|_\infty \cdot \|\xi\|_0 \leqslant \text{const}\|k_c\xi\|_1 \cdot \|\xi\|_0$

We know from (2.4.6) that a bounded H^1-sequence $\{\xi_m\}$ is relatively compact as a C^0-sequence and hence as an H^0-sequence. Since $\|k_c\xi_m\|_1 \leqslant \text{const}\|\xi_m\|_0$, $\{k_c\xi_m\}$ is a relatively compact H^1-sequence. □

2.5.3 Corollary. *The operator* A_c *defined in* (2.5.2) *either has only finitely many eigenvalues, including* 1, *or it has an infinite sequence of eigenvalues* $\neq 1$ *which have* 1 *as the only accumulation point.* 1 *is a spectral value and not an eigenvalue.*

In particular, the tangent space $T_c = T_c\Omega$ *or* $= T_c\Lambda M$ *or* $= T_c H^1_{M_0\times\{p_1\}}$ *possesses an orthogonal decomposition*

$$T_c = T_c^- \oplus T_c^0 \oplus T_c^+$$

into a finite dimensional subspace T_c^-, *spanned by the eigenvectors with negative eigenvalue, a finite dimensional subspace* T_c^0, *spanned by the eigenvectors with eigenvalue zero, and a proper Hilbert space* T_c^+, *spanned by the eigenvectors with positive eigenvalue.*

Proof. This is an immediate consequence of the well-known structure of the spectrum of a compact operator, cf. Dieudonné [1]. □

2.5.4 Definition. Let c be a critical point of $E|P$.

(i) $\dim T_c^-$ is called *index of* c. According to the three cases ΩM, ΛM and $H^1_{M_0\times\{p_1\}}$ we also speak of the *Ω-index of* c, $\text{index}_\Omega\, c$, the *Λ-index of* c, $\text{index}_\Lambda\, c$, and the *focal index of* c, $\text{index}_{foc}\, c$.

(ii) $\dim T_c^0$ is called *nullity of* c, except in the case when $P = \Lambda M$ and $c_0 \neq \text{const}$. In that case the nullity of c is defined to be $\dim T_c^0 - 1$.

Remarks. 1. The reason we define the nullity of a closed geodesic c to be $\dim T_c^0 - 1$ is that the vector field ∂c always belongs to T_c^0. This follows at once from (2.5.1) since $\tilde{R_c}\partial c \equiv 0$. Cf. also (2.5.6) below.

2. A trivial but useful observation is the following: If U is a subspace of $T_c = T_c\Omega M$ or $T_c\Lambda M$ or $T_c H^1_{M_0\times\{p_1\}}$ such that $D^2E(c)|U$ is negative semi-definite, i. e., $\xi \in U$ implies $D^2E(c)(\xi,\xi) \leqslant 0$, then $\dim U \leqslant \text{index}\, c + \dim T_c^0$. If, moreover, $U \cap T_c^0 = 0$, then $\dim U \leqslant \text{index}\, c$.

This is a simple fact of linear algebra: Let

$$pr_- : T_c = T_c^- \oplus T_c^0 \oplus T_c^+ \to T_c^-$$

be the orthogonal projection. Consider

$$pr_- | U : U \to T_c^- .$$

Then $\xi \in \ker pr_- | U$ must belong to $T_c^0 \oplus T_c^+$. That is, since $D^2 E(c)(\xi, \xi) \leqslant 0$, $\xi \in T_c^0$. From

$$\dim U = \dim(\text{image}\, pr_- | U) + \dim(\text{kernel}\, pr_- | U)$$

we prove our claim.

2.5.5 Proposition. *Let ξ be an eigen vector to an eigen value $\lambda \neq 1$ of A_c. Then ξ is a differentiable solution of*

$$\nabla^2 \xi(t) + \frac{1}{1-\lambda} R(\xi, \dot{c})\dot{c}(t) + \frac{\lambda}{1-\lambda} \xi(t) = 0,$$

satisfying the boundary condition

$$\xi \in T_c \Omega M \text{ or } \xi \in T_c \Lambda M \text{ or } \xi \in T_c H^1_{M_0 \times \{p_1\}}.$$

Proof. In $A_c \xi = \lambda \xi$ substitute the expression $A_c = 1 + k_c$, with k_c as in (2.5.2) □

2.5.6 Corollary. *$\xi \in T_c^0$ if and only if $\xi(t)$ is a differentiable Jacobi field along $c(t)$, satisfying the boundary condition*

$$\xi \in T_c \Omega M \text{ or } \xi \in T_c \Lambda M \text{ or } \xi \in T_c H^1_{M_0 \times \{p_1\}}.$$

Note. $\xi \in T_c \Lambda M$ and ξ differentiable means $(\xi(1), \nabla \xi(1)) = (\xi(0), \nabla \xi(0))$. Hence T_c^0 can be identified with the space of the periodic Jacobi fields along c.

Proof. Put $\lambda = 0$ in (2.5.5). □

2.5.7 Examples. (i) Let M be flat, i. e., $\tilde{R}_c = 0$. Then the eigenvalue equation (2.5.5) reads

$$\nabla^2 \xi + \frac{\lambda}{1-\lambda} \xi = 0.$$

For $c \in \Omega M$ or $c \in \Lambda M$ the boundary conditions $\xi(0) = \xi(1) = 0$ or $\xi(0) = \xi(1)$ $\nabla \xi(0) = \nabla \xi(1)$ allow no non-trivial solution for $\lambda < 0$. Thus, index $c = 0$. On the other hand, for $\lambda = 0$, $\dim T_c^0 \Omega = 0$, whereas for $c \in \Lambda M$, $c \neq$ const, $T_c^0 M$ is given by the parallel periodic vector fields ξ, $\xi(0) = \xi(1)$ and $\nabla \xi = 0$. Thus, nullity c is equal to the dimension of the eigenspace for the eigenvalue $+1$ under the linear mapping

$$\overset{1}{\underset{0}{\|}} c : T_0^\perp c \to T_1^\perp c = T_0^\perp c,$$

given by parallel translation of the subspace $T_0^\perp$ of $T_{c(0)} M$ into itself. See (1.10.4) for our notation.

(ii) $M = S^n_\varrho$ = n-dimensional sphere of radius ϱ in $\mathbb{R}^{n+1}$, cf. (1.1.3, ii), (1.11.6). Then

$$R(\xi, \dot{c})\dot{c}(t) = \frac{1}{\varrho^2}(|\dot{c}(0)|^2 \xi(t) - \langle \xi(t), \dot{c}(t)\rangle \dot{c}(t)).$$

A vector field $\xi(t)$ along a non-constant geodesic $c(t)$ splits into a sum $\xi^\top(t) + \xi^\perp(t)$, where $\xi^\top(t) = x(t)\dot{c}(t)$ is tangential to $\dot{c}(t)$ and $\xi^\perp(t)$ is orthogonal to $\dot{c}(t)$:

$$\xi^\perp(t) = \xi(t) - \langle \dot{c}(t), \xi(t)\rangle \dot{c}(t)/|\dot{c}(t)|^2.$$

Writing $\eta(t)$ for the orthogonal component, the eigenvalue equation (2.5.3) splits into the two equations

$$(\top) \qquad \ddot{x}(t) + \frac{\lambda}{1-\lambda} x(t) = 0 \text{ and}$$

$$(\perp) \qquad \nabla^2 \eta(t) + \frac{|\dot{c}(0)|^2/\varrho^2 + \lambda}{1-\lambda} \eta(t) = 0.$$

(ii a) Let c be a critical point of $E|\Omega S^n_\varrho$, i. e., a geodesic from $p \in S^n_\varrho$ to $q \in S^n_\varrho$. $|\dot{c}| = L(c)$ = length of c. ($\top$) has no solutions $x \neq 0$ with $x(0) = x(1) = 0$.

Non-zero solutions η of ($\perp$) with $\eta(0) = \eta(1) = 0$ occur precisely for

$$(|\dot{c}|^2/\varrho^2 + \lambda)/(1-\lambda) = \pi^2 p^2, p = 1, 2, \ldots .$$

i. e. $\qquad \lambda = \lambda_p = (\pi^2 p^2 - |\dot{c}|^2/\varrho^2)/(\pi^2 p^2 + 1).$

The corresponding eigenvectors are

$$\eta(t) = \eta_p(t) = A(t) \sin \pi p t,$$

with $A(t)$ a parallel vector field orthogonal to $\dot{c}(t)$. Thus, the eigenspace belonging to the eigenvalue λ_p has dimension $(n-1)$.

$\lambda_p < 0$ is equivalent to $|\dot{c}| < \varrho \pi p$, where $\varrho\pi$ = length of half a great circle on S^n_ϱ. We see that index $c = k(n-1)$, with k the integer determined by

$$\pi \varrho k < L(c) = |\dot{c}| \leqslant \pi \varrho (k+1).$$

Nullity $c = n-1$, if $|\dot{c}| = L(c) = \pi \varrho k$, k an integer > 0, and $= 0$ otherwise.

(ii b) Let now c be a critical point $\neq$ const of $E|\Lambda S^n_\varrho$. That is, c is a closed geodesic which means it is a q-fold covered great circle. $|\dot{c}| = L(c) = 2\pi q \varrho$.

We see that ($\top$) has no periodic solutions if $\lambda < 0$. For $\lambda = 0$ we get a 1-dimensional space of periodic solutions, i. e., $x(t) = x(0)$ = const.

Non-zero periodic solutions of ($\perp$) can occur only if

$$(|\dot{c}|^2/\varrho^2 + \lambda)/(1-\lambda) = (4\pi^2 q^2 + \lambda)/(1-\lambda) = 4\pi^2 p^2, p = 0, 1, \ldots,$$

i. e., $\qquad \lambda = \lambda_p = 4\pi^2(p^2 - q^2)/(4\pi^2 p^2 + 1).$

An eigenvector for the eigenvalue λ_p is of the form

$$\eta(t) \equiv \eta_p(t) = A(t)\cos 2\pi p t + B(t)\sin 2\pi p t.$$

Here, $A(t)$ and $B(t)$ are parallel vector fields along $c(t)$, orthogonal to $\dot{c}(t)$.

We see that if $p = 0$, the eigenspace for λ_p has dimension $n-1$, whereas for $p > 0$ it has dimension $2(n-1)$.

$\lambda_p < 0$ occurs only for $0 \leqslant p < q$. Hence, index $c = (n-1) + (q-1)2(n-1) = (2q-1)(n-1)$; nullity $c = 2(n-1)$. □

Let $c: [0, a] \to M$ be a geodesic, $\dot{c} \neq 0$, $\dim M = n$. From (1.12.9) we know that the multiplicity k of a conjugate point $c(t_1)$ along $c|[0, t_1]$ is the dimension of the space of Jacobi fields $Y(t)$ along $c(t)$ vanishing at $t = 0$ and $t = t_1$. Here, we allow $k = 0$, in which case $c(t_1)$ is is usually called non-conjugate. For the Jacobi field $Y(t)$ we can assume $\langle Y(t), \dot{c}(t)\rangle = 0$. Clearly, $k \leqslant n-1 = \dim \mathscr{J}_c^{0\perp}$, where $\mathscr{J}_c^{0\perp}$ denotes the space of Jacobi fields vanishing at 0, orthogonal to $\dot{c}$, cf. (1.12.10).

For a later application we need the

2.5.8 Proposition. *Let $c: [0, a] \to M$ be a geodesic, $\dot{c} \neq 0$. Let $c(t_1)$ be a conjugate point of multiplicity $k, 0 \leqslant k \leqslant n-1$.*

(i) *For every sufficiently small neighborhood $I(t_1)$ of t_1 on $[0, a]$, all $c(t)$ with $t \in I(t_1)$, are non-conjugate.*

(ii) *Let Y be a Jacobi field with $Y(0) = Y(t_1) = 0$. Then $\nabla Y(t_1) \in T_{t_1}^{\perp} c$ is orthogonal to every $Y'(t_1)$, $Y' \in \mathscr{J}_c^{0\perp}$.*

Proof. Let $\dim M = n$. In (1.12.3) we had introduced the symplectic form α on the $2n$-dimensional space $\mathscr{J}_c$ of Jacobi fields. $\mathscr{J}_c^{\perp}$ was a $(2n-2)$-dimensional non-degenerate subspace. Now, $\mathscr{J}_c^{0\perp}$ is a so-called Lagrangian subspace of $\mathscr{J}_c^{\perp}$, i. e.,

$$(*) \qquad \alpha(Y, Y') = \langle Y, \nabla Y'\rangle(t) - \langle Y', \nabla Y\rangle(t) = 0$$

for all $Y, Y' \in \mathscr{J}_c^{0\perp}$ and $\dim \mathscr{J}_c^{0\perp} = n-1$.

The values $\nabla Y(t_1)$ of those $Y \in \mathscr{J}_c^{0\perp}$ which vanish at $t = t_1$ form a k-dimensional subspace of the $(n-1)$-dimensional space $T_{t_1}^{\perp} c$ of $T_{c(t_1)} M$. From (*) we get $\langle \nabla Y(t_1), Y'(t_1)\rangle = 0$, if $Y(t_1) = 0$, i. e., we have (ii).

To prove (i) we choose a basis Y_i, $1 \leqslant i \leqslant n-1$, for $\mathscr{J}_c^{0\perp}$. Then the k'th derivative of the determinant $|Y_1(t), \ldots, Y_{n-1}(t)|$ at $t = t_1$ is not zero. □

We now can prove the *Index Theorem of Morse* [2]. It relates the index of a geodesic c, considered as a critical point of $E|\Omega_{c(0)c(1)} M$, to the conjugate points along c. As we shall see in (3.2.12), the right setting for conjugate points is the geodesic flow on the tangent bundle of M. Thus, the following theorem relates properties of critical points of $E|\Omega M$ to properties of the geodesic flow.

2.5.9 Theorem. *Let $c = \{c(t), 0 \leqslant t \leqslant 1\}$ be a critical point of $\Omega = \Omega_{pq} M$. That is, c is a geodesic on M from $c(0) = p$ to $c(1) = q$. Then*

$$\text{index}\, c = \sum_{0<t<1} \{\textit{multiplicity of the conjugate point } c(t)\}$$

$$= \textit{sum of proper conjugate points } c(t_1), 0 < t_1 < 1, \textit{ each counted with its multiplicity } \geqslant 1.$$

Proof. According to (1.12.10), $c(t_1)$ is conjugate to $c(0)$ along $c|[0, t_1]$ of multiplicity k if the space $W(t_1)$ of Jacobi fields $Y(t)$ along $c(t)$ with $Y(0) = Y(t_1) = 0$ has dimension k. Here, we may restrict ourselves to the Jacobi fields $Y(t) \perp \dot{c}(t)$.

Define a mapping

$$\Phi: W(t_1) \to T_c\Omega;\ Y \mapsto \xi_Y$$

by
$$\xi_Y(t) \equiv \xi(t) = \begin{cases} Y(t), & 0 \leqslant t \leqslant t_1, \\ 0, & t_1 \leqslant t \leqslant 1. \end{cases}$$

ξ_Y is a broken Jacobi field with

$$\nabla \xi_Y(t_1 -) - \nabla \xi_Y(t_1 +) = \nabla Y(t_1) \neq 0,$$

unless $Y = 0$.

Clearly, Φ is linear and injective. Moreover, when $t_1 \neq t_1'$, $\Phi W(t_1) \cap \Phi W(t_1') = 0$. Denote by U the subspace of $T_c\Omega$, generated by the $\Phi W(t)$, $0 < t < 1$. Then $\dim U$ $= \sum_{0<t<1} \dim W(t) =$ right hand side of (2.5.9).

We claim that $\dim U \leqslant \text{index}\, c$. We first show that $D^2 E(c)|U \equiv 0$. To see this consider $\xi \in \Phi W(t_1)$, $\xi' \in \Phi W(t_1')$ with $t_1' \leqslant t_1$. Then

$$D^2 E(c)(\xi, \xi') = -\int_0^1 \langle \nabla^2 \xi + R(\xi, \dot{c})\dot{c}, \xi' \rangle (t)\, dt + \langle \nabla \xi, \xi' \rangle (t) \Big|_0^{t_1 -} = 0,$$

since $\xi'(t_1) = \xi'(0) = 0$. Moreover, $U \cap T_c^0 = 0$, since the elements of T_c^0 are the Jacobi fields $Y(t)$ with $Y(0) = Y(1) = 0$, cf. (2.5.6). Our claim therefore follows from remark 2 after (2.5.4).

We conclude the proof by showing that if there is a $\xi^* \in T_c^-$ with $D^2E(c)(\xi^*, U) = 0$ then $\xi^* = 0$. That is, $\dim U = \dim pr_- U = \dim T_c^- = \text{index}\, c$, with $pr_-: T_c \to T_c^-$ the orthogonal projection.

To prove $\xi^* = 0$, we begin by showing that $\xi^*(t)$ can be written as

$$\xi^*(t) = \sum_i w^i(t)\, Y_i(t),$$

where the Y_i, $1 \leqslant i \leqslant n-1$, form a basis for $\mathcal{J}_c^{0\perp}$. First, $\xi^* \in T_c^-$ means that $\xi^*(t) \perp \dot{c}(t)$. Next, consider the case when the $\{Y_i(t_1); 1 \leqslant i \leqslant n-1\}$ are linearly dependant, i. e., $c(t_1)$ is a conjugate point of multiplicity $k > 0$. For $Y \in \mathcal{J}_c^{0\perp}$ with $Y(t_1) = 0$, consider $\xi \in \Phi W(t_1) \subset U$ with $\xi|[0, t_1] = Y|[0, t_1]$, $\xi|[t_1, 1] = 0$. Then

$$0 = D^2 E(c)(\xi, \xi^*) = \langle \nabla \xi, \xi^* \rangle \Big|_0^{t_1 -} = \langle \nabla Y(t_1), \xi^*(t_1) \rangle = 0.$$

That is, according to (2.5.8), $\xi^*(t_1)$ belongs to the space spanned by the $\{Y_i(t_1), 1 \leqslant i \leqslant n-1\}$.

We conclude by computing $D^2E(c)(\xi^*, \xi^*)$. As an element of T_c^-, ξ^* certainly is differentiable. Thus, using partial integration,

$$0 \geqslant D^2E(c)(\xi^*, \xi^*) = -\int_0^1 \langle \nabla^2 \xi^* + R(\xi^*, \dot{c})\dot{c}, \xi^* \rangle (t)\, dt$$

$$= -\int_0^1 \langle \sum_i w^i(\nabla^2 Y_i + R(Y_i, \dot{c})\dot{c}), \sum_j w^j Y_j \rangle (t)\, dt$$

$$-\int_0^1 \sum_{i,j} \dot{w}^i w^j \langle \nabla Y_i, Y_j \rangle (t)\, dt - \int_0^1 \langle \frac{d}{dt}(\sum_i \dot{w}^i Y_i), \sum_j w^j Y_j \rangle (t)\, dt$$

$$= \int_0^1 \sum_{i,j} \dot{w}^i w^j (\langle Y_i, \nabla Y_j \rangle - \langle Y_j, \nabla Y_i \rangle)(t)\, dt$$

$$+\int_0^1 \langle \sum_i \dot{w}^i Y_i, \sum_j \dot{w}^j Y_j \rangle (t)\, dt \geqslant 0.$$

The second last term vanishes, since $\mathscr{J}_c^{0\perp}$ is a Lagrangian subspace, i. e., a subspace of $\mathscr{J}_c^{\perp}$ on which the form α vanishes identically, cf. (1.12.3, iv).

Hence, $D^2E(c)(\xi^*, \xi^*) = 0$, i.e., $\xi^* = 0$. □

Example. We determine the index of a geodesic segment $c = \{c(t), 0 \leqslant t \leqslant 1\}$ of length $L(c) = l$ on S^n_ϱ. The Jacobi equation reads, cf. (2.5.7),

$$(*) \qquad \nabla^2 Y(t) + \frac{l^2}{\varrho^2} Y(t) = 0.$$

Here we have considered only Jacobi fields orthogonal to $\dot{c}(t)$. From (1.12.2) we see that they are the only ones which can vanish at $t = 0$ and at some $t_1 > 0$, unless we have the 0-field.

The solutions of $(*)$ are of the form

$$Y(t) = A(t) \sin(lt/\varrho)$$

with $A(t)$ a parallel vector field.

If $A(t) \neq 0$, $Y(t) = 0$ can occur only for $t_p = \pi\varrho p/l$, p an integer. Each proper conjugate point has multiplicity $n-1$. Thus, index $c = k(n-1)$ where k is the integer determined by

$$\pi\varrho k < l \leqslant \pi\varrho(k+1).$$

Compare this with (2.5.7, ii a). □

As a first consequence of Morse's Index Theorem we prove a result similar to (1.12.13).

2.5.10 Theorem. *Let $c: I \to M$ be a geodesic from $c(0) = p$ to $c(1) = q$. We also consider c as critical point of $E|\Omega_{pq}M$.*

(i) *If there are no conjugate points in the interior or on the boundary of I then there exists a neighborhood $\mathscr{U}(c)$ of c in $\Omega_{pq}M$ such that for all $b \in \mathscr{U}(c)$, $E(b) \geqslant E(c)$, with $E(b) = E(c)$ only for $b = c$.*

(ii) *Let $k > 0$ be the number of conjugate points in the interior of I, each counted with its multiplicity. Then there exist an immersion*

$$F: (B^k, 0) \to (\Omega_{pq}M, c)$$

of the unit ball $B^k = \{x \in \mathbb{R}^k, |x| < 1\}$ such that

$$E(F(x)) \leqslant E(c) \text{ and hence } L(F(x)) \leqslant L(c),$$

for all $x \in B^k$, with equality only for $x = 0$.

Remarks. Part (i) is weaker than (1.12.13, i) in two aspects: First, we compare the energy integral only and not the length – note that from $E(b) \geqslant E(c)$ we cannot always conclude $L(b) \geqslant L(c)$; this we could amend, however, by introducing the functional L on $\Omega_{pq}M$, if $d(p, q) \neq 0$. Then $DL(c) = 0$ and $D^2L(c)$ is, up to a factor, equal to $D^2E(c)$.

Secondly, and this is more serious, in (i) we compare with $E(c)$ only those b which are in some d_{H^1}-neighborhood of c, whereas in (1.12.13, i) we did allow b to be in a d_∞-neighborhood of c in $\Omega_{pq}M$.

On the other hand, part (ii) is definitely a refinement of (1.12.13, ii): $E(c) \geqslant E(b)$ implies $L(c) \geqslant L(b)$ and we can find curves b shorter than c already in every d_{H^1}-neighborhood of c on $\Omega_{pq}M$.

Proof. Under the hypothesis made in (i) we find from (2.5.8) that $D^2E(c)$ is positive definite. Hence, if we choose a sufficiently small natural chart $(\exp_c^{-1}, \mathscr{U}(c))$ around c, the Taylor series for $E(b)$, $b \in \mathscr{U}(c)$, $\xi = \exp_c^{-1} b$ starts with

$$E(b) = E(c) + \frac{1}{2} D^2E(c)(\xi, \xi) + o\|\xi\|_1^2.$$

This is $\geqslant E(c)$, for all ξ, $\|\xi\|_1$ sufficiently small.

The hypothesis made in (ii) implies, according to (2.5.8), that index $c = k$. Let $\{\xi_i, 1 \leqslant i \leqslant k\}$ be an orthonormal basis of T_c^- where ξ_i is an eigenvector for the eigenvalue $-\lambda_i < 0$. For $\delta > 0$ sufficiently small.

$$F: x = (x^1, \ldots, x^k) \in B^k \mapsto \exp_c(\delta \sum_i x^i \xi_i) \in \Omega_{pq}M$$

will be an immersion. From the Taylor expansion

$$E(F(x)) = E(c) - \frac{1}{2}\delta^2 \sum_i \lambda_i (x^i)^2 + o\|\delta\xi\|_1^2$$

we have proved our claim, provided δ is sufficiently small. □

We continue with an Index Theorem for a closed geodesic. First we observe that a closed geodesic c can also be viewed as critical point of $E|\Omega$, $\Omega = \Omega_{c(0)c(0)}M$. Recall from (2.5.4) that we denote by $\mathrm{index}_\Omega\, c$ the index of c as critical point of $E|\Omega$, whereas for index c we also write $\mathrm{index}_\Lambda\, c$.

2.5.11 Proposition. *Let c be a closed geodesic. Then*

$\mathrm{index}_\Omega\, c \leqslant \mathrm{index}_\Lambda\, c \equiv \mathrm{index}\, c.$

Proof. $D^2E(c)$ has the same form, whether we derive it from $E|\Omega M$ or from $E|\Lambda M$. The only difference is that in the first case we restrict it to $T_c\Omega M$, the space of H^1-vector fields $\xi(t)$ along $c(t)$ which vanish for $t=0$ and $t=1$, whereas in the second case we allow $\xi(t)$ with $\xi(0)=\xi(1)$.

Now, as we remarked after (2.5.4), since $T_c^-\Omega$ is a subspace of $T_c\Lambda$ on which $D^2E(c)$ is negative definite, $\mathrm{index}_\Lambda\, c \geqslant \dim T_c^-\Omega = \mathrm{index}_\Omega\, c$. □

Let c be a closed geodesic. We want to write down a formula for the term which has to be added to $\mathrm{index}_\Omega\, c$ to obtain $\mathrm{index}_\Lambda\, c$.

It was introduced by Morse [2] under the name concavity.

As a preliminary step we introduce the following concepts. Recall from (1.12.3) that $\mathscr{J}_c^\perp$ denotes the $(2n-2)$-dimensional space of Jacobi fields along c which are orthogonal to c. $\mathscr{J}_c^{0\perp}$ denotes the $(n-1)$-dimensional subspace of those fields which vanish at $t=0$.

2.5.12 Definition. Let $c(t)$, $0\leqslant t\leqslant 1$, be a closed geodesic. Denote by $\mathscr{J}_c^\perp$ the Jacobi fields $Y(t)$ along $c(t)$ with $\langle \dot c(t), Y(t)\rangle = 0$, cf. (1.12.3). Put $\mathscr{J}_c^\perp \cap T_c\Lambda = \mathscr{J}_{c,\Lambda}^\perp$. I. e., $\mathscr{J}_{c,\Lambda}^\perp$ consists of the Jacobi fields $Y(t)$ with $Y(1)=Y(0)$. $\mathscr{J}_{c,\Lambda}$ contains the space $\mathscr{J}_{c,per}$ of periodic Jacobi fields along $c(t)$ with $\langle \dot c(t), Y(t)\rangle = 0$. $\dim \mathscr{J}_{c,per} = \mathrm{null}\, c$, cf. (2.5.6).Define the *concavity of c*, concav c, by

$$\mathrm{concav}\, c + \mathrm{null}\, c = \dim(\textit{null space plus negative eigenspace of } D^2E(c)\,|\,\mathscr{J}_{c,\Lambda}).$$

Remark. Put $c(0)=c(1)=p$. Consider the subspace $T_0^\perp c = T_1^\perp c$ of T_pM, formed by the vectors orthogonal to $\dot c(0)=\dot c(1)$. $\dim T_0^\perp c = n-1$ if $\dim M = n$. Write $V_h \oplus V_v$ for the direct sum of two copies of $T_0^\perp c$, cf. (3.2.3) for an explanation of this notation. Then we have the isomorphism

$$Y\in \mathscr{J}_c^\perp \mapsto \tilde Y(0) \equiv (Y(0), \nabla Y(0)) \in V_h \oplus V_v.$$

In (1.12.3) we defined on $\mathscr{J}_c^\perp \cong \mathrm{V}_h \oplus \mathrm{V}_v$ the symplectic form α by

$$\alpha(Y,Z) \equiv \alpha(\tilde Y(0), \tilde Z(0)) = \langle Y(t), \nabla Z(t)\rangle - \langle Z(t), \nabla Y(t)\rangle$$

where the right hand side is independent of $t\in S$. We consider the transformation

$$P: V_h\oplus V_v \to V_h \oplus V_v;\;\; \tilde Y(0) = (Y(0), \nabla Y(0)) \mapsto (Y(1), \nabla Y(1)).$$

Then $\alpha(P\tilde Y(0), P\tilde Z(0)) = \alpha(\tilde Y(0), \tilde Z(0))$, i. e., P is a so-called symplectic transformation, cf. (3.1), (3.2) for more on this. With this we can write $\mathscr{J}^{\perp}_{c,per} = \ker(P - id)$, $\mathscr{J}^{\perp}_{c,\Lambda} = \ker pr_h \circ (P - id)$, where $pr_h: V_h \oplus V_v \to V_h$ is the projection.

2.5.13 Proposition. *With the previously introduced notation we define subspaces* T_1 *and* T_2 *of* $T_0^{\perp} c$ *by*

$$T_1 = pr_h \circ \ker(P - id)\,;\; T_2 = V_v \cap \operatorname{im}(P - id)$$

Then $T_0^{\perp}$ *is the orthogonal sum of* T_1 *and* T_2.

Note. Here we have used the canonical identification of V_h and V_v with $T_0^{\perp} c$.

Proof. We claim that $\ker(P - id)$ and $\operatorname{im}(P - id)$ are α-orthogonal complements of each other. Indeed, their dimensions add up to $2n - 2 = \dim(V_h \oplus V_v)$. Moreover, if $\tilde{Y}(0) \in \ker(P - id)$ and $\tilde{W}(0) = P\tilde{Z}(0) - \tilde{Z}(0) \in \operatorname{im}(P - id)$,

$$\alpha(\tilde{Y}(0), \tilde{W}(0)) = \langle Y(0), \nabla Z(1) - \nabla Z(0)\rangle - \langle Z(1) - Z(0), \nabla Y(0)\rangle =$$

$$\{\langle Y(1), \nabla Z(1)\rangle - \langle Z(1), \nabla Y(1)\rangle\} - \{\langle Y(0), \nabla Z(0)\rangle - \langle Z(0), \nabla Y(0)\rangle\} = 0.$$

Hence, if $Y(0) = Y(1) \in T_1$ and $\nabla Z(1) - \nabla Z(0) \in T_2$ we get $\langle Y(0), \nabla Z(1) - \nabla Z(0)\rangle = 0$, i. e., $\langle T_1, T_2\rangle = 0$. Similarly, with $T_1' = V_v \cap \ker(P - id)$ and $T_2' = pr_h \circ \operatorname{im}(P - id)$, we get $\langle T_1', T_2'\rangle = 0$. Thus, $\dim T_1 + \dim T_2 \leqslant n - 1$, $\dim T_1' + \dim T_2' \leqslant n - 1$. On the other hand, $\dim T_1 + \dim T_1' = \dim\ker(P - id)$, $\dim T_2 + \dim T_2' = \dim\operatorname{im}(P - id)$ and therefore, $\dim T_1 + \dim T_2 = n - 1$. □

We now can prove the *Index Theorem for a Closed Geodesic*. In the case that the closed geodesic c is non-degenerate, it is due to Morse [2]. For the case of an arbitrary geodesic see Morse [4] and Klingmann [1]. An entirely different version, where the relation between $\operatorname{index}_\Lambda c$ and $\operatorname{index}_\Omega c$ plays no rôle, is given in Klingenberg [8]: The index is interpreted as the intersection number of a closed curve in the Lagrangian-Grassmann manifold $S\mathbb{U}/S\mathbb{O}$ with its codimension one submanifold. For a special case see (3.2.13). See also Duistermaat [1] for this point of view.

2.5.14 Theorem. *Let* c *be a closed geodesic. Then* $\operatorname{index}_\Lambda c = \operatorname{index}_\Omega c + \operatorname{concav} c \leqslant \operatorname{index}_\Omega c + n - 1$.

Proof. We begin as in the proof of (2.5.9) with the linear mappings $\Phi: W(t_1) \to T_c\Omega M$, $0 < t_1 < 1$. Denote by U the space generated by the $\Phi W(t)$, $0 < t < 1$. $\dim U = \operatorname{index}_\Omega c$.

Let U' be a complement of the space $\mathscr{J}^{\perp}_{c,per}$ in the null space plus negative eigenspace of $D^2E(c)|\mathscr{J}^{\perp}_{c,\Lambda}$. Clearly, $U \cap U' = 0$. Hence, $\dim U + U' = \operatorname{index}_\Omega c + \operatorname{concav} c$. Moreover, $D^2E(c)|U + U' \leqslant 0$. Indeed, we know already $D^2E(c)|U \equiv 0$. If $\xi \in U, \eta \in \mathscr{J}^{\perp}_{c,\Lambda}$, $D^2E(c)(\xi, \eta) = \langle \nabla\eta(1) - \nabla\eta(0), \xi(0)\rangle = 0$, since $\xi(0) = 0$. From the definitions it also follows that under $pr_-: T_c\Lambda \to T_c^-\Lambda$, $U + U'$ is mapped injectively.

To conclude the proof, we therefore only need to show that if $\xi^* \in T_c^- \Lambda$, $D^2E(c)(\xi^*, U+U') = 0$ then $\xi^* = 0$.

To see this let $\eta(t) = Z(t) \in \mathscr{J}_{c,\Lambda}^{\perp}$. With $\eta^- = pr^-(\eta) \in T_c^- \Lambda$ we have

$$0 = D^2E(c)(\eta, \xi^*) = \langle \nabla Z(1) - \nabla Z(0), \xi^*(0) \rangle = 0.$$

That is, $\xi^*(0)$ belongs to the subspace T_1 of $T_0^{\perp} c$, introduced in (2.5.13). Hence, there exists a $Y \in \mathscr{J}_{c,per}^{\perp}$ with $\xi(0) = Y(0)$, i. e., $\zeta = \xi^* - Y \in T_c \Omega$. Since $Y \in$ null space of $D^2E(c)$,

$$0 \geqslant D^2E(c)(\xi^*, \xi^*) = D^2E(c)(\zeta, \zeta).$$

$pr^-(\zeta) \in T_c^- \Lambda$ coincides with $pr^-(\xi^*) = \xi^*$. Moreover, $pr^-(\zeta) \in pr^-(U)$. But by hypothesis, $D^2E(c)(\xi^*, pr^-(U)) = 0$. Therefore, $\xi^* = 0$.

Finally, we observe that an upper bound for concav c is given by the dimension of $T_1 \subset T_0^{\perp} c$. □

We continue with a characterization of the focal index of a geodesic segment $c: I \to M$. Recall from (2.5.4) that c can be viewed as a critical point of $E|H^1_{M_0 \times \{p_1\}}, p_1 = c(1)$, where M_0 is a submanifold of codimension 1 and totally geodesic at $p_0 = c(0)$. $D^2E(c)$ then determines the focal index $\operatorname{index}_{foc} c$.

On the other hand, we have defined in (1.12.20) the concept of focal points. The *Focal Index Theorem* now reads:

2.5.15 Theorem. *Let $c: I \to M$ be a geodesic segment. Then*

$$\begin{aligned}\operatorname{index}_{foc} c &= \sum_{0<t<1} \{\textit{multiplicity of the focal point } c(t)\} \\ &= \textit{sum of focal points } c(t_1), 0 < t_1 < 1, \\ &\quad\ \textit{each counted with its multiplicity} \geqslant 1.\end{aligned}$$

In particular, $D^2E(c) \geqslant 0$ for $E|H^1_{M_0 \times \{p_1\}}$ if c has no focal points in its interior.

Proof. The proof is very similar to the proof of (2.5.9), the main difference being in the definition of the space $W(t_1)$, $0 \leqslant t_1 < 1$. Here we take $W(t_1)$ to be isomorphic to the space of the Jacobi fields $Y(t)$ along $c(t)$ with $\langle Y(t), \dot c(t) \rangle = 0$, and $\nabla Y(0) = 0$, $Y(t_1) = 0$. For each such Jacobi field Y define

$$\xi_Y(t) = \xi(t) = \begin{cases} Y(t), 0 \leqslant t \leqslant t_1, \\ 0, t_1 \leqslant t \leqslant 1. \end{cases}$$

The mapping $\Phi: W(t_1) \to T_c H^1_{M_0 \times \{p_1\}}$ is linearly injective. The subspace U generated by the $\Phi W(t)$, $0 < t < 1$, has dimension equal to the right hand side of (2.5.15). One checks $D^2E(c)|U \equiv 0$; thus, the focal index of c is $\geqslant \dim U$. Equality is obtained just as in (2.5.5), only this time we take a basis Y_i, $1 \leqslant i \leqslant n-1$, of those Jacobi fields Y which satisfy $\nabla Y(0) = 0$, $\langle Y(t), \dot c(t) \rangle = 0$. □

We conclude with a certain generalization of (2.4.16). There we proved essentially the following: Let $\varkappa_0, \varkappa_1$ be real numbers, $0 \leqslant \varkappa_0 \leqslant \varkappa_1$, such that there are no critical points c with $\varkappa_0 \leqslant E(c) \leqslant \varkappa_1$. Then the $-\operatorname{grad} E$ deformation $\phi_s, s \geqslant 0$, deforms $\Lambda^{\varkappa_1} M$ into $\Lambda^{\varkappa_0 -} M$. In particular, all relative homotopy groups vanish,

$$\pi_k = \pi_k(\Lambda^{\varkappa_1} M, \Lambda^{\varkappa_0 -} M) = 0,\ k = 0, 1 \ldots$$

The elements of π_k are the classes of homotopic maps

$$f: (I^k, \partial I^k) \to (\Lambda^{\varkappa_1} M, \Lambda^{\varkappa_0 -} M).$$

In our generalization we do not exclude the existence of critical points c with $\varkappa_0 \leqslant E(c) \leqslant \varkappa_1$ altogether; we only forbid their index from being 'too small'.

2.5.16 Theorem. *Let $\varkappa_0, \varkappa_1$ be real numbers, $0 \leqslant \varkappa_0 < \varkappa_1$ and l an integer > 0. Assume that there are no critical points c of index $< l$ with $\varkappa_0 \leqslant E(c) \leqslant \varkappa_1$ and no critical points at all with E-value $\varkappa_1$. Then*

$$(*) \qquad f: (I^k, \partial I^k) \to (\Lambda^{\varkappa_1} M, \Lambda^{\varkappa_0 -} M),\ k < l,$$

is homotopic to a mapping having its image entirely in $\Lambda^{\varkappa_0 -} M$, i.e.,

$$\pi_k(\Lambda^{\varkappa_1} M, \Lambda^{\varkappa_0 -} M) = 0, \textit{ for } k < l.$$

Notes. 1. A standard result of algebraic topology, cf. Spanier [1], says: If $\pi_k(\Lambda^{\varkappa_1}, \Lambda^{\varkappa_0 -}) = 0$ for $0 \leqslant k < l$ *and* $\pi_1(\Lambda^{\varkappa_0 -}) = 0$ then also $H_k(\Lambda^{\varkappa_1}, \Lambda^{\varkappa_0 -}) = 0$ for $0 \leqslant k < l$ and $\pi_l(\Lambda^{\varkappa_1}, \Lambda^{\varkappa_0 -}) = H_l(\Lambda^{\varkappa_1}, \Lambda^{\varkappa_0 -})$. Thus, in (2.5.10) we actually also prove the vanishing of certain homology groups.

2. In (2.4.18) we showed that in a ϕ-family $\mathscr{A}$ there exist members A which can be pushed under the ϕ_s-deformation into $\mathscr{U} \cup \Lambda^{\varkappa -}$, where $\mathscr{U}$ is a neighborhood of the set $Cr\varkappa$ of critical points of E-value $\varkappa = \varkappa_{\mathscr{A}}$. It should be possible to prove a similar refinement of (2.5.16). More precisely, our methods employed for the proof of (2.5.16) should suffice to show the following: Let $\varkappa_0, \varkappa_1$ be such that there are no critical points c of index $< l$ with E-value $\varkappa_0 < E(c) \leqslant \varkappa_1$ and no critical point at all with E-value $\varkappa_1$. Then, given any neighborhood $\mathscr{U}$ of the set $Cr\varkappa$ of critical points of index $< l$ at E-value $\varkappa_0$, it is possible to deform a mapping f, $(*)$, such that its image belongs to $\mathscr{U} \cup \Lambda^{\varkappa_0 -}$. See the proof of (2.6.A.1) for the special case $l = 2$.

Proof. Let $\varkappa \in [\varkappa_0, \varkappa_1[$ be a critical value. All we need to show is: For a certain open neighborhood $\mathscr{V}$ of the critical set $Cr\varkappa$ at $\{E = \varkappa\}$ in $\Lambda^{\varkappa_1} M$ and a mapping

$$(\dagger) \qquad f_0: (I^k, \partial I^k) \to (\mathscr{V} \cup \Lambda^{\varkappa -}, \Lambda^{\varkappa_0 -})$$

there exists a homotopy

$$(\dagger\dagger) \qquad f_s: (I^k, \partial I^k) \to (\Lambda^{\varkappa_1}, \Lambda^{\varkappa_0 -}),\ 0 \leqslant s \leqslant 1,$$

of f_0 with $\operatorname{im} f_1 \subset \Lambda^{\varkappa -}$. This homotopy in general will be non-standard, i. e., it will not be, up to parameterization, of the form $\phi_s \circ f_0$.

Once we have proved this we can complete the proof of the Theorem as follows: Consider for a mapping f, $(*)$, the ϕ-family $\{\phi_s \circ f(I^k); s \geq 0\}$. Let $\varkappa$ be its critical value. Then $\varkappa < \varkappa_1$. If $\varkappa < \varkappa_0$, we are done. Otherwise, for any prescribed neighborhood $\mathscr{U}$ of $Cr\varkappa$, $\phi_s \circ f$ will be of the form (†), for all $s \geq$ some s_0, cf. (2.4.18). We write f_0 instead of $\phi_s \circ f$, for some $s \geq s_0$. Apply now the homotopy f_s, (††), which shows that our original f, $(*)$ is homotopic to an f' with $\max E|f' < \varkappa$.

To show the existence of the homotopy (††) consider an $c \in Cr\varkappa$. Let $B_\varrho(c)$ be a strongly convex neighborhood of c and denote by u the normal coordinates for $B_\varrho(c)$, based at c. We write $u = (u^-, u^0, u^+)$, in correspondence with the decomposition $T_c\Lambda = T_c^- \oplus T_c^0 \oplus T_c^\perp$ of the tangent space at c, cf. (2.5.3). In these coordinates, $E|B_\varrho(c)$ can be written as

$$E(u) = E(u^-, u^0, u^+) = \varkappa - \frac{1}{2}\sum_i \lambda_i (u_i^-)^2 + \frac{1}{2}\sum_k \lambda_k (u_k^+)^2 + E_{(3)}(u),$$

with $E_{(3)}(u) = o(\|u\|_1^2)$. The $-\lambda_i$, $1 \leq i \leq m = \operatorname{index} c$ are the negative eigenvalues of $D^2E(c)$, while the λ_k, $k \geq m + 2 + \operatorname{null} c$, are the positive eigenvalues of $D^2E(c)$.

We introduce new coordinates $v = (v^-, v^0, v^+)$ by $v^- = u^- - h(u^0, u^+)$, $v^0 = u^0$, $v^+ = u^+$, with $h(0,0) = 0$, $\partial E(h(u^0, u^+), u^0, u^+)/\partial u_i^- = 0$, $1 \leq i \leq m$. Then $\tilde{E}(v) = E(u(v))$ is of the form

$$\tilde{E}(v) = \tilde{E}(v^-, v^0, v^+) = \varkappa - \frac{1}{2}\sum_i \lambda_i (v_i^-)^2 + \frac{1}{2}\sum_k \lambda_k (v_k^+)^2 + \tilde{E}_{(3)}(v)$$

with $\tilde{E}_{(3)}(v) = o(\|v\|_1)^2$ and $\partial \tilde{E}_{(3)}(0, v^0, v^+)/\partial v_i^- = 0$, $1 \leq i \leq m$. That is $\tilde{E}_{(3)}(v)$ contains no terms linear in v^-.

The existence of the implicitly defined functions $h_i(u^0, u^+)$ follows from the fact that the $(m \times m)$-matrix with elements $\partial^2 E(0,0,0)/\partial u_i^- \partial u_j^- = -\lambda_i \delta_{ij}$ is invertible.

We can restrict the domain of the new coordinates v such that the matrix $(\partial^2 E^2(v)/\partial v_i^- \partial v_j^-)$ is negative definite, for all v. Moreover, we have a $\sigma > 0$ and an $\alpha = \alpha(c, \sigma)$ such that $\|v^-\|_1 \geq \alpha \|(v^0, v^+)\|_1$ implies $\tilde{E}(v) \leq \varkappa$, whenever $\|v^-\|_1 \leq \sigma$. We can make α unique by choosing it minimal with this property. Thus, $\tilde{E}(v) < \varkappa$ on the open cone $\{\|v^-\|_1 > \alpha \|(v^0, v^+)\|_1\}$.

We now define the closed neighborhood $C_\sigma(c)$ of c by $\|v^-\|_1 \leq \sigma$, $\|(v^0, v^+)\|_1 \leq \sigma/2\alpha$. For each $v = (v^-, v^0, v^+)$ in the range of $C_\sigma(c)$ with $v^- \neq 0$, the curve $v(t) = (tv^-, v^0, v^+)$, $1 \leq t \leq \sigma/\|v^-\|_1 = t_1$ from $v = v(1)$ to $v(t_1)$ lies inside $C_\sigma(c)$. $\|v^-(t_1)\|_1 > \alpha \|(v^0(t_1), v^+(1))\|_1$, hence, $\tilde{E}(v(t_1)) < \varkappa$. Moreover, $d\tilde{E}(v(t))/dt < 0$, since $\partial \tilde{E}(0, v^0, v^+)/\partial v_i^- = 0$ and $(\partial^2 \tilde{E}(v)/\partial v_i^- \partial v_j^-)$ is negative definite.

On $C_\sigma(c)$ we define the vector field X as the field which in the v-coordinates has the representation $X(v) = (v^-, 0, 0)$. Thus, whenever $e \in C_\sigma(c)$ has a v-coordinate with $v^-(e) \neq 0$, $DE.X(e) < 0$, $D^2E.(X, X) < 0$. Hence, the flow along a non-constant integral curve of X decreases the energy until the energy becomes $< \varkappa$.

Denote by $\frac{1}{2}C_\sigma(c)$ the subset $\{\|v^-\|_1 \leq \sigma/2; \|(v^0, v^+)\|_1 \leq \sigma/4\alpha\}$ of $C_\sigma(c)$. Since $Cr\varkappa$ is compact, a finite subset of the open covering $\{\operatorname{int}\frac{1}{2}C_\sigma(c); c \in Cr\varkappa\}$ will suffice to cover $Cr\varkappa$. We enumerate this set by $\{\mathrm{int}\frac{1}{2}C_{\sigma_r}(c_r), r=1,\ldots,R\}$. We also write briefly $C_r, \frac{1}{2}C_r$ instead of $C_{\sigma_r}(c_r), \frac{1}{2}C_{\sigma_r}(c_r)$.

Let $\gamma_r: \Lambda M \to \mathbb{R}$ be a differentiable function with values in $[0,1]$ such that $\gamma_r|\frac{1}{2}C_r = 1, \gamma_r|\partial C_r = 0$. For each r we define the vector field Y_r on ΛM by $Y_r(e) = \gamma_r(e)X_r(e)$, if $e \in C_r$, and by $Y_r(e)=0$, if $e \in \complement C_r$. Here, X_r is the vector field defined above on C_r.

Denote by $\mathscr{U}$ the union of the $\mathrm{int}\frac{1}{2}C_r, r=1,\ldots,R$. There exists an open neighborhood $\mathscr{V}$ of $Cr\varkappa$ and an $\varrho>0$ such that $d_\Lambda(c',c'')\geq\varrho$, if $c'\in\mathscr{V}, c''\in\complement\mathscr{U}$. Indeed, the distance between a point $c''\in\complement\mathscr{U}$ and $Cr\varkappa$ is $\geq$ some positive constant, independent of the choice of c''. Take now for $\mathscr{V}$, e.g., the points having distance $<\frac{1}{2}$ this constant. It follows, cf. the proof of (2.4.18), that there exists an $\varepsilon>0$ such that the trajectory $\phi_s c'', s\geq 0$, of an $c''\in\complement\mathscr{U}$ with $E(c'')<\varkappa+\varepsilon$ under the $(-\operatorname{grad} E)$-flow has E-value $<\varkappa$ when it enters the neighborhood $\mathscr{V}$.

Consider now an f_0, (†). By applying, if necessary, a deformation ϕ_s we even can assume $\max E|f_0(I^k)\leq\varkappa+\varepsilon$. Beginning with $r=1$, we define a deformation ${}_1\psi_s\circ f_0, s\geq 0$, of f_0 by the integral flow ${}_1\psi_s: \Lambda M\to\Lambda M$ of the vector field Y_1. Whenever $f_0(x)\in\frac{1}{2}C_1$ and the ${}_1v$-coordinate ${}_1v(f_0(x)) = ({}_1v^-(t_0(x)), {}_1v^0(f_0(x)), {}_1v^+(f_0(x)))$ on C_1 satisfies ${}_1v^-(f_0(x))\neq 0, E({}_1\psi_s\circ f_0(x))<\varkappa$, for sufficiently large s. Observe now that the condition ${}_1v^-(f_0(x))\neq 0$ for $x\in f_0^{-1}(\frac{1}{2}C_1)$ is a transversality condition for f_0, since $\dim I^k = k < \operatorname{index} c_1 = \operatorname{codim}\{{}_1v^-=0\}$. Hence, this condition can be met by an arbitrarily small appropriate deformation of f_0.

We thus have the existence of a deformation ${}_1f_0$ of f_0 such that $E({}_1f_0(x))<\varkappa$, whenever $f_0(x)\in\mathrm{int}\frac{1}{2}C_1$. The deformation ${}_1\psi_s$ also operates on $C_1-\frac{1}{2}C_1$. It may have happened that thereby an $f_0(x)$ in this set is deformed into $\complement\mathscr{U}$. We will take care of this at the end. In any case, the E-value of the deformed element ${}_1f_0(x)$ can only have become smaller. In particular, this value will be $<\varkappa+\varepsilon$.

Using the integral flow ${}_2\psi_s: \Lambda M\to\Lambda M$ of the vector field Y_2, we can deform ${}_1f_0$ into ${}_2\psi_s\circ{}_1f_0$ such that, for sufficiently large $s>0, E({}_2\psi_s\circ{}_1f_0(x))<\varkappa$, whenever ${}_1f_0(x)\in\mathrm{int}\frac{1}{2}C_2$ and the negative component ${}_2v^-({}_1f_0(x))$ of the ${}_2v$-coordinate of ${}_1f_0(x)$ is $\neq 0$. This latter condition again can be met for transversality reasons.

Proceeding in this manner, we see that our original f_0, (†), with $\max E|f_0(I^k)<\varkappa+\varepsilon$ can be deformed into an ${}_Rf_0$ such that, for all $x\in I^k, E({}_Rf_0(x))<\varkappa$ or else, ${}_Rf_0(x)\in\complement\mathscr{U}, E({}_Rf_0(x))<\varkappa+\varepsilon$. Now apply upon ${}_Rf_0(I^k)$ a deformation ϕ_{s0} which transforms the image into $\mathscr{V}\cup\Lambda^{\varkappa-}$. Under this deformation any ${}_Rf_0(x)\in\complement\mathscr{U}$ will be transformed into $\Lambda^{\varkappa-}$; thus with $f_1=\phi_{s0}\circ{}_Rf_0$ we get the desired deformation (††). □

2.5 Appendix – The S^1- and the $\mathbb{Z}_2$-action on ΛM

In this appendix we exhibit an additional structure of the Hilbert manifold ΛM of closed H^1-curves, i.e., the existence of a canonical $\mathbb{O}(2)$-action which preserves the

Riemannian metric on ΛM and leaves the function E invariant. This action stems from the canonical $\mathbb{O}(2)$-action on the source S of the elements $\{c: S \to M\}$ in ΛM.

The full importance of this structure for the problem of the existence of many – even infinitely many – geometrically distinct closed geodesics on the compact Riemannian manifold M will not become apparent in our present treatise on Riemannien geometry. Rather, we must refer the reader to Klingenberg [10], [14] for such subjects as the equivariant Morse complex and related topics which can be derived from ΛM. Cf. also the remarks at the end of this appendix. Nevertheless, we will find it convenient in the next chapter to have at our disposal the basic facts of this group action.

2.5.A.1 Definition. Define an S^1-action on ΛM by

$$\chi^{\sim}: S^1 \times \Lambda M \to \Lambda M, (z, c) \mapsto z \,.\, c\,.$$

Here, with $z = e^{2\pi i r} \in S^1$, $z \,.\, c$ is defined by $z \,.\, c(t) = c(t + r)$.

The orbit of $c \in \Lambda M$ under this S^1-action will also be denoted by $S^1 \,.\, c$.

Note. We can say that the elements of $S^1 \,.\, c$ are obtained from c by 'change of the initial point'.

2.5.A.2 Lemma. *The action $\chi^{\sim}$ is continuous but it is not differentiable. On the other hand, for a fixed $z \in S^1$, the mapping*

$$\chi_z^{\sim}: \Lambda M \to \Lambda M; c \mapsto z \,.\, c$$

is an isometry which leaves the function $E: \Lambda M \to \mathbb{R}$ invariant.

Note. Instead of $\chi_z^{\sim}$ we also write $z: \Lambda M \to \Lambda M$.

Proof. Consider the natural charts $(\exp_c^{-1}, \mathcal{U}(c)), (\exp_{z.c}^{-1}, \mathcal{U}(z \,.\, c))$, for $c \in C^{\infty}(S, M)$, $z \in S^1$, cf. (2.3.10). Then the mapping $z: \Lambda M \to \Lambda M$ is represented by the linear isomorphism

$$\xi(t) \in H^1(c^* TM) \mapsto (\xi(t + r)) \in H^1((z \,.\, c)^* TM).$$

From the local representation of E and the Riemannian metric $\langle\ ,\ \rangle_1$ in (2.3) it now follows that $z: \Lambda M \to \Lambda M$ is an isometry, with $E(z \,.\, c) = E(c)$.

It remains to show that $\chi^{\sim}$ is continuous. Since $\chi_z^{\sim}$ is an isometry we have

$$d_\Lambda(c, z \,.\, e) \leqslant d_\Lambda(c, z \,.\, c) + d_\Lambda(c, e);$$

$$d_\Lambda(e, z \,.\, e) \leqslant d_\Lambda(c, z \,.\, c) + 2 d_\Lambda(e, c)\,.$$

Hence, it suffices to show that $d_\Lambda(c, z \,.\, c)$ tends to zero as z tends to $1 \in S^1$, for $c \in C'^{\infty}(S, M)$. But since $\exp_c^{-1}(c) = (\xi(t)) = 0$, the coordinate $\exp_c^{-1}(z \,.\, c)$ of $z \,.\, c$ is given by the horizontal vector field $r \partial t \in H^1(c^* TM)$, where $z = e^{2\pi i r}$. As r goes to 0, so does $r \partial t$.

Finally, we observe that the S^1-action $\chi^{\sim}$ cannot be differentiable. In that case, it would follow that also the composition

$$r \in S \mapsto z = e^{2\pi i r} \in S^1 \mapsto z \,.\, e \in \Lambda M \mapsto (z \,.\, e)(0) = e(r) \in M$$

is differentiable, for every $e \in \Lambda M$. But for a non-differentiable $e: S \to M$, this is false. □

2.5.A.3 Corollary. *If c is a critical point of $E: \Lambda M \to \mathbb{R}$, then every $z \,.\, c$ on the S^1-orbit of c is critical.*

Proof. From (2.5.A.2) follows that $DE(c) = 0$ implies $DE(z \,.\, c) = 0$. □

2.5.A.4 Corollary. *The isotropy subgroup $I^{\sim}(c)$ of c under the S^1-action is either the whole group S^1 – and this is true if and only if $c \in \Lambda^0 M$ = space of constant mappings $c: S \to M$; or else, $I^{\sim}(c)$ is a finite cyclic group.*

Note. Recall that the isotropy group of c is defined as the set of $z \in S^1$ with $z \,.\, c = c$.

Proof. Since $\chi_c^{\sim}: z \in S^1 \mapsto z \,.\, c \in \Lambda M$ is continuous, $I^{\sim}(c) = \chi_c^{\sim -1}(c)$ is closed. $I^{\sim}(c) = S^1$ clearly is equivalent to $c(t) = c(0)$, all $t \in S$, i.e., $c \in \Lambda^0 M$. The only closed subgroups of S^1, different from S^1, are the cyclic subgroups $\mathbb{Z}_m = \{e^{2\pi i l/m}; l = 0, \ldots, m - 1\}$.

2.5.A.5 Definition. We call $c \in \Lambda M - \Lambda^0 M$ an *element of multiplicity m* or *m-fold covered* if $I^{\sim}(c) = \mathbb{Z}_m$. If $m = 1$, i. e., if $I^{\sim}(c) = id$, then c is also called *prime*.

The *m-fold covering* c^m of an element $c \in \Lambda M$ *is defined as* $c^m(t) = c(mt)$.

Remark. $I^{\sim}(c) = \mathbb{Z}_m$ means that $c(t + l/m) = c(t)$, all $t \in S$, all $l \in \mathbb{N}$. For such a c, we can define $c_0(t) = c(t/m)$, $0 \leqslant t \leqslant 1$. c_0 then is called *underlying prime closed curve of c*.

2.5.A.6 Definition. The quotient space $\tilde{\Pi} M$ of ΛM with respect to the S^1-action $\chi^{\sim}$ is called *space of oriented unparameterized closed curves of M*. $\tilde{\Pi} M$ is endowed with the finest topology which makes the projection mapping

$$\tilde{\pi}: \Lambda M \to \tilde{\Pi} M \equiv \Lambda M/_{\chi^{\sim}} S^1$$

continuous. That is, a subset $B \subset \tilde{\Pi} M$ is open if and only if the counter-image of B under $\tilde{\pi}$ is open.

Note. We also can say that $\tilde{\Pi} M$ is the space of orbits in ΛM under the S^1-action $\chi^{\sim}$.

Since $E: \Lambda M \to \mathbb{R}$ is constant on the orbits $S^1 \,.\, c$, we get from E an induced function on $\tilde{\Pi} M$ which we again denote by E. With this we have the

2.5.A.7 Theorem. *The deformation $\phi_s: \Lambda M \to \Lambda M$ of* (2.4.13) *induces a deformation*

$\tilde{\psi}_s : \tilde{\Pi} M \to \tilde{\Pi} M$ *such as to make the following diagram commutative:*

$$\begin{array}{ccc} \Lambda M & \xrightarrow{\ \phi_s\ } & \Lambda M \\ \downarrow{\scriptstyle \tilde{\pi}} & & \downarrow{\scriptstyle \tilde{\pi}} \\ \tilde{\Pi} M & \xrightarrow{\ \tilde{\psi}_s\ } & \tilde{\Pi} M \end{array}$$

Proof. This follows immediately from $\operatorname{grad} E(z \,.\, c) = T_c z \,.\, \operatorname{grad} E(c)$ which is a consequence of (2.5.A.2). □

2.5.A.8 Definition. The *orientation reversing mapping*

$$\theta : \Lambda M \to \Lambda M$$

is defined by $(\theta c)(t) = c(1-t)$. Since $\theta^2 = id$, this defines a $\mathbb{Z}_2$-action on ΛM.

Remark. θ may be viewed as stemming from the involution $\theta : \mathbb{R}^2 \to \mathbb{R}^2; (x, y) \mapsto (x, -y)$ on $\mathbb{R}^2$, restricted to the canonical embedding $t \in S \mapsto (\cos 2\pi t, \sin 2\pi t) \in \mathbb{R}^2$. When we interpret the S^1-action $\chi^{\sim}$ as stemming from the canonical action of $S\mathbb{O}(2) \cong S^1$ on the circle $S \subset \mathbb{R}^2$, the combined action of S^1 and $\mathbb{Z}_2$ can be viewed as an $\mathbb{O}(2)$-action on ΛM.

2.5.A.9 Lemma. *The mapping $\theta : \Lambda M \to \Lambda M$ is an isometry, leaving $E : \Lambda M \to \mathbb{R}$ invariant. Moreover, $z \,.\, \theta c = \theta(\bar{z} \,.\, c)$, for arbitrary $c \in \Lambda M$.*

This shows that θ carries S^1-orbits into S^1-orbits and thus, θ induces an involution on $\tilde{\Pi} M$ which we again denote by θ.

In particular, critical points of E are carried into critical points.

Proof. The representation of θ in the natural coordinates based at c and θc reads $(\xi(t)) \mapsto (\xi(1-t))$. This shows that θ is an isometry and that $E(\theta c) = E(c)$. Finally, if $z = e^{2\pi i r}$, $(z \,.\, \theta c)(t) = (\theta c)(t + r) = c(1 - t - r) = (\bar{z} \,.\, c)(1 - t) = \big(\theta(\bar{z} \,.\, c)\big)(t)$. □

2.5.A.10 Definition. The quotient mapping of ΛM under the $\mathbb{Z}_2$-action generated by $\theta : \Lambda M \to \Lambda M$ is denoted by

$$^{-} : \Lambda M \to \bar{\Lambda} M .$$

$\bar{\Lambda} M$ is called *space of non-oriented parameterized closed curves.*

θ induces an involution also on $\tilde{\Pi} M$ which we denote by the same letter. We denote the corresponding quotient mapping by

$$^{-} : \tilde{\Pi} M \to \Pi M$$

ΠM is called *space of (non-oriented) unparameterized closed curves.* The composition mapping $(^{-}) \circ \tilde{\pi} : \Lambda M \to \Pi M$ will also be denoted by $\pi : \Lambda M \to \Pi M$.

The S^1-action $\chi^{\sim}$ on ΛM determines an equivalence relation on $\bar{\Lambda} M$. We denote the corresponding quotient mapping by $\tilde{\pi}: \bar{\Lambda} M \to \Pi M$ and thus get the commutative diagram

$$\begin{array}{ccc} \Lambda M & \xrightarrow{\ -\ } & \bar{\Lambda} M \\ \Big\downarrow \tilde{\pi} & \searrow \pi & \Big\downarrow \tilde{\pi} \\ \tilde{\Pi} M & \xrightarrow{\ -\ } & \Pi M \end{array}$$

Note. ΠM also can be viewed as space of orbits in ΛM under the $\mathbb{O}(2)$-action described above, i. e., the space of orbits under the combined action of S^1 and $\mathbb{Z}_2$.

The following is an immediate consequence of (2.5.A.9); it constitutes the obvious counterpart to (2.5.A.7). Note that E is defined also on $\bar{\Lambda} M$ and ΠM.

2.5.A.11 Theorem. *The deformation $\phi_s: \Lambda M \to \Lambda M$ of* (2.4.13) *induces deformations*

$$\bar{\phi}_s: \bar{\Lambda} M \to \bar{\Lambda} M \text{ and } \psi_s: \Pi M \to \Pi M$$

such as to make the following diagrams commutative:

$$\begin{array}{ccc} \Lambda M & \xrightarrow{\ \phi_s\ } & \Lambda M \\ \Big\downarrow - & & \Big\downarrow - \\ \bar{\Lambda} M & \xrightarrow{\ \bar{\phi}_s\ } & \bar{\Lambda} M \end{array} ; \qquad \begin{array}{ccc} \Lambda M & \xrightarrow{\ \phi_s\ } & \Lambda M \\ \Big\downarrow \pi & & \Big\downarrow \pi \\ \Pi M & \xrightarrow{\ \psi_s\ } & \Pi M \end{array}$$

Proof. This follows from $\operatorname{grad}(\theta c) = T_c\theta \,.\, \operatorname{grad} E(c)$ and (2.5.A.7). □

Considering the concepts developed in (2.5) for ΛM we see that most of them are compatible with the S^1-action and the $\mathbb{Z}_2$-action. In particular, we have the

2.5.A.12 Lemma. *The index and the nullity of a critical point c of $E: \Lambda M \to \mathbb{R}$ are the same for all elements in the $\mathbb{O}(2)$-orbit $S^1.c \cup S^1.\theta c$ of c.*

Proof. This follows from the canonical isomorphisms between the self-adjoint operators A_c and $A_{z.c}$ and $A_{z.\theta c}$, given by the conjugation with $T_c z$ and $T_c(z\theta)$, respectively. □

We conclude this Appendix by investigating the fixed point set of ΛM under the S^1-action and the $\mathbb{Z}_2$-action described above.

2.5.A.13 Lemma. *The fixed point set $\Lambda^0 M$ of the S^1-action, i. e., the set of constant mappings $c: S \to M$, is a totally geodesic submanifold of ΛM, isometric to M.*

Proof. Let

$$i: M \to \Lambda M; p \mapsto c_p$$

be the canonical inclusion under which we associate to a point p the constant mapping $c_p: S \to p$. c_p is differentiable. The coordinates of the $c \in \mathscr{U}(c_p)$ are sections $\xi(t)$ in $c_p^* \tau$ with $\tau^* c_p \xi(t) \in T_p M$.

With $\Lambda = \Lambda M$, $\Lambda^0 = \Lambda^0 M$ we define

$$T_{c_p}\Lambda^0 = \{\xi_0 \in T_{c_p}\Lambda; \xi_0(t) = \text{const}\}$$

$$T_{c_p}^{\perp}\Lambda^0 = \{\xi \in T_{c_p}\Lambda; \langle \xi, \xi_0 \rangle_1 = \langle \xi, \xi_0 \rangle_0 = 0, \text{ all } \xi_0 \in T_{c_p}\Lambda^0\}$$

Under the canonical identification of $H^1(c_p^* \tau)$ with $T_{c_p}\Lambda M$, the subspace $H^1(\mathbb{O}_{c_p}) \cap T_{c_p}\Lambda^0$ contains the coordinates of the constant curves in $\mathscr{U}(c_p)$. We therefore have with $(\exp_{c_p}^{-1}, \mathscr{U}(c_p))$ a chart satisfying (1.3.5). Thus, $\Lambda^0 = \Lambda^0 M$ is a submanifold.

$\Lambda^0 M$ is totally geodesic since it is the intersection of the fixed point sets of the elements in the 1-parameter family of isometries $z: \Lambda M \to \Lambda M, z \in S^1$, cf. (1.10.15).

2.5.A.14 Lemma. *The fixed point set $\Lambda_\theta M$ of the involution $\theta: \Lambda M \to \Lambda M$ is a totally geodesic submanifold of ΛM, containing the manifold $\Lambda^0 M$ of constant maps. Actually, there exists an isometric embedding*

$$(*) \qquad j: H^1(I, M) \to \Lambda M; \; c \mapsto jc = \begin{cases} c(2t), & 0 \leqslant t \leqslant 1/2 \\ c(2-2t), & 1/2 \leqslant t \leqslant 1 \end{cases}$$

with $\Lambda_\theta M$ as image.

$\Lambda_\theta M$ is transformed into itself under the gradient flow $\phi_s: \Lambda M \to \Lambda M$ on ΛM.

There exists a $\varrho > 0$ such that a non-constant critical point of $E: \Lambda M \to \mathbb{R}$ has d_∞-distance $\geqslant \varrho$ from the submanifold $\Lambda_\theta M$.

Proof. Since $\theta: \Lambda = \Lambda M \to \Lambda$ is an isometry, $\Lambda_\theta = \Lambda_\theta M$ is a totally geodesic submanifold, cf. (1.10.15).

$c_\theta \in \Lambda_\theta$ means $c_\theta(1-t) = c_\theta(t), t \in S$. In the natural chart based at c_θ, the elements in $\Lambda_\theta \cap \mathscr{U}(c_\theta)$ are represented by the vector fields $\xi \in H^1(\mathbb{O}_c)$ satisfying

$$(**) \qquad \xi(t) = \xi(1-t).$$

Now, the $\xi \in T_{c\theta}\Lambda$ satisfying $(**)$ clearly form a closed linear subspace of $T_{c\theta}\Lambda$. This again shows that Λ_θ is a submanifold.

The mapping j, $(*)$, is an embedding. To see that j actually is an isometry one only needs to observe that

$$\langle T_c j \,.\, \xi, T_c j \,.\, \eta \rangle_1 = \tfrac{1}{2}\langle \xi, \eta \rangle_1 + \tfrac{1}{2}\langle \theta\xi, \theta\eta \rangle_1 = \langle \xi, \eta \rangle_1 .$$

$T\theta: T\Lambda \to T\Lambda$ transforms the gradient vector field into itself. Thus, for $c_\theta \in \Lambda_\theta$, $T_{c\theta}\theta \,.\, \operatorname{grad} E(c_\theta) = \operatorname{grad} E(c_\theta)$.

To prove the last statement we consider the injectivity radius $\imath(M)$ of M, cf. (2.1.9). On every closed geodesic c' we have a parameter $t_0, 0 < t_0 \leqslant 1/4$ such that $d(c'(t_0), c'(1-t_0)) = \imath(M)$. Then, if $c_\theta \in \Lambda_\theta$,

$$\imath(M) = d(c'(t_0), c'(1-t_0)) \leqslant d(c'(t_0), c_\theta(t_0)) +$$
$$d(c_\theta(1-t_0), c'(1-t_0)) \leqslant 2 d_\infty(c', c_\theta),$$

cf. (2.3.1) for the definition of d_∞. Thus, $\varrho = \imath(M)/2$ is a positive number such that $d_\infty(c', \Lambda_\theta) \geqslant \varrho$, whenever c' is non-constant critical point of E. This implies also a positive bound for $d_\Lambda(c', \Lambda_\theta)$, since d_Λ is finer than d_∞, cf. (2.5.6). Actually, it is easy to establish the relation $d_\infty(c_1, c_2) \leqslant \sqrt{2}\, d_\Lambda(c_1, c_2)$, for c_1, c_2 in Λ, cf. Klingenberg [10]. □

Remark. As we saw in (2.4.1), (2.4.2), ΛM is a canonically embedded submanifold of $H^1(I, M)$. On the other hand, $H^1(I, M)$ possesses an isometric embedding into ΛM. We have here an example of a Riemannian Hilbert manifold, i. e., $H^1(I, M)$, which can be embedded isometrically into itself. As a submanifold, it has infinite codimension. Clearly, a Riemannian manifold of finite dimension never can be embedded isometrically into itself with codimension > 0.

There can easily be given other examples when a Hilbert manifold can be embedded into itself with codimension > 0. Take e. g. the mapping

$$c \in \Lambda M \mapsto c^m \in \Lambda M,$$

m an integer > 1. In this case, the embedding is not isometric, however.

We conclude with a remark on the problem of the existence of closed geodesics which are geometrically distinct. By this we mean the following: First of all, it means that the geodesics should not just differ by parameterization, i. e., they should represent different elements in ΠM. In addition, however, we want geometrically distinct closed geodesics not to have the same underlying prime closed geodesic. Note that every closed geodesic c gives rise to a whole 'tower' $\{c^m, m = 1, 2, \ldots\}$ of closed geodesics, i. e., the series of its multiple coverings. Two different elements in the same tower should not be viewed as geometrically distinct.

In other words, finding geometrically distinct closed geodesics amounts to finding prime closed geodesics which represent different elements in ΠM. As we mentioned already in the introduction to this appendix, we have to refer the reader to Klingenberg [10] for an exposition of the methods and the results in the problem of finding many geometrically distinct closed geodesics.

2.6 Index and Curvature

In this section we investigate the influence of the curvature on the geometry in the neighborhood of a geodesic. From (1.12) we know already that this geometry is largely determined by the Jacobi fields along the geodesic. We begin with a comparison for the index and the nullity of a pair of geodesics of equal length, cf. (2.6.1). As a corollary we get an estimate for the occurence of the first conjugate point on a geodesic for which the sectional curvature K is bounded by $K_0 \leqslant K \leqslant K_1$, cf. (2.6.2).

For dimension 2, this is a consequence of Sturm's Comparison theorem. As a corollary we obtain the Bonnet-Myers Theorem on the diameter of a manifold with curvature $K \geqslant K_0 > 0$, (2.6.3).

In (2.6.4), we prove the so-called Homotopy Lemma in its general version. Its first application is a simple proof of the v. Mangoldt-Hadamard-Cartan Theorem on manifolds of curvature $K \leqslant 0$,(2.6.6). Next we prove Synge's Theorem, cf. (2.6.7).

In (2.6.9) and (2.6.10) we derive a lower bound for the injectivity radius of a compact, simply connected manifold of positive curvature. This estimate has many implications. In the hypotheses, there is an important difference between the cases $\dim M$ even and $\dim M$ odd. As corollary (2.6.11) we obtain the homotopical version of the Sphere Theorem. We conclude with a computation of the range of the sectional curvature on a distance sphere in complex projective space, cf. (2.6.12).

In the appendix we derive a lower bound for the injectivity radius of a 1/4-pinched, simply connected manifold.

M, M^* etc. continue to denote complete, if not compact Riemannian manifolds. Recall from (1.12.6) the following situation: We have two geodesics

$$c: I \to M;\ c^*: I \to M^*$$

of equal length $|\dot{c}| = |\dot{c}^*|$ and an isometry

$$\iota_0: (T_{c(0)} M, \dot{c}(0)) \to (T_{c^*(0)} M^*, \dot{c}^*(0)).$$

We define, for each $t \in I$, the isometry

$$\iota_t \equiv \overset{t}{\underset{0}{\|}} c^* \circ \iota_0 \circ \overset{0}{\underset{t}{\|}} c: (T_{c(t)} M, \dot{c}(t)) \to (T_{c^*(t)} M^*, \dot{c}^*(t)).$$

Then, according to (1.12.7), the isometries ι_t commute with the covariant derivation ∇, ∇^* along c and c^*, respectively.

We will use this to compare the index of the geodesic c from $c(0) = p$ to $c(1) = q$, considered as critical point of $\Omega = \Omega_{pq} M$, with the index of the geodesic c^* from $c^*(0) = p^*$ to $c^*(1) = q^*$, considered as critical point of $\Omega^* = \Omega_{p^*q^*} M^*$, where M^* is curved more strongly than M, cf. Morse [3], Schoenberg [1].

2.6.1 Lemma. *Let c, c^* be geodesics as above with $|\dot{c}| = |\dot{c}^*| > 0$. Assume that for* *every 2-plane $\sigma(t)$ containing $\dot{c}(t)$ and every 2-plane $\sigma^*(t)$ containing $\dot{c}^*(t)$ we have the relation*

$$K(\sigma(t)) < K(\sigma^*(t)) \quad \{or \leqslant\},\ all\ t \in I.$$

Then

$$(\text{index} + \text{null})\, c \leqslant \text{index}\, c^* \quad \{or\ \text{index}\, c \leqslant \text{index}\, c^*\}$$

Proof. Recall from (1.11.3) and the beginning of (2.5) the formula

$$\langle R_c \xi, \xi \rangle = \langle R(\xi, \dot{c})\, \dot{c}, \xi \rangle = K(\sigma)\, (|\xi|^2 |\dot{c}|^2 - \langle \xi, \dot{c} \rangle^2),$$

with $K(\sigma)$ the sectional curvature of the 2-plane spanned by $\dot{c}, \xi$, provided $\dot{c}, \xi$ are linearly independant. Otherwise, the formula holds by replacing $K(\sigma)$ with 0. Let $\xi \in T_c \Omega_{pq} M$, i.e., ξ is a vector field along c, vanishing for $t = 0$ and $t = 1$. Then $\iota\xi = (\iota_t \xi(t), 0 \leqslant t \leqslant 1)$, is an element of $T_{c^*} \Omega_{p^* q^*} M^*$.

From our hypothesis we get

$$D^2 E(c)(\xi, \xi) = \int_I \langle \nabla \xi, \nabla \xi \rangle (t) - \langle R_c \xi, \xi \rangle (t)\, dt$$

$$> \int_I \langle \nabla \iota\xi, \nabla \iota\xi \rangle (t) - \langle R_{c^*} \iota\xi, \iota\xi \rangle (t)\, dt = D^2 E(c^*)(\iota\xi, \iota\xi).$$

Hence, if U is a subspace of $T_c \Omega M$ with $D^2 E(c)|U \leqslant 0$, then $D^2 E(c^*)|\iota U < 0$. Our claim therefore follows from remark 2 after (2.5.4).

The same arguments prove our claim under the assumption $K(\sigma(t)) \leqslant K(\sigma^*(t))$. □

As an important corollary we get the

2.6.2 Theorem. *Let $c\colon I \to M$ be a geodesic, $|\dot{c}| \neq 0$.*

(i) *Assume that there is a constant K_1 such that $K(\sigma) < K_1$ $\{K(\sigma) \leqslant K_1\}$, for all 2-planes σ tangent to c. If $L(c) \leqslant \pi/\sqrt{K_1}$ $\{< \pi/\sqrt{K_1}\}$, then there are no conjugate points on c. In particular, if $K_1 \leqslant 0$, this is true for every geodesic c.*

(ii) *Assume that there is a constant $K_0 > 0$ such that $K_0 < K(\sigma)$ $\{K_0 \leqslant K(\sigma)\}$, for all 2-planes σ tangent to c. If $L(c) \geqslant \pi/\sqrt{K_0}$ $\{> \pi/\sqrt{K_0}\}$ then* index $c \geqslant n - 1$.

Remark. (2.6.2) should be viewed as a comparison between the index of a geodesic c on the manifold M and the index of geodesics c_1, c_0 of the same length as c on manifolds M_1, M_0 of constant curvature K_1, K_0, respectively.

Proof. (i) Besides $c\colon I \to M$, consider a geodesic $c_1\colon I \to M_1$ on the space M_1 of constant curvature K_1, $L(c_1) = |\dot{c}| = |\dot{c}_1|$. For c_1 we know from (1.12.14) the distribution of the conjugate points which allows us to determine the index c_1. If $L(c) = L(c_1) \leqslant \pi/\sqrt{K_1}$, there are no conjugate points in the interior of c_1; hence, index $c_1 = 0$. The claim now follows from (2.6.1).

(ii) Similarly, consider a geodesic $c_0\colon I \to M_0$ on a space M_0 of constant curvature

$K_0 > 0$. If $L(c_0) \geq \pi/\sqrt{K_0}$, it has at least one conjugate point of multiplicity $n-1$, cf. (1.12.14). Hence, (index + null) $c_0 \geq n-1$. Apply now (2.6.1). □

As an immediate consequence we get the Theorem of Bonnet and Myers, cf. Myers [2].

2.6.3 Theorem. *Let M be a complete Riemannian manifold and assume that the sectional curvature K satisfies $K_0 \leq K$, for some $K_0 > 0$. Then the diameter $d(M)$ of M has the upper bound,*

$$d(M) = \sup_{(p,q) \in M \times M} d(p,q) \leq \pi/\sqrt{K_0}\,.$$

In particular, M as well as its universal covering $\tilde{M}$ is compact. The fundamental group $\pi_1 M$ of M is finite.

Proof. If there were points p, q on M with distance $d(p,q) > \pi/\sqrt{K_0}$, a minimizing geodesic $c = c_{pq}$ from p to q (which does exist according to (2.1.3)) would have index $\geq n-1 \geq 1$, see (2.6.2). But then c cannot be minimizing, cf. (2.5.9), (2.5.10).

Hence, $d(M) \leq \pi/\sqrt{K_0}$. The same is true for the universal Riemannian covering $\tilde{M}$ of $\tilde{M}$. □

We now come to an important tool for the further study of the geometry of a Riemannian manifold. A slightly weaker version of this 'Homotopy Lemma' was proved first in Klingenberg [3].

2.6.4 Lemma. *Let M be complete and the sectional curvature K of M be bounded from above by a constant K_1, $K \leq K_1$.*

Let $c: I \to M$ be a geodesic biangle, i.e., a closed curve formed by two geodesic segments: There exists a $t_0 \in [0,1]$ such that $c|[0,t_0]$ is a geodesic from $c(0) = p$ to $c(t_0) = q$ and $c|[t_0,1]$ is a geodesic from q to p.

Assume that there is given a homotopy of $c = c_1$ in a point curve c_0; i.e. that we have a continuous curve $h(s)$, $0 \leq s \leq 1$, in ΛM from $h(0) = c_0$ to $h(1) = c_1 = c$. Then either, c is a degenerate biangle, i.e., $c(1-t)$, $0 \leq t \leq 1-t_0$, coincides, up to the parameterization, with $c(t)$, $0 \leq t \leq t_0$; or else, there exists a $s_0 \in]0,1]$ with $L(h(s_0)) \geq 2\pi/\sqrt{K_1}$.

Remarks. 1. For most applications it suffices to consider homotopies $h(s)$ which belong to $\Omega_{pp} M$. I.e., each $h(s)$ is a closed curve from $p = c(0)$ back to $p = c(1)$.

2. The geometric meaning of the Homotopy Lemma may be paraphrased by saying that in a homotopy of a non-degenerate geodesic biangle into a point curve there must occur 'long' curves. The example of a short closed geodesic around the waist of a surface shaped like an hour-glass demonstrates this phenomenon; if one wants to deform this closed geodesic into a point curve one must go via relatively long closed curves.

Proof. We put $\pi/\sqrt{K_1} = \varrho$, $\{X \in TM; |X| < \varrho\} = B_\varrho M$. We will show: If there exists a $\delta > 0$ such that $L(h(s)) \leq 2\varrho - 3\delta$, all $s \in [0,1]$, then there exists a continuous mapping

$$\tilde{H}: [0, 1] \times I \to B_\varrho M \subset TM$$

where $\tilde{H}|\{s\} \times I \equiv \tilde{h}(s)$ is an exponential 'lift' of $h(s)$ into $T_{h(s)(0)}M = T_{h(s)(1)}M$. That is, $\tilde{h}(s)$ is a closed curve with $\tilde{h}(s)(0) = \tilde{h}(s)(1) = 0_{h(s)(0)}$, $\exp_{h(s)(0)}\tilde{h}(s)(t) = h(s)(t)$. Actually, $\operatorname{im}\tilde{h}(s) \subset B_{\varrho-\delta}(0_{h(s)(0)})$. Indeed, otherwise we would have from (1.9.2) $L(h(s)) \geqslant 2\varrho - 2\delta$.

Observe that from (2.6.1) we have with $K \leqslant K_1$ that $F = \tau_M \times \exp|B_\varrho M: B_\varrho M \to M \times M$ has maximal rank everywhere. Consider $H: [0, 1] \times I \to M \times M$; $(s, t) \mapsto (h(s)(0), h(s)(t))$. There exists a relatively compact open $M' \subset M$ such that $\operatorname{im} H \subset \operatorname{im} F|B_{\varrho-\delta}M'$. $F|B_{\varrho-\delta}M'$ is a covering projection onto its image, cf. Spanier [1] for this concept. We therefore have the existence of $\tilde{H}$ with $F \circ \tilde{H} = H$, determined by $\tilde{H}(0, t) = (0_{c_0(0)}, 0_{c_0(0)})$. After having shown that $\tilde{H}$ is defined, we consider the lift $\tilde{h}(1) \equiv \tilde{c}$ of the geodesic biangle c. Clearly, $\tilde{c}(t), 0 \leqslant t \leqslant t_0$, is a straight segment in $T_{c(0)}M$, starting from $0_{c(0)}$. Since c is a closed curve, $\tilde{c}(1-t), 0 \leqslant t \leqslant 1 - t_0$, must be the same segment, up to parameterization. Thus, c is degenerate. □

Remark. The hypothesis $K \leqslant K_1$ for the sectional curvature K of M was used only to insure that geodesic segments of length $< \pi/\sqrt{K_1}$ have no conjugate points. We can therefore formulate the

2.6.5 Complement. *Let M be a complete Riemannian manifold with the property that no geodesic segment of length $< \pi/\sqrt{K_1}$ contains a conjugate point. Here we allow $\pi/\sqrt{K_1} = \infty$. Then the conclusions of the Homotopy Lemma (2.6.4) hold.*

In particular, if there do not exist conjugate points on M (which is the case e.g. if $K \leqslant 0$) then a non-degenerate geodesic biangle c on M is not homotopic to a point curve.

Proof. We only have to discuss the last statement. But if it were false, a homotopy $h(s), 0 \leqslant s \leqslant 1$, by closed curves from the biangle $h(1) = c$ to a point curve $h(0) = c_0$ would contain curves of length $\geqslant$ any positive constant, which clearly is impossible. □

This immediately yields the Theorem of v. Mangoldt [1] (for $\dim M = 2$; also proved later by Hadamard [1]) and Cartan [1] (for $\dim M$ arbitrary).

2.6.6 Theorem. *Let M be a complete, simply connected Riemannian manifold and assume that the curvature K of M is $\leqslant 0$. Then, for every $p \in M$,*

$$\exp_p = \exp|T_pM: T_pM \to M$$

is a diffeomorphism.

The same conclusions hold if one only assumes that there are no conjugate points on M.

Proof. According to (2.6.2), $\exp_p$ has bijective tangential everywhere when $K \leqslant 0$.

From the Hopf-Rinow Theorem (2.1.3) – or directly by applying the lifting technique employed in the proof of (2.6.4) – we see that $\exp_p$ is surjective.

Now assume $\exp_p \tilde{q} = \exp_p \tilde{q}' = q$. Then we obtain a geodesic biangle c by

$$c(t) = \begin{cases} \exp_p 2t\tilde{q}, \; 0 \leqslant t \leqslant 1/2 \\ \exp_p (2-2t)\tilde{q}', \; 1/2 \leqslant t \leqslant 1 . \end{cases}$$

Since M is simply connected, c must be degenerate, according to (2.6.5), i.e., $\tilde{q} = \tilde{q}'$. □

Remark. For another proof of (2.6.6) see Kobayashi [2].

The remainder of section (2.6) is devoted to manifolds of positive sectional curvature. We begin with a result due to Synge [1].

2.6.7 Theorem. *Let M be a compact Riemannian manifold of even dimension.*

(i) *Let $c: [0,1] \to M$ be a closed geodesic and assume that $K(\sigma) > 0$ for all 2-planes σ tangential to c. Then either index $c > 0$ or index $c^2 > 0$, where c^2 is the 2-fold covering of c, i.e., $c^2(t) = c(2t), 0 \leqslant t \leqslant 1$.*

(ii) *Let the sectional curvature K be > 0 everywhere. Then M is simply connected, provided M is orientable. Otherwise, $\pi_1 M = \mathbb{Z}_2$.*

Proof. With the notation from the beginning of (1.12) we have for a closed geodesic c on M that

$$(*) \qquad \overset{1}{\underset{0}{\|}} c : T_0^{\perp} c \to T_0^{\perp} c$$

is an element of the odd dimensional orthogonal group $\mathbb{O}(2m-1)$, $\dim T_0^{\perp} c = \dim M - 1 = 2m - 1$. Hence, it has $+1$ or -1 among its eigenvalues. In the second case, $\overset{1}{\underset{0}{\|}} c^2 = \overset{1}{\underset{0}{\|}} c \circ \overset{1}{\underset{0}{\|}} c$ has $+1$ as eigenvalue.

Assume that $X \neq 0$ is an eigenvector of $\overset{1}{\underset{0}{\|}} c$ for the eigenvalue $+1$. We define by $\xi(t) = \overset{t}{\underset{0}{\|}} cX$ a periodic vector field along $c(t)$, i.e., $\xi \in T_c \Lambda M$. Then

$$D^2 E(c)(\xi, \xi) = -\int_I \langle R(\xi, \dot{c})\dot{c}, \xi \rangle (t)\, dt = -\int_I K(\sigma(t)) |\dot{c}|^2 \, dt < 0 ,$$

where $\sigma(t)$ is the plane spanned by $\dot{c}(t), \xi(t)$.

Hence, index $c = \text{index}_\Lambda c > 0$. Otherwise, index $c^2 > 0$. This completes the proof of (i).

For the proof of (ii) let M be orientable. If $\pi_1 M \neq 0$, there exists a closed geodesic c, not homotopic to a point curve, which has minimal E-value in its free homotopy class. But then, index $c = 0$. On the other hand, since M is orientable, $(*)$ belongs to $S\mathbb{O}(2m-1)$, cf. (2.1.A.4). Hence, c has index > 0. Thus, $\pi_1 M = 0$.

Since every non-orientable manifold possesses a 2-fold orientable covering, cf. (2.1.A), we get from $\pi_1 M \neq 0$ and M compact and $K > 0$ that $\pi_1 M = \mathbb{Z}_2$. □

Next we exhibit a close relation between the injectivity radius and the length of a shortest closed geodesic on a manifold with curvature bounded from above.

2.6.8 Proposition. *Let M be compact with sectional curvature $K \leqslant K_1$. Then either the injectivity radius $\iota(M)$ is $\geqslant \pi/\sqrt{K_1}$ or there exists a closed geodesic of length $< 2\pi/\sqrt{K_1}$.*

Proof. Choose $p \in M$ which has minimal distance from its cut locus $C(p)$. From (2.1.11, iii) we know that there exists a closed geodesic c of length $2d(p, C(p))$ if there are no conjugate points on geodesics of length $\leqslant d(p, C(p))$.

If now $\iota(M) = d(p, C(p)) < \pi/\sqrt{K_1}$, this hypothesis is fulfilled according to (2.6.2, i). □

Combining the previous two results we get a lower bound for the injectivity radius $\iota(M)$ of an even dimensional manifold M of positive curvature. This was proved first in Klingenberg [1].

2.6.9 Theorem. *Let M be a compact, simply connected even dimensional Riemannian manifold of positive sectional curvature K. If K_1 is an upper bound for K then $\iota(M) \geqslant \pi/\sqrt{K_1}$; a non-constant geodesic loop, and in particular a closed geodesic on M has length $\geqq 2\pi/\sqrt{K_1}$.*

Remarks. 1. The estimate for $\iota(M)$ is best possible, as is shown by the sphere of constant curvature $K = K_1$.

2. If $\pi_1 M \neq 0$, $0 < K \leqslant K_1$ implies: $\iota(M) \geqslant \pi/2\sqrt{K_1}$. This follows from Synge's Theorem (2.6.7).

Proof. According to (2.6.8) it suffices to derive a contradiction from the assumption that there exists a closed geodesic c of length $2\iota(M) < 2\pi/\sqrt{K_1}$ and c is of minimal length among all non-constant geodesic loops – note that every such geodesic loop c^* starting at $c^*(0) = p^*$ must meet the cut locus $C(p^*)$ of p^* not later than $L(c^*)/2$.

As in the proof of (2.6.7), there exists a periodic vector field ξ along c with $D^2E(c)(\xi, \xi) < 0$. Thus, the family

$$c_s = \{c_s(t) = \exp_{c(t)}(s\xi(t));\ 0 \leqslant t \leqslant 1\},\ 0 \leqslant s \leqslant \varepsilon$$

is a family of closed curves near c_0 with $c_0 = c$, and $L(c_s) < L(c)$, for $0 < s < \varepsilon$, some small $\varepsilon > 0$.

$L(c_s) < 2\iota(M)$ implies that c_s belongs to the ball $B_{\iota(M)}(c_s(0))$ around the initial equal end point $c_s(0)$ of c_s on which normal coordinates can be defined. Thus, for every $s > 0$ there exist a lift $\tilde{c}_s$ of c_s into $T_{c_s(0)}M$, i.e., a closed curve

$$\tilde{c}_s \colon (I, \partial I) \to (T_{c_s(0)}M, 0_{c_s(0)})$$

with $\exp_{c_s(0)} \tilde{c}_s = c_s$. Now, just as in the proof of the Homotopy Lemma (2.6.4), the lift can be extended from $s = s_0 > 0$ to $s = 0$ since $L(c_s) < 2\pi/\sqrt{K_1}$, for all s. But $\tilde{c}_0$ cannot exist, because $c_0 = c$ is a closed geodesic. □

For the case of an odd dimensional manifold an analogous result can be proved only under a much stronger hypothesis: Instead of $0 < K \leqslant K_1$, one must assume $0 < K_1/4 \leqslant K \leqslant K_1$. Here we will consider only the case $0 < K_1/4 < K \leqslant K_1$ which was proved in Klingenberg [3]. For the general case see the Appendix.

Actually, what one needs for the proof is a certain distribution of the conjugate points along a geodesic ray – a distribution which is valid if $0 < K_1/4 < K \leqslant K_1$.

2.6.10 Theorem. *Let M be a simply connected compact Riemannian manifold of dimension n. Assume that the sectional curvature K satisfies $0 < K_1/4 < K \leqslant K_1$. Then the injectivity radius $\iota(M)$ is $\geqslant \pi/\sqrt{K_1}$; a non-constant geodesic loop and, in particular, a closed geodesic has length $\geqslant 2\pi/\sqrt{K_1}$.*

The same conclusions hold if there exists a $K_1 > 0$ such that every geodesic of length $\leqslant \pi/\sqrt{K_1}$ has index 0 while a geodesic of length $\geqslant 2\pi/\sqrt{K_1}$ has index $\geqslant 2$.

Remark. For dim M even, we actually only need $0 < K_0 \leqslant K \leqslant K_1$, cf. (2.6.9). In the appendix we will show that for dim M odd the hypothesis can be weakened to $0 < K_1/4 \leqslant K \leqslant K_1$. It is unknown whether a further weakening is possible. In any case, the conclusion $\iota(M) \geqslant \pi/\sqrt{K_1}$ becomes false if $0 < -\varepsilon + K_1/9 \leqslant K \leqslant K_1, \varepsilon > 0$ arbitrarily small. This is shown by an example due to Berger. In (2.6.12) we give this example in a version suggested by Weinstein [2].

Proof. We first observe that the hypothesis on the index of geodesics from the end of the theorem follows from the assumption $0 < K_1/4 < K \leqslant K_1$. Indeed, according to (2.6.2), a geodesic of length $\leqslant \pi/\sqrt{K_1}$ then has no conjugate points in its interior while a geodesic of length $\geqslant 2\pi/\sqrt{K_1}$ has at least $n-1$ conjugate points in its interior, each counted with its multiplicity. The statement about the index then follows from the Index Theorem (2.5.9), provided $n > 2$.

We will derive a contradiction from the assumption that, under the above hypothesis, there exists a geodesic loop c of length $< 2\pi/\sqrt{K_1}$. Let c_0 be a constant curve. Since $\pi_1 M = 0, c = c_1$ and c_0 can be joined by a curve in ΛM, i.e.,

$$h: I^1 \to \Lambda M; h(0) = c_0, h(1) = c_1.$$

The Homotopy Lemma (2.6.4) implies that

$$\sup E|h(I^1) \geqslant 2\pi^2/K_1,$$

recall the relation $E(c') \geqslant L(c')^2/2$.

On the other hand, if c^* is a closed geodesic with $L(c^*) \geqslant 2\pi/\sqrt{K_1}$,

$$\operatorname{index}_\Lambda c^* \geqslant \operatorname{index}_\Omega c^* \geqslant n-1.$$

Hence, since $\dim M = n \geqslant 3$, there are no closed geodesics c^* in ΛM with $E(c^*) \geqslant 2\pi^2/K_1$ and with $\text{index}_\Lambda c^* = 0$ or 1. The desired contradiction now follows from (2.5.16). $\square$

As an immediated consequence, cf. Klingenberg [4a], we get the

2.6.11 Corollary. *Let M be a compact, n-dimensional Riemannian manifold. Assume that the sectional curvature K satisfies $0 < K_1/4 < K \leqslant K_1$, some $K_1 > 0$; or, more generally, the implied distribution of conjugate points on geodesics of length $2\pi/\sqrt{K_1}$ shall hold. Then $\pi_{k+1} M = 0$ for $0 < k < n-1$. In particular, if $\pi_1 M = 0$, M is a homotopy sphere.*

Remark. The last statement is a standard result of Algebraic Topology, cf. Spanier [1]. A homotopy sphere M of dimension 2 or $\geqslant 5$ is actually homeomorphic to S^n; for $n \geqslant 5$ this is a deep result of Smale [1]. In (2.8) we will prove directly, without using Smale's result, the Sphere Theorem stating that a simply connected Riemannian manifold satisfying $0 < K_1/4 < K \leqslant K_1$ is homeomorphic to the sphere.

Proof. We proceed as in the proof of (2.4.20). Given a mapping

$$f: S^{k+1} \to M$$

we associate with it a mapping

$$F = F(f): (\bar{B}^k, \partial \bar{B}^k) \to (\Lambda M, \Lambda^0 M)$$

and to a $F: (\bar{B}^k, \partial \bar{B}^k) \to (\Lambda M, \Lambda^0 M)$ we associate a $f = f(F)$ such that $f(F(f)) = f$, $F(f(F)) = F$. Moreover, homotopies f_s of $f = f_0$ correspond to homotopies F_s of $F = F_0$.

Let now $\varepsilon > 0$ sufficiently small. Then there are no closed geodesics in $\Lambda^\varepsilon M$. Under our hypothesis we know from (2.6.10) even more: A closed geodesic c has length $\geqslant 2\pi/\sqrt{K_1}$ and hence, index $\geqslant n-1$. Thus, there are no critical points with index $< n-1$ in $\Lambda M - \Lambda^{\varepsilon-} M$. Given $f: S^{k+1} \to M, k < n-1$, the corresponding $F(f)$ will have its image in $\Lambda^\varkappa M$, for some large $\varkappa$. From (2.5.16) it then follows that im $F(f)$ can be deformed into $\Lambda^{\varepsilon-} M$. But then, as we saw already in the proof of (2.4.20), $F(f)$ is homotopic to a constant map. Thus, f is also homotopic to a constant map. $\square$

2.6.12 Example. *Distance spheres in the complex projective space.*

We consider the complex projective space $P\mathbb{E}$, cf. (1.11.14) and (2.2.27). Here we will denote $P\mathbb{E}$ also by M^*.

Choose $p^* \in M^*$. Since the injectivity radius is $= \pi/2$, we get for ϱ, $0 < \varrho < \pi/2$, with

$$F: M = \partial \bar{B}_\varrho(0_{p^*}) \to M^*; \; p \mapsto \exp_{p^*} p$$

an embedding of the sphere of radius ϱ in $T_{p^*} M^*$ into M^*. Its image is the distance sphere $\partial \bar{B}_\varrho(p^*)$ of radius ϱ. By taking on $F(M)$ the induced Riemannian metric g (also denoted simply by $\langle\ ,\ \rangle$), F becomes an isometric embedding.

Choose $p \in F(M) = \partial \bar{B}_\varrho(p^*)$. Let $c_p(t) = \exp_{p^*} tX_0$, $0 \leqslant t \leqslant \varrho$, be the minimizing geodesic from p^* to $\exp_{p^*} p = \exp_{p^*} \varrho X_0$. In (1.10.6) we derived a formula for the second fundamental form $h_{N(p)}$, where $N(p) = \dot{c}_p(\varrho)$ is the outward normal:

$$h_{N(p)}(X, X) = -\langle X(t), \nabla^* X(t)\rangle_{t=\varrho}.$$

Here, $X(t)$ is the Jacobi field along $c_p(t)$ with $X(0) = 0$, $X(\varrho) = X$. We now write h instead of $h_{N(p)}$.

If we take in particular $X = iX_0$ we have from the end of (2.2.27) $X(t) = (\sin 2t/\sin 2\varrho)\, A(t)$, hence,

$$h(iX_0, iX_0) = -2\operatorname{ctg} 2\varrho = -\operatorname{ctg}\varrho + \operatorname{tg}\varrho;$$

similarly, for $X = Y_0, |Y_0| = 1, \langle Y_0, X_0\rangle = \langle Y_0, iX_0\rangle = 0$,

$$h(Y_0, Y_0) = -\operatorname{ctg}\varrho.$$

We determine the sectional curvature $K(\sigma)$ of a 2-plane $\sigma \subset T_pM$ with (1.11.4). The sectional curvature K^* of $TF.\sigma \subset T_{F(p)}M^*$ is given in (1.11.14). Note the change in the notation: An asterisk now denotes objects in $M^* = P\mathbb{E}$ while in (1.11.14) it was used to denote objects associated to $\mathbb{E}^*$. We find

$$K(\sigma) = \begin{cases} 1 + h(iX_0, iX_0)h(Y_0, Y_0) = \operatorname{ctg}^2\varrho, & \text{if } \sigma = \operatorname{span}\{iX_0, Y_0\} \\ 4 + h(Y_0, Y_0)h(iY_0, iY_0) = 4 + \operatorname{ctg}^2\varrho, & \text{if } \sigma = \operatorname{span}\{Y_0, iY_0\}. \end{cases}$$

We claim that these are the extreme values of K. To see this we observe that every 2-plane $\sigma \subset T_pM$ has an intersection with the space orthogonal to $\{X_0, iX_0\}$. Thus, we can represent σ by

$$\sigma = \operatorname{span}\{\cos\alpha\, iX_0 + \sin\alpha\, Y_0', Y_0\}$$

where $|Y_0| = |Y_0'| = 1$, $\langle Y_0, Y_0'\rangle = 0$, $\{Y_0, Y_0'\}$ orthogonal to $\{X_0, iX_0\}$. With this,

$$K(\sigma) = \cos^2\alpha\, K(\operatorname{span}\{Y_0, iX_0\}) + \sin^2\alpha\, K(\operatorname{span}\{Y_0, Y_0'\}) + \\ + 2\cos\alpha\, \sin\alpha \langle R(Y_0, iX_0)\, Y_0', Y_0\rangle.$$

But the coefficient of $2\cos\alpha\,\sin\alpha$ vanishes. This we get from the Gauss equations (1.11.1) and the formula for the curvature tensor R^* of $M^* = P\mathbb{E}$, derived in (1.11.14), i.e.,

$$\langle R^*(Y_0, iX_0)\, Y_0', Y_0\rangle = 0 \quad \text{and}$$

$$\langle R(Y_0, iX_0)\, Y_0', Y_0\rangle = 0 + \langle h(iX_0, Y_0'), h(Y_0, Y_0)\rangle \\ - \langle h(Y_0, Y_0'), h(iX_0, Y_0)\rangle = 0.$$

To see that $K(\sigma)$ assumes its maximum for $\sigma = \operatorname{span}\{Y_0, iY_0\}$ we only have to determine $K(\operatorname{span}\{Y_0, Y_0'\})$. But observe now that $K^*(TF.\operatorname{span}\{Y_0, Y_0'\}) = 1 + 3\cos^2\psi$, with ψ defined at the end of (1.11.14).

We thus have shown the following: If M is isometric to the distance sphere $\partial \bar{B}_\varrho(p^*)$ in $M^* = P\mathbb{E}$, then

$$\min K : \max K = 1/(1 + 4\operatorname{tg}^2 \varrho).$$

We get a closed geodesic c on M by taking the intersection of $F(M) = \partial \bar{B}_\varrho(p^*)$ with one of the 2-spheres $P\mathbb{E}^2$ of constant curvature, passing through $p^* \in P\mathbb{E}$, cf. (2.2.27). Such a 2-sphere is a connected component of the isometry $\sigma_{P\mathbb{E}^2}$ which carries $\partial \bar{B}_\varrho(p^*)$ into itself. Therefore, $\partial \bar{B}_\varrho(p^*) \cap P\mathbb{E}^2$ is a 1-dimensional connected component of $\operatorname{Fix}(\sigma_{P\mathbb{E}^2} | \partial \bar{B}_\varrho(p^*))$ and hence a closed geodesic, see (1.10.15).

We get a parameterization of $F \circ c \subset \partial \bar{B}_\varrho(p^*)$ by

$$F \circ c(t) = \exp_{p^*}(\cos t\ \varrho X_0 + \sin t\ \varrho i X_0),\ 0 \leqslant t \leqslant 2\pi.$$

X_0 is a unit vector in $T_{p^*}M^*$. $\dot{c}(0)$ is the value in $t = \varrho$ of the Jacobi field $Y(t)$ along $c_p(t) = \exp_{p^*} tX_0$ with $Y(0) = 0$, $\nabla Y(0) = iX_0$. Hence, from the end of (2.2.27),

$$L(c) = 2\pi |\dot{c}(0)| = 2\pi \sin \varrho \cos \varrho.$$

In (2.6.A.1) we prove that, for $\min K : \max K \geqslant 1/4$, the length $L(c)$ of a closed geodesic c is $\geqslant 2\pi/\sqrt{\max K}$.

Here we ask whether, for ϱ increasing from 0 to $\pi/2$, the latter estimate ceases to be true. This would mean

$$L(c) = 2\pi \sin \varrho \cos \varrho \leqslant 2\pi/\sqrt{\max K} = 2\pi/\sqrt{4 + \operatorname{ctg}^2 \varrho}.$$

This is equivalent to

$$(1 - \cos^2 \varrho)(3\cos^2 \varrho - 1) \leqslant 0 \quad \text{or} \quad 1/(1 + 4\operatorname{tg}^2 \varrho) \leqslant 1/9.$$

We therefore have shown, cf. Weinstein [2]:

2.6.13 Proposition. *In the complex projective space $P\mathbb{E}$ consider the distance sphere $M = \partial \bar{B}_\varrho(p^*)$ of radius ϱ around $p^* \in P\mathbb{E}$, $0 < \varrho < \iota(P\mathbb{E}) = \pi/2$. For the induced Riemannian metric the sectional curvature K on M has the range*

$$0 < \operatorname{ctg}^2 \varrho \leqslant K \leqslant 4 + \operatorname{ctg}^2 \varrho$$

and the injectivity radius of M is given by

$$\iota(M) = \pi \sin \varrho \cos \varrho.$$

Hence, if ϱ is so large that $\min K : \max K < 1/9$ *then* $\iota(M) < \pi/\sqrt{\max K}$. □

2.6 Appendix – The Injectivity Radius for 1/4-pinched Manifolds.

We present here a strengthened version of (2.6.10) which possesses several important consequences. A first, rather lenghty proof of the result was given in Klingenberg [4].

Cheeger and Gromoll offered a considerable simplification in an unpublished manuscript from 1972. Here we give a proof which uses only the rather basic methods developed in this book; in particular, in contrast to Cheeger and Gromoll, we avoid any appeal to Morse theory and to the technique of approximating a given Riemannian metric by Riemannian metrics with non-degenerate closed geodesics. For the present proof cf. Klingenberg and Sakai [1].

2.6.A.1 Theorem. *Let M be a simply connected compact Riemannian manifold of dimension n. Assume that the sectional curvature K satisfies the relation $K_1/4 \leqslant K \leqslant K_1$, with some $K_1 > 0$. Then the injectivity radius $\iota(M)$ is $\geqslant \pi/\sqrt{K_1}$.*

Proof. We only need to consider the case dim M odd, cf. (2.6.9). For the first part of the proof we only will use dim $M \geqslant 3$.

As in the proof of (2.6.10), we derive a contradiction from the assumption that there exists a closed geodesic c_1 of length $< 2\pi/\sqrt{K_1}$. Choose a point curve c_0. Now define a ϕ-family by taking as its sets A the $f(I)$ where $f\colon I \to \Lambda M$ is a curve from c_0 to c_1. Let $\varkappa$ be the critical value of this ϕ-family. From the Homotopy Lemma (2.6.4) follows $\varkappa \geqslant 2\pi^2/K_1$.

On the other hand, since $K \geqslant K_1/4$, every critical point of length $> 2\pi/\sqrt{K_1}$ or E-value $> 2\pi^2/K_1$ has index $\geqslant n-1 > 1$, see (2.5.11). From (2.5.16) we therefore get $\varkappa = 2\pi^2/K_1$. Moreover, the non-empty compact set $Cr\varkappa$ of critical points of E-value $\varkappa = 2\pi^2/K_1$ contains a non-empty subset $Cr_1\varkappa$ consisting of all the critical points of index 0 or 1. $Cr_1\varkappa$ is closed in $Cr\varkappa$ and therefore it is compact.

In a refinement of (2.5.16) (for the case $k=1$) we show:

Given any open neighborhood $\mathscr{U}_1$ of $Cr_1\varkappa$ there exists a curve $f\colon I \to \Lambda M$ from c_0 to c_1 with im $f \subset \mathscr{U}_1 \cup \Lambda^{\varkappa -}$.

To see this, we consider for every $c \in Cr\varkappa - Cr_1\varkappa$ neighborhoods $C_\sigma(c), \frac{1}{2}C_\sigma(c)$ as constructed in the proof of (2.5.16). Since $Cr\varkappa$ is compact, it can be covered by a finite subset of the open covering formed by $\mathscr{U}_1$ and int$\frac{1}{2}C_\sigma(c), c \in Cr\varkappa - Cr_1\varkappa$. We denote this finite subset by $\mathscr{U}_1$ and int$\frac{1}{2}C_r, r = 1, \ldots, R$.

Denote by $\mathscr{U}$ the union of $\mathscr{U}_1$ and these int$\frac{1}{2}C_r$. As in the proof of (2.5.16), we have an open neighborhood $\mathscr{V} \subset \mathscr{U}$ of $Cr\varkappa$ and an $\varepsilon > 0$ such that a trajectory $\phi_s c'', s \geqslant 0$, from a c'' outside $\mathscr{U}$ with $E(c'') < \varkappa + \varepsilon$ has E-value $< \varkappa$ when it enters $\mathscr{V}$. Now start with an $f_0\colon I \to \mathscr{V} \cup \Lambda^{\varkappa -}, f_0(0) = c_0, f_0(1) = c_1$, $\max E|f_0(I) < \varkappa + \varepsilon$. Then the arguments in the proof of (2.5.16) yield an ${}_Rf_0$, homotopic to f_0 with fixed endpoints, such that either ${}_Rf_0(x) \in \mathscr{U}_1 \cup \Lambda^{\varkappa -}$ or else, ${}_Rf_0(x) \in \complement\mathscr{U}$. Applying to ${}_Rf_0(I)$ a deformation ϕ_{s_0} with sufficiently large $s_0 \geqslant 0$, we get $f = \phi_{s_0}{}_Rf_0$ with $f(I) \subset \mathscr{U}_1 \cup \Lambda^{\varkappa -}$.

Consider now a sequence $\{\mathscr{U}_{1,k}\}$ of neigborhood of $Cr_1\varkappa$ converging towards $Cr_1\varkappa$. For each k choose a $f_k\colon I \to \mathscr{U}_{1,k} \cup \Lambda^{\varkappa -} M, f_k(0) = c_0, f_k(1) = c_1$. There will be a first $a_k, 0 < a_k < 1$, with $c_k = f_k(a_k) \in \bar{\mathscr{U}}_{1,k}$ Clearly, $E(c_k) < \varkappa$. Condition (C) implies these existence of a convergent subsequence of the sequence $\{c_k\}$ which we will denote again by $\{c_k\}$. Thus, $c = \lim c_k \in Cr_1\varkappa$ does exist.

Fix a k. Since $E(f_k(x)) < 2\pi^2/K_1$ for $x \leqslant a_k$, $c_k = f_k(a_k)$ is the final element of a family of curves $f_k(x), 0 \leqslant x \leqslant a_k$, of length $< 2\pi/\sqrt{K_1}$. Therefore, this family can be lifted into a family of closed curves $\tilde{f}_k(x)$ in the total tangent space,

$$\tilde{f}_k(x): (I, \partial I) \to (T_{f_k(x)(0)} M, 0_{f_k(x)(0)}),$$

cf. the proof of the Homotopy Lemma (2.6.4). In particular, we get a lift $\tilde{c}_k$ for $c_k = f_k(a_k)$.

Since the closed geodesic $c = \lim c_k$ obviously does not posses a lift $\tilde{c}$, there must be an obstruction preventing this. But an obstruction to lifting means that the exponential map

$$T_{\dot{c}(0)/2} \exp_{\dot{c}(0)}: T_{\dot{c}(0)/2}(T_{c(0)} M) \to T_{c(1/2)} M$$

has not maximal rank; i.e., $c(1/2)$ is conjugate to $c(0)$ along $c|[0,1/2]$.

The same arguments show that $c(1) = c(0)$ is conjugate to $c(1/2)$ along $c|[1/2,1]$.

Thus, there exists a Jacobi field $Y(t)$ along $c(t)$ with $\langle Y(r), \dot{c}(t)\rangle = 0, |\nabla Y(0)| = 1$ and $Y(0) = Y(1/2) = 0$.

Consider now the geodesic segment $c' = c|[0,1/2]$ from $p = c(0)$ to $q = c(1/2)$. $L(c') = |\dot{c}|/2 = \pi/\sqrt{K_1}$. Since $K \leqslant K_1$, c' possesses no conjugate points in its interior. Thus, the critical element $c' \in \Omega = \Omega_{pq} M$ of $E|\Omega$ has index 0 and nullity $\geqslant 1$.

Denote by M_1 the sphere of constant curvature K_1. Let $c_1'(t), 0 \leqslant t \leqslant 1/2$, be a geodesic from $p_1 \in M_1$ to the antipodal point $q_1 \in M_1$. To c', c_1' the hypotheses of (2.6.1) apply, with $K \leqslant K_1$. Hence, since index $c_1' = 0$, index $c' =$ index c_1'.

The proof of (2.6.1) shows that equality can occur only if $K(t) = K_1, 0 \leqslant t \leqslant 1/2$. Thus, the Jacobi field $Y(t)$ must be of the form

$$Y(t) = V(t) \sin 2\pi t,$$

with $V(t)$ a parallel vector field along $c(t), \langle \dot{c}, V\rangle(t) = 0, |V(t)| = 1, 0 \leqslant t \leqslant 1/2$.

In the same way we get the existence of a non-zero Jacobi field

$$Y'(t) = V'(t) \sin 2\pi t,\ 1/2 \leqslant t \leqslant 1,$$

along $c(t), 1/2 \leqslant t \leqslant 1$, with $\nabla V'(t) = 0, |V'(t)| = 1, \langle \dot{c}, V'\rangle(t) = 0$.

From $Y(t)$ and $Y'(t)$ we construct elements of $T_c \Lambda M$ by

$$\zeta_Y(t) = \begin{cases} Y(t); & 0 \leqslant t \leqslant 1/2 \\ 0; & 1/2 \leqslant t \leqslant 1 \end{cases}$$

$$\zeta_{Y'}(t) = \begin{cases} 0; & 0 \leqslant t \leqslant 1/2 \\ Y'(t); & 1/2 \leqslant t \leqslant 1 \end{cases}$$

Clearly, $\zeta_Y, \zeta_{Y'}$ are linearly independent and do not belong to the space $T_c^0 \Lambda M$ of periodic Jacobi fields along c, cf. (2.5.6). Moreover, $D^2 E(c) \equiv 0$ on the 2-dimensional space U spanned by $\zeta_Y, \zeta_{Y'}$. Since $\text{index}_\Lambda c = 1$, $U \cap T_c^0 \Lambda M$ is 1-dimensional. A certain linear combination $a\zeta_Y + a'\zeta_{Y'}$ must represent a non-zero periodic Jacobi field along $c(t)$, orthogonal to $\dot{c}(t)$. But this means that by replacing, if necessary, $V'(t)$ with $-V'(t)$,

$$V(1/2) = V'(1/2),\ V(0) = V'(1).$$

Thus, $\xi(t) = \Big(\overset{t}{\underset{0}{\|}}\, c\Big)\, V(0),\ 0 \leqslant t \leqslant 1,$

is a non-zero element of $T_c\Lambda M$ with $D^2E(c)(\xi, \xi) < 0$, cf. the proof of (2.6.7).

Consider the orthogonal transformation

$$(*)\qquad \overset{1}{\underset{0}{\|}}\, c: \big(T_{c(0)}M, \dot c(0)\big) \to \big(T_{c(0)}M, \dot c(0)\big).$$

Since M is simply connected, it belongs to the special orthogonal group. Assume now $\dim M$ odd. Then the invariant subspace $T_c^\perp(0)$ orthogonal to $\dot c(0)$ has even dimension. $V(0) \in T_c^\perp(0)$ is an eigenvector for the eigenvalue 1. There must exist a second, linearly independent eigenvector for the eigenvalue 1, say $W(0)$. Then

$$\eta(t) = \overset{t}{\underset{0}{\|}}\, c\ W(0),\ 0 \leqslant t \leqslant 1$$

is an element of $T_c\Lambda M$ with $D^2E(c)(\eta, \eta) < 0$. But this is a contradiction with $\operatorname{index}_\Lambda c = 1$. □

2.7 Comparison Theorems for Triangles

We begin with Rauch's Comparison Theorem (2.7.2). This is a generalization to n dimensions of Sturm's Comparison Theorem. We like to view Rauch's Theorem as a comparison between two infinitesimally slim triangles on manifolds M, M^* with curvature satisfying $K \geqslant K^*$. The two triangles have two long sides and an infinitesimally small angle between them. There is a given correspondence between these two triangles such that the length of corresponding sides and the size of the included angle are the same. Then the length of the infinitesimal side opposite the infinitesimal angle in the triangle on M is $\leqslant$ the infinitesimal side in the triangle on M^*. (2.7.3) is the integrated version of (2.7.2).

In (2.7.4) the concept of a triangle pqr is defined. Associated to pqr there is a so-called Alexandrov triangle $p^*q^*r^*$ on the space M^* of constant curvature K^*. If $K \leqslant K^*$ and $K^* > 0$, one must assume that the sum of the lengths of the sides of pqr is $< 2\pi/\sqrt{K^*}$. $p^*q^*r^*$ is then defined (up to congruence) by having sides of the same length as pqr.

(2.7.6) is the comparison theorem for triangles pqr, $p^*q^*r^*$ in the case the curvature K of M is $\leqslant K^*$. It states, under a certain additional hypothesis, that the angles in $p^*q^*r^*$ are $\geqslant$ the corresponding angles in pqr. If we have the relation $K^* \leqslant K$ then we get Toponogov's Triangle Theorem (2.7.12): The angles in $p^*q^*r^*$ are $\leqslant$ the corresponding angles in pqr.

The proof of Toponogov's Theorem requires extensive preparations. Among these there is a counterpart of Rauch's Comparison Theorem, due to Berger, cf. (2.7.9). Our proof of Toponogov's Theorem then is reduced to the case of a slim triangle, see (2.7.11).

We begin with the following auxiliary result:

2.7.1 Proposition. *Consider a geodesic* $c: [0, t_0] \to M$ *without conjugate points in its interior. Let* $X(t), Y(t)$ *be vector fields along* $c(t)$ *with* $X(0) = Y(0) = 0$, $X(t_0) = Y(t_0)$. *If* Y *is a Jacobi field then*

$$D^2 E(c)(X, X) = \int_0^{t_0} \langle \nabla X, \nabla X \rangle (t) - \langle R(X, \dot{c})\dot{c}, X \rangle (t)\, dt \geq$$

$$\int_0^{t_0} \langle \nabla Y, \nabla Y) \rangle (t) - \langle R(Y, \dot{c})\dot{c}, Y \rangle (t)\, dt = D^2 E(c)(Y, Y)$$

Proof. Consider c as critical point of $E | \Omega_{pq} M$, $p = c(0)$, $q = c(t_0)$. $\xi(t) = X(t) - Y(t)$ then is an element of $T_c \Omega_{pq} M$. Since index $c = 0$, $D^2 E(c)(\xi, \xi) \geq 0$. I.e.,

$$D^2 E(c)(X, X) \geq 2 D^2 E(c)(X, Y) - D^2 E(c)(Y, Y).$$

But

$$D^2 E(c)(X, Y) = D^2 E(c)(Y, X) = \langle \nabla Y, X \rangle (t) \Big|_0^{t_0} = \langle \nabla Y, Y \rangle (t) \Big|_0^{t_0} =$$

$$= D^2 E(c)(Y, Y). \quad \square$$

We can now prove Rauch's Comparison Theorem, see Rauch [1].

2.7.2 Lemma. *Consider* M, M^* *Riemannian manifolds of dimension* n. *Let*

$$c: I \to M;\ c^*: I \to M^*$$

be geodesics of equal length $L(c) = |\dot{c}| = |\dot{c}^*| = L(c^*) > 0$. *Assume that* c^* *has no conjugate points in its interior. For any pair* $\{\sigma(t), \sigma^*(t)\}$ *of 2-planes tangential to* $c(t)$ *and* $c^*(t)$, *respectively, assume* $K(\sigma(t)) \leq K^*(\sigma^*(t))$, *where* K *and* K^* *denote the sectional curvature of* M *and* M^*.

Let $Y(t), Y^*(t)$ *be Jacobi fields along* $c(t)$ *and* $c^*(t)$ *with* $Y(0) = Y^*(0) = 0$, $|\nabla Y(0)| = |\nabla Y^*(0)|$ *and* $\langle \dot{c}(t), Y(t) \rangle = \langle \dot{c}^*(t), Y^*(t) \rangle = 0$. *Then*

$$|Y(t)| \geq |Y^*(t)|, \text{ all } t \in I.$$

In particular, if for all $\sigma(t)$, $K(\sigma(t)) \leq 0$, *then* $|Y(t)| \geq |t \nabla Y(0)|$.

Proof. We can assume $Y^*(t_0) \neq 0$ for $0 < t_0 < 1$. Fix such a t_0. Then we can choose an isometry ι_{t_0} of the form

$$\iota_{t_0}: (T_{c(0)} M, \dot{c}(0), (\overset{0}{\underset{t_0}{\|}} c)\, Y(t_0)) \to \left(T_{c^*(0)} M^*, \dot{c}^*(0), (\overset{0}{\underset{t_0}{\|}} c^*)\, Y^*(t_0) \frac{|Y(t_0)|}{|Y^*(t_0)|} \right)$$

This is possible because for the corresponding pairs of vectors all scalar products are the same. As in the beginning of (2.6) we transport by ι_{t_0} a vector field Y along c into a vector field ιY along c^*.

Using partial integration, the hypothesis on the sectional curvature and (2.7.1), we find with $c_0 = c|[0, t_0]$, $c_0^* = c^*|[0, t_0]$,

$$\frac{1}{2}\frac{d}{dt}|Y|^2(t_0) = \langle Y, \nabla Y\rangle(t)\Big|_0^{t_0} = D^2E(c_0)(Y,Y) \geqslant D^2E(c_0^*)(\iota Y, \iota Y) \geqslant$$

$$\geqslant D^2E(c_0^*)\left(Y^*\frac{|Y(t_0)|}{|Y^*(t_0)|}, Y^*\frac{|Y(t_0)|}{|Y^*(t_0)|}\right) =$$

$$= \langle Y^*, \nabla Y^*\rangle\Big|_0^{t_0}\frac{|Y(t_0)|^2}{|Y^*(t_0)|^2} = \frac{1}{2}\frac{d}{dt}|Y^*|^2(t_0)\frac{|Y(t_0)|^2}{|Y^*(t_0)|^2}.$$

Hence,

$$\frac{d}{dt}\log|Y|^2(t_0) \geqslant \frac{d}{dt}\log|Y^*|^2(t_0).$$

Integration over $[\varepsilon, t]$, $0 < \varepsilon < t < 1$, yields

$$|Y(t)|^2 \geqslant |Y^*(t)|^2\frac{|Y(\varepsilon)|^2}{|Y^*(\varepsilon)|^2}.$$

Since $|\nabla Y(0)| = |\nabla Y^*(0)|$, we get our inequality. The last statement follows from $Y^*(t) = t(\overset{t}{\underset{0}{\|}} c^*)\nabla Y^*(0)$, where M^* has constant curvature $K^* \equiv 0$. □

An integrated version of (2.7.2), together with the Gauss Lemma, yields the

2.7.3 Corollary. *Let M, M^* be Riemannian manifolds of dimension n. Assume for the sectional curvature K and K^* of M and M^* the relation $K \leqslant K^*$.*

Choose $p \in M$, $p^* \in M^*$ and a curve $\tilde{b} = (\tilde{b}(s), s_0 \leqslant s \leqslant s_1)$ in T_pM. Let $\iota\colon T_pM \to T_{p^*}M^*$ be an isometry and put $\iota\tilde{b} = \tilde{b}^*$. Then $b(s) = \exp_p\tilde{b}(s)$, $b^*(s) = \exp_{p^*}\tilde{b}^*(s)$ are curves in M and M^* respectively. Assume that the exponential map $\exp_{p^*}$ is regular for the interior points of the straight segments $\tilde{c}_s^*(t) = t\tilde{b}^*(s)$, $0 \leqslant t \leqslant 1$, from 0_{p^*} to $\tilde{b}^*(s)$.

Then $L(b) \geqslant L(b^*)$, i.e., b is at least as long as b^*.

Remark. As an illustration let M and M^* be the spheres of radius r and r^*, respectively, $r \geqslant r^*$. Choose $p \in M$, $p^* \in M^*$ and an isometry $\iota\colon T_pM \to T_{p^*}M^*$. If b is a curve on M which lies entirely in the interior of the ball $B_{\pi r^*}(p) \subset M$, we take its lift $\tilde{b} = (\exp|T_pM)^{-1}(b)$ in T_pM and transport it into $T_{p^*}M^*$ by ι which yields the curve $\tilde{b}^* = \iota\tilde{b}$. The curve $b^* = \exp_{p^*}b^*$ then lies entirely in $B_{\pi r^*}(p^*)$ and is shorter than b.

Proof. By subdividing, if necessary, $\tilde{b}$ into several pieces and employing an approximation argument we may assume $\tilde{b}(s) \neq 0_p$, all s. The corollary follows once we show $|b'(s)| \geqslant |b^{*\prime}(s)|$.

Let

$$\tilde{b}'(s) = \tilde{b}'_\varrho(s) + \tilde{b}'_\sigma(s)$$

be the radial component and its orthogonal complement, the so-called spherical component. From the Gauss-Lemma (1.9.1) we get the orthogonal decomposition

$$b'(s) = b'_\varrho(s) + b'_\sigma(s),$$

where

$$b'_\varrho(s) = T_{\tilde{b}(s)} \exp_p . \tilde{b}'_\varrho(s);\ b'_\sigma(s) = T_{\tilde{b}(s)} \exp_p . \tilde{b}'_\sigma(s).$$

In the same way we obtain an orthogonal decomposition for $b^{*\prime}(s)$, i.e.,

$$b^{*\prime}(s) = b^{*\prime}_\varrho(s) + b^{*\prime}_\sigma(s).$$

The length of the radial components remains unchanged under the exponential map. Hence we only have to show

$$(*) \qquad |b'_\sigma(s)| \geqslant |b^{*\prime}_\sigma(s)|.$$

To see this we observe that $b'_\sigma(s)$ and $b^{*\prime}_\sigma(s)$ are the values, at $t = 1$, of the Jacobi fields

$$Y_s(t) = T_{t\tilde{b}(s)} \exp_p . (t\tilde{b}'_\sigma(s)),$$

$$Y_s^*(t) = T_{t\tilde{b}^*(s)} \exp_{p^*} . (t\tilde{b}^{*\prime}_\sigma(d))$$

along the geodesics $c_s(t) = \exp_p t\tilde{b}(s)$, $c_s^*(t) = \exp_{p^*} t\tilde{b}^*(s)$ respectively. $(*)$ now follows from (2.7.2). □

We wish to consider triangles on general Riemannian manifolds. Other than in Euclidean geometry or – more generally – in simply connected Riemannian manifolds without conjugate points, cf. (2.6.8), there can exist more than one minimizing geodesic between two points. Therefore, a triangle is not necessarily determined by its corner points. We also will consider generalized triangles.

2.7.4 Definition. (i) A (*geodesic*) *triangle* pqr on a Riemannian manifold M consists of three different points, p, q, r, called *corners* and three *sides*, i.e., minimizing geodesics c_{pq}, c_{qr}, c_{rp} between p and q, q and r, r and p.

(ii) A *generalized triangle* pqr consists of three corners, p, q, r, and two minimizing sides, c_{pq}, c_{pr}, whereas the third side, c_{qr}, is a not necessarily minimizing geodesic from q to r. However, we require the triangle inequality

$$L(c_{qr}) \leqslant L(c_{pq}) + L(c_{pr}) = d(p, q) + d(q, r).$$

In particular, we allow $q = r$, $L(c_{qr}) > 0$.

(iii) The *length* of a triangle pqr, $L(pqr)$, is the sum of the lengths of its sides. The *angle at* p of pqr, $\overset{r}{\underset{q}{\sphericalangle}} p$ or briefly $\sphericalangle p$, is the non-oriented angle between the initial directions of the minimizing c_{pq} and c_{pr} at p. In particular, $0 \leqslant \sphericalangle p \leqslant \pi$.

(iv) Denote by M^* the simply connected space of constant curvature K^* having dimension $= \dim M$. For a (generalized) triangle pqr on M length $L(pqr) < 2\pi/\sqrt{K^*}$ there is defined – uniquely up to congruence – an associated *Alexandrov triangle* $p^*q^*r^*$ on M^*. It is characterized by having same side lengths as pqr, i.e. $d(p^*, q^*) = d(p, q)$, $d(p^*, r^*) = d(p, r)$, $d(q^*, r^*) = d(q, r)$ or $= L(c_{qr})$, if pqr is a generalized triangle.

2.7.5 Remarks. 1. Recall that the typical representatives of a simply connected n-dimensional space M^* of constant curvature K^* are

the sphere $S^n_{1/\sqrt{K^*}}$ of radius $1/\sqrt{K^*}$, if $K^* > 0$,

the Euclidean space $\mathbb{R}^n$, if $K^* = 0$,

the hyperbolic space $H^n_{\sqrt{1/-K^*}}$, if $K^* < 0$,

cf. (1.11.6), (1.11.7).

2. The hypothesis $L(pqr) < 2\pi/\sqrt{K^*}$ is vacuous if $K^* \leqslant 0$; as always we put $a/\sqrt{K^*} = \infty$ if $a > 0$, $K^* \leqslant 0$.

3. The group of isometries of a simply connected space M^* of constant curvature K^* is transitive on the points. Even more, any k-tupel $\{X_i; 1 \leqslant i \leqslant k\}$ of vectors in $T_p \cdot M^*$ can be carried into any other k-tupel $\{Y_i; 1 \leqslant i \leqslant k\}$ of vectors in $T_q \cdot M^*$, provided all possible scalar products coincide, i.e., $\langle X_i, X_j \rangle = \langle Y_i, Y_j \rangle$, $1 \leqslant i, j \leqslant k$. See (1.10.2,3) and (1.11.8).

This shows that any two triangles $p^*q^*r^*$, $p'^*q'^*r'^*$ on M^*, for which corresponding sides have the same length are congruent, i.e., they can be carried by an isometry into each other.

4. For a later application we recall the sinus theorem for triangles on a space M^* of constant curvature K^*. Let a, b, c be the side lengths of a triangle and α, β, γ the angles opposite to these sides. Then

$$\sin\alpha : \sin\beta : \sin\gamma = \sin\sqrt{K^*}\,a : \sin\sqrt{K^*}\,b : \sin\sqrt{K^*}\,c$$

Here, for $K^* < 0$, $\sin\sqrt{K^*}\,a = \sinh\sqrt{-K^*}\,a$. For $K^* = 0$ the right hand side reads $a : b : c$.

We now formulate the Comparison Theorem for Triangles where K is bounded from above:

2.7.6 Theorem. *Let M be a Riemannian manifold where the sectional curvature K is $\leqslant$ some constant K_1. Let pqr be a triangle such that the side from q to r belongs to the* *domain of normal coordinates based at p. Then the angle at p_1 of an associated Alexandrov triangle $p_1 q_1 r_1$ in M_1 is $\geqslant$ the angle at p.*

Note. The hypotheses are satisfied for every triangle pqr if $K \leqslant K_1 \leqslant 0$ and M is simply connected. The assumption $L(pqr) < 2\pi/\sqrt{K_1}$, $K_1 > 0$ does not suffice, cf. (2.6.13) where we gave an example of a closed geodesic of length $< 2\pi/\sqrt{K_1}$. This yields a triangle with all angles $= \pi$, while on the sphere of curvature K_1 a triangle of length $< 2\pi/\sqrt{K_1}$ has all angles $< \pi$.

Proof. Under our hypothesis, the minimizing geodesic $c_{qr}(s) \equiv b(s)$, $0 \leqslant s \leqslant 1$, from q to r possesses a lift $\tilde{b}(s)$ into $B_{\pi/\sqrt{K_1}}(0_p) \subset T_p M$. Put $\tilde{b}(0) = \tilde{q}$, $\tilde{b}(1) = \tilde{r}$.

On M_1 we construct a triangle $p_1 q_1 r_1'$ as follows: Choose $p_1 \in M_1$ and an isometry $\iota: T_p M \to T_{p_1} M_1$. Put $\iota\tilde{b} = \tilde{b}_1$. This is a curve in $B_{\pi/\sqrt{K_1}}(0_{p_1}) \subset T_{p_1} M_1$ from $\tilde{q}_1 = \tilde{b}_1(0)$ to $\tilde{r}_1' = \tilde{b}_1(1)$. $b_1 = \exp_{p_1} \tilde{b}_1$ is a curve from q_1 to r_1'. From our construction follows that the (uniquely determined) minimizing geodesics $c_{p_1 q_1}$ from p_1 to q_1 and $c_{p_1 r'_1}$ from p_1 to r'_1 have the same lengths as the sides c_{pq} and c_{pr}. Moreover, $\underset{q}{\overset{r}{\sphericalangle}} p = \underset{q_1}{\overset{r_1'}{\sphericalangle}} p_1$. For the length of the third side $c_{q_1 r_1'}$ of $p_1 q_1 r_1'$ we have from (2.7.3) the estimate

$$d(q_1, r_1') = L(c_{q_1 r_1'}) \leqslant L(b) = L(c_{qr}).$$

Hence, the angle at p_1 of $p_1 q_1 r_1'$ has to be widened to get a triangle $p_1 q_1 r_1$ with

$$L(c_{q_1 r_1}) = L(c_{qr}). \quad \square$$

We first prove under special assumptions a corresponding Comparison Theorem for a lower bound of the curvature K. In its full generality we will prove it only later, cf. (2.7.12).

2.7.7 Proposition. *Let M be a Riemannian manifold for which the sectional curvature K is bounded from below by a constant K_0. Let pqr be a triangle of length $L(pqr) < 2\pi/\sqrt{K_0}$. Assume that* $\exp_p | B_{\pi/\sqrt{K_0}}(0_p)$ *is of maximal rank everywhere. Let $p_0 q_0 r_0$ be an associated Alexandrov triangle on the space M_0 of constant curvature K_0. Then the angle at p_0 is $\leqslant$ the angle at p.*

Proof. For $\sphericalangle p = \pi$ this certainly is true. Therefore we may assume $\sphericalangle p < \pi$.

Since $K_0 \leqslant K$ and $L(pqr) < 2\pi/\sqrt{K_0}$, the sides of pqr all are shorter than $\pi/\sqrt{K_0}$. We thus can construct on M_0 a triangle $p_0 q_0 r_0'$ with

$$\underset{q_0}{\overset{r_0'}{\sphericalangle}} p_0 = \sphericalangle p;\ d(p_0, q_0) = d(p, q);\ d(p_0, r_0') = d(p, r).$$

The side $b_0 = c_{q_0 r_0'}$ lies entirely in $B_{\pi/\sqrt{K_0}}(p_0)$. It therefore possesses a lift $\tilde{b}_0$ into $B_{\pi/\sqrt{K_0}}(0_{p_0}) \subset T_{p_0} M_0$. We choose an isometry

$$\iota: (T_{p_0} M_0, \dot{c}_{p_0 q_0}(0), \dot{c}_{p_0 r_0'}(0)) \to (T_p M, \dot{c}_{pq}(0), \dot{c}_{pr}(0)).$$

$b = \exp_p \iota\tilde{b}_0$ then is a curve from q to r.

Using (2.7.3) we get

$$d(q_0, r_0') = L(b_0) \geqslant L(b) \geqslant d(q, r) = L(c_{qr}).$$

Hence, the angle at p_0 of $p_0 q_0 r_0'$ must be made smaller to obtain an Alexandrov triangle $p_0 q_0 r_0$ associated with pqr. □

We continue with an analogue of (2.7.1). For the definition of a focal point see (1.12.20).

2.7.8 Proposition. *Consider a geodesic $c\colon [0, t_0] \to M$ without focal points in its interior. Let $X(t)$, be a vector field along $c(t)$ with $X(0) \perp \dot{c}(0)$ and let $Y(t)$ be a Jacobi field along $c(t)$ with $Y(t) \perp \dot{c}(t)$, $\nabla Y(0) = 0$, $Y(t_0) = X(t_0)$. Then*

$$D^2 E(c)(X, X) = \int_0^{t_0} \langle \nabla X, \nabla X \rangle (t) - \langle R(X, \dot{c})\dot{c}, X \rangle (t)\, dt \geqslant$$

$$\int_0^{t_0} \langle \nabla Y, \nabla Y \rangle (t) - \langle R(Y, \dot{c})\dot{c}, Y \rangle (t)\, dt = D^2 E(c)(Y, Y).$$

Proof. Consider c as critical point of $E|H^1_{M_0 \times \{p_1\}}$. Here, M_0 is the in $p_0 = c(0)$ totally geodesic submanifold, obtained by restricting $\exp_{c(0)}$ to a small ball around the origin 0_p of $T_0^{\perp} c = \{X \in T_{p_0} M;\ \langle X, \dot{c}(0) \rangle = 0\}$, cf. (1.10.13, 1).

From (2.5.1, ii) we get a formula for $D^2 E(c)$ which coincides with the formula for $D^2 E(c)$ when c is considered as critical point of $E|\Omega_{p_0 p_1} M$.

Put $X - Y = \xi$. ξ is an element of $T_c H^1_{M_0 \times \{p_1\}}$. Since c contains no focal points in the interior, $D^2 E(c)$ is positive semi-definite, $D^2 E(c)(\xi, \xi) \geqslant 0$, see (2.5.15). That is,

$$D^2 E(c)(X, X) \geqslant 2 D^2 E(c)(Y, X) - D^2 E(c)(Y, Y).$$

Using partial integration and the fact that Y is a Jacobi field we get

$$D^2 E(c)(Y, X) = \langle \nabla Y, X \rangle (t) \Big|_0^{t_0} = \langle \nabla Y, Y \rangle \Big|_0^{t_0} = D^2 E(c)(Y, Y). \quad \square$$

The following Lemma is a counterpart of Rauch's Comparison Theorem (2.7.2), due to Berger [4].

2.7.9 Lemma. *Consider Riemannian manifolds M, M^* of dimension n. Let*

$$c\colon I \to M;\ c^*\colon I \to M^*$$

be geodesics of equal length $L(c) = L(c^*) > 0$. c^* shall contain no focal points in its interior.

For any pair $\{\sigma(t), \sigma^*(t)\}$ of 2-planes containing $\dot{c}(t)$ and $\dot{c}^*(t)$, respectively, let $K(\sigma(t)) \leqslant K^*(\sigma^*(t))$, for all $t \in I$, where K and K^* denote the sectional curvature of M and M^*, respectively.

Let $Y(t), Y^*(t)$ be Jacobi fields along $c(t), c^*(t)$ with $|Y(0)| = |Y^*(0)|$, $\nabla Y(0)$

$= \nabla Y^*(0) = 0$ *and* $\langle \dot c(t), Y(t)\rangle = \langle \dot c^*(t), Y^*(t)\rangle = 0$, *all* $t \in I$. *Then*

$$|Y(t)| \geqslant |Y^*(t)|, \quad all\ t \in I.$$

In particular, if, for all $\sigma(t)$, $K(\sigma(t)) \leqslant 0$, *then* $|Y(t)| \geqslant |Y(0)|$, *all* $t \in \mathbb{R}$.

Proof. We can assume $Y^*(t_0) \neq 0$ for $0 \leqslant t_0 < 1$. Fix such a t_0. Choose an isometry

$$\iota_{t_0}: (T_{c(0)}M, \dot c(0), (\overset{0}{\underset{t_0}{\|}} c)\, Y(t_0)) \to \left(T_{c^*(0)}M^*, \dot c^*(0), (\overset{0}{\underset{t_0}{\|}} c^*)\, Y^*(t_0)\, \frac{|Y(t_0)|}{|Y^*(t_0)|}\right).$$

Such a ι_{t_0} exists since all scalar products of corresponding vectors are the same. We use it to define a mapping ι from vector fields along c into vector fields along c^*, cf. the beginning of (2.6).

Partial integration, the hypothesis on the sectional curvature and (2.7.8) yield, with $c|[0, t_0] = c_0$, $c^*|[0, t_0] = c_0^*$,

$$\frac{1}{2}\frac{d}{dt}|Y|^2(t_0) = \langle \nabla Y, Y\rangle(t)\Big|_0^{t_0} = D^2E(c_0)(Y, Y) \geqslant$$

$$\geqslant D^2E(c_0^*)(\iota Y, \iota Y) \geqslant D^2E(c_0^*)(Y^*, Y^*)\,\frac{|Y(t_0)|^2}{|Y^*(t_0)|^2} = \frac{1}{2}\frac{d}{dt}|Y^*|^2(t_0).$$

Integration over $[0, t_0]$ yields our inequalities. The last statement follows from $Y^*(t) = (\overset{t}{\underset{0}{\|}} c^*)\, Y^*(0)$, if M^* has constant curvature $K^* = 0$. □

The analogue of (2.7.3) reads as follows:

2.7.10 Corollary. *Consider Riemannian manifolds* M, M^* *of dimension* n *with sectional curvatures satisfying* $K \leqslant K^*$.

Let

$$c: I \to M, \quad c^*: I \to M^*$$

be geodesics of equal length $|\dot c| = |\dot c^*| > 0$. $V(t)$, $V^*(t)$ *shall be parallel vector fields along* $c(t)$, $c^*(t)$ *with* $|V(t)| = |V^*(t)| = 1$, $\langle \dot c(t), V(t)\rangle = \langle \dot c^*(t), V^*(t)\rangle$. *Let* $f: I \to \mathbb{R}$ *be a function* $\geqslant 0$ *and consider the curves*

$$b(t) = \exp_{c(t)} f(t)\, V(t), \quad b^*(t) = \exp_{c^*(t)} f(t)\, V^*(t).$$

None of the geodesics

$$c_t^*(s) = \exp_{c^*(t)}(s f(t)\, V^*(t)); \quad 0 \leqslant s \leqslant 1$$

from $c^*(t)$ *to* $b^*(t)$ *shall have a focal point in their interior. Then*

$$L(b) \geqslant L(b^*).$$

Remark. Whereas in (2.7.3) the curves b, b^* were described in polar coordinates, cf. (1.8.16), here we use coordinates of Fermi type, cf. (1.12.1).

Proof. It suffices to show $|\dot{b}(t)| \geqslant |\dot{b}^*(t)|$. By subdividing b into several parts and using, if necessary, an approximation argument, we can restrict ourselves to $t = t_0 \in]0, 1[$ with $f(t_0) > 0$.

Consider the splitting of the derivative $(f(t)\,V(t))^{\cdot}$ of the vector field $f(t)\,V(t)$ along $c(t)$ in its horizontal and its vertical part, i.e.,

$$(f(t)\,V(t))^{\cdot} = (f(t)\,V(t))^{\cdot}_h + (f(t)\,V(t))^{\cdot}_v = \dot{c}(t) + \dot{f}(t)\,V(t).$$

The first component stems from the derivation of $\tau \circ f(t)\,V(t) = c(t)$; the second is $= \nabla(f(t)\,V(t))$.

This splitting yields

$$\begin{aligned}\dot{b}(t) &= T_{f(t)V(t)} \exp_{c(t)} . \dot{c}(t) + T_{f(t)V(t)} \exp_{c(t)} \dot{f}(t)\,V(t) = \\ &= \text{(briefly) } \dot{b}(t)_h + \dot{b}(t)_v .\end{aligned}$$

Now fix a t_0. Then $\dot{b}(t_0)_h$ is the value for $s = 1$ of a Jacobi field $Y_{t_0}(s)$ along the geodesic $c_{t_0}(s) = \exp_{c(t_0)} sf(t_0)\,V(t_0)$. $Y_{t_0}(s)$ is obtained by the 1-parameter family $(c_{t_0,t}(s), 0 \leqslant t \leqslant 1)$ of geodesics of the form

$$c_{t_0,t}(s) = F(s, t) = \exp_{c(t)}(sf(t_0)\,V(t)).$$

Note: $c_{t_0,t_0}(s) = c_{t_0}(s)$.

Indeed, we have

$$Y_{t_0}(s) = \frac{\partial F}{\partial t}(s, t)|_{t=t_0} = T_{sf(t_0)V(t_0)} \exp_{c(t_0)} . \dot{c}(t_0)$$

thus,

$$Y_{t_0}(1) = \dot{b}(t_0)_h,\ Y_{t_0}(0) = \dot{c}(t_0).$$

Moreover,

$$\nabla Y_{t_0}(0) = \frac{\nabla}{\partial s}\frac{\partial}{\partial t} F(s, t)|_{0,t_0} = \frac{\nabla}{\partial t}\frac{\partial}{\partial s} F(s, t)|_{t_0,0} = \frac{\nabla}{\partial t} f(t_0)\,V(t)|_{t_0} = 0.$$

The $c'_{t_0}(s)$-orthogonal component of $Y_{t_0}(s)$ is given by

$$Y^{\perp}_{t_0}(s) = Y_{t_0}(s) - a(s)\,c'_{t_0}(s);\ a(s) = \frac{\langle c'_{t_0}(s), Y_{t_0}(s)\rangle}{|c'_{t_0}(s)|^2}.$$

Since $\nabla Y_{t_0}(0) = 0$ also $a'(0) = 0$ and $\nabla Y^{\perp}_{t_0}(0) = 0$. Since $a(s)\,c'_{t_0}(s)$ is a Jacobi field it follows that $a(s) = a(0) = \text{const}$. With $c'_{t_0}(0) = f(t_0)\,V(t_0)$, $Y_{t_0}(0) = \dot{c}(t_0)$ we get

$$a(s) = a(0) = \langle \dot{c}(t_0), V(t_0)\rangle / f(t_0).$$

Therefore we can write $\dot{b}(t_0) = \dot{b}(t_0)_h + \dot{b}(t_0)_v$ as

$$\dot{b}(t_0) = Y^{\perp}_{t_0}(1) + T_{f(t_0)V(t_0)} \exp_{c(t_0)} . [-\langle \dot{c}(t_0), V(t_0)\rangle + \dot{f}(t_0)]\,V(t_0).$$

Here, the second component is radial in $f(t_0)\,V(t_0) \in T_{c(t_0)}M$ while the first one is spherical, i.e., orthogonal to the vector $f(t_0)\,V(t_0)$.

In exactly the same way we can write

$$\dot{b}^*(t_0) = Y_{t_0}^{*\perp}(1) + T_{f(t_0)V^*(t_0)} \exp_{c^*(t_0)} \cdot [-\langle \dot{c}^*(t_0), V^*(t_0)\rangle + \dot{f}(t_0)]\, V^*(t_0).$$

Under our assumption, the radial components of $\dot{b}(t_0)$ and $\dot{b}(t_0)$ have the same length. To the spherical components we apply (2.7.5), $|Y_{t_0}^{\perp}(1)| \geq |Y_{t_0}^{*\perp}(1)|$. Hence, $|\dot{b}(t_0)| \geq |\dot{b}^*(t_0)|$. □

We now prove the Triangle Comparison Theorem for a lower bound of the curvature and for 'slim' triangles.

2.7.11 Proposition. *Let M be a Riemannian manifold and K_0 a lower bound for the sectional curvature K of M. Let pqr be a triangle on M for which the side c_{qr} is short in the following sense: Let K_1 be an upper bound for K on a compact domain of M containing all minimizing geodesics between points on the sides of pqr. One can take the set of $u \in M$ having distance $\leq L(pqr)/2$ from p. In addition,*

(i) *If $K_0 \leq 0$, $L(c_{qr}) = d(q, r) \leq \pi/2\sqrt{K_1}$.*

(ii) *If $K_0 > 0$, then there shall exist an $\varepsilon > 0$ such that $L(pqr) = 2\pi/\sqrt{K_0} - 4\varepsilon$ and*

$$d(q, r) \leq \inf\left\{\varepsilon, \frac{\pi}{2\sqrt{K_0}}, \frac{\sin\sqrt{K_0}\,\varepsilon}{\sqrt{K_0}} \sin\left(\frac{\pi}{2}\frac{\sqrt{K_0}}{\sqrt{K_1}}\right)\right\}.$$

Let $p_0q_0r_0$ be an associated Alexandrov triangle on the simply connected space M_0 of constant curvature K_0. Then $\measuredangle q_0 \leq \measuredangle q$, $r_0 \leq \measuredangle r$.

Proof. Since the assumptions are symmetric in q and r it suffices to show that $\measuredangle q_0 \leq \measuredangle q$. For $K_1 \leq 0$ this follows from (2.7.7). We can therefore assume $K_1 > 0$.

1. Consider first the case $K_0 \leq 0$. On M_0 we construct a triangle $p_0q_0r_0'$ with

$$d(q_0, p_0) = d(q, p),\ d(q_0, r_0') = d(q, r),\ \overset{r_0'}{\underset{p_0}{\measuredangle}} q_0 = \measuredangle q.$$

We can describe $b_0(t) = c_{p_0r_0'}(t)$ as

$$b_0(t) = \exp_{c_0(t)}\big(f(t)\,V_0(t)\big).$$

Here, $V_0(t)$ is a parallel unit vector field along the geodesic

$$c_0(t) = c_{p_0q_0}(t),\ 0 \leq t \leq 1$$

from $p_0 = c_0(0)$ to $q_0 = c_0(1)$.

$$0 \leq f(t) = d\big(c_0(t), b_0(t)\big) \leq f(1) = d(q, r) \leq \pi/2\sqrt{K_1}.$$

Hence, there are no focal points in the interior of the minimizing geodesics

$$\exp_{c_0(t)}(sf(t)\,V_0(t)),\ 0 \leqslant s \leqslant 1,$$

from $c_0(t)$ to $b_0(t)$.

We can write $r = \exp_q(f(1)\,V(1))$, $|V(1)| = 1$. Denote by $c(t)$, $0 \leqslant t \leqslant 1$, the geodesic $c_{pq}(t)$ from $p = c(0)$ to $q = c(1)$. With the parallel vector field $V(t) = (\overset{t}{\underset{1}{\|}} c)\,V(1)$ we obtain with

$$b(t) = \exp_{c(t)} f(t)\,V(t),\ 0 \leqslant t \leqslant 1,$$

a curve from $p = b(0)$ to $r = b(1)$. Using (2.7.10) we get

$$d(p, r) \leqslant L(b) \leqslant L(b_0) = d(p_0, r_0').$$

Thus, the angle at q_0 in $p_0 q_0 r_0'$ must be made smaller to obtain an Alexandrov triangle $p_0 q_0 r_0$ associated with pqr.

2. Now consider the case $K_0 > 0$ and hence $K_1 > 0$. Since $d(p, q) \leqslant L(pqr)/2 = \pi/\sqrt{K_0} - 2\varepsilon$ we can consider on the sphere M_0 of radius $1/\sqrt{K_0}$ the triangle $p_0 q_0 r_0'$ with $d(q_0, p_0) = d(p, q)$, $d(q_0, r_0') = \mathrm{d}(q, r)$, $\overset{r_0'}{\underset{p_0}{\sphericalangle}}\, q_0 = \sphericalangle q$.

Since $d(q_0, r_0') \leqslant \varepsilon$ we have from the triangle inequality

$$L(p_0 q_0 r_0') \leqslant \left(\frac{\pi}{\sqrt{K_0}} - 2\varepsilon\right) + (\varepsilon) + \left(\frac{\pi}{\sqrt{K_0}} - 2\varepsilon + \varepsilon\right) = \frac{2\pi}{\sqrt{K_0}} - 2\varepsilon.$$

In particular, $\sphericalangle q < \pi$. As in 1. we can describe $b_0(t) = c_{p_0 r_0'}(t)$ in the form

$$b_0(t) = \exp_{c_0(t)}(f(t)\,V_0(t)).$$

Here, $V_0(t)$ is a parallel vector field along $c_0(t) = c_{p_0 q_0}(t)$, $0 \leqslant t \leqslant 1$, $|V_0(t)| = 1$, $f(0) = 0$, $f(1) = d(q, r) > 0$.

If $d(p_0, r_0') \leqslant \pi/2\sqrt{K_0}$, the curve $b_0(t)$ does not go beyond the equator of p_0; $f(t)$ then is monotonic increasing and we can complete our arguments as under 1.

It remains to discuss the case

$$\frac{\pi}{2\sqrt{K_0}} < d(p_0, r_0') < \frac{\pi}{\sqrt{K_0}} - \varepsilon.$$

Let $t_0 \in]0, 1[$ be such that $f(t_0)$ is maximal. This is the value where $b_0(t)$ meets the equator of p_0, $d(p_0, b_0(t_0)) = \pi/2\sqrt{K_0}$.

Denote by α_0 the angle between $\dot{c}_0(t)$ and $V_0(t)$ and by γ_0 the angle in p_0 between $\dot{c}_0(0)$ and $\dot{b}_0(0)$. From the sinus theorem, cf. (2.7.5,4), we get

$$\sin\sqrt{K_0}\, f(t_0) = \frac{\sin\sqrt{K_0}\, f(t_0)}{\sin\sqrt{K_0}\,\dfrac{\pi}{2\sqrt{K_0}}} = \frac{\sin\gamma_0}{\sin(\pi - \alpha_0)} = \frac{\sin\gamma_0}{\sin\alpha_0} =$$

$$= \frac{\sin \sqrt{K_0}\, d(q_0, r_0')}{\sin \sqrt{K_0}\left(\frac{\pi}{\sqrt{K_0}} - d(p_0, r_0')\right)}$$

$$\leqslant \frac{\sin \sqrt{K_0}\, d(q, r)}{\sin \sqrt{K_0}\, \varepsilon} \leqslant \frac{\sqrt{K_0}\, d(q, r)}{\sin \sqrt{K_0}\, \varepsilon} \leqslant \sin\left(\frac{\pi}{2} \frac{\sqrt{K_0}}{\sqrt{K_1}}\right).$$

Here, the first inequality follows from $\pi/\sqrt{K_0} - d(p_0, r_0') > \varepsilon$ and $0 < \sqrt{K_0}\,\varepsilon < \pi/2$. The second inequality is obvious and the third one is one of our assumptions.

Hence, as soon as we are sure that the arguments of sin on the left hand side and on the right hand side lie in the interval $[0, \pi/2]$ we can conclude

$$f(t) \leqslant f(t_0) \leqslant \pi/2\sqrt{K_1}$$

and thus complete the proof as under 1.

Clearly, $\frac{\pi}{2} \frac{\sqrt{K_0}}{\sqrt{K_1}} \in \left[0, \frac{\pi}{2}\right]$. It thus remains to show $\sqrt{K_0}\, f(t_0) \leqslant \pi/2$. This will follow from $\gamma_0 \leqslant \pi/2$.

Now, γ_0 is also the angle at the antipodal point $\bar{p}_0$ of p_0 in the triangle $\bar{p}_0 q_0 r_0'$. From our assumptions we have

$$d(q_0, r_0') \leqslant \varepsilon;\ d(q_0, \bar{p}_0) \geqslant 2\varepsilon;\ \varepsilon < d(\bar{p}_0, r_0') < \pi/2\sqrt{K_0}.$$

Calling r_0' the north pole of the sphere M_0 of curvature K_0 we see that $\bar{p}_0$ and q_0 both are on the northern hemisphere. If $\gamma_0 > \pi/2$, the points on the half great circle from $\bar{p}_0$ through q_0 to p_0 would have distance $\geqslant d(r_0', \bar{p}_0) > \varepsilon$ from r_0' which contradicts $d(r_0', q_0) \leqslant \varepsilon$. Hence, $\gamma_0 \leqslant \pi/2$, which completes the proof. □

After these preparations we can easily prove Toponogov's Triangle Theorem, see Toponogov [4]. The basic ideas for the proof go back to Alexandrov [1].

Our proof differs in several essential details from Toponogov's proof; we use a manuscript of H. Karcher. See also Gromoll, Klingenberg und Meyer [1], Cheeger and Ebin [1].

2.7.12 Theorem. *Let M be a Riemannian manifold and K_0 a lower bound for the sectional curvature K of M. Consider a generalized triangle pqr on M where the side c_{qr} from q to r is not necessarily minimizing but $L(c_{qr}) \leqslant d(q, p) + d(p, r)$. Assume, moreover, $L(c_{qr}) \leqslant \pi/\sqrt{K_0}$, which is no restriction if $K_0 \leqslant 0$. Then we have the following.*

(i) *$L(pqr) \leqslant 2\pi/\sqrt{K_0}$ (which is non-trivial only for $K_0 > 0$).*

(ii) *If $L(pqr) < 2\pi/\sqrt{K_0}$, then there exists on the simply connected space M_0 of constant curvature K_0 an associated Alexandrov triangle – uniquely determined up to congruence. For the angles at q_0 and r_0 we have $\measuredangle q_0 \leqslant \measuredangle q$; $\measuredangle r_0 \leqslant \measuredangle r$.*

(iii) *If $L(pqr) = 2\pi/\sqrt{K_0}$ (which can occur only for $K_0 > 0$) and all sides are shorter* *than* $\pi/\sqrt{K_0}$ *then* pqr *forms a closed geodesic. If one of the minimizing sides* c_{pq} *or* c_{pr} *has length* $=\pi/\sqrt{K_0}$, *the two remaining sides together form a minimizing geodesic of length* $\pi/\sqrt{K_0}$, *i.e.,* pqr *is a geodesic biangle. In either case, there exist points on* M *having distance* $\pi/\sqrt{K_0}$.

Note. The example of the real projective space of constant curvature $K_0 = 1$ where $d(p,q) = d(p,r) = \pi/2$, $q = r$ and c_{qr} is a closed geodesic of length π shows that (iii) cannot be strengthened.

Proof. We can exclude the trivial case in which the side c_{qr} coincides with the side c_{qp} followed by the side c_{pr}.

1. Assume $L(pqr) < 2\pi/\sqrt{K_0}$. Choose a subdivision $q = m^0, m^1, \ldots, m^N = r$ of c_{qr} so fine that for the triangles $pm^i m^{i+1}$, $0 \leqslant i < N$, the hypotheses made in (2.7.11) are fulfilled. In particular, we take for K_1 an upper bound of K on the compact subset of M formed by the points with distance $\leqslant L(pqr)$ from p. For m^i, m^{i+1} sufficiently close to each other, the arc from m^i to m^{i+1} on c_{qr} will be minimizing.

Note: If $m \in c_{qr}$ and we consider the sides $c_{pq}, c_{pm}, c_{qm} =$ arc of c_{qr} from q to m, we obtain a generalized triangle. Indeed, from

$$d(p,q) + d(p,m) + L(c_{mr}) \geqslant d(p,q) + d(p,r) \geqslant L(c_{qr})$$

we get the desired triangle inequality

$$d(p,q) + d(p,m) \geqslant L(c_{qr}) - L(c_{mr}) = L(c_{qm}).$$

Let $p_0 m_0^i m_0^{i+1}$ be an Alexandrov triangle in M_0, associated with $pm^i m^{i+1}$. We fit together the triangles $p_0 m_0^0 m_0^1 = p_0 q_0 m_0^1$ and $p_0 m_0^1 m_0^2$ along their common side $p_0 m_0^1$ to obtain a planar quadrangle $p_0 q_0 m_0^1 m_0^2$. By planar we mean that it lies in a 2-dimensional subspace of constant curvature M_0.

From (2.7.11) we know that at the corners $m_0^0 = q_0, m_0^1, m_0^2$ this quadrangle is convex, i.e., all angles her are $\leqslant \pi$. The triangle inequality for the possibly generalized triangle pqm^2, with side $c_{qm^2} =$ arc on c_{qr} from q to m^2, tells us that $p_0 q_0 m_0^1 m_0^2$ is convex also at p_0. We therefore obtain from this quadrangle a triangle, denoted by $p_0 q_0 m_0^2$, (although, as we shall see, the point m_0^2 here will in general be different from the point m_0^2 in the quadrangle) by 'bending inwards' the convex corner at m_0^1 and straightening it out. This process obviously makes the angles at q_0 and m_0^2 smaller. $p_0 q_0 m_0^2$ is an Alexandrov triangle associated with pqm^2 with side c_{qm^2}.

Proceeding in this manner we put $p_0 q_0 m_0^2$ and $p_0 m_0^2 m_0^3$ together at the common side $p_0 m_0^2$ to obtain a planar quadrangle $p_0 q_0 m_0^2 m_0^3$. It again is convex, as follows from (2.7.11) for the corners m_0^2 and from the triangle inequality of the possibly generalized triangle pqm^3 with the side $c_{qm^3} =$ arc from q to m^3 on c_{qr}. Bending this quadrangle in at m_0^3 we get a triangle denoted by $p_0 q_0 m_0^3$ which is an Alexandrov triangle associated with pqm^3. Its angles satisfy the inequality

$$\mathop{\measuredangle}\limits_{p_0}^{m_0^3} q_0 \leqslant \mathop{\measuredangle}\limits_{p}^{m^3} q = \measuredangle q.$$

Proceeding in this manner until $m^N = r$ we get (ii) for the angle at q and also for the angle at r.

2. Assume now $L(pqr) = 2\pi/\sqrt{K_0}$. If one of the minimizing sides has length $= \pi/\sqrt{K_0}$, say, $L(c_{pq}) = \pi/\sqrt{K_0}$, then

$$L(c_{pr}) + L(c_{rq}) = \pi/\sqrt{K_0} = d(p, q),$$

i.e., c_{pr}, followed by c_{rq}, is a minimizing geodesic from p to q; pqr is a geodesic biangle.

If $L(pqr) = 2\pi/\sqrt{K_0}$ and all sides are shorter than $\pi/\sqrt{K_0}$, we consider the point $m \in c_{qr}$ opposite p. That is, if we denote by c_{qm} and c_{rm} the arcs on c_{qr} from q to m and r to m, respectively,

$$d(p, q) + L(c_{qm}) = d(p, r) + L(c_{rm}) = \pi/\sqrt{K_0}.$$

We then get two generalized triangles pqm and prm — for the triangle inequality, see the note in part 1 of the proof.

If $d(p, m) = \pi/\sqrt{K_0}$ we just saw that $\sphericalangle q = \sphericalangle r = \pi$. Thus, we may extend in pqr the side from p to q beyond q towards q' near m and the side from p to r beyond r towards r' near m such as to obtain a triangle $pq'r'$ where all sides are shortest geodesics between its end points and all sides have length $< \pi/\sqrt{K_0}$, while $L(pq'r') = 2\pi/\sqrt{K_0}$. Consider now the point m' on $c_{pq'}$, opposite r'. Let $c_{r'm'}$ be a minimizing geodesic from r' to m'. We thus obtain two non-generalized triangles $pr'm'$ and $qr'm'$. If $d(r', m') = \pi/\sqrt{K_0}$ we can conclude $\sphericalangle p = \pi$. That is, pqr is a closed geodesic.

It thus remains to show that $d(p, m) < \pi/\sqrt{K_0}$ cannot occur. The same argument will show that $d(r', m') < \pi/\sqrt{K_0}$ also cannot occur. To see this we apply (ii) to the generalized triangles pqm and prm, where the side from q to m is c_{qm} and the side from r to m is c_{rm}. Associated Alexandrov triangles $p_0 q_0 m_0$ and $p_0 r_0 m_0$ can be fitted together to yield a planar quadrangle $p_0 q_0 m_0 q_0$ of length $2\pi/\sqrt{K_0}$ = length of a great circle on M_0.

From (ii) we know that $p_0 q_0 m_0 r_0$ is convex at q_0, m_0, r_0. Denote by $\bar{p}_0$ the antipodal point of p_0 on M_0. Then we have the three distinguished half circles $c(q_0)$, $c(m_0)$, $c(r_0)$ on the sphere M_0 from p_0 to $\bar{p}_0$ which run through q_0, m_0, r_0, respectively. Since $d(q_0, \bar{p}_0) = \pi/\sqrt{K_0} - d(q_0, p_0) = d(q_0, m_0)$, $m_0 q_0 \bar{p}_0$ is an isoscles triangle. That is, the angles at $\bar{p}_0$ and m_0 are the same. Our constructions show that this can happen only if these angles both are $= \pi/2$ and $d(q_0, \bar{p}_0) = d(q_0, m_0) = \pi/2\sqrt{K_0}$. Similarly, we get $d(r_0, \bar{p}_0) = d(r_0, m_0) = \pi/2\sqrt{K_0}$. Thus, $L(c_{qr}) \geq d(q_0, r_0) = \pi/\sqrt{K_0}$ – a contradiction of our assumption that all sides in pqr are shorter than $\pi/\sqrt{K_0}$.

3. It remains to derive a contradiction from $L(pqr) > 2\pi/\sqrt{K_0}$. Since $L(c_{qr}) \leq \pi/\sqrt{K_0}$ this implies

$$d(r,p) + d(p,q) > \pi/\sqrt{K_0}.$$

We can choose p' on c_{rp} and q' on c_{rq} such that $p'q'r$ with $c_{q'r}$ = restriction of c_{qr} to the arc from q' to r is a (possibly degenerate) triangle of length $L(p'q'r) = 2\pi/\sqrt{K_0}$. We have the freedom to make sure that all sides are shorter than $\pi/\sqrt{K_0}$. Thus, according to (iii), $p'q'r$ is a closed geodesic. In particular, $\sphericalangle r = \pi$. Similarly, $\sphericalangle q = \pi$. But then, p' must belong to the extension of c_{rq} beyond q which means that pqr is a closed geodesic which coincides with the closed geodesic $p'q'r$ of length $2\pi/\sqrt{K_0}$ – a contradiction. □

2.8 The Sphere Theorem

In this section we will investigate the problem of to what extent restrictions on the range of the sectional curvature K of a compact Riemannian manifold M yield information on the topology (or even geometry) of M. We will always assume K to be strictly positive – for the case that K is negative we already proved the Theorem of v. Mangoldt-Hadamard-Cartan (2.6.6).

The restriction on K will therefore be of the form

(*) $\quad 0 < K_0 \leqslant K \leqslant K_1.$

By a change of scale, i. e., by multiplying the Riemannian metric g of M with K_1, we may assume $K \leqslant 1$. Thus, in the hypothesis (*), only the ratio $\delta = K_0/K_1$ plays a rôle for questions in which the scale does not matter. Manifolds M satisfying (*) are also called *δ-pinched,* with $\delta = K_0/K_1$.

We begin with Toponogov's Rigidity Theorem (2.8.1) which states that a manifold for which the diameter assumes the upper bound given by Bonnet-Myers' Theorem (2.6.3) is isometric to the sphere. As a consequence we get the characterization of the sphere as the only simply connected manifold of constant positive curvature, cf. (2.8.2).

In (2.8.4) we show that if the diameter of M is $>$ (1/2) times the upper bound given by Bonnet-Myers, then it has the homotopy type of the sphere. (2.8.5) is the Sphere Theorem stating that a simply connected manifold with δ-pinched curvature, $\delta > 1/4$, is homeomorphic to a sphere.

We add a rigidity theorem, (2.8.7), due to Berger, which states that a (1/4)-pinched simply connected manifold with minimal diameter is isometric to one of the symmetric spaces of rank 1. If the diameter of such a manifold is bigger than the possible minimum then it is homeomorphic to the sphere again, cf. (2.8.10).

We start with a Rigidity Theorem, due to Toponogov [4].

2.8.1 Theorem. *Let M be a compact Riemannian manifold for which the sectional curvature K satisfies $0 < K_0 \leqslant K$. According to (2.6.5), the diameter $d(M)$ of M is*

$\leqslant \pi/\sqrt{K_0}$. *If now* $d(M) = \pi/\sqrt{K_0}$ *then* M *is isometric to the sphere* $S^n_{1/\sqrt{K_0}}$ *of constant curvature* $K = K_0$.

The same conclusions hold if there exists on M *a geodesic triangle in the sense of* (2.7.4) *of length* $2\pi/\sqrt{K_0}$.

Proof. Choose points p, q on M of maximal distance $d(p, q) = d(M) = \pi/\sqrt{K_0}$. Let $r \in M, r \neq p, r \neq q$. Let pqr be a triangle having the corners p, q, r. $L(pqr) \leqslant 2\pi/\sqrt{K_0}$, see (2.7.12). Hence

$$\pi/\sqrt{K_0} = d(p, q) \leqslant d(p, r) + d(r, q) = L(pqr) - d(p, q) \leqslant \pi/\sqrt{K_0}.$$

From (2.7.12) we therefore know that the sides c_{pr} and c_{rq} of pqr together form a minimizing geodesic from p to q. Every half-geodesic starting from p meets q at the distance $\pi/\sqrt{K_0}$. Therefore, every such half-geodesic is minimizing until distance $\pi/\sqrt{K_0}$; the cut locus $C(p)$ of p consists of q only.

Let $c: I = [0, 1] \to M$ be a minimizing geodesic from p to q. We can view c as a critical point of $E|\Omega, \Omega = \Omega_{pq} M$. Choose a parallel vector field $V(t)$ along $c(t), |V(t)| = 1, \langle \dot{c}(t), V(t) \rangle = 0$. Put

$$\xi(t) = \sin \pi t \, V(t).$$

Then ξ is an element of $T_c \Omega$. Since c has no conjugate points in its interior, $D^2 E(c)$ is positive semi-definite. Using $-K \leqslant -K_0$ and $|\dot{c}|^2 = \pi^2/K_0$ we find

$$0 \leqslant D^2 E(c)(\xi, \xi) = \int_0^1 (\pi^2 \cos^2 \pi t - K(t) \pi^2 \sin^2 \pi t / K_0) \, dt$$

$$\leqslant \pi^2 \int_0^1 (\cos^2 \pi t - \sin^2 \pi t) \, dt = 0.$$

Hence, the sectional curvature $K(t)$ of the 2-plane spanned by $\{\dot{c}(t), V(t)\}$ is $= K_0$. $\xi(t)$ actually is a Jacobi field, and it is identical with a Jacobi field on the sphere $M_0 = S^n_{1/\sqrt{K_0}}$ of constant curvature $K_0, n = \dim M$.

Choose on M_0 a pair (p_0, q_0) of antipodal points. Pick an isometry $\iota: T_p M \to T_{p_0} M_0$. Then

$$\phi = \exp_{p_0} \circ \iota \circ (\exp_p | B_{\pi/\sqrt{K_0}}(0_p))^{-1}: M - \{q\} \to M_0 - \{q_0\}$$

is an isometry. This follows from the fact that Jacobi fields along half-geodesics emanating from p are carried into Jacobi fields along half-geodesics emanating from p_0 in such a way that scalar products remain preserved. Cf. also Cartan's Theorem (1.12.8).

ϕ can be extended to a diffeomorphism $\bar{\phi}: M \to M_0$ by mapping q into q_0. $\bar{\phi}$ even is an isometry at q. To see this we consider a small ball $\bar{B}_\varrho(q)$ around q. Let $r, \bar{r}$ be diametral points on $\partial \bar{B}_\varrho(q)$, i. e., points which are joined by a minimizing geodesic passing through q. The image of this geodesic under $\bar{\phi}$ must be minimizing, since ϕ was an isometry. The image also passes through q_0, hence, it is a geodesic.

If there exists on M a geodesic triangle of length $2\pi/\sqrt{K_0}$ then we have from (2.7.12) that $d(M) \geqslant \pi/\sqrt{K_0}$. □

With (1.7.1) and the estimate on the injectivity radius from (2.6.10) it is now easy to prove Hopf's Theorem on the uniqueness of a simply connected space of constant positive curvature, cf. H. Hopf [1].

We wish to add that one can also give a more elementary proof which uses not much more than the monodromy principle, together with the fact that a space of constant curvature locally is isometric to the sphere, cf. Kobayashi and Nomizu [1].

2.8.2 Theorem. *Let M be a simply connected compact manifold of constant curvature $K_0 > 0$. Then M is isometric to the sphere $M_0 = S^n_{1/\sqrt{K_0}}$.*

Proof. This will follow from (2.8.1) once we show that $d(M) = \pi/\sqrt{K_0}$. But from (2.6.10) we know that $\pi/\sqrt{K_0} \leqslant d(M)$ while (2.6.3) tells us $d(M) \leqslant \pi/\sqrt{K_0}$. □

We continue with a technical result.

2.8.3 Proposition. *Choose $p \in M$ be such that the distance function from $p, r \in M \mapsto d(p, r) \in \mathbb{R}$, has a local maximum at q. If X is a vector $\neq 0$ in $T_q M$, then there exists a minimizing geodesic $c = c_{pq}: I \to M$ from p to q with $\langle \dot{c}(1), X\rangle \leqslant 0$, i.e., the angle between X and $-\dot{c}(1)$ is $\leqslant \pi/2$.*

Proof. Consider the geodesic $c_X(s) = \exp_q sX$. For every small $|s|$ we choose a minimizing geodesic $c_s: I \to M$ from p to $c_X(s)$. Let $\{s_m\}$ be a sequence of positive numbers with limit 0. By going, if necessary, to a subsequence which we will again denote by $\{s_m\}$, we may assume that the c_{s_m} have a limit geodesic $c: I \to M$ from p to q. Simply observe that the $\dot{c}_{s_m}(0)$ are in a compact subset of $T_p M$.

For large m, we can write $c_{s_m}(t) = \exp_{c(t)} \xi_m(t)$, where $\xi_m(t)$ is a vector field along $c(t)$ with $\xi_m(0) = 0$, $\xi_m(1) = s_m X$. Since

$$2E(c_m) = L(c_{s_m})^2 = d(p, c_X(s_m))^2 \leqslant d(p, q)^2 = L(c)^2 = 2E(c),$$

we get from (2.3.20), or from the proof of (1.12.12), that if m is sufficiently large then

$$0 \geqslant DE(c) \,.\, \xi_m = \int_0^1 \langle \dot{c}(t), \nabla \xi_m(t)\rangle\, dt =$$

$$\int_0^1 \frac{d}{dt} \langle \dot{c}(t), \xi_m(t)\rangle\, dt = \langle \dot{c}(1), \xi_m(1)\rangle = s_m \langle \dot{c}(1), X\rangle. \quad □$$

If in (2.8.1) the diameter $d(M)$ of M is allowed to vary in the interval $]\pi/2\sqrt{K_0}, \pi/\sqrt{K_0}]$, M will not necessarily be isometric to a sphere of constant curvature, of course. But M still is homeomorphic to the sphere. This was proved by Grove and Shiohama [1].

Here we only show that under the above hypothesis M has vanishing homotopy groups in dimensions 1 to $n-1, n = \dim M$. As we remarked after (2.6.11), M is then actually homeomorphic to S^n, except possibly for $n = 3$ or 4.

2.8.4 Theorem. *Let* M *be compact,* $\dim M = n$. *Assume* $0 < K_0 \leqslant K$. *If* $\pi/2\sqrt{K_0} < d(M) \leqslant \pi/\sqrt{K_0}$, *then there exists a* $q \in M$ *such that every non-trivial geodesic* c *from* q *to* q *has length* $> \pi/\sqrt{K_0}$. *Hence, according to* (2.6.2), *index* $c \geqslant n-1 \geqslant 1$.

It follows that $\Omega = \Omega_{qq} M$ *is connected and* $\pi_k \Omega = \pi_{k+1} M = 0$ *for* $0 < k < n-1$.

Proof. We can assume $d(M) < \pi/\sqrt{K_0}$, see (2.8.1). Choose p and q on M with maximal distance $d(p, q) = d(M)$. We will derive a contradiction from the existence of a geodesic c from q to q of length $0 < L(c) \leqslant \pi/\sqrt{K_0}$.

For such a c we consider the generalized triangle pqr with $q = r$ where $c_{qr} = c$. According to (2.8.3) we can choose the side $c_{pq} = c_{pr}$ such that $\measuredangle q \leqslant \pi/2$. The triangle inequality is valid since

$$d(p, q) + d(p, r) = 2d(p, q) > \pi/\sqrt{K_0} \geqslant L(c_{qr}).$$

We claim $L(pqr) < 2\pi/\sqrt{K_0}$. Otherwise, according to (2.7.12), $L(pqr) = 2\pi/\sqrt{K_0}$. Since pqr is not a closed geodesic, $L(c_{qr}) = \pi/\sqrt{K_0}$, but then, $2d(M) = L(c_{pq}) + L(c_{pr}) = \pi/\sqrt{K_0}$ which was excluded.

Let $p_0 q_0 r_0$ be an Alexandrov triangle associated with pqr on the sphere M_0 of constant curvature K_0. $p_0 q_0 r_0$ is an isosceles triangle, with q_0, r_0 having distance $> \pi/2\sqrt{K_0}$ from p_0. But for such a triangle $\measuredangle q_0 = \measuredangle r_0 > \pi/2$ while, according to (2.7.12), $1 \measuredangle q_0 \leqslant \measuredangle q \leqslant \pi/2$.

Assume Ω were not connected, i. e., $\pi_1 M \neq 0$. Then there exists a not-null homotopic geodesic of index 0 in $\Omega = \Omega_{qq} M$, see (2.4.19). Thus, $\pi_1 M = 0$. The remainder follows from (2.5.16). □

We now come to the *Sphere Theorem* in Riemannian geometry.

Rauch [1] was the first to study successfully the problem of whether a δ-pinched M with some $\delta < 1$ – while certainly not necessarily isometric to a sphere of constant curvature – at least is homeomorphic to such a sphere.

Subsequent work by the author, cf. Klingenberg [1], pushed down the value of δ until Berger [1] and – independently – Toponogov [1] proved that if M is δ-pinched with $\delta > 1/4$ and of even dimension, then its universal covering $\tilde{M}$ is homeomorphic to the sphere.

The restriction to the case of dim M even at that time was necessary because one had a good lower bound for the injectivity radius only for this case, see (2.6.9). When the author, cf. Klingenberg [3], was able to extend this lower bound to odd-dimensional δ-pinched manifolds, $\delta > 1/4$, cf. (2.6.10), the restriction to manifolds of even dimension was removed and the Sphere Theorem was proved in its full generality.

2.8.5 Theorem. *Let M be a compact, simply connected Riemannian manifold of dimension n. Assume that M is δ-pinched with $\delta > 1/4$, i.e., there exists a K_1 such that*

$$0 < K_1/4 < K \leqslant K_1 .$$

Then M is homeomorphic to the sphere S^n.

Remark. At least for n even, the hypothesis cannot be weakened to 1/4-pinched: The complex projective spaces $P^m\mathbb{C}$ of real dimension $2m > 2$ are 1/4-pinched, and not homeomorphic to S^{2m}, cf. (1.11.14).

Proof. By replacing the metric g on M with the metric $g^* = K_1 g$ we get $K^* \leqslant 1$, cf. the formula for K in (1.11.3). Thus we may assume $K_1 = 1$.

Since M is compact there exists a K_0 with $0 < 1/4 < K_0 \leqslant K \leqslant 1$. Choose p, q on M with maximal distance $d(p, q) = d(M)$. From (2.6.9) and (2.6.10) we have

$$\pi/2\sqrt{K_0} < \pi \leqslant d(M) = d(p, q) \leqslant \pi/\sqrt{K_0} .$$

We define

$$\bar{B}(p) = \{r \in M;\, d(r, p) \leqslant d(r, q)\};\ \bar{B}(q) = \{r \in M;\, d(r, q) \leqslant d(r, p)\},$$

$$E(p, q) = \bar{B}(p) \cap \bar{B}(q).$$

These sets can be defined for every M. In our case we will show that $\bar{B}(p)$ and $\bar{B}(q)$ are homeomorphic to the two hemispheres of a pair of antipodal points p_0, q_0 on S^n and that their intersection is homeomorphic to the *equator* $E(p_0, q_0) \cong S^{n-1}$ of (p_0, q_0).

We claim

$$\bar{B}(p) \subset \bar{B}_{\pi/2\sqrt{K_0}}(p),\ \bar{B}(q) \subset \bar{B}_{\pi/2\sqrt{K_0}}(q).$$

Indeed, assume e. g. $d(q, r) > \pi/2\sqrt{K_0}$. Consider a triangle pqr where according to (2.8.3) we can assume $\measuredangle q \leqslant \pi/2$. If $d(p, q) = \pi/\sqrt{K_0}$, we have from (1.8.1) that M is isometric to the sphere $M_0 = S^n_{1/\sqrt{K_0}}$ and thus $d(p, r) < \pi/2\sqrt{K_0}$. Otherwise, we have directly from (2.7.12) that c_{pr} and c_{rq} together form a minimizing geodesic from p to q of length $\pi/\sqrt{K_0}$, thus, $d(p, r) < \pi/2\sqrt{K_0}$.

It remains to discuss the case $d(M) = d(p, q) < \pi/\sqrt{K_0}$. Since pqr is not a closed geodesic we have from (2.7.12) $L(pqr) < 2\pi/\sqrt{K_0}$. We thus can construct on M_0 an associated Alexandrov triangle $p_0 q_0 r_0$. Here, the sides $c_{q_0 r_0}, c_{q_0 p_0}$ both are longer than $\pi/2\sqrt{K_0}$. I. e., $c_{p_0 r_0}$ has distance $< \pi/2\sqrt{K_0}$ from the antipodal point $\bar{q}_0$ of q_0 on M_0. Since $\measuredangle q_0 \leqslant \measuredangle q \leqslant \pi/2$, $L(c_{p_0 r_0}) = d(p, r)$ must be $< \pi/2\sqrt{K_0}$.

We have thus proved our claim. Since $\pi/2\sqrt{K_0} < \pi$, normal coordinates based at p and at q are valid in the open balls $B_\pi(p)$ and $B_\pi(q)$. $B_\pi(p) \cup B_\pi(q) = M$, i. e., M can be covered by two cell-like coordinate neighborhoods. It now is a standard result in the theory of topological manifolds that such an M must be homeomorphic to S^n.

In our situation we are able to exhibit such a homeomorphism in a particularly simple and geometric manner. This we shall do now.

Consider the functions

$$\Phi_p : r \in B_\pi(p) - \{p\} \mapsto d(p, r) \in \mathbb{R}^+,$$

$$\Phi_q : r \in B_\pi(q) - \{q\} \mapsto d(q, r) \in \mathbb{R}^+.$$

They are differentiable. From what we just saw,

$$E(p, q) = \bar{B}(p) \cap \bar{B}(q) \subset B_\pi(p) \cap B_\pi(q).$$

For $r \in B_\pi(p) \cap B_\pi(q)$, the minimizing geodesics c_{rp} and c_{rq} from r to p and q clearly have different initial direction. If we consider the gradient fields $\operatorname{grad}\Phi_p, \operatorname{grad}\Phi_q$, derived from Φ_p, Φ_q with the Riemannian metric, then $(\operatorname{grad}\Phi_p - \operatorname{grad}\Phi_q)(r) \neq 0$, for $r \in B_\pi(p) \cap B_\pi(q)$.

It follows that the differentiable function

$$\Phi = \Phi_q - \Phi_p : B_\pi(p) \cap B_\pi(q) \to \mathbb{R}$$

has 0 as a regular value. According to (1.3.8), $E = E(p, q) = \Phi^{-1}(0)$ then is a submanifold of codimension 1. $T_r E$ is formed by the $Y \in T_r M$ with

$$\langle \operatorname{grad}\Phi_p, Y \rangle = \langle \operatorname{grad}\Phi_q, Y \rangle$$

To every unit vector $X_0 \in T_p M$ we associate the intersection $r(X_0)$ of the geodesic $c_{X_0}(t), 0 \leqslant t \leqslant \pi$, with E, where $\dot{c}_{X_0}(0) = X_0$. Put $\Phi_p \circ r = \psi$. We define a homeomorphism

$$h_p : (\bar{B}_1(0_p), \partial \bar{B}_1(0_p)) \to (\bar{B}(p), \partial \bar{B}(p))$$

by $h_p(0_p) = p$, and, if $\tilde{p} \neq 0_p$, $h_p(\tilde{p}) = \exp_p\left(\psi\left(\frac{\tilde{p}}{|\tilde{p}|}\right)\tilde{p}\right)$.

In the same way we define a homeomorphism

$$h_q : (\bar{B}_1(0_q), \partial \bar{B}_1(0_q)) \to (\bar{B}(q), \partial \bar{B}(q))$$

Note: $\partial \bar{B}(p) = \partial \bar{B}(q) = E(p, q)$.

The conclusion of the proof consists now of an application of the following Lemma. □

2.8.6 Lemma. *Let M be a topological manifold of dimension n. Denote by $\bar{B}^n$ the closed unit ball in $\mathbb{R}^n$. Let*

$$h_p, h_q : (\bar{B}^n, \partial \bar{B}^n) \to (M, E)$$

be homeomorphisms such that

$$h_p(\bar{B}^n) \cup h_q(\bar{B}^n) = M,$$

$$h_p(\bar{B}^n) \cap h_q(\bar{B}^n) = h_p(S^{n-1}) = h_q(S^{n-1}) = E.$$

Then M is homeomorphic to S^n.

Proof. Consider on $S^n = \{\sum_0^n x_i^2 = 1\} \subset \mathbb{R}^{n+1}$ the two half spheres $HS_+^n = S^n \cap \{x_0 \leqslant 0\}$, $HS_-^n = S^n \cap \{x_0 \geqslant 0\}$. $HS_+^n \cap HS_-^n$ is the sphere S^{n-1} in the hyperplane $\{x_0 = 0\}$. We take for $\bar{B}^n$ the ball $\{\sum_1^n x_1^2 \leqslant 1\}$ in $\{x_0 = 0\}$. Thus, $\partial \bar{B}^n = S^{n-1}$.

In (1.1.3, ii) we defined the stereographic projections $u_\pm$. The mappings

$$j_+ = u_+ | HS_+^n : HS_+^n \to \bar{B}^n ; j_- = u_- | HS_-^n : HS_-^n \to \bar{B}^n$$

then are homeomorphisms with $j_+ | S^{n-1} = j_- | S^{n-1} = id | S^{n-1}$.

Under our assumptions,

$$f = h_q^{-1} \circ (h_p | S^{n-1}) : S^{n-1} \to S^{n-1}$$

is a homeomorphism. This allows us to define the radial extension

$$F : \bar{B}^n \to \bar{B}^n$$

by $F(0) = 0$, $F(x) = |x| f(x/|x|)$, for $x \neq 0$.

We now get a homeomorphism $\phi : S^n \to M$ by $\phi | HS_+^n = h_p \circ j_+ ; \phi | HS_-^n = h_q \circ F \circ j_-$. Indeed, since on $S^{n-1} = HS_+^n \cap HS_-^n$ we have $F = h_q^{-1} \circ h_p$, ϕ is well defined, bijective and continuous. But then, since M and S^n are compact, ϕ^{-1} is also continuous. □

The sphere Theorem (2.8.5) can be complemented by two statements concerning the case of 1/4-pinched manifolds. They both follow from Berger [2] if one uses Klingenberg [4], cf. (2.6.A.1).

We first prove a counterpart of (2.8.1), i. e., a rigidity theorem for manifolds with minimal diameter.

2.8.7 Theorem. *Let M be a compact Riemannian manifold, simply connected such that $0 < K_1/4 \leqslant K \leqslant K_1$. From* (2.6.A.1) *we know that the diameter of M is $\geqslant \pi/\sqrt{K_1}$. If in this case $d(M) = \pi/\sqrt{K_1}$, then M is isometric either to the sphere of constant curvature $K = K_1$ or to one of the projective spaces over the complex numbers, the quaternions or the Cayley numbers.*

Remark. We have defined only the projective space over the complex numbers. For the other projective spaces, cf. Kobayashi and Nomizu [1] or Wolf [1]. In our proof we will show that under the hypotheses of (2.8.7), either M is the sphere of constant curvature K_1, or it is locally symmetric and the curvature operator possesses just two eigenvalues, K_1 and $K_1/4$. That M must then be of the type just mentioned follows from the classification of symmetric spaces l. c..

Proof. We can assume $K_1 = 1$. Given $p \in M$ and a unit vector X_0 in T_pM, the geodesic $c(t) = \exp_p tX_0$ yields, for $t = \pi$, a point q with $d(p, q) = d(M) = \pi$.

By $\mathscr{J}_1^\perp$ we denote the space of the Jacobi fields $Y(t)$ along $c(t)$ with $Y(0) = Y(\pi) = 0$, $\langle Y(t), \dot c(t)\rangle = 0$. We claim that $Y(t) = \sin t A(t)$ with $\nabla A(t) = 0$. Indeed, compare c with a geodesic c^* of length π on the unit sphere M^* of curvature $K^* = 1$. Index $c^* = 0$. From (2.6.1) we then have that index $c = 0$ and hence our claim.

We now consider the space $\mathscr{J}_{1/4}^\perp$ formed by the Jacobi fields $Y(t)$ along $c(t)$ satisfying $Y(0) = 0$, $\langle Y(t), \dot c(t)\rangle = 0$ and $Y(\pi) \neq 0$.

In our notation we anticipate already that eventually we will show that the $Y \in \mathscr{J}_{1/4}^\perp$ are of the form $\sin\frac{t}{2} B(t)$ with $\nabla B(t) = 0$.

We can assume that $\mathscr{J}_1^\perp$ is not all of $\mathscr{J}_c^{0\perp}$. Otherwise, M is isometric to the sphere of constant curvature $K_1 = 1$.

Now fix a $Y \in \mathscr{J}_{1/4}^\perp$, $Y \neq 0$. Put $Y(\pi) = B \in T_q M$. $B \neq 0$. Consider the geodesic $\tilde c(t) = t\dot c(0)$ on $T_p M$. We can write $Y(t)$ in the form $Y(t) = T_{\tilde c(t)} \exp_p . (t \nabla Y(0))$, cf. (1.12.5). Here, $t \nabla Y(0)$ is considered as an element of $T_{\tilde c(t)} T_p M$. Put $\pi \nabla Y(0) = \tilde B \in T_{\tilde c(\pi)} T_p M$. Let $\tilde h(s)$ be the circle of radius π in $T_p M$ with origin 0_p and $\tilde h'(0) = \tilde B$. Then $h(s) = \exp_p \tilde h(s)$ is a curve in M with distance π from p.

For every small $\varepsilon > 0$ let $k_\varepsilon(s)$ be the uniquely determined minimizing geodesic with $|k_\varepsilon'(0)| = 1$, $k_\varepsilon(s_\varepsilon) = h(\varepsilon)$, $k_\varepsilon(0) = q$, $d(q, h(\varepsilon)) = s_\varepsilon$. We show that all points $k_\varepsilon(s)$ have distance π from p. Indeed, otherwise there is a $s_0, |s_0| < s_\varepsilon$, where $d(p, k_\varepsilon(s))$, $|s| \leqslant s_\varepsilon$, assumes its minimum $< \pi$. The triangles $pk_\varepsilon(s_\varepsilon)k_\varepsilon(s_0)$ and $pqk_\varepsilon(s_\varepsilon)$ have at $k_\varepsilon(s_0)$ the angle $\pi/2$. The associated Alexandrov triangles on the sphere M^* of constant curvature $K^* = 1/4$ must, at the corresponding points, have an angle $\leqslant \pi/2$, cf. (2.7.12). But this clearly is impossible for the angle at r^* of a triangle $p^*q^*r^*$ on M^* with $d(p^*, q^*) = \pi$, $d(p^*, r^*) < \pi$.

As $\varepsilon \to 0$, the tangent vector $k_\varepsilon'(0)$ goes towards $h'(0) = B = Y(\pi)$. We therefore put $k_\varepsilon'(0) = B_\varepsilon$ and $B_0 = B$.

Denote by $B_\varepsilon(t)$ the parallel field along $c(t)$ generated by B_ε and consider the curve $b_\varepsilon(t) = \exp_{c(t)} s_\varepsilon \sin\frac{t}{2} B_\varepsilon(t)$, $0 \leqslant t \leqslant \pi$. $L(b_\varepsilon) \geqslant \pi$, since $b_\varepsilon(0) = p$, $b_\varepsilon(\pi) = h(\varepsilon)$. Similarly, define on the sphere M^* of constant curvature $K^* = 1/4$ the curve $b_\varepsilon^*(t) = \exp_{c^*(t)} s_\varepsilon \sin\frac{t}{2} B_\varepsilon^*(t)$, $0 \leqslant t \leqslant \pi$. This is a geodesic of length π. From (2.7.10) we have $L(b_\varepsilon) \leqslant L(b_\varepsilon^*)$, thus, $L(b_\varepsilon) = L(b_\varepsilon^*)$. But this only can hold if $Y_\varepsilon(t) = \sin\frac{t}{2} B_\varepsilon(t)$ is a Jacobi field. Since $\lim\limits_{\varepsilon \to 0} B_\varepsilon = B$, $Y(t) = \sin\frac{t}{2} B(t)$, with $B(t) \equiv B_0(t)$.

In particular, the sectional curvature of the 2-plane spanned by $\dot c(t)$ and $B(t)$ is 1/4. Thus, the elements in $\mathscr{J}_1^\perp$ and $\mathscr{J}_{1/4}^\perp$ correspond to the eigenvectors of the operator $R_{\dot c(t)}$ for the eigenvalue 1 and 1/4, respectively. This shows that $\mathscr{J}_1^\perp$ and $\mathscr{J}_{1/4}^\perp$ are orthogonal. In addition, this splitting $\mathscr{J}_c^{0\perp} = \mathscr{J}_1^\perp \perp \mathscr{J}_{1/4}^\perp$ extends to all values $t \in \mathbb{R}$.

We can now conclude from Cartan's Theorem (1.12.8, ii) that the reflection $\sigma_p: B_\pi(p) \to B_\pi(p)$ at the origin of p is an isometry. Recall from (2.2.1) that σ_p

$= \exp_p \circ \iota_p \circ (\exp_p | B_\pi(p))^{-1}$, where $\iota_p : T_p M \to T_p M$ is the reflection on 0_p. It follows that, for every geodesic $c(t), |t| < \pi$, with $c(0) = p, |\dot{c}(0)| = 1$, the mapping $\iota_t : T_{c(t)} M \to T_{c(-t)} M$, constructed from ι_p with the help of parallel translations, carries a parallel vector field $A(t)$ along $c(t)$ into the field $-A(-t)$ along $c(-t)$. Finally, the curvature operator $R_{\dot{c}(t)}$ possesses the $Y_1(t) \in \mathcal{J}_1^{\perp}$ and the $Y_{1/4}(t) \in \mathcal{J}_{1/4}^{\perp}$ as eigenvectors with eigenvalue 1 and 1/4, respectively. Thus, (†) in (1.12.8) is satisfied.

The proof of (2.8.7) can now be completed as follows: A simply connected locally symmetric space is symmetric. Apply the classification of symmetric spaces with just the two positive eigenvalues $\lambda, \lambda/4$ for the curvature operator. □

For the discussion of a (1/4)-pinched manifold M with diameter $> \pi\sqrt{\max K}$ we need a number of auxiliary results, cf. Berger [2].

2.8.8 Proposition. *Consider a $p \in M$ and linearly independent unit vectors $\{X, Y_0, Y_1\}$ in $T_p M$ with $\langle X, Y_0 \rangle = \langle X, Y_1 \rangle = 0$. Assume that for all 2-planes σ in $T_p M$ the sectional curvature $K(\sigma)$ satisfies $0 < K_0 \leqslant K(\sigma) \leqslant 1, K_0 < 1$. Denote by σ_ϕ the 2-plane spanned by $Z_i = \cos\phi\, X + \sin\phi\, Y_i, i = 0,1, 0 < \phi \leqslant \pi/2$. Assume that $K(\sigma_i) = K_0$, where σ_i is the plane spanned by $\{X, Y_i\}, i = 0,1$. If now $\phi < \pi/2$, then $K(\sigma_\phi) < 1$.*

Proof. Consider the self-adjoint operator $R_X : Y \mapsto R(Y, X)X$.

The smallest eigenvalue of R_X is $\geqslant K_0$. Thus, we have from our assumptions that $Y_i, i = 0, 1$, are eigenvectors of R_X with the eigenvalue K_0. Similarly, X is an eigenvector of R_{Y_i} with the eigenvalue K_0. In particular, $\langle R_{Y_i} X, Y_j \rangle = 0$ for $i \neq j$.

Since every vector in the plane spanned by $\{Y_0, Y_1\}$ is an eigenvector of R_X with eigenvalue K_0, we can describe the plane σ_ϕ also as the span of two vectors $Y_i' = \cos\phi' X + \sin\phi' Y_i'$ with $\langle Y_0', Y_1' \rangle = 0$. Therefore, it suffices to consider the case $\langle Y_0, Y_1 \rangle = 0$.

With $K(\sigma_{\pi/2}) = K_1$ and $|Z_0|^2 |Z_1|^2 - \langle Z_0, Z_1 \rangle^2 = (1 + \cos^2\phi)\sin^2\phi$ we get

$$
\begin{aligned}
&(1 + \cos^2\phi)\sin^2\phi\, K(\sigma_\phi) = \\
&\langle R(\cos\phi\, X + \sin\phi\, Y_0, \cos\phi\, X + \sin\phi\, Y_1)(\cos\phi\, X + \sin\phi\, Y_1), (\cos\phi\, X + \sin\phi\, Y_0)\rangle \\
&= \sin^2\phi \cos^2\phi\, K_0 + \\
&+ 2\sin\phi\cos\phi \langle R(Y_0, \cos\phi\, X + \sin\phi\, Y_1)(\cos\phi\, X + \sin\phi\, Y_1), X\rangle \\
&+ \sin^2\phi \langle R(Y_0, \cos\phi\, X + \sin\phi\, Y_1)(\cos\phi\, X + \sin\phi\, Y_1), Y_0\rangle \\
&= 2\sin^2\phi\cos^2\phi\, K_0 + \sin^4\phi\, K_1 \leqslant \sin^2\phi\,(2\cos^2\phi\, K_0 + \sin^2\phi) \\
&< \sin^2\phi\,(1 + \cos^2\phi), \text{ since } K_0 < 1. \quad \square
\end{aligned}
$$

2.8.9 Proposition. *Let p, q be in M such that $d(p, q) = d(M) =$ diameter of M. Fix a unit vector $X \in T_p M$. Assume that there is no minimizing geodesic $c(t), 0 \leqslant t \leqslant d$*

$= d(p, q)$ *from* p *to* q *such that* $\langle \dot{c}(0), X\rangle > 0$. *Then there exist at least two minimizing geodesics* c_0, c_1 *from* p *to* q *with* $\langle \dot{c}_0(0), X\rangle = \langle \dot{c}_1(0), X\rangle = 0$.

Proof. From (2.8.3) we know that there must be at least one such geodesic, say c_1. Consider the vector $X - \varepsilon \dot{c}_1(0), \varepsilon \geqslant 0$. There exists a minimizing geodesic c_ε from p to q with $\langle \dot{c}_\varepsilon(0), X - \varepsilon \dot{c}_1(0)\rangle \geqslant 0$. Hence, since $\langle \dot{c}_\varepsilon(0), X\rangle \leqslant 0, \langle \dot{c}_\varepsilon(0), \dot{c}_1(0)\rangle \leqslant 0$. Let c_0 be a limit geodesic for the family c_ε, as $\varepsilon \to 0$. Then $\langle \dot{c}_0(0), \dot{c}_1(0)\rangle \leqslant 0$. Thus, $c_0 \neq c_1$ and $\langle \dot{c}_0(0), X\rangle = 0$. □

2.8.10 Theorem. *Let* M *be a compact, simply connected Riemannian manifold and assume that the sectional curvature* K *of* M *satisfies* $0 < K_1/4 \leqslant K \leqslant K_1$. *If the diameter* $d(M)$ *of* M *is* $> \pi/\sqrt{K_1}$ *then* M *is homeomorphic to the sphere.*

Proof. We can restrict ourselves to the case $K_1 = 1$. According to (2.8.1) we can assume that $d(M) < 2\pi$. Let p, q be points of maximal distance $d(p, q) = d(M)$. $\exp_r | B_\pi(0_r)$ is injective, for every $r \in M$, cf. (2.6.A.1). Hence, if $M = B_\pi(p) \cup B_\pi(q)$ we can proceed as in the proof of the sphere theorem (2.8.8).

We will derive a contradiction from the assumption that there is a $r \in M$ with $d(p, r) \geqslant \pi, d(q, r) \geqslant \pi$. We may assume just as well that $d(p, r) = d(q, r) = \pi$. Let $c_{pr}(t), 0 \leqslant t \leqslant \pi$, be a minimizing geodesic from p to r. According to (2.8.3) there exists a minimizing geodesic $c_{pq}(t), 0 \leqslant t \leqslant d = d(M)$, from p to q with $\langle \dot{c}_{pr}(0), \dot{c}_{pq}(0)\rangle \geqslant 0$.

We consider the triangle pqr with c_{pq}, c_{pr} as two of its sides and an associated Alexandrov triangle $p^* q^* r^*$ on the sphere M^* of constant curvature $K^* = 1/4$.

The points p^*, q^*, having distance π from r^*, lie on the equator of r^*. Hence $\pi/2 = \measuredangle p^* \leqslant \measuredangle p \leqslant \pi/2$, i. e., $\measuredangle p = \pi/2$. Moreover, $\pi/2 < \measuredangle r^* < \pi$.

We claim that the minimizing geodesics c_t from r to the points $c_{pq}(t), 0 < t < d$, on c_{pq} all have length π and meet c_{pq} orthogonally.

To see this we consider a minimizing geodesic c_t from r to $c_{pq}(t)$ and the minimizing geodesic c_t^* of length π from r^* to $c_{p^*q^*}(t)$. For small $t > 0$, we get from (2.7.10) (using the description of c_t^* by Fermi coordinates along $c_{r^*p^*}$)

$$L(c_t) = d(r, c_{pq}(t)) \leqslant L(c_t^*) = \pi.$$

If $L(c_t) < \pi$, a glance at the Alexandrov triangles $p^* c_{pq}(t)^* r^*, q^* c_{pq}(t)^* r^*$ associated with $pc_{pq}(t)r, qc_{pq}(t)r$ tells us that the angles at $c_{pq}(t)$ must be $> \pi/2$; but they also are $\leqslant$ the angles at $c_{pq}(t)$ in $pc_{qp}(t)r, qc_{pq}(t)r$ which together add up to π. Thus we prove our claim.

The occurence of the equality sign in all steps of the proof of (2.7.10) implies even more, i. e., the sectional curvature of the 2-plane σ_0 in $T_r M$, spanned by $\dot{c}_{rp}(0)$ and the parallel translate $(\overset{\pi}{\underset{0}{\|}} c_{pr})\dot{c}_{pq}(0)$ of $\dot{c}_{pq}(0)$, is 1/4. In addition, we have a minimizing geodesic c_{rq} from r to q of length π with $\dot{c}_{rq}(0)$ in σ_0.

Our proof also shows that every minimizing geodesic from q to p at p is orthogonal to the initial direction $\dot{c}_{pr}(0)$ of a fixed c_{pr}. Therefore, (2.8.9) applies and we have, besides c_{pq}, a second minimizing geodesic c'_{pq} from p to q. $(\overset{\pi}{\underset{0}{\|}} c_{pr})\dot{c}'_{pq}(0)$ does not belong to the plane σ_0 in $T_r M$. Hence, there must also be a second minimizing geodesic c'_{rq} from r to q. Put $\dot{c}_{rp}(0) = X$, $\dot{c}_{qr}(\pi) = Z_0$, $\dot{c}'_{qr}(\pi) = Z_1$. Then $0 < \langle X, Z_0 \rangle = \langle X, Z_1 \rangle < 1$ since the angle between X and $-Z_i$ is the same as the angle at r^* in the Alexandrov triangle $p^* q^* r^*$, where $\pi < d(p^*, q^*) < 2\pi$. The curvature of the 2-plane σ_i spanned by $\{X, Z_i\}$ is 1/4.

We therefore have from (2.8.8) that the curvature of the 2-plane $\tilde{\sigma}$, spanned by $\{Z_0, Z_1\}$ is < 1. But this now is a contradiction. To see this we consider the triangle formed by $m = c_{qr}(\pi/2)$ and q and r where the side from q to r shall be given by c'_{qr}. Since Z_0, Z_1 are linearly independent, the side c'_{qr} belongs to $B_\pi(m)$. We thus can apply (2.7.6) with M_1 = sphere of curvature $K_1 = 1$. It follows that in an associated Alexandrov triangle $m_1 q_1 r_1$ the angle at m_1 is $= \pi =$ angle at m. Thus, the sectional curvature of the plane spanned by the tangent vectors Z_0, Z_1 is $= 1$. □

Remarks. 1. The Sphere Theorem (2.8.5) only determines the structure of the underlying topological manifold M. Since it is known that there exist non-equivalent differentiable structures on a sphere S^n, at least for n sufficiently large, one may ask whether the homeomorphism $S^n \to M$ cannot actually be chosen to be a diffeomorphism.

So far, this has been proved only under stronger pinching hypotheses. The best version of the so-called *Differentiable Sphere Theorem* reads: A simply connected 0.76-pinched n-dimensional manifold M is diffeomorphic to S^n with the standard differentiable structure, cf. Grove, Karcher and Ruh [1].

2. There is no example known of a Riemannian Metric on an exotic sphere (i. e., a sphere with a non-standard differentiable structure) with strictly positive curvature. It remains an open problem whether the pinching in the Differential Sphere Theorem can be improved to $> 1/4$. There exists an example of an exotic 7-sphere with curvature $\geqslant 0$, see Gromoll and Meyer [2].

3. (2.8.5), (2.8.7), (2.8.10) together give a complete classification of the homeomorphism type of simply connected 1/4-pinched manifolds. Nothing comparable is known for δ-pinched manifolds with $0 < \delta < 1/4$.

If one restricts oneself to simply connected homogeneous manifolds with $K > 0$, a complete classification is known. First, there are the symmetric spaces of rank one, i. e., the sphere and the projective spaces over the complex and quaternionic numbers and the Cayley projective plane. Among the so-called normal homogeneous spaces, there are two more, in dimension 7 and 13, respectively. The pinching of these spaces is $\delta = 1/37$ and $\delta = 16/29 \cdot 37$, respectively; cf. Berger [3], Eliasson [1] and Heintze [1].

There are three more "exceptional" non-normal homogeneous spaces with $K > 0$, all having even dimensions. They can be described as the manifold of flags in the complex, quaternionic and Cayley projective plane, see Wallach [1].

In addition, there exists an infinite series of homogeneous spaces in dimension 7, cf.

Aloff and Wallach [1]. That there are no other homogeneous spaces with strictly positive curvature was proved by Berard-Bergery [1].

Recently, Eschenburg [2] has shown that for any space $M = M_{pq}$ in the infinite series of 7-dimensional homogeneous spaces of Alloff and Wallach, there exists an infinite series $\{M_n\}$ of compact, simply connected 7-dimensional manifolds all topologically distinct and non-homotopic to a homogeneous space, such that the pinching δ_n of M_n converges to the pinching of M.

Weinstein [2] has given an upper bound for the number of homotopy types of compact, even dimensional manifolds with strictly positive curvature K. He derives his result from the lower bound for the injectivity radius of such a manifold, cf. (2.6.9).

By an ingeneous application of Topogonov's Triangle Theorem, Gromov [3] could show that the sum $\sum b_i$ of the Betti numbers b_i of an n-dimensional compact manifold M with $K \geqslant 0$ is bounded by a universal constant $c(n)$, only depending on n. If M is the torus, $\sum b_i = 2^n$, and this quite possibly is the maximal value for $\sum b_i$. However, Gromov's constant $c(n)$ is very much larger. In any case, this shows that 'most' compact manifolds do not admit a metric with $K \geqslant 0$.

4. When considering compact manifolds M for which the curvature K changes its sign, the pinching number min K: max K carries little, if any, information. Usually, one takes the diameter $d(M)$ into consideration.

Gromov's Theorem on Almost Flat Manifolds states: Let M be compact, n-dimensional and $|K| \leqslant \varepsilon_n d(M)^2$, with sufficiently small $\varepsilon_n > 0$. Then M has a finite covering by a compact nilmanifold, i. e., a manifold which is a compact quotient of a nilpotent Lie group. Cf. Gromov [2]. A detailed proof with considerable improvement on Gromov's ε_n is given in Buser and Karcher [1]. In the latter paper, a number of auxiliary results are derived which are of interest not only for their contents but also for the methods employed in the proofs.

5. In conclusion, we draw attention to relations between the diameter $d(M)$ and the volume $\operatorname{vol}(M)$ of a compact Riemannian manifold M with $-1 \leqslant K < 0$. Employing methods of Cheeger [1], Gromov [1] has shown:

(i) If dim $M \geqslant 8$, then $\operatorname{vol}(M) \geqslant \gamma_n(1 + d(M))$, with γ_n a constant only depending on n.

(ii) If dim $M \neq 3$, the number of diffeomorphism types of M with vol $M \leqslant$ some constant is finite.

2.9 Non-compact Manifolds of Positive Curvature

In this section we study the geometry of non-compact, complete manifolds M with sectional curvature $K \geqslant 0$. As always in chapter 2, we assume dim $M < \infty$.

In the case dim $M = 2$, the classical investigations stem from Cohn-Vossen [1], [3]. The case of arbitrary dimension and $K > 0$ was studied in Gromoll and Meyer [1]. The extension to the case $K \geqslant 0$ is due to Cheeger and Gromoll [1]; cf. also the exposition given in Cheeger and Ebin [1].

Geodesics are – unless stated otherwise – parameterized by arc length.

2.9.1 Definition. A point p of a Riemannian manifold M is called a *pole* if the exponential mapping $\exp_p: T_pM \to M$ has maximal rank everywhere.

Note. If M is simply connected, p being a pole means that $\exp_p: T_pM \to M$ is a diffeomorphism.

2.9.2 Example. *The paraboloid of revolution.* By this we mean the surface $M = \{x^2 + y^2 = 2z\}$ in $\mathbb{R}^3$. The poles of M and, more generally, the poles of the other connected non-compact surfaces of second order in $\mathbb{R}^3$ have been determined by v. Mangoldt [1].

We use the fact that the geodesics on a surface of revolution in appropriate coordinates can be described in a particularly simple way, cf. (3.5.21). The appropriate coordinates on M, outside the point (0,0,0) where M intersects the axis of revolution $\{x = y = 0\}$, are of the form

$$(u_1, u_2) \in \mathbb{R}^+ \times \mathbb{R} \mapsto (r(u_1)\cos u_2, r(u_1)\sin u_2, r(u_1)^2/2)$$

with $$\sqrt{1 + r^2(u_1)}\, dr(u_1)/du_1 = 1\,.$$

Of course, one has to restrict u_2 to an open interval of length $\leqslant 2\pi$ to obtain an injective mapping. The curves $\{u_1 > 0; u_2 = \text{const}\}$ are called *meridians* while the curves $\{u_1 = \text{const} > 0\}$ are the *parallel circles*.

The line element of M in these coordinates becomes $ds^2 = du_1^2 + r(u_1)^2\, du_2^2$.

The coordinates $(\dot u_1(t), \dot u_2(t))$ of the tangent vector field along a geodesic are characterized by

$$\dot u_1^2 + r(u_1)^2 \dot u_2^2 = 1; \quad r(u_1)^2 \dot u_2 = \gamma$$

If $\gamma = 0$, $u_2(t) = \text{const}$, i. e., the geodesic lies on a meridian and passes through the point (0,0,0) not in the domain of our coordinates.

Now let $\gamma \neq 0$. Using the symmetries of M we can restrict ourselves to the case $\gamma > 0$ and $u_2(0) = 0$. $r(t)$ is bounded from below by $\gamma > 0$.

$$\dot r(t)^2 = \big(r(t)^2 - \gamma^2\big)/r(t)^2\big(r(t)^2 + 1\big).$$

This shows that there is exactly one t-value, say t_0, where $r(t_0)$ assumes its minimum $r(t_0) = \gamma$. $\dot u_1(t_0) = 0, \dot u_2(t_0) = 1/\gamma$. $\dot u_1(t)$ and $\dot r(t)$ are negative, if $t < t_0$; $\dot u_1(t)$ and $\dot r(t)$ are positive, if $t > t_0$.

We can describe u_2 as function of r by

$$u_2 = u_2(r; \gamma) = \int_{r(0)}^{r} \gamma\sqrt{r^2 + 1}\,/\big(r\sqrt{r^2 - \gamma^2}\big)\, dr\,.$$

Here, $\sqrt{r^2 + 1}$ has to be taken with the $+$-sign, while the sign of $\sqrt{r^2 - \gamma^2}$ has to be taken equal to the sign of $\dot r$, see above.

Let now in $u_2(r;\gamma)$ the parameter γ vary. Then we get a variation of our geodesic c by geodesics with the same initial point as c. A conjugate point $c(t_1)$ of $c(0)$ along $c|[0,t_1]$ will occur when the function

$$J(r;\gamma)=\frac{\partial u_2(r;\gamma)}{\partial\gamma}=\int_{r(0)}^{r} r\sqrt{r^2+1}/(r^2-\gamma^2)\sqrt{r^2-\gamma^2}\,dr$$

has a zero. This can happen only if the integral changes its sign, i. e., if $t_0>0$. That is to say, $\dot u_1(0)<0$; i. e., the tangent vector $\dot c(0)$ forms an angle $\mu(\dot c(0))<0$ with the parallel circle $\{u_1=r(0)\}$.

As γ decreases towards 0, μ moves towards $-\pi/2$ and the value $t_1>0$ of the conjugate point of $c(0)$ moves closer to 0. Among all geodesics starting from the same point p, the one moving downwards on the meridian is the one where the conjugate point occurs earliest.

For each unit vector $X\in T_pM$ denote by $t_1(X)$ the first conjugate point of the geodesic c_X with $\dot c_X(0)=X$. $t_1(X)$ assumes its minimum when $\mu(X)=-\pi/2$, and it goes to $+\infty$, as $\mu(X)$ goes to $-\pi$ and 0. For $\mu(X)\in[0,\pi]$, c_X has no conjugate points. □

2.9.3 Proposition. *Let M be non-compact. For every $p\in M$, there exists at least one ray $c(t)$, $t\geqslant 0$, starting from p. That is, $c(t)$ is a geodesic such that $c|[0,t]$ is minimizing between its end points, for all $t\geqslant 0$.*

Proof. There exists a sequence $\{p_n\}$ on M with $\lim d(p,p_n)=\infty$. By c_n denote a minimizing geodesic from p to p_n. We can assume that the sequence $\{X_n=\dot c_n(0)\}$ is convergent with limit $X\in T_pM$. Let c be the geodesic with $\dot c(0)=X$. For every fixed $t^*>0$, $c_n|[0,t^*]$ will converge to $c|[0,t^*]$. As in the second part of the proof of (2.1.5), we can conclude that $c|[0,t^*]$ is a minimizing geodesic from $p=c(0)$ to $c(t^*)$. □

We now return to manifolds with strictly positive curvature. The following Lemma is at the basis of all further results on the topological structure of such manifolds.

2.9.4 Lemma. *Let M be complete and $K>0$. For every half geodesic $c(t)$, $t\geqslant 0$, there exists a $t_0\geqslant 0$ such that $t^*\geqslant t_0$ implies that given an X_0, $|X_0|=1$, $\langle X_0,\dot c(0)\rangle=0$, there exists a vector field $X(t)$, $0\leqslant t\leqslant t^*$ with $X(0)=X_0$, $X(t^*)=0$ and*

$$(*)\qquad D^2E(c|[0,t^*])(X,X)=\int_0^{t^*}\langle\nabla X,\nabla X\rangle(t)-\langle R(X,\dot c)\dot c,X\rangle(t)\,dt<0.$$

Using (2.7.8) *we find from* (2.9.4) *that every Jacobi field $Y(t)$ along $c|[0,t^*]$ with $\nabla Y(0)=0$, $\langle Y(t),\dot c(t)\rangle=0$ has a zero before t^*. In particular – but this is a weaker statement – the focal index* (*cf.* (2.5.4)) *of every sufficiently long half-geodesic on M is $\geqslant n-1$.*

Proof. For fixed $t^*>0$, and $X_0\in T_0^{\perp}c$, $|X_0|=1$, we consider the vector field

$$X(t) = \begin{cases} (\overset{t}{\underset{0}{\|}} c)\, X_0\,; & 0 \leqslant t \leqslant t^*/2 \\ \frac{1}{t}(t^* - t)\,(\overset{t}{\underset{0}{\|}} c)\, X_0\,; & t^*/2 \leqslant t \leqslant t^*. \end{cases}$$

For $0 \leqslant t \leqslant t^*/2$, the second integrand is the sectional curvature of the plane spanned by $\{\dot{c}(t), X(t)\}$. There exists a positive function $k(t)$ which bounds this curvature from below, independent of the choice of X_0. Thus,

$$D^2 E(c\,|\,[0, t^*])(X, X) < \frac{7}{3t^*} - \int_0^{t^*/2} k(t)\, dt$$

For t^* sufficiently large, the right hand side becomes < 0. □

2.9.5 Corollary. *Under the hypothesis of* (2.9.4), *for every geodesic* c *there exists a* $t^* > 0$ *such that* $\operatorname{index}_\Omega c\,|\,[-t^*, t^*] \geqslant n-1$.

Proof. Simply form, for every $X_0 \in T_0^\perp c$, a vector field X along $c\,|\,[-t^*, t^*]$ with $X(-t^*) = X(t^*) = 0$, $X(0) = X_0$, by combining the construction in the proof of (2.9.4) on $c\,|\,[0, t^*]$ with the corresponding construction on $c(-t)$, $0 \leqslant t \leqslant t^*$. □

2.9.6 Definition. A set C in a Riemannian manifold M is called *totally convex* if it is non-empty and closed and if for any two points p and q in C, every geodesic from p to q lies entirely in C.

The topological importance of totally convex sets becomes clear from the following

2.9.7 Lemma. *Let* C *be a totally convex subset of* M. *Then the inclusion* $C \hookrightarrow M$ *is a homotopy equivalence.*

Note. What we will show is that $\pi_k(M, C) = 0$ for all k. The conclusion then is a standard result of homotopy theory, cf. Spanier [1].

Proof. We will restrict ourselves to the case where C is a submanifold with $\partial C = \emptyset$. Choose $p_1 \in C$ and consider the space $P = H^1_{C \times \{p_1\}} M$, cf. (2.4.2). Recall from (2.4.3) that the critical points of $E\,|\,P$ are the geodesics c with $\dot{c}(0)$ orthogonal to $T_{c(0)} C$. Since C is totally convex, such a geodesic must be the constant curve p_1.

Let $\bar{B}^k$ be the closed unit ball in $\mathbb{R}^k$. For each $p \in \partial \bar{B}^k$, *let* $c_p(t)$, $0 \leqslant t \leqslant 1$, be the straight segment from $\tilde{p}_0 = (1, \ldots, 0) \in \partial \bar{B}^k$ to p. To $f\colon (\bar{B}^k, \partial \bar{B}^k) \to (M, C)$ we associate a mapping $F = F(f)\colon (\partial \bar{B}^k, \tilde{p}_0) \to (P, p_1)$ by $F(p)(t) = f \circ c_p(t)$. As in the proof of (2.4.20), a homotopy F_s of $F = F_0$ determines a homotopy f_s of $f = f_0$. As we saw above, $\phi_s \circ F(\partial \bar{B}^k)$, for $s \to \infty$, comes arbitrarily close to the constant curve p_1. Thus, f is homotopic to zero. □

Remark. As for the general case, where C is not a manifold with $\partial C = \emptyset$, we again consider the subset $P = H^1_{C \times \{p_1\}}$ of $H^1(I, M)$. P need not be a submanifold. Thus, the arguments of (2.4) do not apply immediately. However, it is not difficult to show that there is no obstruction in deforming a mapping $F = F(f): (\partial \bar{B}^k, \tilde{p}_0) \to (P, p_1)$ into small neighborhoods of the constant curve p_1. We shall not enter into the details, since we shall need (2.5.7) only in the case where C is a submanifold with $\partial C = \emptyset$. For a survey article on convex sets in Riemannian manifolds see Walter [1].

For non-compact manifolds M satisfying the property in (2.9.4), and, in particular, satisfying $K > 0$, we can construct compact totally convex sets. For this construction we need the following extension of (2.9.4).

2.9.8 Lemma. *Let $C \subset M$ be compact, M non-compact and satisfying the property of* (2.9.4). *Then there exists a compact $D \supset C$ such that every geodesic $c_0(s)$, $s_0 \leqslant s \leqslant s_1$, in C is concave with respect to all $q \in \complement D$. That is, the function $s \in [s_0, s_1] \mapsto d(c_0(s), q)$ does not assume a relative minimum at an interior point s^* of $[s_0, s_1]$.*

Proof. Assume that there exist sequences $\{p_n\}$ in C, $\{q_n\}$ in $\complement C$, $d(p_n, q_n) \to \infty$, such that a minimizing geodesic c_n from p_n to q_n does not satisfy the property (*) of (2.9.4). We can assume that $\{p_n\}$ has a limit $p \in C$ and that the sequence $\{X_n = \dot{c}_n(0)\}$ of initial directions converges to an $X \in T_p M$. The geodesic c with $\dot{c}(0) = X$ satisfies (*), (2.9.4), with some $t^* > 0$. But then, also the c_n must satisfy (*), (2.9.4), with some $t_n^* > t^* + a$, $a > 0$, if only n is sufficiently large.

Now, the property (*) in (2.9.4) implies that the function $d(q, c_0(s))$, $s_0 \leqslant s \leqslant s_1$, cannot assume its infinimum at a value $s^* \in]s_0, s_1[$. If this were the case, consider a minimizing geodesic c from $c_0(s^*)$ to q. $\dot{c}(0) \perp c_0'(s^*)$. (2.9.4) then yields that $d(c_0(s^* + \varepsilon), q) < d(c_0(s^*), q)$, for sufficiently small $|\varepsilon| \neq 0$. □

We can now describe the basic construction of Cheeger, Gromoll and Meyer l. c., which shows the complement $\complement B_c$ of the open half space B_c of a geodesic ray c to be totally convex. Here,

$$B_c = \bigcup_{t > 0} B_t(c(t)),$$

where $B_t(c(t)) = \{q \in M; d(c(t), q) < t\}$.

2.9.9 Lemma. *Let M satisfy the property* (*) *in* (2.9.4) *and assume that $c(t)$, $t \geqslant 0$, is a ray on M. Then $\complement B_c$ is totally convex.*

The same conclusion holds if $K \geqslant 0$.

Note. The property (*) in (2.9.4) and the property $K \geqslant 0$ are independent of each other: (*) can be valid even when there 2-planes σ occur with $K(\sigma) < 0$. Only globally, sections σ with $K(\sigma) > 0$ must dominate those with $K(\sigma) < 0$. On the other hand, if $K \equiv 0$, (*) does not hold.

Proof. $\complement B_c \neq \emptyset$, since $c(0) \notin B_c$. Suppose that there exists a geodesic $c_0(s)$, $s_0 \leqslant s \leqslant s_1$, with $p_0 = c_0(s_0) \in \complement B_c$, $p_1 = c_0(s_1) \in \complement B_c$, but $q = c(s^*) \in B_c$ some $s^* \in]s_0, s_1[$. For large t_0, $q \in B_{t_0}(c(t_0))$ and $t_0 \leqslant t^*$ implies $B_{t_0}(c(t_0)) \subset B_{t^*}(c(t^*))$

(triangle inequality). Let C be a compact set containing c_0. For every $t^* > t_0$, the interior point $c_0(s^*)$ on c_0 is closer to $c(t^*)$ than the end points $c_0(s_0)$, $c_0(s_1)$. This is a contradiction of (2.9.8)

If $K \geqslant 0$, then we apply (2.7.12) as follows: For large $t > 0$, consider the possibly generalized triangles $p_0 q c(t)$ and $p_1 q c(t)$ where the sides $p_0 q$ and $q p_1$ are given by $c_0|[s_0, s^*]$ and $c_0|[s^*, s_1]$. In associated Alexandrov triangles $p_0^* q^* c(t)^*$ and $p_1^* q^* c(t)^*$ of the Euclidean space M^*, the angles at q^* together are $\leqslant \pi$. If we therefore put these two triangles together at their common side $q^* c(t)^*$ we get a quadrangle which is convex at q^*. But this contradicts the fact that $p_0^* c(t)^*$ and $p_1^* c(t)^*$ each are longer than $q^* c(t)^*$ by some $\varepsilon > 0$, uniformly for all large t. □

(2.9.9) now yields the existence of compact totally convex sets.

2.9.10 Theorem. *Let M be non-compact and satisfy* (∗) *in* (2.9.4) *or else, let $K \geqslant 0$. For every compact set $E \subset M$, there exists a compact, totally convex set C containing E.*

Remark. It follows that there exists a sequence $\{C_i\}$ of totally convex compact sets with $C_i \subset \operatorname{int} C_{i+1}$, $\bigcup_i C_i = M$.

Proof. Call a geodesic $c(t)$, $t \geqslant 0$, $|\dot{c}| = 1$, a *ray, starting from E*, if, for all $t \geqslant 0$, $d(c(t), E) = t$. Such rays exist: Consider a sequence $\{q_n\}$ in $\complement E$ with $\{d_n = d(q_n, E)\}$ going to infinity. Let c_n be a geodesic with $c_n(0) \in E$, $c_n(d_n) = q_n$. The sequence $\{\dot{c}_n(0)\}$ has an accumulation point X. The geodesic c with $\dot{c}(0) = X$ then is a ray starting from E.

Define now C as the intersection of the $\complement B_c$, where c is a ray starting from E. C is totally convex and contains E. C is also compact since otherwise, a sequence $\{q_n\}$ in $C \cap \complement E$ with $\{d(q_n, E)\}$ going to infinity, would yield, just as above, a ray c starting from E and $c \subset \complement B_c$. □

A point $p \in M$ is called *simple* if $\{p\}$ is totally convex, i. e., if there exists no non-constant geodesic in M, starting and ending in p.

2.9.11 Theorem. *Let M be non-compact and satisfy* (∗) *in* (2.9.4). *This is satisfied in particular if $K > 0$. Then the set S of simple points on M is open and, for every totally convex $C \subset M$, $S \cap C \neq \emptyset$. It follows that $\pi_i M = 0$, for all i.*

Proof. Let $\{p_n\}$ be a sequence of non-simple points with limit p. According to (2.9.10), there exists a compact, totally convex C with $p_n \in C$, all n. For each n, we have a non-constant geodesic c_n, starting and ending in p_n, c_n lying entirely in C. It follows that the sequence $\{c_n\}$ has an accumulation point c, which is a geodesic starting and ending in p. Since $\{L(c_n)\}$ is bounded away from 0, $c \neq$ const. Thus, S is open.

Let $E \neq \emptyset$, compact. From (2.9.10) we have a compact totally convex $C \supset E$. Choose $D \supset C$ as in (2.9.8). For $q \in \complement D$, let $p \in C$ have maximal distance from q. If p were not simple, there would exist a non-constant geodesic $c_0(s)$, $s_0 \leqslant s \leqslant s_1$ in C

starting and ending in p. Since

$$d(q, c_0(s)) \leqslant d(q, c_0(s_0)) = d(q, c_0(s_1)),$$

this is a contradiction to (2.9.8).

$\pi_i M = 0$ now follows from (2.9.7). □

2.9.12 Example. *The paraboloid of revolution.* We continue with (2.9.2). $p_0 = (0, 0, 0)$ is the only pole, but p_0 possesses an open neighborhood formed by simple points. We show: If $p \neq p_0$ and $u_2(p) = 0$ then the totally convex set C, constructed from $E = \{p\}$ as in (2.9.10), is the set $\{u_1 \leqslant u_1(p)\} \cup \{p_0\}$. Indeed, the only ray $c(t), t \geqslant 0$ emanating from p is given by $\{(t + u_1(p), 0); t \geqslant 0\}$. For sufficiently large t, there exists, for every value of u_2, a point $(u_1(t, u_2), u_2)$, on $\partial \bar{B}_t(c(t))$ having distance t from $c(t)$. Except for $u_2 = \pi$, where there are two minimizing geodesics of length t from $(u_1(t, u_2), \pi)$ to $(u_1(p) + t, 0)$, there is only one such minimizing geodesic. The initial vector $X(t, u_2)$ of such a geodesic forms an angle $\mu(t, u_2) \in]0, \pi/2[$ with the parallel circle. As $t \to \infty, \mu(t, u_2) \to \pi/2$, since a geodesic for which the initial direction X forms an angle $\neq \pi/2$ with the parallel circle meets the opposite meridian.

Since the $X(t, u_2)$, for $t \to \infty$, all become vectors tangent to the meridian, their foot points all have the same u_1-value $u_1(p)$.

The following theorem is a considerable sharpening of (2.9.11): A non-compact manifold M with $K > 0$ is homeomorphic to the Euclidean space. The idea is to use a continuous filtration $\{C_t, t \geqslant 0\}$ of totally convex compact sets C_t with $C_0 = \{p\}$ a simple point and to define a crooked exponential mapping $T_p M \to M$ by means of broken half geodesics starting from p. With more technical work this construction can be smoothed out to yield a diffeomorphism from $T_p M$ to M.

In particular, our construction will show that M is $(n-2)$*-connected at infinity*, $n = \dim M$. This means that for every compact set $E \subset M$ there exists a compact set $C \supset E$ with $\pi_k(\complement C) = 0, 0 \leqslant k \leqslant n-2$. Together with (2.9.11) then also follows by differential topological methods that M is diffeomorphic to $\mathbb{R}^n$, provided n is not 3 or 4, cf. Stallings [1].

2.9.13 Theorem. *Let M be non-compact with $K > 0$. Then there exists a simple point $p \in M$ and family $\{C_t, t \geqslant 0\}$ of compact totally convex sets with $C_0 = \{p\}$. $C_t \subset C_{t'}$ if $t \leqslant t'$ and C_t contains the metric ball $B_t(p)$. For every $t^* > 0$ we have an integer $k = k(t^*) \geqslant 0$ such that C_{t^*} is simply covered by k-fold broken geodesics from p to ∂C_{t^*} where each of the k geodesics segments is minimizing between its end points. This permits to define a homeomorphism from $C_{t^*} \subset M$ to $\bar{B}_{t^*}(0_p) \subset T_p M$.*

Note. In case p is a pole, we will obtain unbroken geodesics. However, in general there need not exist a pole on M.

Proof. Choose some $p' \in M$. Let $c'(t), t \geqslant 0$. be a ray starting from p'. For every $t_0 \geqslant 0$ we define $c'_{t_0}(t)$ by $c'(t + t_0)$. Then $B_{c'_{t_0}}$, is defined as before (2.9.9). The intersection C'_{t_0} of the complements of the $B_{c'_{t_0}}$, for all rays c' starting from p', is then compact and totally convex, cf. the proof of (2.9.10). Since

$$d(q, c'(t)) \geqslant d(p', c'(t)) - d(p', q) = t - d(p', q),$$

$B_t(p') \subset C'_{t'}$ if $t \leqslant t'$. Moreover, if $t \leqslant t'$,

$$C'_t = \{q \in C'_{t'};\, d(q, \partial C'_{t'}) \geqslant t' - t\}$$

with $q \in \partial C'_t$ if and only if $d(q, \partial C'_{t'}) = t' - t$.

Fix some $t_0 > 0$. Then $\partial C'_{t_0} \neq \emptyset$ and there exists $p \in C'_{t_0}$ having maximal distance d_{t_0} from $\partial C'_{t_0}$. p has maximal distance $d_t = d_{t_0} + t - t_0$ from $\partial C'_t$ also for all other $t \geqslant t_0$. We claim: If X is a unit vector in $T_p M$ then there exists a ray $c(t), t \geqslant 0$, starting from p with $\langle X, \dot{c}(0)\rangle \geqslant 0$.

To see this we proceed similar as in the proof of (2.8.3). Actually, p may be viewed as a point having maximal distance from infinity since all points $\neq p$ on the geodesic $b(s) = \exp_p(sX)$, $0 \leqslant s$ small, are closer to $\partial C'_t$ than p is, for all large t. We take a sequence $\{s_n > 0\}$ with $\lim s_n = 0$. Fix some $t > 0$. Let c_n be a minimizing geodesic from $b(s_n)$ to $\partial C'_t$. Then $d(b(s_n), \partial C'_t) \leqslant d(p, \partial C'_t)$ implies $\langle b'(s_n), \dot{c}_n(0)\rangle \geqslant 0$, for large n. If c_t is a geodesic in the limit set of the c_n then $\langle X, \dot{c}_t(0)\rangle \geqslant 0$. Let c be in the limit set of the c_t, for t going to ∞. Then c is a ray with $\langle X, \dot{c}(0)\rangle \geqslant 0$.

Define C_t, $t \geqslant 0$, just as above with the help of geodesic rays starting from p. Our claim implies $C_0 = \{p\}$, since all geodesic segments $b(s)$ starting from p look concave from ∂C_t, t large.

Fix some $t^* > 0$. Given $\varepsilon > 0$, there exists a $\delta = \delta(t^*) > 0$ with the following property: If $0 \leqslant t < t' \leqslant t^*$ and $t' - t < \delta$ then every $q' \in \partial C_{t'}$ has distance $< \varepsilon$ from ∂C_t. Indeed, otherwise we would have sequences $\{t_n\}, \{t'_n\}$ in $[0, t^*]$ with $t_n < t'_n$ and $\lim t_n = \lim t'_n = t_0$ and sequences $\{q'_n \in \partial C_{t'_n}\}, \{q_n \in \partial C_{t_n}\}$ with $\lim q'_n = q' \in \partial C_{t_0}$, $\lim q_n = q \in \partial C_{t_0}$, such that $d(q'_n, q_n) = d(q'_n, \partial C_{t_n}) \geqslant$ some $\eta > 0$. If c is a geodesic in the limit set of the minimizing geodesics from q'_n to q_n, for $n \to \infty$, c belongs to ∂C_{t_0} and has length $\geqslant \eta > 0$. But ∂C_{t_0} cannot contain a non-constant geodesic c since, for large t, c looks concave from $q \in \partial C_t$ and thus, not all points on c can have distance $t - t_0$ from ∂C_t.

According to (1.9.11) we have a $\varkappa = \varkappa(t^*) > 0$ such that $B_\varkappa(q)$ is strongly convex, for all $q \in C_{t^*}$. Let $\delta > 0$ be such that, for $0 \leqslant t < t' \leqslant t^*$, $t^* - t < \delta$, $q \in \partial C_{t'}$ has distance $< \varkappa/3$ from ∂C_t. Fix t_0, t_1 with $0 \leqslant t_0 < t_1 \leqslant t^*$, $t_1 - t_0 < 3\delta/2$. Put $C_{t_0} = C_0$, $C_{t_1} = C_1$. We claim: Every $q \in C_1 - C_0$ possesses a unique $h(q) \in \partial C_0$ having minimal distance d_q from q and the mapping $h: C_1 - C_0 \to \partial C_0$ is continuous. Moreover, if $h(q) \neq q$, that is, $q \notin \partial C_0$, we have the geodesic c with $c(0) = q$, $c(d_q) = h(q)$. Then $h(c(-t)) = h(q)$ for all $t \geqslant 0$ with $c(-t) \in B_\varkappa(q)$ and c meets ∂C_1 at some $c(-t)$ with $t < \varkappa/2$.

The uniqueness of h follows from the fact that two different points r_0, r_1 on ∂C_0 with $d(q, r_0) = d(q, r_1) = d_q = d(q, \partial C_0)$ belong to $B_\varkappa(q)$, their mid point therefore has distance $< d_q$ from q and also belongs to C_0. Moreover, since d_q depends continuously on q, we get by the same argument the continuity of h. Finally, if $b(s), s \geqslant 0$, is a non-constant geodesic in $B_\varkappa(q) \cap C_0$ with $b(0) = h(q)$ then

$\langle b'(0), \dot{c}(d_q)\rangle > 0$. Hence, $d(c(-t), b(s)) > d(c(-t), h(q))$ for $t \geq 0, c(-t) \in B_\varkappa(q)$, $s > 0$.

We now can define a homeomorphism $H: \partial C_{t'} \times [0, 1] \to C_{t'} - C_t$, for all $0 < t < t' \leq t^*$ with $t' - t < \delta$ as follows: Put $C_t = C$, $C_{t'} = C'$. Choose $t_0 < t$ with $t' - t_0 < 3\delta/2$. For every $q' \in \partial C'$ define $h(q') \in \partial C_{t_0}$ as above and denote by $c_{q'}(t)$, $0 \leq t \leq d_{q'} = d(q', h(q'))$ the minimizing geodesic from $h(q')$ to q'. $c_{q'}$ meets ∂C at a point $q = c_{q'}(t_{q'})$. The mapping $q' \in \partial C' \mapsto q \in \partial C$ is a homeomorphism since it is continuous and bijective. Define now $H(q', s)$ by $c_{q'}(s t_{q'})$.

To conclude the proof, choose a subdivision $0 = t_0 < t_2 < \ldots < t_{2k} = t^*$ of $[0, t^*]$ with $t_{2k} - t_{2k-2} < \delta$. Then through every $q \in C_{t^*} - \{p\}$ there passes exactly one k-fold broken geodesic from p to ∂C_{t^*}: Put $(t_{2i} + t_{2i-2})/2 = t_{2i-1}$, $C_{t_j} = C_j$. If $q \in \partial C_{t^*} = \partial C_{2k}$, there is a unique geodesic from $q = q_{2k}$ to a $q_{2k-2} \in \partial C_{2k-2}$, determined as above with ∂C_{2k} and ∂C_{2k-3}. Continue by going from q_{2k-2} to $q_{2k-4} \in \partial C_{2k-4}$, until $q_2 \in \partial C_2$ from where we have a unique minimizing geodesic to p. □

A geodesic c, $\|\dot{c}\| = 1$, is called a *line* if $d(c(t), c(t')) = |t - t'|$, for all t, t' in $\mathbb{R}$. Cohn-Vossen [2] showed that on a manifold with strictly positive curvature there are no lines, cf. (2.9.5). However, if M is non-compact, no complete geodesic remains in a bounded domain:

2.9.14 Proposition. *Let M be non-compact with $K > 0$. Let C be a compact, totally convex set. Then there exists a $L = L(C) > 0$ such that every geodesic segment in C has length $\leq L$. It follows that for every non-constant geodesic c on M, $d(c(0), c(t))$ goes to infinity for $t \to \infty$. In particular, M contains no closed geodesics.*

Proof. The conclusions are clear from the first statement. We derive a contradiction from the existence of a sequence $\{c_n: [-n, n] \to C\}$ of geodesic segments of length $2n$. We may assume that the $\dot{c}_n(0)$ converge to an X. Put $\exp(tX) = c(t) \in C$. Let A be the compact closure of $c(\mathbb{R})$. Choose q outside C such that every geodesic segment in C looks concave from q, cf. (2.9.8). Let $p \in A$ have minimal distance from q. There exists a sequence $\{t_n\}$ such that $\lim c(t_n) = p$ and $\lim \dot{c}(t_n) = \bar{X}$.

For the geodesic $\bar{c}(t) = \exp_p(t\bar{X})$ we then have $d(q, \bar{c}(t)) \geq d(q, \bar{c}(0))$, for $|t|$ small, which is the desired contradiction. □

We now investigate the consequence of the existence of a line on a necessarily non-compact manifold M with $K \geq 0$. Cohn-Vossen [3], for $\dim M = 2$, and Toponogov [3], for arbitrary dimension, have shown that then M is isometric to a product $M' \times \mathbb{R}$ where M' has curvature ≥ 0, cf. (2.9.16). By refining the methods of Toponogov, Cheeger and Gromoll [3] have constructed for every non-compact manifold M with non-negative curvature the so-called soul S which is a totally convex compact submanifold of M such that M is diffeomorphic to the total space of the normal bundle of S. If $K > 0$, S reduces to a point cf. (2.9.13). Using these constructions, one gets a short proof of the Cohn-Vossen-Toponogov Splitting Theorem (2.9.16).

We will describe the essential steps of their constructions later on, beginning with (2.9.17).

We start with a preliminary result.

2.9.15 Lemma. *Let M be a manifold with $K \geqslant 0$ and assume that there exists on M a line $c_0(t)$, $t \in \mathbb{R}$. Then there exists, for every $p \in M$, a unique line $c = c_p$ and a unique $t_p \in \mathbb{R}$ with $c_p(t_p) = p$ and $d(c(t), c_0(t)) = const = d(p, c_0) = inf\, d(p, c_0(t))$. If $c \neq c_0$, that is, if $p \notin c_0$, every geodesic from $p = c_p(t_p)$ to $c_0(t_p)$ meets c_0 and c orthogonally.*

Proof. We first assume $p \notin c_0$. Let $p_0 = c_0(t_p)$ be such that $d(p, c(t_p)) = d(p, c)$. Let $\{t_n^+\}$ and $\{t_n^-\}$ be sequences going to $+\infty$ and $-\infty$, respectively. Put $c_0(t_n^+) = p_n^+$, $c_0(t_n^-) = p_n^-$. Let $\bar{c}$ be an arbitrary geodesic from p to p_0 which meets c_0 orthogonally. Such a geodesic does exist: take e. g. a minimizing geodesic from p to p_0.

We take n so large that minimizing geodesics c_n^+ from p to p_n^+ and c_n^- from p to p_n^- as well as $c_0 | [t_p, t_n^+]$, $c_0 | [t_n^-, t_p]$ all are longer than $\bar{c}$. Denote by $pp_0p_n^+$ and $pp_0p_n^-$ the possibly generalized triangles where the side from p to p_0 is the geodesic $\bar{c}$. With these triangles we associate Alexandrov triangles $p^*p_0^*p_n^{+*}$ and $p^*p_0^*p_n^{-*}$ in the Euclidean space. Then we have from (2.7.12) that the angles at p^* and p_0^* are $\leqslant$ than the angles at p and p_0. In particular, the angles at p_0^* in the Alexandrov triangles are $\leqslant \pi/2$.

We now put together the two Alexandrov triangles at their common side $p^*p_0^*$ and obtain a planar quadrangle $p_n^{-*}p_0^*p_n^{+*}p^*$. At p_0^*, the angle is $\leqslant \pi$. This is also true for the angle at p^* because the triangle inequality for $p_n^- p_n p$ shows that the sides $p_n^{-*}p^*$ and $p^*p_n^{+*}$ together have length $\geqslant$ than the two sides $p_n^{-*}p_0^*$ and $p_0^*p_n^{+*}$.

Therefore, the quadrangle is convex. By straightening out the corner at p_0^* we obtain an Alexandrov triangle $p_n^{-*}p_n^{+*}p^*$, associated to $p_n^- p_n^+ p$. The distance from p^* to the opposite side in this triangle is $\leqslant d(p, p_0)$. Hence, as $n \to \infty$, the angle at p^* in $p_n^{-*}p_n^{+*}p^*$ goes to π and so does the angle at p in $p_n^- p_n^+ p$.

Let $X^+ = \lim \dot{c}_n^+(0)$, $X^- = \lim \dot{c}_n^-(0)$. Then we have just shown $X^- = -X^+$. We define the geodesic $c = c_p$ by $\dot{c}_p(t_p) = X^+$.

The geodesic $\bar{c}$ from $p \in c(t_p)$ to $p_0 = c_0(t_p)$ is orthogonal to $c = c_p$ in its initial point, since in $p^*p_0^*p_n^{+*}$ and $p^*p_0^*p_n^{-*}$ the angles at p^* both go to $\pi/2$.

Let now $p' = c(t_p')$ be an arbitrary point on $c = c_p$ with $t_p' > t_p$. Let $p_0' \in c_0$ have minimal distance from p' and let $\bar{c}'$ be a minimizing geodesic from p' to p_0'. By $\bar{c}$ we now denote a minimizing geodesic from $p = c(t_p)$ to $p_0 = c_0(t_p)$. Let $\{p_n^+\}$ be a sequence on c_0, as above. Consider the triangle $pp'p_n^+$, n large. Then we have that in an associated Alexandrov triangle $p^*p'^*p_n^{+*}$ the angle at p^* goes to 0 and the angle at p'^* goes to π, as $n \to \infty$. Therefore, the angle in p' in $pp'p_n^+$ also goes to π, the sides $c_n'^+$ from p' to p_n^+ converge to $c(t + t_p')$, $t \geqslant 0$. $\bar{c}$ and $c(t + t_p)$, $t \geqslant 0$, as well as $\bar{c}'$ and $c(t + t_p')$, $t \geqslant 0$, form an angle $\pi/2$. Hence, $d(p, p_0) = d(p', p_0') =$ length of the sides $p^*p_0^*$ and $p^*p_0'^*$. That is, $d(c_0(t), c(t)) =$ const for $t \geqslant t_p$. The same argument with p_n^+ replaced by p_n^- yields $d(c_0(t), c(t)) =$ const, for all $t \in \mathbb{R}$.

Finally, $c(t)$ is a line, since $c|[t'^-, t'_p]$, for every $t'^- < t'_p$, and $c|[t'_p, t'^+]$, for every $t'_p < t'^+$, are minimizing geodesic segments between its end points. This follows from the construction of $c(t+t'_p)$, $t \geq 0$, and $c(-t+t'_p)$, $t \geq 0$, as limit of miminizing geodesic segments. $\square$

We are now ready to prove the *Splitting Theorem of Cohn-Vossen* [3] *and Toponogov* [3].

2.9.16 Theorem. *Let M have curvature ≥ 0. Then M is isometric to a Riemannian product $M' \times \mathbb{R}^k$, some $k, 0 \leq k \leq n = \dim M$, where M' does not contain a line.*

Proof. Clearly, it suffices to show that the existence of one line $c_0(t)$, $t \in \mathbb{R}$, on M implies a Riemannian splitting $M = M' \times \mathbb{R}$.

For every $t_0 \in \mathbb{R}$ we define M_{t_0} to be the intersection of the two totally convex sets $\complement B_{c_0|[t_0,\infty[}$ and $\complement B_{c_0|]-\infty,t_0]}$, cf. (2.9.9). We first show that $M_{t_0} = \exp_{c_0(t_0)} T^\perp_{t_0} c_0 \subset T_{c_0(t_0)} M$ where $T^\perp_{t_0} c_0$ consists of the vectors orthogonal to $\dot{c}_0(t_0)$.

Indeed, let $X \in T^\perp_{t_0} c_0$. Put $c_0(t_0) = p_0$, $\exp_{p_0} X = p$. From (2.9.15) we then get a parallel line c to c_0 with $c(t_0) = p$, $\dot{c}(t_0)$ orthogonal to the final direction of the geodesic $\bar{c}(t) = \exp_{p_0}(tX/|X|)$, $0 \leq t \leq |X|$, where we assume $|X| \neq 0$. Then $c(t_0)$ is the only point of c which belongs to M_{t_0}. To see this simply note that – with the notation of the proof (2.9.15) – the sides $p_0^* p_n^{+*}$ and $p^* p_n^{+*}$ in an Alexandrov triangle $p^* p_0^* p_n^{+*}$, associated with $p p_0 p_n^+$, differ arbitrarily little in length from each other, as $n \to \infty$, since the angles at p_0^* and p^* both converge to $\pi/2$. Thus, $c(t+t_0)$, $t > 0$, belongs to $B_{c_0|[t_0,\infty[}$. Similarly, $c(t+t_0)$, $t < 0$, belongs to $B_{c_0|]-\infty,t_0]}$.

It follows that M_{t_0} is a totally geodesic submanifold of codimension 1. Indeed, the criterion (1.3.5) for a submanifold is satisfied if we introduce Fermi coordinates around $p = c(t_0)$ along c with radius $\varrho > 0$, cf. (1.12.1). Then the intersection of the domain of these coordinates with M_{t_0} has range $\{t_0\} \times B_\varrho(0)$.

If M_{t_0} and M_{t_1} have a point p in common then they coincide. To see this let $p = c(t_0)$, c a parallel line to c_0. Then $T_p M_{t_0} = T^\perp_{t_0} c = T_p M_{t_1}$.

We can therefore define a mapping $\phi_{t_1-t_0}: M_{t_0} \to M_{t_1}$ by associating with $p = c(t_0) \in M_{t_0}$ the point $c(t_1) \in M_{t_1}$. Since parallels have constant distance from each other, this is an isometry. Put $M_0 = M'$. Then

$$(*) \qquad (p', t) \in M' \times \mathbb{R} \mapsto \phi_t(p') \in M$$

is the desired isometry of the Riemannian product $M' \times \mathbb{R}$ with M. $\square$

Note. If $F: M' \to M$ is the embedding with $\operatorname{im} F = M_0$, we have the normal bundle ν_F of F. The mapping $(*)$ above can be viewed as the mapping $\exp_F: T^\perp F \to M$ of the total space $T^\perp F$ of ν_F, cf. (1.12.15).

We continue with the study of general non-compact manifolds M with $K \geq 0$. Our goal is to show the existence of a so-called soul for M. This is a totally geodesic totally convex compact submanifold $S \subset M$ of codimension ≥ 1. S contains all the relevant homotopical structure of M. More precisely, the normal bundle of S is diffeomorphic to M. This, however, we will not prove here but refer the reader to the paper of Cheeger and Gromoll [1].

As a first step towards the construction of a soul we study *locally convex subsets* of M. These are sets C such that, for each $p \in \bar{C}$, there exists a $\varrho = \varrho(p) > 0$ such that, for any of its points, $C \cap B_\varrho(p)$ contains, in addition, the minimizing geodesic between these points.

For the following discussion, the hypothesis $K \geqslant 0$ is not needed.

2.9.17 Lemma. *Let C be locally convex, connected, $C \neq \emptyset$. Then there exists a totally geodesic connected submanifold $N \subset C$ with $C \subset \bar{N}$.*

Proof. Let k, $0 \leqslant k \leqslant n = \dim M$, denote the largest integer such that the collection $\{N_\alpha\}_{\alpha \in A}$ of smoothly embedded k-dimensional submanifolds N_α of M with $N_\alpha \subset C$ is not empty. Put $\bigcup_\alpha N_\alpha = N$. Let $p \in N$. That is, $p \in N_\alpha$ for some $\alpha \in A$. There is a neighborhood $U \subset N_\alpha \cap B_{\varrho(p)/2}(p)$ of p in N_α and a $\delta < \varrho(p)/2$ such that the exponential map, restricted to the set of vectors orthogonal to U and of length $< \delta$, is a diffeomorphism onto a neighborhood V_δ of p in M. In order to show that N is a submanifold, it will suffice to show that $N \cap V_\delta = U$. In fact, if $q \in (C \cap V_\delta) - U$ and $q' \in U$ closest to q, the minimizing geodesic from q to q' is perpendicular to U. Then, for a sufficiently small open neighborhood U' of q in U, the minimizing geodesic from q to any $q'' \in U'$ intersects U' transversally. It follows that the cone $\{\exp_q(tX); X \in T_q M, |X| < \varrho(q), \exp_q X \in U', 0 < t < 1\}$ is a $(k+1)$-dimensional smooth submanifold of M which is contained in C by convexity. But this contradicts the definition of k, i. e., $q \notin N$. By the convexity of C and the existence of V_δ, it is immediate that N is totally geodesic.

Next, let $p \in C \cap \bar{N}$, $p' \in B_{\varrho(p)/4}(p) \cap C$ and $q \in B_{\varrho(p)/4}(p) \cap N$. Let c be the geodesic in M such that $c|[0, \varepsilon]$ is the minimizing curve from q to p'. Then $c|[0, \varepsilon[\subset N$, hence, $p' \in \bar{N}$. If $p' \notin N$, then $c(t) \notin C$, all $\varepsilon < t < \varrho(p)/4$.

To see this, let W be a sufficiently small $(k-1)$-dimensional hypersurface of $B_{\varrho(p)/4}(p) \cap N$ through q which is transversal to c. Let $\bar{\varepsilon}, 0 < \bar{\varepsilon} < \varepsilon + \varrho(p)/4$ and $\bar{p} = c(\bar{\varepsilon}) \in C$. The cone $V = \{\exp(tX); X \in T_{\bar{p}} M, |X| < \varrho(p), \exp_{\bar{p}} X \in W, 0 < t < 1\}$ is a smooth k-dimensional submanifold of M and $V \subset C$, since C is convex. Hence, $V \subset N$ by construction of N. In particular, taking $\bar{p} = p'$, we have $c|[0, \varepsilon[\subset N$, i. e., $p' \in \bar{N}$. If $p' \notin N$, then the construction shows that no $c(t)$, $\varepsilon < t < \varrho(p)/4$, can belong to C.

From our last statement it follows in particular that $C \subset \bar{N}$. To see that N is connected we observe that any two points q_0 and q_1 in N can be joined by a piecewise broken geodesic c^* belonging to $C \subset \bar{N}$. But then N cannot consist of two disjoint open sets since that would make the domain on which c is defined non-connected. □

For a closed, locally convex set C we define the *tangent cone* by

$$T_p C = \{X \in T_p M; \exp_p(tX/|X|) \in N; \text{ some positive } t < \varrho(p)\} \cup 0_p$$

Let $\hat{T}_p C$ denote the linear subspace of $T_p M$, generated by $T_p C$. Clearly, if

$p \in N$, $T_p C \subset \hat{T}_p C = T_p N$. If, on the other hand, $p \in \partial C = C - N$, then by the proof of (2.9.17) there exists a $q \in B_{\varrho(p)/4}(p) \cap N$ and the geodesic $c(t)$, $0 \leqslant t \leqslant \varepsilon = d(p, q)$ from p to q belongs to N for $0 < t$. While $\dot{c}(0) \in T_p C$, $-\dot{c}(0) \notin T_p C$. Hence, in this case $\hat{T}_p C \neq T_p C$ Moreover, if $q' \in B_{\varrho(p)/4} \cap N$ and, for sufficiently small $t > 0$, c_t denotes the minimizing geodesic from $c(t)$ to q', then $\lim_{t \to 0} \dot{c}_t(0) = \dot{c}_0(0)$.

Since N is totally geodesic, $T_{c(t)} N$ is invariant under parallel translation along c. In particular, $\hat{T}_p C \subset (\| c)^{-1} T_q N$. However, as in the proof of (2.9.17), if we form the geodesic cone at p based on a small hypersurface W of N, transversal to c at q, we see that $T_p C - \{0\}$ is open in $(\| c)^{-1} T_q N$. Therefore, $\hat{T}_p C = (\| c)^{-1} T_q N$ and

$$\dim T_p C = \dim \hat{T}_p C = \dim N.$$

Moreover, if $p' \in B_{\varrho(p)/4}(p) \cap \partial C$; $q \in B_{\varrho(p)/4}(p) \cap N$ and $c'(t)$, $0 \leqslant t \leqslant \varepsilon'$, is the minimizing geodesic from p' to q, then $\hat{T}_p C = (\| c)^{-1} \circ (\| c') \hat{T}_{p'} C$.

It follows that $\hat{T}_p C$ varies continuously with p.

Using these preliminary results we now show:

2.9.18 Proposition. *Let C be closed, non-empty locally convex. Suppose that there is $q \in \operatorname{int} C$ and $p \in \partial C$ and a minimizing geodesic $c(t)$, $0 \leqslant t \leqslant d$, from q to p such that $d = d(q, \partial C) = \inf\{d(q, p'); p' \in \partial C\}$. Then the tangent cone $T_p C - \{0\}$ is a* half space,

$$H = \{X \in T_p M; \langle X, -\dot{c}(d) \rangle > 0\}.$$

Proof. Take $s < d$ such that $d(c(s), p) < \varrho(p)/2$. Then $c | [s, d]$ is a minimizing geodesic from $c(s)$ to ∂C and hence

$$\bar{B}_{d-s}(c(s)) \cap \partial C = p.$$

It follows that $T_p C - \{0\} \supset H$. Conversely, let $q' \in B_{\varrho(p)}(p) \cap C$, $q' \neq p$, and let $\bar{c}$ denote the minimal geodesic from p to q'. If $\langle \dot{\bar{c}}(0), -\dot{c}(d) \rangle < 0$ then $-\dot{\bar{c}}(0)$ points into $B_{d-s}(c(s))$. Hence, by (2.9.16) we would have $p \in \operatorname{int} C$. Since, as we have previously remarked, $T_p C - \{0_p\}$ is open in $\hat{T}_p C$, we get $T_p C - \{0_p\} = H$. □

Note. A halfspace $H \subset \hat{T}_p C$, $p \in \partial C$, with $T_p C \subset \bar{H}$ is called a *supporting half space* of C at p. One can show that $T_p C - \{0_p\}$ is the intersection of all supporting half spaces at p, cf. Cheeger and Gromoll [1].

The next lemma will yield the basic tool for constructing minimal totally convex sets.

2.9.19 Lemma. *Let $K \geqslant 0$. Let C be a closed, totally convex set with $\partial C \neq \emptyset$. Define $\psi : C \to \mathbb{R}$ by $\psi(q) = d(q, \partial C)$. Then ψ is (weakly)* concave *in the sense that, for any geodesic segment $c(t)$, $t \in J$, in C, we have*

$$\psi \circ c(at + a't') \geqslant a\psi \circ c(t) + a'\psi \circ c(t'), \quad \textit{whenever } a + a' = 1;\ a, a' \geqslant 0.$$

In particular, $\psi \circ c$ has no strict minimum on an interior point.

Proof. Choose s an interior point of J. Let $p \in \partial C$ have minimal distance $d_s = \psi \circ c(s)$ from $q = c(s)$. Let $c_s(t), 0 \leqslant t \leqslant d_s = d(q, p)$ be a minimizing geodesic from q to p. It will suffice to show that

(*) $\quad \psi \circ c(t) \leqslant d_s - (t-s)\cos\alpha = \text{(briefly) } l_s(t)$ for all $t \in [s-\delta, s+\delta]$,

for some small $\delta > 0$. Here $\alpha = \sphericalangle(\dot{c}(s), \dot{c}_s(0))$. For $|t_0 - s| \neq 0$, small, assume now $(t_0 - s)\cos\alpha \geqslant 0$. Let $\tilde{c}(t), 0 \leqslant t \leqslant t_s$ be the perpendicular from $c(t_0)$ to c_s, with $\tilde{c}(t_s) = c_s(t_0')$. We then have

(†) $\quad t_0' \geqslant (t_0 - s)\cos\alpha \geqslant 0.$

To see this we put $c(s) = q, c_s(t_0') = p_0, c(t_0) = r_0$ and compare the triangle qp_0r_0 with an associated Alexandrov triangle $q^*p_0^*r_0^*$ in the Euclidean space M^*. Then $\sphericalangle q^* \leqslant \alpha$ (if $\alpha \leqslant \pi/2$) or $\leqslant \pi - \alpha$ (if $\alpha > \pi/2$) and $\sphericalangle p_0^* \leqslant \pi/2$. The length t_0' of the side $q^*p_0^*$ therefore is $\geqslant (t_0 - s)\cos\alpha$ = length of the side in a triangle $\tilde{q}\tilde{p}_0\tilde{r}_0$ with $\sphericalangle \tilde{q} = \alpha, \sphericalangle \tilde{p}_0 = \pi/2$, length $\tilde{q}\tilde{r}_0 = |t_0 - s|$.

Let now $V(t)$ be the parallel unit vector field along $c_s(t)$ with $V(t_0') = -\dot{\tilde{c}}(t_s)/|\dot{\tilde{c}}(t_s)|$. Then the curve

$$b_{t_0}(t) = \exp_{c_s(t+t_0')}(t_s V(t + t_0')), \; 0 \leqslant t \leqslant d_s - t_0'$$

starts from $c(t_0)$ and its end point does not belong to $\operatorname{int} C$, because the vector $V(d_s) \perp \dot{c}_s(d_s)$ does not belong to the supporting halfspace of C at $p = c_s(d_s) \in \partial C$, cf. (2.9.18). We get an upper bound for the length of b_{t_0} by the length $d_s - t_0'$ of the curve $b_{t_0}^*$ in the Euclidean space M^* which is constructed analogously to b_{t_0}, cf. (2.7.10). Note that in (2.7.10) the symbol * has been used differently. Therefore, using (†), $\psi \circ c(t_0) \leqslant d_s - t_0' \leqslant l_s(t_0)$.

Let now $(t_0 - s)\cos\alpha < 0$. $V(t)$ shall be the unit parallel field along $c_s(t)$, orthogonal to $c_s(t)$, with $\langle \dot{c}(s), V(0) \rangle = \sin\alpha$. Consider the curve

$$b_{t_0}(t) = \exp_{c_s(t)}((t_0 - s)\sin\alpha\, V(t)), \; 0 \leqslant t \leqslant d_s$$

As above, $L(b_{t_0}) \leqslant d_s$. The distance from $c(t_0)$ to $b_{t_0}(0)$ is $\leqslant -(t_0 - s)\cos\alpha$. Since $b_{t_0}(d_s) \notin \operatorname{int} C$ we again get $\psi \circ c(t_0) \leqslant l_s(t_0)$. □

We are now ready to prove the existence of a soul for a manifold of non-negative curvature.

2.9.20 Theorem. *Let M be a complete, non-compact manifold with $K \geqslant 0$. Then M contains a compact totally convex totally geodesic submanifold S without boundary,* $\dim S < \dim M$. *Such a manifold is called a* soul *of M.*

Notes. Starting from any compact, totally convex subset C of M we will construct a soul $S \subset C$. In general, S will depend on the choice of C. However, we know from (2.9.7) that the inclusion $S \hookrightarrow M$ induces an isomorphism for the homotopy groups. Thus, all souls of M have the same dimension.

If $K > 0$, every soul of M is a point, cf. (2.9.11).

Any compact manifold S with $K \geqslant 0$ arises as the soul of some non-compact M with $K \geqslant 0$. Just take $M = S \times \mathbb{R}$.

Proof. Let C be a compact totally convex set. In (2.9.10) we constructed such a C, containing an arbitrarily given point of M. If $\partial C \neq \emptyset$, we put, for $a \geqslant 0$,

$$C^a = \{p \in C; d(p, \partial C) \geqslant a)\}, \quad C_0 = \bigcap_a \{C^a; C^a \neq \emptyset\}.$$

C_0 is totally convex since $C^a \neq \emptyset$ is totally convex. Indeed, if $c(t)$, $t \in J$, is a geodesic segment having its end points in C^a, we have from (2.9.19) $d(c(t), \partial C) \geqslant a$ for all $t \in J$, i. e., $c(t) \in C^a$.

Moreover, $\dim C_0 < \dim C$, since, for $p_0 \in C_0$, $d(p_0, \partial C) = \sup\{d(p, \partial C); p \in C\}$. Here, the dimension of a totally convex set is the dimension of the maximal totally geodesic submanifold contained in it, cf. (2.9.17).

If $\partial C_0 \neq \emptyset$ we iterate this process and construct from C_0 the totally convex set $C_1 = (C_0)_0$. After at most $n = \dim M$ such steps we arrive at a totally convex set $S \neq \emptyset$ with with $\partial S = \emptyset$. □

Remark. Let $F: S \to M$ be the canonical embedding, $\nu_F: T^\perp F \to S$ its normal bundle. The mapping $\exp_F: T^\perp F \to M$, cf. (1.12.15), in general need not be a diffeomorphism. However, Cheeger and Gromoll [1] showed that a modification of $\exp_F$, where the normal lines are being mapped into broken geodesics, yields a homeomorphism from $T^\perp F$ onto M which can be modified to give a diffeomorphism $T^\perp F \to M$. For $K > 0$, cf. also (2.9.13).

As was observed by Greene and Wu [1] the diffeomorphism of the total normal space of $S \subset M$ with M is based on the construction of a strictly convex so-called exhaustion function $\psi: M \to \mathbb{R}$, where $\{\psi \leqslant \text{const}\}$ are convex subsets of M.

That the total space $T^\perp F$ of the normal bundle of the soul S in M is diffeomorphic to M also follows from general results of differential topology, at least if $\dim M \geqslant 5$, cf. Stallings [1].

2.9.21 Examples. Let M^* be a complete Riemannian manifold and H a subgroup of the group of isometries of M^* such that $M = M^*/H$ is a manifold. Then the quotient mapping

$$F: M^* \to M = M^*/H$$

is an isometric submersion, cf. (1.11.9). From (1.11.12) we know that M has curvature $K \geqslant 0$ if M^* has curvature $K^* \geqslant 0$.

Take now for H a closed subgroup of a compact Lie group G, acting on some $\tilde{M}$ by isometries. Let $\tilde{M}$ have curvature $\tilde{K} \geqslant 0$. G, considered as symmetric space, also has curvature $\geqslant 0$, cf. (2.2.25), (2.2.26). On the Riemannian product $M^* = G \times \tilde{M}$ with curvature $K^* \geqslant 0$ we take the free H-action by letting H operate on the first factor by left translations. Then $M = (G \times \tilde{M})/H$ has curvature $\geqslant 0$.

For example, let $G = S\mathbb{O}(n+1)$, $H = S\mathbb{O}(n)$ acting on $\tilde{M} = \mathbb{R}^n$ by rotations. Then $M = (S\mathbb{O}(n+1) \times \mathbb{R}^n)/S\mathbb{O}(n)$ is the total tangent space TS^n of S^n. Hence, TS^n admits a metric with $K \geqslant 0$. For $n \geqslant 2$, the soul S of TS^n is uniquely determined and is given by $(S\mathbb{O}(n+1) \times \{0\})/S\mathbb{O}(n) = S^n$.

Chapter 3: Structure of the Geodesic Flow

This chapter contains material which is usually not presented in a course on Riemannian geometry. Starting from the basic facts on Hamiltonian systems, we introduce the co-geodesic flow on the dual tangent bundle. Its periodic orbits which correspond to the closed geodesics, will be among our main topics.

The case of a surface is treated in great detail, and here, (3.5) is devoted entirely to the geodesic flow on the ellipsoid. In (3.6) we give a number of results on closed geodesics on n-dimensional spheres with a non-standard metric. (3.7) gives a complete proof of the famous Lusternik-Schnirelmann Theorem of the Three Closed Geodesics. The last two sections are devoted to the structure of the geodesic flow on manifolds of non-positive curvature. Here in particular, the basic results on topological transitivity of the flows and the density of periodic orbits on manifolds of negative curvature are derived entirely by the methods prepared in the earlier chapters of this book.

3.1 Hamiltonian Systems

We begin with some basic results of (linear) symplectic geometry.

3.1.1 Definition. *A symplectic vector space* (V, α) is a finite dimensional real vector space V, endowed with a non-degenerate skew-symmetric bilinear form α. We also simply write V instead of (V, α).

Note. For the mapping

$$\alpha : V \times V \to \mathbb{R}$$

we therefore have

$$\alpha(\xi, \eta) = -\alpha(\eta, \xi); \; \alpha(a\xi + a'\xi', \eta) = a\alpha(\xi, \eta) + a'\alpha(\xi', \eta).$$

Non-degenerate means that for every $\xi \neq 0$ in V there exists an η with $\alpha(\xi, \eta) \neq 0$.

A subspace $U \subset V$ is called *non-degenerate*, if $(U, \alpha \mid U)$ is a symplectic space. U is called (*totally*) *isotropic*, if $\alpha \mid U \equiv 0$.

Every 1-dimensional subspace of V is isotropic. If $\dim V = 2n$, n is the maximum of the dimension of an isotropic subspace; take e.g. the space spanned by the elements

$\varepsilon_i, 1 \leqslant i \leqslant n$, of a symplectic basis, see (3.1.2). An isotropic subspace of maximal dimension is also called *Lagrangian*.

Just as for a symmetric bilinear form, there also exist distinguished bases for a vector space V with skew symmetric form α.

3.1.2 Proposition. *A symplectic space* (V, α) *possesses a* symplectic basis $\{(\varepsilon_i, \varepsilon_{i+n}); 1 \leqslant i \leqslant n\}$, *satisfying*

$$\alpha(\varepsilon_i, \varepsilon_{j+n}) = \delta_{ij}; \ \alpha(\varepsilon_i, \varepsilon_j) = \alpha(\varepsilon_{i+n}, \varepsilon_{j+n}) = 0.$$

In particular, $\dim V$ *is even.*

Proof. Start with any $\varepsilon_1 \neq 0$. Then there is an $\varepsilon_{1+n} \in V$ with $\alpha(\varepsilon_1, \varepsilon_{1+n}) = 1$. $(\varepsilon_1, \varepsilon_{1+n})$ generates a 2-space V^2. The space $V^{2\perp} = \{\xi \in V; \alpha(\xi, \varepsilon_1) = \alpha(\xi, \varepsilon_{1+n}) = 0\}$ is complementary to V^2 and non-degenerate. Now proceed by induction. □

A linear mapping (or transformation)

$$P: V \to V$$

of a symplectic space $V = (V, \alpha)$ is called *symplectic* if $\alpha(P\xi, P\eta) = \alpha(\xi, \eta)$, for all ξ, η in V.

3.1.3 Proposition. *A symplectic* $P: V \to V$ *is an automorphism. If* ϱ *is a (not necessarily real) eigenvalue of* P, *so are* $\bar{\varrho}, \varrho^{-1}, \bar{\varrho}^{-1}$.

Proof. Let $P\xi = 0$. Then $0 = \alpha(P\xi, P\eta) = \alpha(\xi, \eta)$, for all $\eta \in V$, implies $\xi = 0$. Thus $P \in GL(V)$.

Since α is non-degenerate, we have the isomorphism

$$L_\alpha: V \to V^*; \ \xi \mapsto \alpha(\xi, \ )$$

from V onto its dual space V^*. $\alpha(P\xi, \eta) = \alpha(\xi, P^{-1}\eta)$, i.e., $L_\alpha . P = {}^tP^{-1} L_\alpha$ where ^{t} denotes the transpose of a mapping. Thus, ${}^tP^{-1}$ is conjugate to P.

Let ϱ be an eigenvalue of P. Since P is real, $\bar{\varrho}$ is also an eigenvalue. From what we just saw, $\varrho^{-1}, \bar{\varrho}^{-1}$ are also eigenvalues. □

Remark. The proof also shows that $\det P = \pm 1$. In fact, $\det P = +1$, as follows from our next result. This will concern the decomposition of V into P-invariant subspaces, P a given fixed symplectic transformation.

To simplify the statements we complexify the real vector space V, i.e., we consider the space $V_{\mathbb{C}} = V \otimes \mathbb{C}$. Often we will again simply write V instead of $V_{\mathbb{C}}$. If $P: V \to V$ is a (real) symplectic transformation, $P\bar{\xi} = \overline{P\xi}$.

The extension of the form α to $V_{\mathbb{C}}$ shall be given by

$$\alpha: (\xi, \eta) \in V_{\mathbb{C}} \times V_{\mathbb{C}} \mapsto \alpha(\xi, \bar{\eta}) \in \mathbb{C}$$

where the right hand side is to be computed using bilinearity.

A subspace U of $V_{\mathbb{C}}$ is called *real* if $\bar{U} = U$.

3.1.4 Lemma. *Let $P: V \to V$ be a (real) symplectic transformation. Consider the spaces*

$$V(\varrho) = V_P(\varrho) = \{\xi \in V; \text{ there exists an integer } k \geqslant 0 \text{ with } (P\varrho^{-1} - 1)^k \xi = 0\}.$$

Then $V_{\mathbb{C}}$ is the α-orthogonal sum of non-degenerate subspaces of the form

$$(*) \qquad \begin{cases} V(\varrho) \oplus V(\bar{\varrho}^{-1}), & \text{if } \varrho\bar{\varrho} \neq 1 \\ V(\varrho), & \text{if } \varrho\bar{\varrho} = 1. \end{cases}$$

Here ϱ runs through the eigenvalues of P. In particular, $\det P = 1$.

Proof. Let $\varrho\bar{\sigma} \neq 1$. We claim

$$\alpha(V(\varrho), V(\sigma)) = 0.$$

To see this we denote by $V_j(\tau)$, j an integer $\geqslant 0$, the subspace of $V(\tau)$ which is annihilated by $(P\tau^{-1} - 1)^j$. For $\xi \in V_1(\varrho)$, $\eta \in V_1(\sigma)$ we have

$$\alpha(\xi, \eta) = \alpha(P\xi, P\eta) = \varrho\bar{\sigma}\alpha(\xi, \eta) = 0.$$

Assume we know already that

$$\alpha(V_{r-1}(\varrho), V_s(\sigma)) = \alpha(V_r(\varrho), V_{s-1}(\sigma)) = 0.$$

For $(r, s) = (1, 2)$ or $= (2, 1)$ we have just proved this. Then we find for $\xi \in V_r(\varrho)$, $\eta \in V_s(\sigma)$ that

$$\alpha((P\varrho^{-1} - 1)\xi, P\eta) = \alpha(\xi, (P\sigma^{-1} - 1)\eta) = 0.$$

Hence,

$$0 = \alpha(\xi, \eta) - \alpha(P\xi, P\eta) = (1 - \varrho\bar{\sigma})\alpha(\xi, \eta).$$

Finally, for $\xi \in V(\varrho)$, $\xi \neq 0$, there exists a ξ^* with $\alpha(\xi, \xi^*) \neq 0$. We may assume $\xi^* \in V(\bar{\varrho}^{-1})$. If $\varrho\bar{\varrho} \neq 1$, $V(\bar{\varrho}^{-1}) \neq V(\varrho)$ and $V(\varrho) \oplus V(\bar{\varrho}^{-1})$ is non-degenerate. If $\varrho\bar{\varrho} = 1$, $V(\varrho)$ is non-degenerate. □

3.1.5 Corollary. *Let $P: V \to V$ be a symplectic transformation. Then we have for V the decomposition*

$$V = (V_s \oplus V_u) \oplus V_{ce}$$

into P-invariant subspaces. V_s is called the stable subspace of *V (w.r.t.P). It is generated by the $V(\varrho)$ with $|\varrho| < 1$. V_u, the* unstable subspace of *V (w.r.t.P), is generated by the $V(\varrho)$, $|\varrho| > 1$. V_{ce}, the* center subspace of *V (w.r.t.P), is generated by the $V(\varrho)$ with $|\varrho| = 1$. V_s and V_u are isotropic. $V_s \oplus V_u$ as well as V_{ce} are non-degenerate.*

Proof. Immediate from (3.1.4). □

Note. For a finer discussion of a symplectic transformation, leading to a complete classification of the conjugacy classes in the symplectic group, cf. Klingenberg [8].

3.1.6 Definition. A symplectic transformation $P: V \to V$ is called *hyperbolic* if P has no eigenvalue ϱ with $|\varrho| = 1$, i.e., $V = V_s \oplus V_u$. P is called *elliptic* if all eigenvalues of P lie on the unit circle, i.e., $V = V_{ce}$.

Note. If $\dim V = 2$, P either is elliptic or else it is hyperbolic. For $\dim V > 2$, there are various other possibilities, i.e., $\dim(V_s \oplus V_u) = 2p \neq 0$, $\dim V_{ce} = 2q \neq 0$, $2p + 2q = \dim V$.

We now come to the concept of a symplectic manifold. Essentially this is a differentiable manifold N where for each $q \in N$, $T_q N$ is a symplectic space.

Recall from (1.4.11) the definition of a 2-form α on N. In a chart (v, N') of N, α has a representation of the form

$$\sum_{i,j} \alpha_{ij}(v)\, dv^i \wedge dv^j .$$

Here, the $\alpha_{ij}(v)$ are differentiable functions. The differential $d\alpha$ of α is determined from the differential of its representation. $d\alpha = 0$, i.e., α closed, implies that locally, α is the boundary $d\beta$ of a 1-form β, but globally this need not be true. See Abraham and Marsden [1], Kobayashi and Nomizu [1] for more on this.

3.1.7 Definition. A *symplectic manifold* (N, α), or briefly N, is a differentiable manifold N endowed with a closed, non-degenerate 2-form α.

Note. For every $q \in N, (T_q N, \alpha(q))$ is a symplectic space. That α is non-degenerate implies in particular that $\dim N$ is even, say $2n$. Then the nth exterior power α^n of α defines a non-vanishing form of maximal dimension on N, a so-called *volume form*. For $\dim N = 2$ a symplectic manifold (N, α) is the same as a surface with a volume form.

The most important example of a symplectic manifold is given by the cotangent bundle.

3.1.8 Proposition. *Let $N = T^*M$ be the total space of the cotangent bundle τ_M^* of a differentiable manifold. T^*M carries a canonical symplectic structure given by the 2-form $\alpha^* = -d\theta$. Here, the 1-form θ is defined by*

$$\eta \in T_{X^*} T^*M \mapsto \langle X^*, T\tau_M^* \eta \rangle \in \mathbb{R} .$$

Note. In the definition of θ we use the following commutative diagram, representing the tangent functor T applied to τ_M^*:

$$\begin{array}{ccc} TT^*M & \xrightarrow{\tau_{T^*M}} & T^*M \\ \Big\downarrow T\tau_M^* & & \Big\downarrow \tau_M^* \\ TM & \xrightarrow{\tau_M} & M \end{array} ; \qquad \begin{array}{ccc} (u, v, \dot{u}, \dot{v}) & \longmapsto & (u, v) \\ \Big\downarrow & & \Big\downarrow \\ (u, \dot{u}) & \longmapsto & (u) \end{array}$$

Proof. Clearly, α^* is closed. To see that it is non-degenerate we write down the local representation of θ w.r.t. a chart (u, M') of M, cf. the second diagram above. There we have written u instead of $u \circ \tau_M^*$ for the first set of coordinates on T^*M'. Then $\theta(\eta)$ is given by $\langle v, \dot{u} \rangle$ (the pointed brackets denote the pairing between a space and its dual), i.e., θ is represented by $\sum_i v^i du^i$. Hence, $\alpha^* = -d\theta$ has the representation $\sum_i du^i \wedge dv^i$ which shows that α^* is non-degenerate. □

Remark. We see even more. All charts of the canonical atlas of T^*M, derived from an atlas of M, are *symplectic charts*. By this we mean that the representation of the symplectic 2-form α with respect to the chart has 'diagonal form' $\sum_i du^i \wedge dv^i$. In other words, a chart is symplectic if the canonical basis vector fields, determined by this chart (in (1.4.4) we called this the Gauss frame of the chart) form a symplectic basis for α, simultaneously for all points in the domain of the chart.

There exists a symplectic atlas (i.e., an atlas all of whose charts are symplectic) for every symplectic manifold (N, α). This is the content of the so-called *Darboux Lemma*. See e.g. Abraham and Marsden [1] for a simple proof. Since we later consider only the case $N = T^*M$ with its canonical symplectic structure α^*, we do not need this Lemma – in our case we already have a symplectic atlas.

We now come to the main topic of this section.

3.1.9 Definition. A *Hamiltonian system* (N, α, H) is a symplectic manifold (N, α) endowed with a differentiable function $H: N \to \mathbb{R}$. Its *Hamiltonian vector field* ζ_H is defined by

$$\zeta_H = L_\alpha^{-1} DH, \text{ i.e., } DH = \alpha(\zeta_H, \ ).$$

The integral curves $t \in \mathbb{R} \mapsto \phi_t q \in N$ of ζ_H, are called the *Hamiltonian flow*.

Remark. To obtain the local representation of the Hamiltonian vector field in a symplectic chart of N we write the symplectic coordinates in the form (u, v) so that α is represented by $\sum_i du^i \wedge dv^i$. Let (ξ_H, η_H) be the corresponding decomposition of the Hamiltonian field ζ_H. Then the identity $\alpha(\zeta_H, \ ) = DH$ reads

$$-\sum_i du^i \eta_H^i + \sum_i dv^i \xi_H^i = \sum_i \frac{\partial H}{\partial u^i} du^i + \sum_i \frac{\partial H}{\partial v^i} dv^i,$$

i.e.,

$$\xi_H^i = H_{v^i}, \eta_H^i = -H_{u^i}.$$

The differential equation for the Hamiltonian flow therefore becomes

$$\frac{du^i}{dt} = H_{v^i}(u, v); \quad \frac{dv^i}{dt} = -H_{u^i}(u, v).$$

3.1.10 Proposition. *Let* $\phi_t: N \to N$ *be the Hamiltonian flow of the system* (N, α, H). *Then*

(i) $H(\phi_t q) = \text{const.}$

(ii) $\phi_t^* \alpha = \alpha$, *i.e.,* $\alpha(T\phi_t . \xi, T\phi_t . \eta) = \alpha(\xi, \eta)$ *does not depend on* t, *for arbitrary* ξ, η.

Proof. (i) follows from

$$\frac{d}{dt} H(\phi_t q) = DH . \frac{d}{dt} \phi_t q = \alpha(\zeta_H, \zeta_H) = 0 .$$

For the proof of (ii) we refer to the literature, e.g., Abraham and Marsden [1]. It makes use of some simple techniques on forms which we don't have at our disposal. However, for the case of the geodesic flow (3.1.12), which is our main subject, we will offer in (3.1.16) a proof of (ii), drawing on Riemannian geometry. □

3.1.11 Example. *The co-geodesic flow.* Let M be a Riemannian manifold. Consider the symplectic manifold (T^*M, α^*) with $\alpha^* = -d\theta$. As Hamiltonian function take

$$E^*(X^*) = \tfrac{1}{2} g^*(X^*, X^*) .$$

Here, g^* is the scalar product on $T_p^* M$, induced from the scalar product g on $T_p M$. The flow of the system (T^*M, α^*, E^*) is called the *co-geodesic flow*.

From the Riemannian metric g on M we get a bundle isomorphism

$$\begin{array}{ccc} TM & \xrightarrow{L_g} & T^*M \\ \downarrow{\scriptstyle \tau_M} & & \downarrow{\scriptstyle \tau_M^*} \\ M & \xrightarrow{id} & M \end{array} \quad ; \quad \begin{array}{ccc} (u^i, \dot u^i) & \longmapsto & (u^i, \sum_k g_{ik}(u) \dot u^k) \\ \downarrow & & \downarrow \\ (u^i) & \longmapsto & (u^i) \end{array}$$

Here, $L_g | T_p M$ is given by $X \mapsto g(X, \ )$, cf. the beginning of (1.11).

By α we denote the pull-back of the canonical symplectic form α^* on T^*M, i.e., $\alpha(\xi, \eta) = \alpha^*(TL_g . \xi, TL_g . \eta)$. The pullback E of E^* yields the kinetic energy

$$E: TM \to \mathbb{R};\ X \mapsto \tfrac{1}{2} g(X, X) = \tfrac{1}{2} g^*(L_g X, L_g X) .$$

3.1.12 Definition. Let M be a Riemannian manifold. The *geodesic (Hamiltonian) system (associated with M)* is the Hamiltonian system (TM, α, E). Its flow $\phi_t: TM \to TM$ is called *geodesic flow*.

The justification for this terminology is given by the following

3.1.13 Proposition. *Let* M *be a complete Riemannian manifold.*

(i) *The flow lines* $\phi_t X$ *of the geodesic system* (TM, α, E) *are defined for all* $t \in \mathbb{R}$. *The projection* $\tau_M \phi_t X$, $t \in \mathbb{R}$, *of such a flow line is the geodesic* $c_X(t)$ *determined by* $\dot c_X(0) = X$. $\phi_t X = \dot c_X(t)$.

(ii) *The non-constant periodic flow lines* $\phi_t X$, i.e., $\phi_\omega X = X$, *some* $\omega > 0$, *and* $|X| \neq 0$, *are in* 1 : 1 *correspondence with the non-constant closed geodesics on* M. *And hence, according to* (2.4.3), *they are, up to parameterization, in* 1 : 1 *correspondence with the non-constant critical points of the functional* $E: \Lambda M \to \mathbb{R}$ *on the Hilbert manifold* ΛM *of closed curves on* M.

Proof. To show that $\tau_M \phi_t X$ is the geodesic $c_X(t)$ on M, with $\dot{c}_X(0) = X$ we use local coordinates. A more conceptual proof of this fact will be given in (3.1.15).

Let $g_{ik}(u)$ be the (i, k)-coefficient of the symmetric matrix representing the Riemannian metric g in $T_u U$, where U is the range of a chart (u, M'), cf. (1.8.2, ii). From the proof of (1.6.5) we then know that a curve $t \mapsto (u^i(t))$ is a geodesic precisely if it satisfies the equations

$$(*) \qquad \frac{d^2 u^k}{dt^2}(t) + \sum_{i,j} \Gamma_{ij}^k(u(t)) \frac{du^i}{dt}(t) \frac{du^j}{dt}(t) = 0 .$$

Here, the $\Gamma_{ij}^k(u)$ are the coefficients of the Christoffel symbols, cf. (1.8.12).

The function $E^*: T^* M \to \mathbb{R}$ now reads $E^*(u, v) = \frac{1}{2} \sum_{i,k} g^{ik}(u) v^i v^k$. The $g^{ik}(u)$ are the elements of the metric tensor g^* on $T_u^* U$, i.e., the solutions of the equations $\sum_j g_{ij}(u) g^{jk}(u) = \delta_i^k$.

The Hamiltonian equations of the system $(T^* M, \alpha^*, E^*)$ now become

$$\frac{du^i}{dt} = E_{v^i}^*(u, v) = \sum_k g^{ik}(u) v^k ; \quad \frac{dv^i}{dt} = -E_{u^i}^*(u, v) = -\frac{1}{2} \sum_{l,m} \frac{\partial g^{lm}(u)}{\partial u^i} v^l v^m .$$

Inserting the relation $v^l = \sum_r g_{rl}(u) \dfrac{du^r}{dt}$ from the first equation into the second one we get $(*)$.

If $\phi_\omega X = X$, then $c(t), 0 \leq t \leq 1$, with $\dot{c}(0) = \omega X/|X|$, is a critical point of $E: \Lambda M \to \mathbb{R}$. Conversely, a critical point $c \neq$ const of E determines the periodic orbits $\phi_t X$, with $X = a\dot{c}(0) \neq 0$. □

To relate the system (TM, α, E) further with the geometry of the tangent bundle $\tau_M: TM \to M$ we recall from (1.5.10), (1.9.12) the decomposition

$$T_X TM = T_{Xh} TM \oplus T_{Xv} TM$$

of the tangent space $T_X TM$ into its horizontal and its vertical subspace. Each of these subspaces is canonically isomorphic to $T_p M, p = \tau_M X \in M$. This isomorphism is used to define a Riemannian metric on TM. We also freely use this isomorphism to identify elements $\xi_h \in T_{Xh} TM$ and $\xi_v \in T_{Xv} TM$ with their images in $T_p M$.

We first give a description of the symplectic form α.

3.1.14 Proposition. *The symplectic* 2*-form* α *of the geodesic system* (TM, α, E) *can be* *described by the formula*

$$(\dagger) \qquad \alpha(\xi, \eta) = \langle \xi_h, \eta_v \rangle - \langle \eta_h, \xi_v \rangle .$$

Here, the elements ξ *and* η *of* $T_X TM$ *are decomposed into their horizontal and vertical components,* $\xi = (\xi_h, \xi_v)$, $\eta = (\eta_h, \eta_v)$. $\langle\ ,\ \rangle$ *denotes the (Riemannian) scalar product on* $T_p M, p = \tau_M X$.

Note. Compare this with (1.12.3, ii).

Proof. Both sides in $(\dagger)$ are defined intrinsically. Therefore, to prove $(\dagger)$ it suffices to verify it in special coordinates. Let $X \in TM$. Put $\tau_M X = p$. We consider normal coordinates (u, M') based at p. Thus, $u(p) = 0$, $g_{ik}(0) = \delta_{ik}$, $\Gamma_{ij}^k(0) = 0$, cf. (1.8.16). The isomorphism $L_g \colon T_p M \to T_p^* M$ is then represented by $v^i = \dot{u}^i$. Since α^* has in the symplectic coordinates (u, v) the representation $\sum_i du^i \wedge dv^i$, α in $T_X TM$ is given by $\sum_i du^i \wedge d\dot{u}^i$.

Let X be represented by $(0, \dot{u})$. Since the $\Gamma(0)$'s are 0, $T_{Xh} TM$ and $T_{Xv} TM$ have the representation

$$\{0, \dot{u}\} \times \mathbb{E} \times \{0\} \quad \text{and} \quad \{0, \dot{u}\} \times \{0\} \times \mathbb{E},$$

$\mathbb{E} =$ model of M, cf. the proof of (1.5.10).

Let $(X, \dot{X}) \in \mathbb{E} \times \mathbb{E}$, $(Y, \dot{Y}) \in \mathbb{E} \times \mathbb{E}$ be the principal parts of the representation of ξ and η in $T_X TM$. The evaluation of α on the pair then yields $\langle X, \dot{Y} \rangle - \langle Y, \dot{X} \rangle$. But this is just the right hand side of $(\dagger)$. □

3.1.15 Lemma. *The Hamiltonian vector field* ζ_E *of the geodesic system* (TM, α, E) *is given by* $\zeta_E(X) = (X, 0) \in T_{Xh} TM \oplus T_{Xv} TM$. *It follows that the geodesic flow lines* $\phi_t X$ *project under* $\tau_M \colon TM \to M$ *onto geodesics.*

Proof. Clearily, it suffices to verify this in special coordinates. As such we again take the normal coordinates based at $p = \tau_M X$. Let $(\dot{u}^i)$ be the principal part of the representation of X. Then we have from the proof of (3.1.13) that the representation $(E_{v^i}^*(u, v), -E_{u^i}^*(u, v))$ of ζ_{E^*} for $u = 0$ reads $(v^i, 0) = (\dot{u}^i, 0)$. But this is just the principal part of the representation of $(X, 0) \in T_X TM$.

The defining equation $\nabla \dot{c}(t) = 0$ of a geodesic $c(t)$ now reads

$$\frac{d}{dt} \dot{c}(t) = (\dot{c}(t), 0) = \zeta_E(\dot{c}(t)).$$

Thus, $\dot{c}(t) = \phi_t \dot{c}(0)$. □

Recall from (1.9.12) that TM carries a canonical Riemannian metric. With this we have

3.1.16 Proposition. *A geodesic flow line* $\phi_t X$, $t \in \mathbb{R}$, *is a geodesic in* TM. *Let* $E(X) = a > 0$. *Then this geodesic lies in the codimension one submanifold* $E^{-1}(a)$.

That is, $\phi_t X$, $t \in \mathbb{R}$, is also a geodesic on this submanifold with its induced Riemannian metric.

Proof. $\tau_M: TM \to M$ is a Riemannian submersion, cf. (1.11.9) and the subsequent example. According to (3.1.15), $\tau_M \phi_t X$, $t \in \mathbb{R}$ is a geodesic on M and $\phi_t X$, $t \in \mathbb{R}$, is a horizontal lift of this geodesic. Now apply (1.11.11).

If $a > 0$, a is a regular value of $E: TM \to \mathbb{R}$, since $DE(X).(0, X) = \alpha(\zeta_E(X), (0, X)) = \langle X, X \rangle = 2a > 0$. Thus, according to (1.3.8), $E^{-1}(a)$ is a submanifold. $E(\phi_t X) = E(X)$, see (3.1.10). □

We continue with a characterization of Jacobi fields.

3.1.17 Lemma. *The Jacobi fields $Y(t)$ along a geodesic $c(t)$ on M are in 1 : 1 correspondence with the flow-invariant vector fields $\eta(t)$ along the corresponding flow line $\phi_t \dot{c}(0) = \dot{c}(t)$. The correspondence is given by*

$$Y(t) \in T_{c(t)} M \leftrightarrow \eta(t) = \big(Y(t), \nabla Y(t)\big) \in T_{\dot{c}(t)h} TM \oplus T_{\dot{c}(t)v} TM.$$

Proof. Consider a vector $\eta_0 = (\eta_{0h}, \eta_{0v}) \in T_{\dot{c}(0)h} TM \oplus T_{\dot{c}(0)v} TM$. Then $T\phi_t . \eta_0$, $t \in \mathbb{R}$, is a flow-invariant vector field $\eta(t)$ along $\dot{c}(t)$. We write

$$\eta(t) = \big(\eta_h(t), \eta_v(t)\big) \in T_{\dot{c}(t)h} TM \oplus T_{\dot{c}(t)v} TM.$$

Let $k(s)$, $|s|$ small, be a curve in TM with $k'(0) = \eta_0$. Then $\eta(t) = \frac{\partial}{\partial s} \phi_t k(s)|_0$.

From the description of the Hamiltonian vector field ζ_E in (3.1.15) we get

$$\frac{\partial}{\partial t} \tau_M \phi_t k(s) = T\tau_M . \frac{\partial}{\partial t} \phi_t k(s) = \phi_t k(s).$$

Observe now that $t \mapsto \tau_M \phi_t k(s)$ is a geodesic, for each fixed s. Hence, according to (1.12.4),

$$\frac{\partial}{\partial s} \tau_M \phi_t k(s)|_0 = T\tau_M . \frac{\partial}{\partial s} \phi_t k(s) = \eta_h(t)$$

is a Jacobi field $Y(t)$ along $c(t)$. Moreover,

$$\eta(t) = \frac{\partial}{\partial s} \phi_t k(s)|_0 = \big(T\tau_M . \frac{\partial}{\partial s} \phi_t k(s)|_0, K . \frac{\partial}{\partial s} \phi_t k(s)|_0\big) =$$

$$= \big(\eta_h(t), \frac{\nabla}{\partial s} \frac{\partial}{\partial t} \tau_M \phi_t k(s)|_0\big) = \big(\eta_h(t), \nabla \eta_h(t)\big).$$

In the last equation we made use of (1.5.8). □

We now can prove (3.1.10, ii) for the geodesic system (TM, α, E).

3.1.18 Corollary. *In the geodesic system (TM, α, E), $\alpha(T\phi_t . \xi, T\phi_t . \eta) = \alpha(\xi, \eta)$, all ξ, η in some $T_X TM$, all $t \in \mathbb{R}$.*

Proof. According to (3.1.14), (3.1.17) we may write

$$\alpha(T\phi_t \,.\, \xi, T\phi_t \,.\, \eta) = \langle X(t), \nabla Y(t) \rangle - \langle Y(t), \nabla X(t) \rangle$$

where $X(t)$, $Y(t)$ are Jacobi fields along the geodesic $\tau_M \phi_t X$, $t \in \mathbb{R}$, on M. That this expression does not depend on $t \in \mathbb{R}$ follows from (1.12.3, ii). $\square$

For a general Hamiltonian system (N, α, H), the level sets $\{H = \text{const}\}$ are invariant under the flow, see (3.1.10, i). Whenever the constant is a regular value of H, $\{H = \text{const}\}$ is a submanifold of codimension 1 in N. In general, the flows on two different level sets are not isomorphic. However for the geodesic system (TM, α, E) we have the following

3.1.19 Proposition. *In the system (TM, α, E), the restrictions of the geodesic flow to any two level hypersurfaces $\{E = a > 0\}$ and $\{E = b > 0\}$ are isomorphic, if we reparameterize the flow. That is,*

$$X \in E^{-1}(a) \leftrightarrow \sqrt{\frac{b}{a}}\, X \in E^{-1}(b)$$

and

$$\phi_t X \in E^{-1}(a) \leftrightarrow \phi_t \sqrt{\frac{b}{a}}\, X \in E^{-1}(b)\,.$$

Proof. Clear from $E\left(\sqrt{\frac{b}{a}}\, X\right) = \frac{b}{a}\, E(X)$. $\square$

For an investigation of the structure of the geodesic flow $\phi_t\colon TM \to TM$ it therefore suffices to consider the restriction $\phi_t | T_1 M$ to the unit tangent bundle $T_1 M$. In this way, the projected curves $\tau_M \phi_t X$, $X \in T_1 M$, are always parameterized by arc length.

3.2 Properties of the Geodesic Flow

We continue to consider complete – if not compact – Riemannian manifolds of finite dimension. We want to study in more detail the geodesic flow on the tangent bundle of such a manifold. While many of the subsequent constructions and results also hold for general Hamiltonian flows, quite a few are valid only for the geodesic flow. Since we are interested mainly in Riemannian geometry we take full advantage of the special situation at hand.

We start with a useful splitting of the tangent bundle $\tau_{T_1 M}$ of $T_1 M$.

3.2.1 Proposition. *The tangent bundle $\tau_{T_1 M}\colon TT_1 M \to T_1 M$ of $T_1 M$ possesses a $T\phi_t$-invariant splitting*

$$\tau_{T_1 M} = \tau^1_{T_1 M} \oplus \bar{\tau}_{T_1 M}$$

into two subbundles. Here, the 1*-dimensional fibre* $T^1_{X_0} T_1 M$ *of* $\tau^1_{T_1 M}$ *over* $X_0 \in T_1 M$ *is generated by the flow vector* $\zeta_E(X_0) = (X_0, 0)$. *The fibre* $\bar{T}_{X_0} T_1 M$ *of* $\bar{\tau}_{T_1 M}$ *over* $X_0 \in T_1 M$ *is the* α*-orthogonal complement in* $T_{X_0} TM$ *of the* 2*-dimensional subspace* $T^2_{X_0} TM$ *generated by the vectors* $(X_0, 0) \in T_{X_0 h} TM$ *and* $(0, X_0) \in T_{X_0 v} TM$. $T^2_{X_0} TM$ *and hence also* $\bar{T}_{X_0} T_1 M$ *are nondegenerate.*

Putting

$$\bar{T}_{X_0} T_1 M \cap T_{X_0 h} TM = \bar{T}_{X_0 h} T_1 M$$

$$\bar{T}_{X_0} T_1 M \cap T_{X_0 v} TM = \bar{T}_{X_0 v} T_1 M$$

we get a splitting of the fibre $\bar{T}_{X_0} T_1 M$ *over* X_0 *of* $\bar{\tau}_{T_1 M}$ *into two totally isotropic (or Lagrangian) subspaces, called the* horizontal *and* vertical *subspaces, respectively. This yields a decomposition of* $\bar{\tau}_{T_1 M}$ *into its horizontal and its vertical subbundles*

$$\bar{\tau}_{T_1 M} = \bar{\tau}_{T_1 Mh} \oplus \bar{\tau}_{T_1 Mv}.$$

Note. The decomposition of $\bar{\tau}_{T_1 M}$ into its horizontal subbundle $\bar{\tau}_{T_1 Mh}$ and vertical subbundle $\bar{\tau}_{T_1 Mv}$ is not $T\phi_t$-invariant.

Proof. $\tau^1_{T_1 M}$ clearly is $T\phi_t$-invariant. Denote by $\tau^2_{T_1 M}$ the subbundle of τ_{TM} where the fibre over $X_0 \in T_1 M$ given by $T^2_{X_0} TM$.

$T^2_{X_0} TM$ is a non-degenerate subspace of $T_{X_0} TM$ since

$$\alpha((X_0, 0), (0, X_0)) = \langle X_0, X_0 \rangle = 1,$$

cf. (3.1.14). Therefore, the α-orthogonal subspace $\bar{T}_{X_0} TM$ of $T^2_{X_0} TM$ is also non-degenerate.

Since

$$T\phi_t(X_0, 0) = (\phi_t X_0, 0),\ T\phi_t(0, X_0) = (t\phi_t X_0, \phi_t X_0)$$

we see that $T\phi_t T^2_{X_0} TM = T^2_{\phi_t X_0} TM$. Hence $\bar{T}_{X_0} T_1 M$ is transported under $T\phi_t$ into $\bar{T}_{\phi_t X_0} T_1 M$. Finally, we have from (3.1.14) that the horizontal and the vertical subspace of $\bar{T}_{X_0} T_1 M$ are Lagrangian subspaces. □

3.2.2 Complement. *Let* $X_0 \in T_1 M$. *Consider the immersion*

$$f \equiv f_{X_0} : t \in \mathbb{R} \mapsto \phi_t X_0 \in T_1 M.$$

Then, from the bundle $\bar{\tau}_{T_1 M}$ *over* $T_1 M$ *we obtain the induced bundle* $\upsilon \equiv \upsilon_{X_0} : V = V_{X_0} \to \mathbb{R}$, *cf.* (1.3.10). *That is, we have the following bundle morphism*

$$\begin{array}{ccc} V & \xrightarrow{\quad F \quad} & \bar{T}T_1 M \\ \Big\downarrow \upsilon & & \Big\downarrow \bar{\tau}_{T_1 M} \\ \mathbb{R} & \xrightarrow{\quad f \quad} & T_1 M \end{array}$$

where $F \equiv F_{X_0}$, *restricted to a fibre* V_t *over* $t \in \mathbb{R}$ *of* υ, *is an isomorphism.*

Note. The identification of V_t with $\bar{T}_{\phi_t X_0} T_1 M$ under $F | V_t$ (or briefly F) induces on V_t a symplectic structure. The non-degenerate 2-form on V_t will also be denoted by α. In addition, the splitting $\bar{\tau}_{T_1 M} = \bar{\tau}_{T_1 Mh} \oplus \bar{\tau}_{T_1 Mv}$ induces a splitting $\upsilon = \upsilon_h \oplus \upsilon_v$ of υ into a *horizontal* and a *vertical subbundle*. For the fibre V_t we write this splitting into two Lagrangian subspaces in the form $V_t = V_{t,h} \oplus V_{t,v}$. $V_{t,h}$ and $V_{t,v}$ are the *horizontal* and the *vertical* subspace of V_t, respectively.

The 1-parameter group $\bar{T}\phi_t = T\phi_t | \bar{T}T_1 M$, $t \in \mathbb{R}$, of diffeomorphisms of $\bar{T}T_1 M$ induces on the total space V of υ a 1-parameter group of *fibre preserving diffeomorphisms* P_t, $t \in \mathbb{R}$. For each $t_0 \in \mathbb{R}$, all $t \in \mathbb{R}$,

$$P_t | V_{t_0} : V_{t_0} \to V_{t+t_0}$$

is a symplectic linear transformation.

The P_t-invariant sections $\eta(t)$ in υ, $\eta(t) = P_t \eta(0)$, correspond under the canonical isomorphism F to the vector fields $(Y(t), \nabla Y(t)) \in \bar{T}_{\phi_t X_0 h} T_1 M \oplus \bar{T}_{\phi_t X_0 v} T_1 M$ where $Y(t)$ is a Jacobi field, orthogonal to the underlying geodesic $\tau \phi_t X_0$.

Finally, υ carries a Riemannian metric, i.e., the riemannian metric induced by F from the Riemannian metric on $\bar{T}T_1 M$, considered as submanifold of codimension 2 of the Riemannian manifold TTM, cf. (1.5.12).

Proof. Everything is an immediate consequence of (3.2.1). □

In case, $\phi_t X_0$, $t \in \mathbb{R}$, is periodic, i.e., $\phi_\omega X_0 = X_0$, for some $\omega > 0$, we get the following analogue of (3.2.2):

3.2.3 Proposition. *Let $\phi_t X_0$, $0 \leqslant t \leqslant \omega$, be a periodic geodesic flow line. Then the mapping $f \equiv f_{X_0} : \mathbb{R} \to T_1 M$; $t \mapsto \phi_t X_0$ yields a mapping $f \equiv f_{X_0} : S_\omega = \mathbb{R}/\omega\mathbb{Z}$ of the circle $S_\omega = [0, \omega]/\{0, \omega\}$ into $T_1 M$. This mapping induces from the bundle $\bar{\tau}_{T_1 M}$ the bundle*

$$\upsilon : V \to S_\omega$$

having the same structure as the bundle υ defined in (3.2.2). *In particular, since the fibre V_0 over $0 \in S_\omega$ is identified with the fibre V_ω over $\omega \in S_\omega$, $P_\omega : V_0 \to V_0$ is a symplectic transformation, called the* linear Poincaré mapping, *associated with the periodic orbit $\phi_t X_0$, $0 \leqslant t \leqslant \omega$.* □

Remark. From the note after (3.2.2), $P_\omega : V_0 \to V_0$ can be described as follows: We identify V_0 via $F | V_0$ with $\bar{T}_{X_0} T_1 M$. Write $\eta \in \bar{T}_{X_0} T_1 M$ as $\eta = (\eta_h, \eta_v) \in \bar{T}_{X_0 h} \oplus \bar{T}_{X_0 v}$. Let $Y(t)$ be the Jacobi field along $c_{X_0}(t) = \tau_M \phi_t X_0$ with $\langle Y(t), \dot{c}_{X_0}(t) \rangle = 0$ and $Y(0) = \eta_h$, $\nabla Y(0) = \eta_v$. Then $P_\omega \eta = (Y(\omega), \nabla Y(\omega))$.

We now come to the definition of the non-linear Poincaré mapping associated with a periodic geodesic flow line. This mapping can be defined for a periodic flow line of arbitrary Hamiltonian systems (N, α, H), (even for periodic flow lines of general dynamical systems), provided $DH \neq 0$ on the flow line. Here, we shall employ the special structure of the geodesic systems to make the necessary constructions in an intrinsic and canonical manner, i.e., free of choices which are not dictated by the circumstances. This is true in particular for the construction of the local flow-transversal hypersurface given below.

Recall from (1.9.12) that TM has a canonical Riemannian metric. A flow line $\phi_t X_0$, $t \in \mathbb{R}$, in $T_1 M$ is a geodesic in $T_1 M$, see (3.1.16). In (1.12.11) we had defined Fermi coordinates along a geodesic.

3.2.4 Definition. Let $X_0 \in T_1 M$. Then we define, for every sufficiently small $\varrho > 0$, the *local flow-transversal hypersurface at* X_0, $\Sigma_\varrho(X_0)$ or simply $\Sigma(X_0)$ as the image under $\exp_{X_0}$ of the ϱ-Ball $\bar{B}_\varrho(0_{X_0}) \cap \bar{T}_{X_0} T_1 M$ in $\bar{T}_{X_0} T_1 M$. $\exp_{X_0}$ denotes the restriction of the exponential map in $TT_1 M$ to $T_{X_0} T_1 M$ (or to some open neighborhood of the origin 0_{X_0} of $T_{X_0} T_1 M$, if we are not sure whether $TT_1 M$ is geodesically complete).

Note. In general, $\phi_t \Sigma_\varrho(X_0)$ will not – even near to $\phi_t X_0$ – coincide with $\Sigma_\varrho(\phi_t X_0)$, since $\phi_t: T_1 M \to T_1 M$ does not operate as isometry. Only the tangent space of $\Sigma_\varrho(X_0)$ at X_0, i.e. $\bar{T}_{X_0} T_1 M$, is carried under $T\phi_t$ into $\bar{T}_{\phi_t X_0} T_1 M$, i.e. the tangent space of $\Sigma_\varrho(\phi_t X_0)$ at $\phi_t X_0$.

3.2.5 Proposition. *For $\varrho > 0$ sufficiently small, the symplectic form α on TM, restricted to $\Sigma_\varrho(X_0)$, induces on $\Sigma_\varrho(X_0)$ the structure of a symplectic submanifold of codimension* 2.

Proof. Consider $\alpha \mid T_{X_0} \Sigma_\varrho(X_0) = \alpha \mid \bar{T}_{X_0} T_1 M$. From (3.2.1) we know that this restriction is non-degenerate. Hence, this is true by continuity also for $\alpha \mid T_X \Sigma_\varrho(X_0)$ if X is sufficiently near X_0. □

The existence of the Poincaré mapping is now a consequence of the following Lemma.

3.2.6 Lemma. *Let $\phi_t X_0, 0 \leqslant t \leqslant \omega$, be a periodic geodesic flow line. Let $\Sigma = \Sigma_\varrho(X_0)$ be the small flow-transversal hypersurface. Then there exist neighborhoods Σ_0, Σ_ω of X_0 on Σ and a differentiable function $\delta: \Sigma_0 \to \mathbb{R}$ with $\delta(X_0) = 0$ such that the mapping*

$$(*) \qquad \mathscr{P}: X \in \Sigma_0 \mapsto \phi_{\omega + \delta(X)}(X) \in \Sigma_\omega$$

is a symplectic diffeomorphism. That is to say, $\mathscr{P}$ is a diffeomorphism which preserves the symplectic structure on $\Sigma_0 \subset \Sigma$ and $\Sigma_\omega \subset \Sigma$.

3.2.7 Definition. The *Poincaré mapping* (associated with the periodic orbit $\phi_t X_0, 0 \leqslant t \leqslant \omega$) is the mapping $\mathscr{P}$ defined in (3.2.6).

Remarks. 1. We will prove the Lemma in the case that ω is the prime period of the flow line $\phi_t X_0$, $t \in \mathbb{R}$. I.e., ω is the smallest positive number with $\phi_\omega X_0 = X_0$. Every other period is of the form $m\omega$, m an integer > 1. The Poincaré mapping associated with $\phi_t X_0, 0 \leqslant t \leqslant m\omega$, is the mth power $\mathscr{P}^m$ of the Poincaré mapping $\mathscr{P}$ associated with $\phi_t X_0, 0 \leqslant t \leqslant \omega$.

The existence of a prime period follows from the observation that there exists an $\varepsilon = \varepsilon(X_0) > 0$ such that $t \in [0, \varepsilon] \mapsto \phi_t X_0 \in T_1 M$ is injective.

2. The Poincaré mapping gives information about the flow $\phi_t: T_1 M \to T_1 M$ in a neighborhood of the periodic orbit. For instance, a fixed point X of $\mathscr{P}$, $\mathscr{P} X = X$, describes the fact that $\phi_t X, 0 \leqslant t \leqslant \omega + \delta(X)$, is a periodic orbit near the given one, $\phi_t X_0, 0 \leqslant t \leqslant \omega$. More generally, if $X \in \Sigma_0$ is a periodic point of $\mathscr{P}$ of period m, m an integer $\geqslant 1$, (i.e., $\mathscr{P}^m X = X$, where we assume that $\mathscr{P}^m$ is defined for X) then this means that the flow line $\phi_t X$ starting at X is periodic of period approximately equal to $m\omega$.

If $X \in \Sigma_0$ is such that $\mathscr{P}^m X$ is defined for $m = 1, 2, \ldots$ and $\lim_{m \to \infty} \mathscr{P}^m X = X_0$ then it means that the flow line $\phi_t X$, $t \in \mathbb{R}$, approaches the flow line $\phi_t X_0$, $t \in \mathbb{R}$, asymptotically as $t \to \infty$.

Note, however, that $\mathscr{P}$ does not suffice to give a complete description of the flow near $\phi_t X_0, 0 \leqslant t \leqslant \omega$. E.g., $\mathscr{P}$ does not tell us how often $\phi_t X, 0 \leqslant t \leqslant \omega + \delta(X)$, has turned around the periodic orbit $\phi_t X_0, 0 \leqslant t \leqslant \omega$, through X_0, when going from Σ_0 to Σ_ω. If we want to talk of such turning around, we need a frame of reference along the periodic orbit $\phi_t X_0, 0 \leqslant t \leqslant \omega$, of course. At least if $\dim M = 2$, and hence $\dim \bar{T}_{\phi_t X_0} T_1 M = 2$, the vertical line and the horizontal line in $\bar{T}_{\phi_t X_0} T_1 M$ provide such a frame. In this case, we can count the number of times that the variation vector field $T\phi_t . \eta$, determined by a family $\phi_t X_s, 0 \leqslant s \leqslant 1$, of nearby orbits, crosses the vertical line. Cf. (3.2.13)for more on this.

Proof of the Lemma (3.2.6). Let $\omega > 0$ be a prime period. Then there exists $\eta > 0$ and a neighborhood U of X_0 in $T_1 M$ such that

$$\phi_\omega: U \to \phi_\omega U$$

is a diffeomorphism and

$$\phi_t U \cap U = \emptyset, \quad \text{for} \quad \eta < t < -\eta + \omega.$$

This follows from $\phi_t X_0 \neq X_0$ for $0 < t < \omega$ and the continuity of the flow mapping

$$(t, X) \in \mathbb{R} \times T_1 M \mapsto \phi_t X \in T_1 M.$$

Choose $\Sigma \subset \phi_\omega^{-1} U \cap U$ and $\varepsilon > 0$ sufficiently small. Then

$$\psi:]-\varepsilon + \omega, \omega + \varepsilon[\times \Sigma \to T_1 M; \ (t, X) \mapsto \phi_t X$$

will have its image in U. Since

$$T_{(\omega, X_0)} \psi: \mathbb{R} \times T_{X_0} \Sigma \to \mathbb{R} \times T_{X_0} \Sigma$$

has maximal rank there exists $\varrho > 0$ small and an $\varepsilon > 0$, possibly smaller than the one chosen above, such that

$$\psi:]-\varepsilon + \omega, \omega + \varepsilon[\times \Sigma_\varrho(X_0) \to \operatorname{im} \psi$$

is a (bijective) diffeomorphism. $X_0 \in \operatorname{im} \psi$ implies that some open neighborhood Σ_ω

of X_0 in $\Sigma = \Sigma_\varrho(X_0)$ belongs to $\operatorname{im}\psi$. Define now

$$\chi: \Sigma_\omega \to \chi(\Sigma_\omega) = \Sigma_0;\ \gamma: \Sigma_\omega \to \mathbb{R}$$

by

$$\psi(\omega + \gamma(X'), \chi(X')) \equiv \phi_{\omega+\gamma(X')}\chi(X') = X'$$

Then χ^{-1} is the desired mapping $\mathscr{P}$ with $\delta = \gamma \circ \chi^{-1}$. Indeed, if we put $\chi(X') = X$, $\phi_{\omega+\delta(X)}(X) = \chi^{-1}(X) \in \Sigma_\omega$.

It remains to show that $\mathscr{P}$ is a symplectic diffeomorphism, i.e., $\alpha(T\mathscr{P}\,.\,\xi, T\mathscr{P}\,.\,\eta) = \alpha(\xi, \eta)$. To see this we write the tangential of $\mathscr{P}$ in the form $T\psi = T_1\psi + T_2\psi$, with $\mathscr{P}X = \psi(\omega + \delta(X), X)$. Here, $T_1\psi = T\psi\,|\,T(]-\varepsilon+\omega, \omega+\varepsilon[)$; $T_2\psi = T\psi\,|\,T\Sigma_0$.

Since $T_1\psi\,.\,\xi$, $T_1\psi\,.\,\eta$ are linearly dependent,

$$\alpha(T_1\psi\,.\,\xi, T_1\psi\,.\,\eta) = 0.$$

From $\alpha(d\phi_t X/dt, \ ) = DE(\phi_t X)$

we get

$$\alpha(T_1\psi\,.\,\xi, \ ) = T\delta\,.\,\xi\, DE.$$

Hence, since $E\,|\,\Sigma = \text{const}$,

$$\alpha(T_1\psi\,.\,\xi, T_2\psi\,.\,\eta) = T\delta\,.\,\xi\, DE\,.\,(T_2\psi\,.\,\eta) = 0.$$

Therefore

$$\alpha(T\psi\,.\,\xi, T\psi\,.\,\eta) = \alpha(T_2\psi\,.\,\xi, T_2\psi\,.\,\eta) = \alpha(\xi, \eta),$$

where we use that

$$T_2\psi\,|\,T_X\Sigma: \xi \mapsto T_X\phi_{\omega+\delta(X)}\,.\,\xi$$

is a symplectic transformation. □

The relation between $\mathscr{P}$ and P_ω is given by the

3.2.8 Proposition. *Let $\phi_t X_0, 0 \leqslant t \leqslant \omega$, be a periodic geodesic flow line. Then the linear Poincaré mapping $P_\omega: V_0 \to V_0$, introduced in* (3.2.3), *corresponds via the canonical isomorphism $F: V_0 \to \bar{T}_{X_0}T_1M$ to the tangential $T_{X_0}\mathscr{P}$ of $\mathscr{P}$ at the origin X_0 of the local transversal hypersurface $\Sigma_\varrho(X_0)$.*

Proof. Note that $T_{X_0}\mathscr{P}$ carries $T_{X_0}\Sigma_0$ into $T_{X_0}\Sigma_\omega = T_{X_0}T_1M$. □

As we saw in (3.1.13), the periodic flow lines of $\phi_t: T_1M \to T_1M$ are in 1:1 correspondence with the non-constant critical points of the functional $E: \Lambda M \to \mathbb{R}$ on the Hilbert manifold ΛM of closed curves. The next Theorem is a first example of a relation between properties of a periodic flow line and properties of the corresponding critical point.

3.2.9 Theorem. *Let $\phi_t X_0, 0 \leqslant t \leqslant \omega$, be a periodic orbit. Then the dimension of the*

eigenspace for the eigenvalue 1 *of the associated linear Poincaré mapping* P_ω *is equal to the nullity of the corresponding critical point* $c(t) = \tau_M \phi_{\omega t} X, 0 \leqslant t \leqslant 1$ *of* $E: \Lambda M \to \mathbb{R}$.

Proof. $P_\omega \eta = \eta$ means that the Jacobi field $Y(t)$ satisfying $F\eta(t) = (Y(t), \nabla Y(t))$ is periodic, $(Y(\omega), \nabla Y(\omega)) = (Y(0), \nabla Y(0))$. According to (2.5.6), $\xi(t) = Y(\omega t), 0 \leqslant t \leqslant 1$, then is in the null space of $D^2 E(c)$ and not of the form $a\dot{c}(t), a \neq 0$.

Conversely, every element in the null space of $D^2 E(c)$, not of the form $a\dot{c}(t), a \neq 0$, is a periodic Jacobi field, orthogonal to $\dot{c}(t)$, and therefore it determines an eigenvector of P_ω for the eigenvalue 1. □

We now come to a particularly important class of geodesic flow lines.

3.2.10 Definition. The flow line $\phi_t X_0, t \in \mathbb{R}$, in $T_1 M$ is called *hyperbolic* if the induced bundle $\upsilon: V \to \mathbb{R}$ has a P_t-invariant splitting

$$\upsilon = \upsilon_s \oplus \upsilon_u$$

into two subbundles.

$$\upsilon_s: V_s \to \mathbb{R}; \ \upsilon_u: V_u \to \mathbb{R}$$

such that $P_t | V_s$ operates as contraction and $P_t | V_u$ operates as expansion. By this we mean the following:

There exist numbers $a \geqslant 1$ and $b > 0$ such that, for all $t \geqslant 0$ and all $t_0 \in \mathbb{R}$

$$(\dagger) \qquad \begin{aligned} &|P_t \xi| \leqslant a |\xi| e^{-bt}, \ \text{if } \xi \in V_{t_0,s} = \upsilon_s^{-1}(t_0) \\ &|P_t \eta| \geqslant a^{-1} |\eta| e^{bt}, \ \text{if } \eta \in V_{t_0,u} = \upsilon_u^{-1}(t_0). \end{aligned}$$

Here, $|\ |$ denotes the norm derived from the Riemannian metric on υ.

$V_{t_0,s}$ and $V_{t_0,u}$ are called *stable* and *unstable* subspace of V_{t_0}, respectively.

We also call the underlying geodesic $c_{X_0}(t) = \tau_M \phi_t X_0$ of a hyperbolic flow line *hyperbolic*. In the case of a periodic flow line $\phi_t X_0, 0 \leqslant t \leqslant \omega$, we call the underlying closed geodesic $c_{X_0}(t), 0 \leqslant t \leqslant \omega$, as well as the closed geodesic $c(t) \equiv c_{X_0}(\omega t), 0 \leqslant t \leqslant 1$, considered as critical point of $E: \Lambda M \to \mathbb{R}$, *hyperbolic*.

3.2.11 Proposition. *Let* $\phi_t X_0, t \in \mathbb{R}$, *be hyperbolic. Then the subspaces* $V_{t_0,s}, V_{t_0,u}$ of $V_{t_0} = \upsilon^{-1}(t_0)$ *are totally isotropic (or Lagrangian).*

Proof. Let $\xi_0, \xi_1 \in V_{0,s}$. Then, for all $t \in \mathbb{R}, |\alpha(\xi_0, \xi_1)| = |\alpha(P_t \xi_0, P_t \xi_1)|$. But as $t \to \infty$, the norm of $P_t \xi_0, P_t \xi_1$ goes to zero. Hence, $\alpha(\xi_0, \xi_1) = 0$. Likewise for elements η_0, η_1 of $V_{0,u}$, as $t \to -\infty$. □

In the case of a periodic orbit, hyperbolicity is determined by the linear Poincaré mapping:

3.2.12 Lemma. *Let* $\phi_t X_0, 0 \leqslant t \leqslant \omega$, *be periodic. The associated flow line* $\phi_t X_0, t \in \mathbb{R}$, *is hyperbolic if and only if the linear Poincaré mapping* P_ω *is hyperbolic, i.e.,* P_ω *only has eigenvalues* ϱ *with* $|\varrho| < 1$ *or* $|\varrho| > 1$.

Proof. From the definition (†), (3.2.10), it follows at once, that, for a hyperbolic flow line, P_ω must be hyperbolic.

To prove the converse we see from (3.1.5) that, for P_ω hyperbolic, the fibre V_0 over $t=0=\omega \in S_\omega$ has a decomposition

$$V_0 = V_{0,s} \oplus V_{0,u}$$

into P_ω-invariant subspaces $V_{0,s}$ and $V_{0,u}$. These are the spaces generated by the $V(\varrho)$ with $|\varrho|<1$ and $|\varrho|>1$, respectively. A similar decomposition holds for $V_{t_0} = P_{t_0} V_0 = V_{t_0,s} \oplus V_{t_0,u}$, with $V_{t_0,s} = P_{t_0} V_{0,s}$, $V_{t_0,u} = P_{t_0} V_{0,u}$. Indeed $P_\omega | V_{t_0}$ is given by $(P_{t_0}|V_0) \circ (P_\omega|V_0) \circ (P_{t_0}|V_0)^{-1}$. We therefore have a splitting

$$\upsilon = \upsilon_s \oplus \upsilon_u$$

of $\upsilon: V \to S_\omega$ into two P_t-invariant subbundles, called *stable* and *unstable subbundle,* respectively. We claim that this is precisely the splitting of the bundle υ given by the hyperbolicity of the complete flow line $\phi_t X_0$, $t \in \mathbb{R}$.

To show this we first prove the existence of $a_0 \geqslant 1$, $b>0$ such that

$$\text{(††)} \quad \begin{aligned} &|P_\omega^k \xi| \leqslant a_0 |\xi| e^{-kb\omega}, \quad \xi \in V_{t_0,s}; \\ &|P_\omega^k \eta| \geqslant a_0^{-1} |\eta| e^{bk\omega}, \quad \eta \in V_{t_0,u}, \text{ for all integers } k \geqslant 0. \end{aligned}$$

If we had an orthonormal basis of eigenvectors of P_ω, (††) would be clear immediately, with $a_0 = 1$ and b such that $e^{-b\omega} \geqslant |\varrho|$, for every eigenvalue ϱ of P_ω with $|\varrho|<1$. For the general case we proceed as follows.

First, we approximate P_ω by a symplectic $\tilde{P}_\omega$ such that $V_{t_0,s}$ has a basis consisting entirely of eigenvectors of $\tilde{P}_\omega$. $\tilde{P}_\omega$ still may be chosen hyperbolic. Take $\tilde{b}>0$ be such that $e^{-\tilde{b}\omega} \geqslant |\tilde{\varrho}|$, for every eigenvalue $\tilde{\varrho}$ of $\tilde{P}$ with $|\tilde{\varrho}|<1$. Then

$$|\tilde{P}_\omega \xi| \leqslant |\xi| e^{-\tilde{b}\omega} \quad \text{or} \quad |\tilde{P}_\omega \eta| \geqslant |\eta| e^{\tilde{b}\omega}$$

for eigenvectors ξ, η of $\tilde{P}_\omega$.

Secondly, introduce on V_{t_0} a scalar product $\langle\,,\rangle'$ such that a basis of eigenvectors of $\tilde{P}_\omega$ becomes orthonormal. There exists $a_0 \geqslant 1$ such that

$$a_0^{-1} \langle \zeta, \zeta \rangle \leqslant \langle \zeta, \zeta \rangle' \leqslant a_0 \langle \zeta, \zeta \rangle, \quad \text{all } \zeta \in V_{t_0}.$$

Here, $\langle\,,\rangle$ is the given Riemannian scalar product on V_{t_0}. We thus have (††), with P_ω replaced by $\tilde{P}_\omega$ and b replaced by $\tilde{b}$.

For $b>0$ with $0<\tilde{b}-b$ arbitrarily small, if only $\tilde{P}_\omega$ is sufficiently close to P_ω, we therefore get (††).

To derive (†) in (3.2.10) we choose $a_1>1$ such that

$$\text{(§)} \qquad a_1^{-1} |\zeta| e^{bt_0} \leqslant |P_{t_0} \zeta| \leqslant a_1 |\zeta| e^{-bt_0}$$

simultaneously for all t_0, $0 \leqslant t_0 < \omega$. Write $t \geqslant 0$ in the form $t = k\omega + t_0$ with k integer $\geqslant 0$, $0 \leqslant t_0 < \omega$. Then we have with $P_t = P_{k\omega} \circ P_{t_0}$ from (††) and (§), if $\xi \in V_{t_0,s}$,

$$|P_t\xi| \leqslant a_0 |P_{t_0}\xi| e^{-bk\omega} \leqslant a_0 a_1 |\xi| e^{-bt};$$

similary for $\eta \in V_{t_0,u}$. □

We can now prove the *Index Theorem for a hyperbolic closed geodesic,* cf. Klingenberg [7].

3.2.13 Theorem. *Let* $\phi_t X_0, 0 \leqslant t \leqslant \omega$, *be a hyperbolic periodic orbit of the flow* $\phi_t: T_1 M \to T_1 M$. *Let* $c(t) = \tau_M \phi_{\omega t} X_0, 0 \leqslant t \leqslant 1$, *be the corresponding critical point of* $E: \Lambda M \to \mathbb{R}$.

Then nullity $c = 0$, *i.e., c is non-degenerate, and*

$$(*) \qquad \operatorname{index} c = \sum_{0 \leqslant t < \omega} \dim (P_t V_{0,s} \cap V_{t,v})$$

Here, $P_t V_{0,s} = V_{t,s}$ *is the stable subspace of the fibre* V_t *of* υ *over t and* $V_{t,v}$ *is the vertical subspace of* V_t.

3.2.14 Remarks. 1. Since we know already that index $c < \infty$, cf. (2.5.3), (2.5.4), this formula shows that $V_{t,s} \cap V_{t,v} = 0$ for all $t \in S_\omega$ with the possible exception of finitely many t's.

2. It would have been possible to replace on the right hand side the stable subspace $V_{t,s}$ by the unstable subspace $V_{t,u}$.

3. We wish to point out a close analogy of (3.2.13) with the Morse Index Theorem (2.5.9) for a geodesic segment $c(t), 0 \leqslant t \leqslant 1$, from $c(0) = p$ to $c(1) = q$ where c is considered as critical point of $E: \Omega = \Omega_{pq} \to \mathbb{R}$. For that purpose we first reinterpret the statement that $c(t_1/\omega)$, $t_1 > 0$, is conjugate to $c(0)$ along $c|[0, t_1/\omega]$, of multiplicity $k \geqslant 0$, cf. (1.12.9).

In fact, this statement can be written in the form

$$\dim (P_{t_1} V_{0,v} \cap V_{t_1,v}) = k.$$

To see this we consider for the Jacobi field $Y(t)$ along $c_{X_0}(t) = \tau_M \phi_t X_0$ with $\langle Y(t), \dot{c}_{X_0}(t)\rangle = 0$ and $Y(0) = Y(t_1) = 0$ the correspondling P_t-invariant sections $\eta(t)$ in the bundle $\upsilon: V \to \mathbb{R}$. Then $\eta(0) \in V_{0,v}, \eta(t_1) \in V_{t_1,v}$. Hence, these $\eta(t)$ are those which for $t = t_1$ belong to $P_{t_1} V_{0,v} \cap V_{t_1,v}$.

The Morse Index Theorem (2.5.9) therefore can be written in the form

$$\operatorname{index} c = \sum_{0 < t < \omega} \dim (P_t V_{0,v} \cap V_{t,v})$$

which is very similar to the formula in (3.2.13). We will employ this analogy to carry over most of the arguments from the proof of (2.5.9) into the present case.

Proof of (3.2.13). Since P_ω is hyperbolic, 1 is not an eigenvalue. Hence, there are no non-zero periodic Jacobi fields along the corresponding closed geodesic c, orthogonal to c. From (2.5.6) now follows nullity $c = 0$.

As in the proof of (2.5.9) we construct for each $t_1, 0 \leqslant t_1 \leqslant \omega$, an injective linear mapping

$$\Phi: W(t_1) = V_{t_1,s} \cap V_{t_1,v} \to T_c\Lambda = T_c\Lambda M.$$

as follows: Let $\eta(t) \in P_t V_{0,s} = V_{t,s}$ be a P_t-invariant section in v_s with $\eta(t_1) \in W(t_1)$. Put $(P_\omega - id)^{-1}\eta(t) = \zeta(t)$. $\zeta(t)$ also is a P_t-invariant section in v_s. Let $Y(t), Z(t)$ be the corresponding Jacobi fields along $c_{X_0}(t) = \tau_M \phi_t X_0$, i.e. $Y(t) = (F\eta(t))_h$, $Z(t) = (F\zeta(t))_h$. Then

$$(T\phi_\omega - id)(Z(t), \nabla Z(t)) = (Y(t), \nabla Y(t))$$

and $Y(t_1) = 0$ imply

$$Z(t_1+\omega) - Z(t_1) = 0; \ \nabla Z(t_1+\omega) - \nabla Z(t_1) = \nabla Y(t_1).$$

We now define $\Phi(\eta)$ to be the element $\xi \equiv \xi_Y \in T_c\Lambda$, with Z determined from η as above, by

$$\xi(t) = \begin{cases} Z(\omega t + \omega); & 0 \leqslant t \leqslant t_1/\omega \\ Z(\omega t); & t_1/\omega \leqslant t \leqslant 1. \end{cases}$$

$\xi_Y(t), 0 \leqslant t \leqslant 1$, is differentiable, with the exception of $t = t_1/\omega$ where

$$\nabla \xi_Y(t_1/\omega -) - \nabla \xi_Y(t_1/\omega +) = \omega \nabla Y(t_1) \neq 0,$$

unless $\eta(t) \equiv 0$, i.e., $Y(t) \equiv 0$.

The mapping $\Phi: W(t_1) \to T_c\Lambda$ thus defined is clearly linear injective. Moreover, if $t_1 \neq t_1'$, $\Phi W(t_1) \cap \Phi W(t_1') = \{0\}$. Denote by U the subspace of $T_c\Lambda$, generated by the $\Phi W(t), 0 \leqslant t < \omega$. Then, with $\xi \in \Phi W(t_1)$, $\xi' \in \Phi W(t_1')$,

$$D^2E(c)(\xi,\xi') = \omega(\langle \nabla Y(t_1), Y'(t_1)\rangle - \langle Y(t_1), \nabla Y'(t_1)\rangle) = 0,$$

since $F(V_{t,s})$ is Lagrangian. Hence $D^2E(c)|U \equiv 0$. Since nullity $c = 0$, index $c \geqslant \dim U$, cf. the remark after (2.5.4). In particular, $\dim U < \infty$. $\dim U$ is equal to the right hand side of $(*)$ in (3.2.13).

It remains to be shown that an element $\xi^* \in T_c^-\Lambda$ with $D^2E(c)(\xi^*, pr_- U) = 0$ must be 0. As in the proof of (2.5.9) we choose a basis Y_i, $1 \leqslant i \leqslant \dim M - 1$, of the space $\mathscr{I}_s^\perp$ of stable Jacobi fields $Y(t)$ along $c_{X_0}(t)$. That is, $Y \in \mathscr{I}_s^\perp$ means $\langle \dot{c}_X(t), Y(t)\rangle = 0$ and $(Y(t), \nabla Y(t)) \in F(V_{t,s})$. We show that an element with the above properties can be written as

$$(\S) \qquad \xi^*(t) = \sum_i w^i(t)\, Y_i(\omega t), 0 \leqslant t \leqslant 1.$$

To see this, consider $\xi_Y \in U$, determined by $\eta(t_1) \in W(t_1)$, i.e., $(Y(t_1), \nabla Y(t_1)) = F\eta(t_1)$. Then

$$0 = D^2E(c)(\xi_Y, \xi^*) = \langle \nabla \xi_Y(t_1/\omega -) - \nabla \xi_Y(t_1/\omega +), \xi^*(t_1/\omega)\rangle =$$
$$= \omega \langle \nabla Y(t_1), \xi^*(t_1/\omega)\rangle.$$

This shows the validity of (§), with appropriate $w^i(t)$, $1 \leqslant i \leqslant \dim M - 1$.

With the expression (§) for $\xi^*(t)$ we show, by the same computation which we employed in the proof of (2.5.9), that $D^2E(c)(\xi^*, \xi^*) = 0$ and therefore $\xi^* = 0$. Here we use the fact that $\mathcal{J}_s^\perp$ is a Lagrangian subspace in the vector space $\mathcal{J}_c^\perp$ of Jacobi fields $Y(t)$ along $c_{X_0}(t)$, orthogonal to $\dot{c}_{X_0}(t)$. $\mathcal{J}_s^\perp$ corresponds under the canonical isomorphism F to the P_t-invariant sections of the stable bundle $v_s\colon V_s \to S_\omega$. □

As an important consequence we have the

3.2.15 Corollary. *Let c be a hyperbolic closed geodesic, i.e., the corresponding periodic orbit $\phi_t X_0, 0 \leqslant t \leqslant \omega$, in T_1M is hyperbolic. Let c^m be the m-fold covering of c, m an integer $\geqslant 1$. That is, $c^m(t) = c(mt), 0 \leqslant t \leqslant 1$. Then*

$$\operatorname{index} c^m = m \operatorname{index} c.$$

Note. (3.2.15) was proved, with a different method, by Bott [2].

Proof. $V_{t+\omega,s}$ and $V_{t,s}$ correspond under F to the same subspace $T_{\phi_t X_0 s} T_1 M$ of $T_{\phi_t X_0} T_1 M$ and so do $V_{t+\omega,v}$ and $V_{t,v}$. Therefore, according to (3.2.13),

$$\operatorname{index} c^m = \sum_{0 \leqslant t < m\omega} \dim(V_{t,s} \cap V_{t,v}) = m \sum_{0 \leqslant t < \omega} \dim(V_{t,s} \cap V_{t,v})$$

$= m \operatorname{index} c$. □

Important examples of hyperbolic geodesics are those for which the sectional curvatures $K(\sigma)$ of all 2-planes σ tangent to that geodesic are negative. To prove this we need another result of symplectic geometry.

3.2.16 Proposition. *Let $V = (V, \alpha)$ be a symplectic vector space. Let there be given a splitting $V = V_h \oplus V_v$ into two Lagrangian subspaces and assume that we have for V_h as well as for V_v an isomorphism with an Euclidean vector space $\bar{V}$ with scalar product $\langle\, ,\, \rangle$. We freely identify V_h and V_v with $\bar{V}$.*

Now let the 2-form α be given by

$$\alpha(\xi, \eta) = \langle \xi_h, \eta_v \rangle - \langle \eta_h, \xi_v \rangle.$$

Then the Lagrangian subspaces L of (V, α) which project under $V \to V_h$ onto V_h are in 1:1 correspondence with the self-adjoint operators S_L of $(\bar{V}, \langle\, ,\, \rangle)$.

More precisely, such a Lagrangian subspace L can be represented by

$$L = \{(\xi_h, S_L . \xi_h) \in V_h \oplus V_v;\ \ \xi_h \in V_h \text{ arbitrary}\};$$

$S_L\colon \bar{V} \to \bar{V}$ then is self-adjoint.

Conversely, given a self-adjoint operator S of $(\bar{V}, \langle\, ,\, \rangle)$, define

$$L_S = \{(\xi, S . \xi) \in V_h \oplus V_v;\ \xi \in V_h \text{ arbitrary}\}.$$

Then L_S is Lagrangian. Finally, $S_{L_S} = S$, $L_{S_L} = L$.

Proof. If L is Langrangian, we have

$$\alpha(\xi, \eta) = \langle \xi_h, S_L . \eta_h \rangle - \langle \eta_h, S_L . \xi_h \rangle = 0$$

for all ξ_h, η_h in V_h. Hence, S_L is self-adjoint. Conversely, the same formula shows that the space L_S defined above from a self-adjoint S is Lagrangian. □

We now can prove the relation between hyperbolicity and negative sectional curvature which we announced above. The proof is due essentially to Anosov, cf. Anosov and Sinai [1].

3.2.17 Theorem. *Let $\phi_t X_0$, $t \in \mathbb{R}$, be a geodesic flow line on $T_1 M$ and $c_{X_0}(t) = \tau_M \phi_t X_0$ the underlying geodesic on M.*

Assume that there exist positive constants k_0, k_1 such that, for all 2-planes σ tangent to the geodesic $c_{X_0}(t)$, the sectional curvature $K(\sigma)$ satisfies

$$-k_0^2 \leqslant K(\sigma) \leqslant -k_1^2 < 0.$$

Then $c_{X_0}(t)$ and $\phi_t X_0$, $t \in \mathbb{R}$, are hyperbolic.

More precisely, the constants $a \geqslant 1$ and $b > 0$ in (3.2.10) can be chosen $a = k_0/k_1$, $b = k_1$, and we also obtain the following lower bounds for the contraction and upper bounds for the expansion, for all $t \geqslant 0$

$$(*) \qquad \begin{aligned} &\frac{k_1}{k_0} |\xi| e^{-k_0 t} \leqslant |P_t \xi| \leqslant \frac{k_0}{k_1} |\xi| e^{-k_1 t}, && \text{if } \xi \in V_{t_0,s} \\ &\frac{k_0}{k_1} |\eta| e^{k_0 t} \geqslant |P_t \eta| \geqslant \frac{k_1}{k_0} |\eta| e^{k_1 t}, && \text{if } \eta \in V_{t_0,u} \end{aligned}$$

Proof. If $k_0 = k_1$, i.e., if M has constant curvature $K = -k_0^2 =$ (briefly) $-k^2$, we can write down explicitely the stable and unstable fields as

$$\xi_s(t) = e^{-tk} \xi_s(0); \; \xi_u(t) = e^{tk} \xi_u(0).$$

Indeed, the Jacobi equations in this case read $\nabla^2 Y(t) - k^2 Y(t) = 0$ from which we see that we get the above solutions.

Assume $k_1 < k_0$. Choose a real τ. Let $S_\tau \colon V_{\tau,h} \to V_{\tau,v} \cong V_{\tau,h}$ be a self-adjoint operator with spectrum in $]k_1, k_0[$. From (3.2.16) we know that S_τ determines a Lagrangian subspace $L_{S_\tau} =$ (briefly) L_τ in $V_\tau = V_{\tau,h} \oplus V_{\tau,v}$.

For $t \in \mathbb{R}$, the flow translate $P_{t-\tau} L_\tau$ of L_τ is a Lagrangian subspace in V_t, which we also denote by $L_\tau(t)$. Whenever $L_\tau(t)$ projects onto the horizontal subspace $V_{t,h}$ of V_t it determines a self-adjoint operator $S_\tau(t)$, cf. (3.2.16). Note: $S_\tau(\tau) = S_\tau$.

From the identity

$$P_{t-\tau}(\eta_h(\tau), S_\tau \eta_h(\tau)) = (\eta_h(t), S_\tau(t) \, . \, \eta_h(t)),$$

for any given $\eta_h(\tau) \in V_{\tau,h}$, we see that $\eta_h(t)$ is a Jacobi field $Y(t)$ with $\nabla Y(t) = S_\tau(t) \, . \, Y(t)$. Hence, from the Jacobi equation $\nabla^2 Y(t) + R_{\dot c(t)} \, . \, Y(t) = 0$ we obtain for the operator $S_\tau(t)$ the following first order differential equation of Riccati type:

$$(\S) \qquad \nabla S_\tau(t) + S_\tau(t) \, . \, S_\tau(t) + R_{\dot c(t)} = 0.$$

Here, t has to be restricted to those values where $S_\tau(t)$ can be defined, i.e., where $L_\tau(t)$ projects onto $V_{t,h}$. We will see in a moment that this is the case for all $t \geqslant \tau$.

In fact, we will show that for all $t \geqslant \tau$ the operator $S_\tau(t)$ has its spectrum in $[k_1, k_0]$ which clearly implies that $L_\tau(t)$ projects onto $V_{t,h}$.

To prove this we denote by $\lambda(t)$ and $\mu(t)$ the smallest and the largest eigenvalue of $S_\tau(t)$. Let $t_0 > \tau$ be the first t-value where $\mu(t_0)$ assumes the value k_0. Then we get from (§) for $\dot{\mu}(t_0)$ the estimate

$$\dot{\mu}(t_0) \leqslant -k_0^2 + \sigma(t_0) \leqslant 0$$

where $\sigma(t_0)$ is the largest eigenvalue of $-R_{\dot{c}(t_0)}$ which, by hypothesis, is $\leqslant k_0^2$. And whenever there is a $t_0' > t_0$ with $\mu(t_0') > k_0$ we find by the same argument $\dot{\mu}(t_0') < 0$. Thus, it is always the case that $\mu(t) \leqslant k_0$.

Similarly, if $t_1 > \tau$ is the first value with $\lambda(t_1) = k_1$, we find $\dot{\lambda}(t_1) \geqslant -k_1^2 + \varrho(t_1) \geqslant 0$ where $\varrho(t_1)$ is the smallest eigenvalue of $R_{\dot{c}(t_1)}$ which is $\geqslant k_1^2$: and if there should be a $t_1' > t_1$ with $\lambda(t_1') < k_1$ we find $\dot{\lambda}(t_1') > 0$. Thus, $\lambda(t) \geqslant k_1$ for all $t \geqslant \tau$.

We continue by showing that if $\tau' \leqslant \tau$ and $S_{\tau'}(t)$, $S_\tau(t)$ are self-adjoint operators defined as above from some $S_{\tau'}$, S_τ, then, for all $t \geqslant \tau$,

$$(\dagger) \qquad |S_\tau(t) - S_{\tau'}(t)| \leqslant |S_\tau - S_{\tau'}(\tau)| \, e^{-2k_1(t-\tau)}$$

To prove (†) we put $S_\tau(t) - S_{\tau'}(t) = R(t)$. From (§) we obtain by an easy computation for $R(t)$ the differential equation

$$(\S\S) \qquad \nabla R(t) = -S_\tau(t) \, . \, R(t) - R(t) \, . \, S_{\tau'}(t) \, .$$

We put

$$R(t) = Q(t) \, . \, (S_\tau - S_{\tau'}(\tau)) \, . \, Q'(t)$$

with

$$\nabla Q(t) = -S_\tau(t) \, . \, Q(t); \; \nabla Q'(t) = -Q'(t) \, . \, S_{\tau'}(t)$$

and initial conditions $Q(\tau) = Q'(\tau) = E =$ identity operator. Then (§§) is satisfied. Now,

$$\frac{d}{dt} |Q(t)|^2 = -\langle 2S_\tau(t) \, . \, Q(t), Q(t) \rangle \leqslant -2k_1 |Q(t)|^2,$$

hence, $|Q(t)| \leqslant e^{-k_1(t-\tau)}$. Similarly for $|Q'(t)|$. This yields (†).

It follows that the limit of $S_\tau(t)$, for $\tau \to -\infty$, exists and is a well defined self-adjoint operator $S_u(t)$ with spectrum in $[k_1, k_0]$. With the help of $S_u(t)$ we define, for all $t \in \mathbb{R}$, the Lagrangian subspace

$$V_{t,u} = \{(\eta_h, S_u(t) \, . \, \eta_h), \quad \text{all} \quad \eta_h \in V_{t,h}\} \, .$$

Then, $V_{t,u}$ is P_t-invariant, i.e., $P_{t_0} V_{t,u} = V_{t+t_0,u}$. To see this consider, for large negative τ, an element

$$\eta^\tau(t) = (\eta_h(t), S_\tau(t) \, . \, \eta_h(t))$$

which, for $\tau \to -\infty$, yields an element $\eta(t) \in V_{t,u}$. Applying P_{t_0} to $\eta^\tau(t)$ we get the element

$$\eta^\tau(t+t_0) = (\eta_h(t+t_0), S_\tau(t+t_0) \,.\, \eta_h(t+t_0))$$

which, for $\tau \to -\infty$, becomes $\eta(t+t_0) = P_{t_0}\eta(t)$.

We conclude the proof by writing down a number of inequalities which are immediate consequences of our previous estimates:

$$(\alpha) \qquad 2k_0|\eta_h(t)|^2 \geqslant 2\langle \eta_h(t), S_u(t) \,.\, \eta_h(t)\rangle = \frac{d}{dt}|\eta_h(t)|^2 \geqslant 2k_1|\eta_h(t)|^2$$

$$(\beta) \qquad k_0^2|\eta_h(t)|^2 \geqslant |S_u(t) \,.\, \eta_h(t)|^2 = |\nabla\eta_h(t)|^2 \geqslant k_1^2|\eta_h(t)|^2.$$

By integrating (α), from t_0 to $t+t_0 \geqslant t_0$ we get

$$(\gamma) \qquad |\eta_h(t_0)|^2 e^{2k_0t} \geqslant |\eta_h(t+t_0)|^2 \geqslant |\eta_h(t_0)|^2 e^{2k_1t}$$

Applying (β), (γ) and (β) again we get

$$(\delta) \qquad \left\{\begin{aligned} &\frac{k_0^2}{k_1^2}|\nabla\eta_h(t_0)|^2 e^{2k_0t} \geqslant k_0^2|\eta_h(t_0)|^2 e^{2k_0t} \geqslant k_0^2|\eta_h(t+t_0)|^2 \geqslant \\ &\quad \geqslant |\nabla\eta_h(t+t_0)|^2 \geqslant k_1^2|\eta_h(t+t_0)|^2 \geqslant \\ &\quad \geqslant k_1^2|\eta_h(t_0)|^2 e^{2k_1t} \geqslant \frac{k_1^2}{k_0^2}|\nabla\eta_h(t_0)|^2 e^{2k_1t} \end{aligned}\right.$$

Since $k_1 \leqslant k_0$, we find that from (γ) and (δ) for a P_t-invariant $\eta(t)$ in $V_{t,u}$

$$\frac{k_0}{k_1}|\eta(t_0)|e^{k_0t} \geqslant |P_t\eta(t_0)| \geqslant \frac{k_1}{k_0}|\eta(t_0)|e^{k_1t},$$

which is the second of the relations $(*)$.

Define $V_{t,s} \subset V_t$ to be $V^-_{-t,u} \subset V^-_{-t}$ where $v^-: V^- \to \mathbb{R}$ is the bundle induced by the immersion $t \in \mathbb{R} \mapsto \phi_{-t}X_0 \in T_1M$. Then we get for a P_t-invariant vector field $\xi(t)$ in $V_{t,s}$, i.e., for the P_t-invariant vector field $\eta^-(-t)$ in $V^-_{-t,u}$, the estimate

$$\begin{aligned} \frac{k_0}{k_1}|P_t\xi(t_0)|e^{k_0t} &= \frac{k_0}{k_1}|P_{-t}\eta^-(-t_0)|e^{k_0t} \geqslant |\xi(t_0)| = |\eta^-(-t_0)| \geqslant \\ &\geqslant \frac{k_1}{k_0}|P_{-t}\eta^-(-t_0)|e^{k_1t} = \frac{k_1}{k_0}|P_t\xi(t_0)|e^{k_1t}, \end{aligned}$$

which is the first of the relations $(*)$. □

3.3 Stable and Unstable Motions

We consider complete Riemannian manifolds M of finite dimension. A geodesic flow line $\phi_t X_0$, $t \in \mathbb{R}$, of the system (TM, α, E) describes the motion of a point $p_0 = \tau_M X_0$ on M which at the time $t = 0$ is pushed on with the initial velocity vector X_0. While $\tau_M \phi_t X_0$ gives the position of the point at the time t, $\phi_t X_0$ is its velocity vector at the time t.

Recall from (3.1.16) that a geodesic flow line $\phi_t X_0$, $t \in \mathbb{R}$, on $T_1 M$, is a geodesic on the Riemannian manifold $T_1 M$. To describe a neighborhood of such a flow line we use a covering of this line by Fermi coordinates, based on subarcs of this geodesic. We need a certain uniform size of the domain of Fermi coordinates, since in general such a geodesic will not be periodic. Therefore, we begin with the following

3.3.1 Proposition *Let M be a compact Riemannian manifold. Then there exist positive constants λ_M and ϱ_M such that, for every geodesic $\phi_t X_0$, $t \in \mathbb{R}$, on the Riemannian manifold $T_1 M$, the restriction to intervals $J \subset \mathbb{R}$ of length λ_M, can serve as a basis for a domain of Fermi coordinates with radius ϱ_M.*

Proof. Recall from the definition (1.12.1) of Fermi coordinates that the starting point is a geodesic segment – in our case a segment of the form $\phi_t X_0$, $t \in I$ – and the family of normal spaces to that geodesic – in our case these normal spaces are the $\bar{T}_{\phi_t X_0} T_1 M$, $t \in I$, defined in (3.2.1). The tangential of the exponential mapping defined on $TT_1 M$ (or at least on some open neighborhood $\tilde{T}T_1 M$ of the 0-section $\cong T_1 M$ in $TT_1 M$) is of maximal rank at the origin $0_{\phi_t X_0}$ of $T_{\phi_t X_0} T_1 M$. Hence, a suitable restriction of $\exp: \tilde{T}T_1 M \to T_1 M$ to a set of the form $\bar{B}_\varrho(\phi_J X_0)$, J some interval of I, will be an injective diffeomorphism. Here $\bar{B}_\varrho(\phi_J X_0)$ is the union of the $\bar{B}_\varrho(0_{\phi_t X_0})$, $t \in J$, i.e., the balls of radius ϱ around the origin of $\bar{T}_{\phi_t X_0} T_1 M$.

Now use the compactness of $T_1 M$ to conclude that there exist $\lambda_M > 0$, $\varrho_M > 0$ such that these coordinates are well defined, whenever J has length $\leqslant \lambda_M$ and ϱ is $\leqslant \varrho_M$ – independently of the choice of X_0 in $T_1 M$. $\square$

Now let λ_M and μ_M positive numbers as in (3.3.1).

Fix a flow line $\phi_t X_0$, $t \in \mathbb{R}$. Consider $X \in \Sigma_\varrho(X_0)$, $\varrho \leqslant \varrho_M$. If $\phi_t X$, for $t \geqslant 0$, stays in the domain of the Fermi chart based at $\phi_t X_0$, $0 \leqslant t \leqslant \lambda_M$ then this means that there is a uniquely determined differentiable function $\delta(X; t)$ such that $\delta(X_0; t) = \delta(X; 0) = 0$ and $\phi_{t+\delta(X;t)} X \in \Sigma_{\varrho M}(\phi_t X_0)$, $0 \leqslant t \leqslant \lambda_M$. Continuing in this manner from $X'_0 = \phi_{\lambda_M} X_0$ with $X' = \phi_{\lambda_M + \delta(X;\lambda_M)} X$ we may repeat this process, possibly for all $t \in [\lambda_M, 2\lambda_M]$, using the Fermi coordinates based at $\phi_t X_0$, $\lambda_M \leqslant t \leqslant 2\lambda_M$.

We use this construction to define the concept of stability for a geodesic flow line.

3.3.2 Definition. Let M be a compact Riemannian manifold or – more generally – a complete Riemannian manifold with the property that there exist positive constants λ_M, ϱ_M such that, for every geodesic of length λ_M on $T_1 M$, there exist Fermi coordinates of radius ϱ_M.

A geodesic flow line $\phi_t X_0$, $t \in \mathbb{R}$, on $T_1 M$ is called *stable*, if for every ϱ, $0 < \varrho \leqslant \varrho_M$, there exists a ϱ_0, $0 < \varrho_0 \leqslant \varrho$ such that the following holds: Let $\upsilon \equiv \upsilon_{X_0}: V \equiv V_{X_0} \to \mathbb{R}$ be the bundle asociated to $\phi_t X_0$. For $X \in \Sigma_{\varrho M}(X_0)$ let $\xi(X) = F^{-1} \circ (\exp_{X_0} | \bar{B}_{\varrho M}(0_{X_0}))^{-1}(X)$ be its canonical coordinate in V_0.

For every $X \in \Sigma_{\varrho 0}(X_0)$ there shall exist a differentiable function $t \in \mathbb{R} \mapsto \delta(X; t) \in \mathbb{R}$ with $\delta(X; 0) = 0$, $\delta(X_0; t) = 0$ and a differentiable section $\eta(t; \xi(X))$ in υ such that, for all $t \in \mathbb{R}$

$$\phi_{t+\delta(X;t)} X = \exp_{\phi_t X_0} \circ F \circ \eta(t; \xi(X)) \in \Sigma_\varrho(\phi_t X_0)$$

We also say briefly that every flow line $\phi_t X$, $t \in \mathbb{R}$, starting from a X sufficiently near X_0, shall remain for all $t \in \mathbb{R}$ in the tubular neighborhood of $\phi_t X_0$, $t \in \mathbb{R}$, of radius ϱ.

A geodesic $c \neq$ const is called *stable* if the corresponding flow line $\phi_t \dot{c}(0)/|\dot{c}(0)|$ is stable.

Remark. If $\phi_t X$ with $X \in \Sigma_{\varrho M}(X_0)$ permits a description of the form (*) then this description is unique. Indeed, this follows from the existence of Fermi coordinates of radius ϱ_M based at geodesics of length λ_M by the continuation process described above.

If a flow line $\phi_t X$, $t \in \mathbb{R}$ stays near $\phi_t X_0$, $t \in \mathbb{R}$, such as to allow a description of the form (*) (3.3.2), its first order approximation has the coordinate description $P_t \xi \in V_t$, with $\xi = \xi(X)$. We use this to derive a necessary condition for the stability of $\phi_t X_0$, $t \in \mathbb{R}$, in the case that this is a periodic orbit.

3.3.3 Lemma. *Let $\phi_t X_0$, $t \in \mathbb{R}$, be periodic with period $\omega > 0$. If $\phi_t X_0$, $t \in \mathbb{R}$, is stable, then the associated linear Poincaré mapping P_ω is compact, i. e., $|P_\omega^m|$ is bounded for all $m \in \mathbb{N}$*

Note. See (3.3.4) for a characterization of a compact symplectic transformation.

Proof. Using the coordinate description $\eta(t; \xi)$ of $\phi_t X$, with $\xi = \xi(X)$, we have from $\eta(\omega; s\xi) = s P_\omega \xi + o(s)$

$$\eta(m\omega; s\xi) = \eta(\omega; \eta((m-1)\omega; s\xi)) = s P_\omega^m \xi + o(s), \text{ all } m \in \mathbb{Z}, s \geqslant 0.$$

Since $|\eta(m\omega; s\xi)|$ is small with $|s|$ small uniformly for all $m \in \mathbb{Z}$, $o(s)/s \to 0$ for $s \to 0$, uniformly for all $m \in \mathbb{Z}$. Therefore, $|P_\omega^m \xi|$ is bounded for all $m \in \mathbb{Z}$ which proves our claim. □

As a consequence of (3.3.3) we introduce the

3.3.4 Definition. A periodic flow line $\phi_t X_0$, $0 \leqslant t \leqslant \omega$, on $T_1 M$ is called *infinitesimally stable* if the associated linear Poincaré mapping P_ω is compact. This means that P_ω can be represented as orthogonal sum of 2-dimensional rotations and reflections. In particular, all eigenvalues of P_ω are on the unit circle and P_ω therefore is elliptic in the sense of (3.1.6). We shall use this name also for the associated closed geodesic $c_{X_0}(t) = \tau_M \phi_t X_0, 0 \leqslant t \leqslant \omega$ as well as its reparameterization $c(t) = c_{X_0}(\omega t), 0 \leqslant t \leqslant 1$.

To find sufficient conditions for the stability of a geodesic flow line is very difficult. Let us restrict ourselves to periodic flow lines only. If $\dim M = 2$ then there is a result of Kolmogorov-Arnold-Moser which guarantees stability under a generically satisfied hypothesis, the so-called twist condition. For details see Arnold and Avez [1] and Moser [1].

If, however, $\dim M > 2$, then it seems likely that generically there are no stable periodic flow lines as defined in (3.3.2). Rather, one can expect that periodic flow lines are stable only in the following weakened sense: For $\Sigma_\varrho(X_0)$, ϱ small, there exist a set $\Sigma_\varrho^0(X_0)$ in $\Sigma_\varrho(X_0)$ of positive measure such that the flow lines $\phi_t X$, $t \in \mathbb{R}$, starting at $X \in \Sigma_\varrho^0(X_0)$ remain in a small tubular neighborhood of $\phi_t X_0$, $t \in \mathbb{R}$, whereas the other flow lines leave such neighborhoods, as t goes to $+\infty$ or $-\infty$. The measure of $\Sigma_\varrho^0(X_0)$, divided by the measure of $\Sigma_\varrho(X_0)$, is < 1 but it approaches 1 as ϱ goes to zero. For details we refer to Moser [1]. Moser considers general Hamiltonian systems, but the geodesic flow does not behave in any special way with respect to stability. See also Klingenberg [10] for this case.

For the remainder of this section we will give conditions for the existence of hyperbolic and non-hyperbolic – if not infinitesimally stable – closed geodesics. Recall from (3.1.10) that the linear Poincaré mapping P_ω, considered as symplectic transformation of V_0, determines for V_0 the decomposition

$$(V_{0,s} \oplus V_{0,u}) \oplus V_{0,ce}$$

into P_ω-invariant subspaces. $\dim V_{0,ce} = 2q$ is even. Let $\dim M = n+1$. Then $\dim V_0 = 2n$. Thus, whenever $n > q > 0$, P_ω is neither hyperbolic nor elliptic but something in between.

Thet 'most unstable' case is the hyperbolic one. In fact, we have the

3.3.5 Theorem. *Let $\phi_t X_0$, $t \in \mathbb{R}$, be a hyperbolic geodesic flow line on $T_1 M$. Then it is* totally unstable *in the following sense: For every sufficiently small $\varrho > 0$, all elements $X \neq X_0$ in the flow-transversal hypersurface $\Sigma_{\varrho_0}(X_0)$, with arbitrarily small ϱ_0, $0 < \varrho_0 \leqslant \varrho$, have a flow line $\phi_t X$, $t \in \mathbb{R}$, which leaves the tubular neighborhood of radius ϱ of $\phi_t X_0$.*

Proof. Consider $X \neq X_0$ in $\Sigma_{\varrho_0}(X_0)$, some small $\varrho_0 > 0$. Let $\xi = \xi(X)$ be its coordinate in V_0. Write $\xi = (\xi_0, \xi_u) \in V_{0,s} \oplus V_{0,u}$. We first consider the case $\xi_u \neq 0$. Then (†) (3.2.10) yields the estimate

$$|P_t \xi| = |(P_t \xi_u, P_t \xi_s)| \geqslant a^{-1} |\xi_u| e^{bt} - a |\xi_s| e^{-bt} \geqslant a' |\xi| e^{bt}$$

for all $t \geqslant$ some $t_0 > 0$, some $a' > 0$.

Assume that $\phi_t X$, $t \in \mathbb{R}$, would remain in a small tubular neighborhood of $\phi_t X_0$, $t \in \mathbb{R}$. We then could describe $\phi_t X$ by its coordinate $\eta(t; \xi)$ in the bundle υ of $\phi_t X_0$, cf. (3.3.2). In particular, $|\eta(t; \xi)|$ is bounded. For $|\eta(t; \xi)|$ we have a differential equation of the form

$$|\eta(t;\xi)|^{\cdot} = b(t)|\eta(t;\xi)| + o(\eta).$$

Here, $b(t)$ is determined by $|P_t\xi|^{\cdot} = b(t)|P_t\xi|$. In our case, $b(t) \geqslant$ some $2b_0 > 0$, if t is sufficiently large. Therefore, $|\eta(t;\xi)|^{\cdot} > b_0|\eta(t;\xi)|$, which contradicts the boundedness of $|\eta(t;\xi)|$.

If $\xi = (\xi_s, 0) \neq 0$, consider $|P_{-t}\xi|$ for $t \to \infty$. From (†), (3.2.10) we get, by replacing in the first inequality $P_t\xi$ by $P_{-t}\xi$, $|P_{-t}\xi| \geqslant a^{-1}|\xi|e^{bt}$. It follows from the differential equation for $\eta(-t;\xi)$ that with $|P_{-t}\xi|$, $t \to \infty$, unbounded, also $|\eta(-t;\xi)|$, $t \to \infty$, is unbounded. □

(3.3.5) allows a slight generalization to partially hyperbolic flow lines. These represent a generalization of hyperbolic flow lines, as defined in (3.2.10). Recall from (3.2.2) the definition of the bundle $\upsilon: V \to \mathbb{R}$, associated with such a flow line.

3.3.6 Definition. The flow line $\phi_t X_0$, $t \in \mathbb{R}$, on $T_1 M$ is called *partially hyperbolic* if the associated bundle υ permits a P_t-invariant splitting

$$\upsilon = \upsilon_s \oplus \upsilon_u \oplus \upsilon_{ce}$$

into subbundles with the following properties:

Denote by $V_t = (V_{t,s} \oplus V_{t,u}) \oplus V_{t,ce}$ the splitting of the fibre V_t over t. Then $\dim V_{t,s} = \dim V_{t,u} = p > 0$, hence, $\dim V_{t,ce} = 2n - 2p \geqslant 0$, where $n + 1 = \dim M$.

Moreover, there exist numbers $a \geqslant 1$ and $b > 0$ such that, for all $t \geqslant 0$ and every $t_0 \in \mathbb{R}$,

$$|P_t\xi| \leqslant a|\xi|e^{-bt}, \qquad \text{if } \xi \in V_{t_0,s}$$

$$|P_t\eta| \geqslant a^{-1}|\eta|e^{bt}, \qquad \text{if } \eta \in V_{t_0,u}$$

The subbundles υ_s, υ_u and υ_{ce} are called the *stable*, the *unstable* and the *center bundle* respectively.

With this, (3.3.5) possesses the following generalization:

3.3.7 Theorem. *Let $\phi_t X_0$, $t \in \mathbb{R}$, be a partially hyperbolic flow line on $T_1 M$. Then it is not stable.*

More precisely, if $V_0 = (V_{0,s} \oplus V_{0,u}) \oplus V_{0,ce}$ is the splitting of the fibre V_0 over 0 of the bundle υ and $\dim V_{0,s} = \dim V_{0,u} = p > 0$, then, for every sufficiently small $\varrho > 0$, the flow $\phi_t X$, $t \in \mathbb{R}$, of every X with coordinate $\xi(X)$ not in $V_{0,ce}$ and $|\xi(X)|$ arbitrarily small, will not remain in the tubular neighborhood of radius ϱ around $\phi_t X_0$, $t \in \mathbb{R}$.

Proof. As in (3.3.5), we consider the decomposition $\xi = (\xi_s, \xi_u, \xi_{ce}) \in V_{0,s} \oplus V_{0,u} \oplus V_{0,ce}$. If $\xi_u \neq 0$, $\phi_t X$ leaves the tabular neighborhood for $t \to \infty$. If $\xi_u = 0$ but $\xi_s \neq 0$, then this is the case for $t \to -\infty$. □

As a consequence of (3.2.17) we now get the following

3.3.8 Theorem. *Let M be a compact manifold with strictly negative sectional curvature K. Then every geodesic flow line on $T_1 M$ is totally unstable in the sense of* (3.3.5). *The same is true if, more generally, M is complete and there exist positive constants k_0, k_1 such that $-k_0^2 \leqslant K(\sigma) \leqslant -k_1^2$, for all 2-planes σ in M.*

Proof. According to (3.2.17), the hypotheses of (3.3.5) are satisfied. □

It would be false to expect, however, that positive sectional curvature implies that the geodesic flow lines are stable. Only under special additional assumptions one can show that a flow line in such a case is at least not hyperbolic. Here is a first result in this direction. For others see (3.3.13), and (3.3.17).

3.3.9 Theorem. *Let $\phi_t X_0, 0 \leqslant t \leqslant \omega$, be a periodic flow line on $T_1 M$. Let $c(t) = \tau_M \circ \phi_{\omega t} X_0, 0 \leqslant t \leqslant 1$, be the corresponding critical point of E on ΛM. That is, c is a closed geodesic.*

If c has index 0 *and if the curvature $K(\sigma)$ of all 2-planes σ tangent to c is strictly positive then $\phi_t X_0, 0 \leqslant t \leqslant \omega$, and c are not hyperbolic.*

Proof. Let $K_0 > 0$ be a lower bound for the values of $K(\sigma)$, σ tangential to $c(t), 0 \leqslant t \leqslant 1$. Then, for a sufficient large integer m (e.g., $mL(c) = m\omega > \pi/\sqrt{K_0}$), $c^m(t) = c(mt), 0 \leqslant t \leqslant 1$, considered as critical point of $\Omega = \Omega_{c(0),c(0)}$, has $\operatorname{index}_\Omega \geqslant n-1 \geqslant 1$, cf. (2.6.2, ii). Hence, according to (2.5.14), $\operatorname{index}_\Lambda c^m \geqslant 1$. But then c cannot be hyperbolic, because in that case $\operatorname{index}_\Lambda c^m = m \operatorname{index}_\Lambda c = 0$, see (3.2.15). □

3.3.10 Corollary. *Let M be compact and not simply connected. Assume that $K > 0$ on M. Then in every connected component Λ' of ΛM, consisting of non null homotopic closed curves, there exists a closed geodesic c of minimal E-value. Such a c has index* 0 *and is not hyperbolic.*

Proof. The existence of c is shown in (2.4.19). Since $E(c)$ is the infimum of $E|\Lambda'$, (2.5.10), (2.5.11) show that index $c = 0$. Apply now (3.3.9). □

We continue with a connection between infinitesimal stability (or instability) of a periodic geodesic flow line and the isometry group of M. Note that any isometry of M induces an isometry on TM and $T_1 M$. Thus, an isometry carries intrinsically defined objects like the horizontal and the vertical subspaces of $\bar{T}_X T_1 M$ into itself. The same is true for the stable and unstable subspace of a hyperbolic flow line.

The following result is proved in Thorbergsson [1].

3.3.11 Theorem. *Let $\phi_t X_0, 0 \leqslant t \leqslant \omega$, be a periodic geodesic flow line on $T_1 M$. Let $c(t) = \tau_M \phi_{t\omega} X_0, 0 \leqslant t \leqslant 1$, be the associated closed geodesic on M.*

(i) *Assume that the Riemannian manifold M possesses a* 1*-parameter group of isometries which does not operate as the identity on every point $c(t)$ of c. Whenever* index c + nullity $c > 0$, *c is not hyperbolic.*

(ii) *Assume that on M we have a finite cyclic group $\mathbb{Z}_m$ of isometries such that the* *closed geodesic c contains the full $\mathbb{Z}_m$-orbit of the point $c(0)$. Let c be hyperbolic. Then* index c *is divisible by the number of different elements of the $\mathbb{Z}_m$-orbit through $c(0)$.*

Proof. Under the hypotheses of (i), we have along $c(t)$ a non-zero periodic Jacobi field $Y(t)$. Indeed, the 1-parameter group defines a 1-parameter family of closed geodesics containing c. Such a family determines a Jacobi field, see (1.12.4), and this Jacobi field must be periodic from the way it has been constructed.

If $Y(t) \neq a\dot{c}(t)$ then we know from (2.5.6) that nullity $c > 0$, hence, c is not hyperbolic, see (3.2.13). So let now $Y(t) = a\dot{c}(t)$. That means that $c(t)$ is, up to parameterization, an orbit of the isometric $\mathbb{R}$-action. And so is $\phi_t X_0$, for the induced isometric $\mathbb{R}$-action on $T_1 M$. This means that the bundle $v: V \to S_\omega$, associated with $\phi_t X_0$, $0 \leqslant t \leqslant \omega$, also permits an isometric $\mathbb{R}$-action which operates on the basis S_ω.

Let now $\phi_t X_0$, $0 \leqslant t \leqslant \omega$, be hyperbolic and in particular nullity $c = 0$. Then the stable subspaces $V_{t,s}$ of V_t, $0 \leqslant t \leqslant \omega$, must be carried into themselves by the $\mathbb{R}$-action on V, and the same is true for the vertical subspaces $V_{t,v}$. Thus, if we had $V_{t,s} \cap V_{t,v} \neq 0$ for $t = t_0$, then it would hold for all t. But this would imply – cf. (3.2.13) – that index $c = \infty$ which is impossible. Hence, if c is hyperbolic, index $c = 0$ is the only possibility.

For the proof of (ii) let $c(j/k), j = 0, 1, \ldots, k-1$, be the orbit of an isometric $\mathbb{Z}_m$-action, with m divisible by k. We may assume index $c > 0, k > 1$. Let $\phi_t X_0$, $0 \leqslant t \leqslant \omega$, $X_0 = \dot{c}(0)/|\dot{c}(0)|$ be the corresponding periodic orbit and $v: V \to S_\omega$ the associated bundle. The induced $\mathbb{Z}_m$-action carries the intrinsically defined subspaces $V_{t,s}$ and $V_{t,v}$ into themselves. Hence, if $V_{t,s} \cap V_{t,v} \neq 0$ for $t = t_0$, the same is true for the $k-1$ spaces $V_{t,s}$ and $V_{t,v}$ with $t = t_0 + j\omega/k, j = 1, \ldots, k-1$, and all have the same dimension. Hence, according to (3.2.13), index c is a multiple of k. □

3.3.12 Examples. As an example of case (i) in (3.3.11), we consider a surface of revolution, cf. the example after (1.10.9). According to (1.10.10), we get a closed geodesic as orbit of the isometry group whenever the distance function from the axis of revolution has a stationary value. If the distance is a minimum, with non-vanishing second derivative, i. e., if the parallel circle is a waist, the sectional curvature along the closed geodesic is negative. Hence, in this case the closed geodesic is hyperbolic. If the distance function has a critical value of different type then the orbit is not hyperbolic. Cf. also (3.3.13).

As an example of the case (ii) in (3.3.11) we consider a 2-dimensional manifold $\bar{M}$ which is the projective plane with an arbitrary Riemannian metric. The universal covering M of $\bar{M}$ then is the 2-sphere with a Riemannian metric which permits an isometric $\mathbb{Z}_2$-action without fixed points such that $M/\mathbb{Z}_2 = \bar{M}$. As an example, take for M an ellipsoid $\{\sum_{0}^{2} x_i^2/a_i = 1\}$ in $\mathbb{R}^3$, cf. (1.10.6). The $\mathbb{Z}_2$-action shall be the reflection on the origin of $\mathbb{R}^3$.

On $\bar{M}$ we have a closed geodesic $\bar{c}$ which has minimal E-value among all closed curves on $\bar{M}$ which are not homotopic to a point curve, cf. (2.4.19). In the case that $\bar{M}$

has positive curvature we know from (3.3.10) that $\bar{c}$ is elliptic. In any case, index $\bar{c} = 0$.

For every closed geodesic $\bar{c}'$ on $\bar{M}$ which is not homotopic to zero, we have a closed geodesic c' on M which under the isometric submersion $M \to \bar{M} = M/\mathbb{Z}_2$ is a 2-fold covering of $\bar{c}'$. Thus c' contains the $\mathbb{Z}_2$-orbit of any of its points. If c' is hyperbolic, its index therefore must be even. On the other hand, in (3.4.3) we will show that a non-degenerate elliptic closed geodesic on M has odd index.

The following generalizes the example of a waist on a surface of revolution. Compare also (3.3.11, i).

3.3.13 Lemma. *Let $c = c(t), 0 \leqslant t \leqslant 1$, be a closed geodesic which is the orbit of an isometric $\mathbb{R}$-action on M. Then c is hyperbolic if and only if $K(\sigma) < 0$, for all 2-planes σ tangent to c.*

Proof. Let c be hyperbolic. According to (3.3.11) we have index $c = 0$ and the stable subspace $V_{t,s} \subset V_t$ does not intersect the vertical subspace $V_{t,v} \subset V_t$. From (3.2.16) we then know that $V_{t,s}$ in Fermi coordinates can be represented in the form

$$V_{t,s} = \{(\eta_h(t), S_t \eta_h(t)) \in V_{t,h} \oplus V_{t,v}\}$$

where $\dot{\eta}_h(t) = S_t \eta_h(t), \ddot{\eta}_h(t) = -R_{\phi_t X_0} \cdot \eta_h(t)$

S_t is self-adjoint. Its eigenvalues $\lambda(t)$ therefore are real. Since $c(t)$ is the orbit of an isometric $\mathbb{R}$-action, the $\lambda(t)$ do not depend on t. Hence, if $-\lambda$ is such an eigenvalue we have for a corresponding eigenvector $\eta_s(t) \neq 0$

$$\eta_s(t) = e^{-\lambda t}(\eta_h(0), -\lambda \eta_h(0)) \in V_{t,h} \oplus V_{t,v}$$

with $-\lambda < 0$. Thus, the sectional curvature of the plane spanned by $\dot{c}(t), e^{-\lambda t}\eta_h(0)$ is equal to $-\lambda^2$.

The converse is contained in (3.2.17). □

Note. Actually, we have shown a little more than $K(\sigma) < 0$ for a hyperbolic closed geodesic c which is an orbit of a 1-parameter group of isometries: There exists a basis for the space $\mathscr{J}_c^{\perp}$ of Jacobi fields, formed by stable and unstable fields of the form $Y_s(t) = Y_s(0)e^{-\lambda t}$; $Y_u(t) = Y_u(0)e^{\lambda t}, \lambda > 0$.

For symmetric spaces, the linear Poincaré mapping has a particularly simple form, as was observed by Ziller [1].

3.3.14 Theorem. *Let M be a symmetric space,* dim $M = n+1$. *Let $c(t), 0 \leqslant t \leqslant 1$, be a closed geodesic on M and $\phi_t X_0, 0 \leqslant t \leqslant \omega = L(c) = |\dot{c}(0)|$, the corresponding periodic geodesic flow line on $T_1 M$. Then the linear Poincaré mapping $P_\omega: V_0 \to V_0$ has all eigenvalues equal to* 1.

More precisely, the self-adjoint curvature operator $R_{X_0}: T_0^{\perp} c \to T_0^{\perp} c$ on the subspace of $T_{c(0)} M$ orthogonal to $\dot{c}(0)$ has all eigenvalues $\lambda \geqslant 0$. Let $A_i, 1 \leqslant i \leqslant n$, be an orthonormal basis for $T_0^{\perp} c$, consisting of eigenvectors of R_{X_0}. Then we get a symplectic basis for V_0 by

$$\varepsilon_i = (A_i, 0) \in V_{0,h} \oplus V_{0,v}; \ \varepsilon_{i+n} = (0, A_i) \in V_{0,h} \oplus V_{0,v}.$$

P_ω leaves each of the non-degenerate 2-dimensional subspaces, generated by $(\varepsilon_i, \varepsilon_{i+n})$, invariant. The operation of P_ω on such a subspace is given by

$$\begin{pmatrix} 1 & \omega \\ 0 & 1 \end{pmatrix} \quad \text{or} \quad \begin{pmatrix} 1 & 0 \\ 0 & 1 \end{pmatrix}$$

depending on whether the eigenvalue λ_i of A_i is $=0$ or >0.

Proof. From (2.2.10) we know that $K(\sigma) \geqslant 0$ for all 2-planes σ tangent to c. Hence, all eigenvalues λ of R_{X_0} are $\geqslant 0$.

Let A be one of the A_i, $1 \leqslant i \leqslant n$. If $R_{X_0} A = 0$ then we have from (2.2.10) that the corresponding Jacobi field ${}_A Y_1(t)$ is periodic, whereas

$$({}_A Y_0, \nabla_A Y_0)(0) = (0, A); ({}_A Y_0, \nabla_A Y_0)(\omega) = (\omega A, A).$$

Thus, P_ω operates on $\{(A, 0), (0, A)\}$ as $\begin{pmatrix} 1 & \omega \\ 0 & 1 \end{pmatrix}$.

If $R_{X_0} A = \lambda A$ with $\lambda > 0$, we see from (2.2.10) that both Jacobi fields, ${}_A Y_0$ and ${}_A Y_1$, have a t-value where the covariant derivative vanishes. According to (2.2.14), both therefore are generated by isometries and hence periodic. This means that P_ω operates as identity on $(A, 0)$ as well as on $(0, A)$. □

Remark. A closed geodesic c on a symmetric space is not stable in the sense of (3.3.2), unless the sectional curvature $K(\sigma)$ for σ tangent to c is strictly positive. Indeed, if there is a flat parallel strip along c then there are geodesics which run away from c as two lines in Euclidean space do, which start from the same point but have linearly independent initial directions. Locally at least this is the picture.

The only symmetric spaces of strictly positive sectional curvature are the ones of rank 1, i. e., the spheres and the various projective spaces over the reals, the complex numbers, the quarternions or the Cayley numbers, cf. (2.8.7).

We conclude this section with an investigation of closed geodesics on δ-pinched manifolds, with $\delta \geqslant 1/4$, cf. the introduction to (2.8) for this concept. Preliminary results were obtained by Thorbergsson [1]. The improvements given below are due to Ballmann, Thorbergsson and Ziller [1].

Our first result exhibits a gap in the range of the length of closed geodesics on a positively curved manifold, the gap depending on the pinching number $\delta = \min K : \max K$. It also may be viewed as a characterization of the sphere.

3.3.15 Theorem. *Let M be a compact, simply connected Riemannian manifold. Suppose that the sectional curvature K satisfies $0 < K_1/4 \leqslant K_0 \leqslant K \leqslant K_1$, with certain constants K_0, K_1. Then M carries a closed geodesic c with length $L(c) \in [2\pi/\sqrt{K_0}, 4\pi/\sqrt{K_1}[$ if and only if M is isometric to the sphere M_0 of constant curvature $K = K_0$, and in this case, $L(c) = 2\pi/\sqrt{K_0}$. In particular, if $L(c) < 4\pi/\sqrt{K_1}$, actually $L(c) \leqslant 2\pi/\sqrt{K_0}$.*

Proof. We can assume $K_1 = 1$, cf. the proof of (2.8.5). Then the diameter $d(M)$ is $\leqslant \pi/\sqrt{K_0} \leqslant 2\pi$ and the injectivity radius $\iota(M)$ is $\geqslant \pi$, see (2.6.10).

It suffices to consider the case $K_0 < 1$. Let $c(t)$, $0 \leqslant t \leqslant \omega$, be a closed geodesic of length $L(c) = \omega$ with $2\pi/\sqrt{K_0} \leqslant \omega < 4\pi$. Put $c(0) = p, c(\pi) = q, c(\omega - \pi) = r$. With $c_{pq} = c|[0, \pi]$; $c_{qr} = c|[\pi, \omega - \pi]$, $c_{rp} = c|[\omega - \pi, \pi]$ we get a possibly generalized triangle pqr – note that $L(c_{qr}) = \omega - 2\pi < 2\pi = L(c_{pq}) + L(c_{rp})$. Then from (2.7.12, i) we get $\omega = L(c) = L(pqr) \leqslant 2\pi/\sqrt{K_0}$.

Assume $L(c) = \omega = 2\pi/\sqrt{K_0} < 4\pi$. Hence, $1/4 < K_0 < 1$. All sides in pqr have length $< \pi/\sqrt{K_0}$ and we get from (2.7.12, iii) that $d(M) \geqslant \pi/\sqrt{K_0}$, i. e., according to (2.8.1), M is isometric to the sphere M_0 of curvature K_0. □

With this we now can prove:

3.3.16 Theorem. *Let M be a simply connected compact Riemannian manifold,* $\dim M = n$. *Assume that the sectional curvature K is bounded by $0 < K_0 \leqslant K \leqslant K_1$, with $\delta = K_0 : K_1 \geqslant 1/4$.*

(i) *If c is hyperbolic with $\sqrt{\delta} \geqslant (n-1+m)/(2n-2)$, $0 \leqslant m \leqslant n-2$, then* index $c > n-1+m$. *In particular, for $\delta \geqslant 1/4$, a closed geodesic c of index $n-1$ is not hyperbolic.*

(ii) *Let c be hyperbolic with $L(c) < 4\pi/\sqrt{K_1}$ and $\sqrt{\delta} \geqslant (2n-2)/(3n-3-l) \geqslant 2/3$, $0 \leqslant l \leqslant n-2$. Then* index $c < 3n-3-l$. *In particular, if $\delta \geqslant 4/9$, a closed geodesic of index $3n-3$ is not hyperbolic.*

Remark. (i) is proved without using (3.3.15) while (ii) uses essentially that a geodesic of length $< 4\pi/\sqrt{K_1}$ actually has length $\leqslant 2\pi/\sqrt{K_0}$. Special cases of (3.3.16) are due to Thorbergsson [1] while the full result – and actually several further results describing the center subspace of the Poincaré mapping of c in the sense of (3.1.5) – are proved in Ballmann, Thorbergsson and Ziller [1].

Proof. We may assume $K_1 = 1$ and hence $K_0 = \delta$. Then we know from (2.6.A.1) that $L(c) \geqslant 2\pi$. If $L(c) = 2\pi$, we know from (2.8.7) that c cannot be hyperbolic. We thus may assume $L(c) > 2\pi$.

To prove (i) we note that, for large k,

$$L(c^{k(n-1)}) = k(n-1)L(c) \geqslant (2k(n-1)\sqrt{\delta} + 2)(\pi/\sqrt{\delta}),$$

where c^m denotes the m-fold covering of c. A geodesic segment of length $\pi/\sqrt{\delta}$ contains at least $(n-1)$ conjugate points, cf. (2.6.2). Thus, if c is hyperbolic, we get with index $c^m = m$ index c

$$\text{index}\,(c^{k(n-1)}) = k(n-1)\,\text{index}\,c \geqslant (n-1)[2k(n-1)\sqrt{\delta} + 1].$$

Here, $[a]$ denotes the greatest integer $\leqslant a$. Hence

$$\text{index}\,c \geqslant \frac{1}{k}[2k(n-1)\sqrt{\delta} + 1] \geqslant \frac{1}{k}[k(n-1+m) + 1] > n-1+m,$$

if $\sqrt{\delta} \geqslant (n-1+m)/(2n-2)$.

For the proof of (ii) we have from (3.3.15) that $L(c) \leqslant 2\pi/\sqrt{\delta}$. Hence, just as above,

$$L(c^{k(n-1)}) = k(n-1) L(c) \leqslant 2\pi k(n-1)/\sqrt{\delta}.$$

If the restriction of c to a segment of length π has $(n-1)$ conjugate points, c is not hyperbolic. We therefore have the estimate

$$\operatorname{index}(c^{k(n-1)}) = k(n-1) \operatorname{index} c < (n-1)[2k(n-1)/\sqrt{\delta}].$$

Hence, if $1/\sqrt{\delta} \leqslant (3n-3-l)/(2n-2)$,

$$\operatorname{index} c < \frac{1}{k}[k(3n-3-l)] = 3n-3-l. \quad \square$$

We now can show, cf. Ballmann, Thorbergsson and Ziller [1]:

3.3.17 Theorem. *Let M be a compact, simply connected Riemannian manifold of dimension n. Assume that the sectional curvature K of M satisfies $K_1/4 < K \leqslant K_1$, for some $K_1 \geqslant 0$. Then there exists on M a simple, non-hyperbolic closed geodesic c of index $n-1$ and length in the interval $[2\pi/\sqrt{K_1}, 4\pi/\sqrt{K_1}[$.*

Proof. We can restrict ourselves to the case $K_1 = 1$. Put $\min K = \delta$. Then $\delta > 1/4$. From (2.8.5) we know that M is homeomorphic to the sphere S^n. Thus, there exists a differentiable $f: S^n \to M$, not homotopic to a constant mapping. The proof of (2.4.20) yields a closed geodesic c at which the associated mapping $F(f): (D^{n-1}, \partial D^{n-1}) \to (\Lambda M, \Lambda^0 M)$ remains hanging. (2.5.16) shows that we may assume $\operatorname{index} c \leqslant n-1$. (2.6.9) implies $L(c) \geqslant 2\pi > \pi/\sqrt{\delta}$. Hence, according to (2.6.2), $\operatorname{index} c = n-1$, and $L(c) \leqslant 2\pi/\sqrt{\delta} < 4\pi$. But then, c must be simple, i. e., without self-intersections. $\square$

3.4 Geodesics on Surfaces

In this section we consider 2-dimensional Riemannian manifolds, also called briefly surfaces. Often, we will assume the surface to be compact, to simplify the exposition, although the results also might be true for complete or even non-complete surfaces.

We begin by exposing a close relation between the parity of the index of a closed geodesic and the stability behaviour of the corresponding periodic orbit. At its basis is the observation that, in the index formula (2.5.14), for surfaces the concavity term is closely related to the orientability of the geodesic, see (3.4.2) below.

At the end we consider convex surfaces, having the property that they permit a free isometric $\mathbb{Z}_2$-action. This amounts to considering Riemannian metrics of positive curvature on the real projective plane. In (3.7) we will extend these results – with more elaborate methods – to general surfaces of genus 0.

We conclude by showing that a surface with a waist possesses infinitely many closed geodesics.

For a surface, the linear Poincaré mapping P_ω associated to a periodic flow line $\phi_t X_0, 0 \leqslant t \leqslant \omega$, operates on a 2-dimensional symplectic space V_0. Hence, P_ω has at most two different eigenvalues. If the eigenvalues of P_ω are real and not $+1$ or -1, P_ω and the flow line are called *hyperbolic*, otherwise they are called *elliptic*, cf. (3.1.6). If $+1$ is an eigenvalue, P_ω and the flow line also are called *parabolic*. By *properly elliptic* we mean the case that P_ω has an eigenvalue $\varrho, |\varrho| = 1, \varrho \neq 1$.

If P_ω is hyperbolic, V_0 splits into two P_ω-invariant 1-dimensional subspaces $V_{0,s}$ and $V_{0,u}$, called *stable* and *unstable* space, cf. (3.1.5). If, on the other hand, P_ω is not hyperbolic, V_0 does not possess a P_ω-invariant subspace, unless $+1$ or -1 is an eigenvalue of P_ω. In the latter case, V_0 has a basis such that P_ω is represented by a matrix of the form

$$\begin{pmatrix} 1 & \varepsilon \\ 0 & 1 \end{pmatrix} \quad \text{or} \quad \begin{pmatrix} -1 & \varepsilon \\ 0 & -1 \end{pmatrix}, \text{ with } \varepsilon \in \{0, 1\}.$$

Let now P_ω have non-real eigenvalues $\varrho = e^{i\phi}, \varrho = e^{-i\phi}, \phi \neq 0, \pi$. Then P_ω has in an appropriate basis the matrix representation

$$\begin{pmatrix} \cos\phi & -\sin\phi \\ \sin\phi & \cos\phi \end{pmatrix}.$$

In the terminology of (3.3.4), P_ω and the flow line $\phi_t X_0, 0 \leqslant t \leqslant \omega$, are then called *infinitesimally stable*. This name is applied also to the case where $P_\omega = \pm\, id$.

Observe that $|\text{trace } P_\omega| > 2$ characterizes hyperbolicity. $|\text{trace } P_\omega| < 2$ characterizes the case where P_ω has non-real eigenvalues.

3.4.1 Definition. Let $c(t), 0 \leqslant t \leqslant 1$, be a closed geodesic on M. c is called *orientable* if the induced bundle $\upsilon: V \to S_\omega$ over S_ω, cf. (2.3.4), is orientable in the sense of (2.1.A.2).

Note. Consider the isometry

$$(*) \qquad \overset{\omega}{\underset{0}{\|}} c : T_0^\perp c \to T_\omega^\perp c = T_0^\perp c,$$

cf. the proof of (2.6.7). Then c is orientable if and only if the transformation $(*)$ has determinant $+1$, i. e., belongs to the special orthogonal group. This follows from (2.1.A.3).

The next result is taken from Klingenberg [11]. Similar results were obtained by Hedlund [1].

3.4.2 Theorem. *Let M be a surface and c a closed geodesic on M. If $c = c(t), 0 \leqslant t \leqslant 1$, is non-degenerate and not hyperbolic (i. e., properly elliptic) then* index c *is odd or even, depending on whether c is orientable or not.*

More precisely, if k is the number of conjugate points of $c(0)$ which occur before $c(1)$ then, for c orientable, index $c = k$ *or* $k+1$, *depending on whether k is odd or even. For c non-orientable,* index $c = k$ *or* $k+1$ *depending on whether k is even or odd.*

The same conclusions hold if c is degenerate with maximal nullity $= 2$.

Remarks. 1. According to (2.5.14), index c is the sum of the Ω-index of c and the concavity of c. Now, $\text{index}_\Omega c$ is equal to the number k of conjugate points of $c(0)$ which occur before $c(1)$, cf. (2.5.9). Our Theorem therefore may be interpreted also as a statement on the concavity of a non-degenerate elliptic closed geodesic on a surface.

2. The Theorem brings together properties of the two aspects of a closed geodesic, i. e., the index, which is a feature of a critical point of $E: \Lambda M \to \mathbb{R}$, and the ellipticity, which is a property of the corresponding periodic orbit in the geodesic flow. Cf. (3.2.12) for another example for such a relation.

Proof. We first consider the case where c is orientable. Put $\dot{c}(0)/|\dot{c}(0)| = X_0$. Then $\phi_t X_0, 0 \leqslant t \leqslant \omega = |\dot{c}(0)| = L(c)$, is the periodic orbit of the geodesic flow, associated with c. We write $\tau_M \phi_t X_0 = c_{X_0}(t), 0 \leqslant t \leqslant \omega$.

Let $\upsilon: V \to S_\omega$ be the associated vector bundle, cf. (3.2.2).

Since c_{X_0} is orientable we can choose an orthonormal parallel basis $\{E_0(t), E_1(t)\}$ for $T_{c_{X_0}(t)} M$ with $E_0(t) = \dot{c}_{X_0}(t)$, $\{E_0(\omega), E_1(\omega)\} = \{E_0(0), E_1(0)\}$. A symplectic basis $\{\varepsilon_h(t), \varepsilon_v(t)\}$ for $V_t = \upsilon^{-1}(t)$ then is given by

$$\varepsilon_h(t) = (E_1(t), 0);\ \varepsilon_v(t) = (0, E_1(t)).$$

A $T\phi_t$-invariant section $\zeta(t)$ in the universal covering $\tilde{\upsilon}: \tilde{V} \to \mathbb{R}$ of υ can be written as

$$\zeta(t) = z(t)\varepsilon_h(t) + \dot{z}(t)\varepsilon_v(t),$$

where $z(t)$ is a real valued function satisfying the Jacobi equation

$$\ddot{z}(t) + K(t)z(t) = 0,$$

with $K(t) = K \circ c(t)$. That is, $Z(t) = z(t)E_1(t)$ is a Jacobi field along $c_{X_0}(t)$, orthogonal to $\dot{c}_{X_0}(t)$.

Denote by $\xi(t)$ and $\eta(t)$ the two basic $T\phi_t$-invariant sections in $\tilde{\upsilon}$ with initial condition $\xi(0) = \varepsilon_h(0), \eta(0) = \varepsilon_v(0)$. Then the Poincaré mapping P_ω possesses the matrix representation

$$\begin{pmatrix} p & q \\ r & s \end{pmatrix} \equiv \begin{pmatrix} x(\omega) & y(\omega) \\ \dot{x}(\omega) & \dot{y}(\omega) \end{pmatrix}$$

with $ps - qr = 1, |p+s| \leqslant 2$. The latter relation is the condition that c is non-hyperbolic. That c is non-degenerate means $p + s \neq 2$.

The concavity of c can be computed as follows: Let $Z(t) = z(t)E_1(t)$ be a Jacobi field with $Z(\omega) = Z(0)$, but $\nabla Z(\omega) - \nabla Z(0) \neq 0$. We can even assume $\nabla Z(\omega) - \nabla Z(0) = E_1(0)$. Then

$$\langle \nabla Z(\omega) - \nabla Z(0), Z(0) \rangle = \langle E_1(0), Z(0) \rangle = z(0).$$

Hence, according to (2.5.12), concav $c = 1$ if $z(0) \leqslant 0$; otherwise, concav $c = 0$.

We put $z(t)\varepsilon_h(t) + \dot{z}(t)\varepsilon_v(t) = \zeta(t)$. Then $P_\omega \zeta(0) - \zeta(0) = \eta(0)$ yields the relation

$$(2 - p - s)z(0) = -q, \text{ i. e., } (2 - x(\omega) - \dot{y}(\omega))z(0) = -y(\omega).$$

Assume first that c is non-degenerate. Then $2 - p - s > 0$. If k is even, i. e., if $\eta(t)$ crosses the vertical axis an even number of times in the interval $0 < t < \omega$, then $y(\omega) \geqslant 0$. Note that each time where $\eta(t_0) \in V_{t_0 v}$, i. e., $y(t_0) = 0$ and $\dot{y}(t_0) \neq 0$, we have $\nabla \eta(t_0) = \dot{y}(t_0)\varepsilon_h(t_0)$. That is to say, $\eta(t)$ always crosses the vertical axis transversally and clockwise. Now, $y(\omega) \geqslant 0$ means $z(0) \leqslant 0$, i. e., concav $c = 1$.

If, on the other hand, k is odd, $y(\omega) \leqslant 0$. $y(\omega) = 0$ is excluded since in that case $ps = 1$, $s = \dot{y}(\omega) > 0$ would imply under our assumptions $p = s = 1$. Hence, concav $c = 0$.

Assume now nullity $c = 2$. Then $\eta(\omega) = \eta(0)$, i. e., k is odd and concav $c = 0$.

Let us now consider the case where c is not orientable. Then $(\overset{\omega}{\underset{0}{\|}} c_{X_0})E_1(0) = -E_1(0)$. The matrix representation of P_ω in this case reads

$$\begin{pmatrix} p & q \\ r & s \end{pmatrix} = -\begin{pmatrix} x(\omega) & y(\omega) \\ \dot{x}(\omega) & \dot{y}(\omega) \end{pmatrix}$$

Thus, as compared to the orientable case, all signs are reversed. With this we easily complete the proof. □

3.4.3 Corollary. *On an orientable surface M, a non-degenerate elliptic closed geodesic or a closed geodesic with nullity 2 has odd index.*

Proof. Simply observe that on an orientable surface every closed geodesic is orientable. Now apply (3.4.2). □

In (3.4.2) we had excluded the case nullity $c = 1$. In fact, in this case, one cannot make a general statement on the parity of index c. Even index $c = 0$ is possible. This latter case, however, permits an interesting characterization due to Poincaré [1] which we now present. The original proof of Poincaré is based on a geometric argument.

3.4.4 Theorem. *An orientable closed geodesic $c = c(t), 0 \leqslant t \leqslant 1$, on a surface has index 0 if and only if there are no conjugate points of $c(0)$ on the half geodesic $c(t), t \geqslant 0$.*

It follows that all multiple coverings c^m of such a geodesic c also have index 0 for all m. If c is non-hyperbolic, nullity $c^m = 1$.

Proof. If, for some $t_0 > 0$, $c(t_0)$ is conjugate to $c(0)$ along $c|[0, t_0]$, index $c^m > 0$ for all sufficiently high iterates c^m of c. Indeed, whenever m is an integer $> t_0$, $c^m(t) = c(mt), 0 \leqslant t \leqslant 1$, has a conjugate point in its interior and hence, index $c^m > 0$.

Consider first the case when c is hyperbolic. Then the formula index $c^m = m$ index c, cf. (3.2.15), proves our theorem.

Now let c be non-hyperbolic, index $c = 0$. From (3.4.3) we then know that nullity $c = 1$. Denote by $\eta^*(0)$ a non-zero eigenvector for the eigenvalue 1 of the Poincaré mapping P_ω.

Consider the $T\phi_t$-invariant section $\eta^*(t) = T\phi_t . \eta^*(0), 0 \leqslant t \leqslant \omega$, in υ. Since index $c = 0, \eta^*(t)$ never meets the vertical axis $V_{t,v}, 0 \leqslant t \leqslant \omega$. By periodicity, this is also true for all $t \geqslant 0$ in the universal covering $\tilde{\upsilon}$ of υ. It follows that the section $\eta(t) = T\phi_t . \eta(0), \eta(0) = \varepsilon_v(0) \in V_{0,v}$, also never again meets $V_{t,v}, t > 0$. Thus, $c(0)$ has no conjugate points on $c|[0, \infty[$.

Conversely, assume that $c(0)$ has no conjugate points on $c|[0, \infty[$. That means that $\eta(t)$ with $\eta(0) = \varepsilon_v(0) \in V_{0,v}$ never again meets $V_{t,v}$, for $t > 0$. In particular, since the vertical axis is crossed transversally and clockwise, $\varepsilon_v(0)$ is not eigenvector of P_ω or of any iterate $P_\omega^m = P_{m\omega}$. We can assume c non-hyperbolic. Thus, nullity $c \leqslant 1$.

The case nullity $c = 0$ and c non-hyperbolic cannot occur, cf. (3.4.3). If nullity $c = 1$, consider a non-zero eigenvector $\eta^*(0)$ of P_ω for the eigenvalue 1. $(P_\omega - id) V_0$ is spanned by $\eta^*(0) \notin V_{0,v}$. Therefore, $(P_\omega - id) V_0$ does not contain $\varepsilon_v(0)$ and hence, concav $c = 0$, i. e., index $c = 0$. □

3.4.5 Corollary. *Let c be an orientable closed geodesic. Then either index $c^m = 0$ and nullity $c^m \leqslant 1$, for all m, or else, index c^m is unbounded, as m goes to infinity.*

Proof. We know from (3.4.4) that the first case is equivalent to index $c = 0$ and to the fact that there are no conjugate points on the geodesic $c(t), t \geqslant 0$. Thus, if this is not true, let $m_0 \geqslant 1$ be an integer so large that $c|[0, m_0]$ contains a conjugate point in its interior. Then $c|[0, km_0]$, k an integer $\geqslant 1$, contains at least k conjugate point in its interior, i. e., index $c^{km_0} \geqslant k$. □

If c is orientable and hyperbolic, both cases, index c even and index c odd, can occur. We have the following condition for one of these two possibilities to occur.

3.4.6 Lemma. *Let c be a hyperbolic, orientable closed geodesic. Then index c is even if and only if the two real eigenvalues ϱ, $1/\varrho$ of the associated Poincaré mapping P_ω are positive.*

Remark. As we will see, ϱ and $1/\varrho$ positive and index $c = 2l$ is equivalent to saying that the flow-invariant stable section $\eta_s(t)$ in the bundle υ completes l full rotations in the frame $\varepsilon_h(t), \varepsilon_v(t)$ which we introduced in the proof of (3.4.2).

Proof. We write a stable section $\eta_s(t)$ in the form $\eta_s(t) = y_s(t)\varepsilon_h(t) + \dot{y}_s(t)\varepsilon_v(t)$. Let $|\varrho| < 1$. Then $\eta_s(\omega) = \varrho\eta_s(0)$. If index $c = 0$ then $\eta_s(t)$ never meets $V_{t,v}$. Thus, $\varrho > 0$. If index $c = k > 0$ then $\varrho > 0$ or $\varrho < 0$ depending on whether k is odd or even, cf. (3.2.13). □

Remarks. 1. It would be false to assume, however, that on an orientable surface every hyperbolic closed geodesic has even index. Poincaré [2] has given an example of a convex surface obtained from slightly perturbing the metric of an ellipsoid with pinching 1/16 where the shortest closed geodesic has index 1 and is hyperbolic.

Another example can be constructed along the following lines: Start with an ellipsoid

$$M = \{x_0^2/a_0 + x_1^2/a_1 + x_2^2/a_2 = 1\}.$$

$0 < a_0 < a_1 < a_2$, $|1 - a_i|$ small, cf. (1.10.16). As we will see in (3.5.16), a parameterization of the ellipse of middle length in the (x_0, x_2)-plane yields a hyperbolic closed geodesic of index 2. Now take a narrow ribbon around this ellipse on M and identify the points which correspond to each other under the symmetry $(x_0, x_1, x_2) \mapsto (-x_0, x_1, -x_2)$. This ribbon can be embedded isometrically in $\mathbb{R}^3$ to yield a narrow cylinder with curvature approximately $= 1$ and diameter approximately $= 1$. To produce a convex surface, close up this cylinder by two caps with maximal curvature approximately 4, similar to half spheres of radius 1/2. One thus obtains a convex surface with $\min K : \max K = 1/4 - \varepsilon$, for some small $\varepsilon > 0$, which carries on the original ribbon a closed hyperbolic geodesic of index 1. For details of this construction cf. Ballmann, Thorbergsson and Ziller [1].

2. One may ask under what conditions a slight perturbation of the metric transforms an elliptic closed geodesic c on an orientable surface M into a hyperbolic closed geodesic. Such a penomenon is called a *bifurcation*. Poincaré [1] was the first one to point out the importance of bifurcations for the study of periodic orbits.

Here we only observe the following: Let M_0 be an orientable surface. We also write $M_0 = (M, g_0)$ where M denotes the underlying differentiable manifold and g_0 the Riemannian metric on M. Let now g_σ, $-\varepsilon \leqslant \sigma \leqslant \varepsilon$, $\varepsilon > 0$, be a 1-parameter family of Riemannian metrics on M, i. e., a curve $\sigma \in [-\varepsilon, \varepsilon] \mapsto g_\sigma \in \mathscr{G} M$ in the space of $\mathscr{G} M$ of Riemannian metrics on M, cf. (1.10.2). Assume that c_σ is a closed geodesic on $M_\sigma = (M, g_\sigma)$ which depends continuously on σ. For $\sigma < 0$, c_σ shall be non-degenerate elliptic and for $\sigma > 0$ it shall be hyperbolic. What can be said for c_0, i. e., for the bifurcation point of the family c_σ?

Under a so-called generic bifurcation, the critical points c_σ of $E_\sigma : \Lambda M_\sigma \to \mathbb{R}$ are all non-degenerate. Thus, for $\sigma < 0$, we know from (3.4.3) that index c_σ is odd. But then, according to (3.4.5), the eigenvalues $(\varrho_\sigma, \varrho_\sigma^{-1})$ of the Poincaré mapping associated to c_σ, must be negative when $\sigma > 0$. Thus, while for $\sigma < 0$, $\varrho_\sigma^{-1} = \bar{\varrho}_\sigma$, i.e., the eigen values $\{\varrho_\sigma, \varrho_\sigma^{-1}\}$ are complex conjugates of modulo 1, for $\sigma = 0$ they must read $\{-1, -1\}$.

For a discussion of the generic bifurcations for a Hamiltonian system with two degrees of freedom see Meyer [1].

Morse [2] has made the following interesting observation: Let M be an ellipsoid, not too different from a sphere, but with no two axes of equal length. Then the only prime closed geodesics – up to a high E-value – are the ellipses, obtained from intersecting the ellipsoid with a coordinate 2-plane. We present here a simple proof which was first given in Klingenberg [10].

3.4.7 Lemma. *Let* $M = M(a_0, \ldots, a_n) = \{\sum_0^n x_i^2/a_i = 1\} \subset \mathbb{R}^{n+1}$ *be an n-dimensional ellipsoid.*

Given any large $\varkappa > 2\pi^2$ *there exists an* $\varepsilon = \varepsilon(\varkappa) > 0$ *such that on every* $M(a_0, \ldots, a_n)$ *with* $a_0 < \ldots < a_n$ *and* $|1 - a_i| < \varepsilon$, *the only closed geodesics with E-*

value $<\varkappa$ *are the ones which lie in one of the* $n(n+1)/2$ *coordinate planes, i.e., are simple or multiple coverings of one of the principle ellipses.*

Proof. $x_i(s), 0 \leqslant i \leqslant n$, is a geodesic on M, parameterized by arc length, if and only if the second (Euclidean) derivative is orthogonal to M, cf. (1.10.3). In our case, $M^* = \mathbb{R}^{n+1}$ and the defining equation $\nabla c'(s)/ds = 0$ of a geodesic may be written as $pr \circ c''(s) = 0$, i. e.,

$$(*) \qquad x_i''(s) + \lambda(s) x_i(s)/a_i = 0$$

To determine $\lambda(s)$ we twice differentiate the equation $\sum_i x_i(s)^2/a_i = 1$. We find

$$\lambda(s) = \Big(\sum_i (x_i'(s)^2/a_i)\Big) \Big/ \Big(\sum_i x_i(s)^2/a_i^2\Big)$$

If $a_i = 1, i = 0, \ldots, n$, then $M = M(a_0, \ldots, a_n)$ is the sphere of constant curvature $K = 1$. In this case, we have for every geodesic $x_i(s), 0 \leqslant i \leqslant n$, on M the following property: Let $x_k(s) \not\equiv 0$. $x_k(s)$ has a zero, say $x_k(s_0) = 0$. Then all other zeros of $x_k(s)$ occur at $s = x_0 + q\pi, q \in \mathbb{Z}$.

Fix now $\varkappa > 2\pi^2$. Let q^* be an integer with $\sqrt{2\varkappa} < q^*\pi + \pi/2$. The continuous dependence of the zeros of a solution $x_i(s)$ of $(*)$ from the coefficient $\lambda(s)/a_i$ shows: There exists an $\varepsilon = \varepsilon(q^*) > 0$ with the following property: For every ellipsoid $M = M(a_0, \ldots a_n)$ with $|1 - a_i| < \varepsilon, i = 0, \ldots, n$, a geodesic $x_i(s), 0 \leqslant i \leqslant n$, on M for which $x_k(s)$ is not identically zero, $x_k(s)$ has a zero, say s_0, and there are q^* subsequent zeros $s_1, \ldots, s_{q^*}$ satisfying

$$s_0 + q\pi - \pi/2 < s_q < s_0 + q\pi + \pi/2, \ 1 \leqslant q \leqslant q^*.$$

We now choose $M = M(a_0, \ldots, a_n)$ with $|1 - a_i| < \varepsilon$ and $a_0 < \ldots < a_n$. Let $x_i(s), 0 \leqslant i \leqslant n, 0 \leqslant s \leqslant \omega$, be a closed geodesic on M with E-value $\leqslant \varkappa$. Hence, $\omega < q^*\pi + \pi/2$. We want to derive a contradiction from the assumption that there are at least three i-values, say $i = j, k, l$, where $x_i(s) \not\equiv 0$. We can assume $j < k < l$, thus, $\lambda(s)/a_j > \lambda(s)/a_k > \lambda(s)/a_l$.

There exists a s_0 with $x_k(s_0) \neq 0, x_l(s_0) = 0$. Indeed, if $x_k(s_0) = x_l(s_0)$ then for the zero s_1 of $x_l(s)$ which follows s_0, we have $x_k(s_1) \neq 0$. This is seen by comparing the zeros of the solutions of $(*)$, for $i = k$, with the zeros of the solutions of $(*)$, for $i = l$, using $\lambda(s)/a_k > \lambda(s)/a_l$.

We associate with a solution $u_i(s)$ of $(*)$ the curve $U_i(s) = (u_i(s), u_i'(s)) \in \mathbb{R}^2$. With this, we now consider the linear transformation

$$P_k: \mathbb{R}^2 \mapsto \mathbb{R}^2; \ U_k(s_0) \mapsto U_k(s_0 + \omega).$$

Here, $U_k(s) = (u_k(s), u_k'(s))$ and $u_k(s)$ is solution of $(*)$, for $i = k$.

P_k is area preserving since $u_k(s) v_k'(s) - v_k(s) u_k'(s) = \text{const}$. Moreover, $X_k(s_0) = (x_k(s_0), x_k'(s_0))$ is an eigenvector of P_k for the eigenvalue 1. Now choose the vector $U_k(s_0) \in \mathbb{R}^2$ such that $U_k(s_0) = \pm X_l(s_0) = (0, \pm x_l'(s_0))$ and $(X_k(s_0), U_k(s_0))$ form a positively oriented basis. Then

$$(P_k - 1)\,U_k(s_0) = \lambda X_k(s_0).$$

The sign of λ is an invariant of P_k. Indeed, note that the most general positively oriented basis (A, B) of $\mathbb{R}^2$ with A an eigenvector of P_k for the eigenvalue 1 is of the form

$$A = a X_k(s_0),\ B = b X_k(s_0) + d U_k(s_0)$$

with $ad > 0$. We find

$$(P_k - 1)\,B = \lambda dA/a,$$

i. e., $\operatorname{sign}\lambda = \operatorname{sign}\lambda\, d/a$.

To determine the sign of λ we compare the zeros of $u_k(s)$, which is a solution of $(*)$ for $i = k$, with the zeros of $x_l(s)$, which is a solution of $(*)$ for $i = l$. $u_k(s_0) = x_l(s_0) = 0$. The next following zero of $u_k(s)$ comes before the next zero of $x_l(s)$. This goes on for all the following q^* zeros of $u_k(s)$, since the zeros of both, $u_k(s)$ and $x_l(s)$, occur in the disjoint intervals $]s_0 + q\pi - \pi/2, s_0 + q\pi + \pi/2[$, $q = 0, \dots q^*$, exactly once.

Hence, since $x_l(s_0 + \omega) = x_l(s_0)$, $x_l'(s_0 + \omega) = x_l'(s_0)$, the λ in $P_k U_k(s_0) - U_k(s_0) = \lambda X_k(s_0)$ is positive.

Let now s_0 be such that $x_j(s_0) = 0$, $x_k(s_0) \neq 0$ (this will in general be another s_0 as before). We define, just as above, a linear mapping

$$Q_k : \mathbb{R}^2 \to \mathbb{R}^2$$

by associating with $V_k(s_0) = (v_k(s_0), v_k'(s_0))$ the vector $V_k(s_0 + \omega) = (v_k(s_0 + \omega), v_k'(s_0 + \omega))$. Again, $X_k(s_0)$ is eigenvector for the eigenvalue 1, because Q_k is conjugate to P_k by an area preserving transformation.

Determine $V_k(s)$ by $V_k(s_0) = \pm X_j(s_0)$ such that $(X_k(s_0), V_k(s_0))$ is a positively oriented basis. By comparing the zeros of $v_k(s)$ with the zeros of $x_j(s)$, $s \geqslant s_0$, we find that the zeros of $v_k(s)$ come after the zeros of $x_j(s)$, when we consider the interval $[s_0, s_0 + \omega]$. Therefore, in

$$(Q_k - id)\,V_k(s_0) = \mu X_k(s_0),$$

μ must be negative. This is the desired contradiction. □

In (3.5) we will discuss the geodesic flow for the 2-dimensional ellipsoid in much more detail. Here we continue with some more results on general *convex surfaces*. We use the term 'convex surface' synonymous with a Riemannian manifold M of dimension 2, simply connected and of strictly positive curvature K. A classical result of H. Weyl and H. Lewy (for the case of a real analytic metric) and A. D. Alexandrov, A. V. Pogorelov and L. Nirenberg (for the differentiable case, working with quite different methods) states, that such a manifold always can be embedded as a convex surface in $\mathbb{R}^3$. Cf. Klingenberg [9] for references.

For the following result see Klingenberg [2].

3.4.8 Theorem. *Let M be a convex surface. Then a geodesic loop c on M has length $\geqslant 2\pi/\sqrt{\max K}$. $L(c) = 2\pi/\sqrt{\max K}$ can hold only if M is the sphere of constant curvature.*

Proof. Put $\max K = K_1$. $L(c) \geqslant 2\pi/\sqrt{K_1}$ follows from (2.6.9). Thus only the case $L(c) = 2\pi/\sqrt{K_1}$ remains to be discussed.

We first show that the loop $c = c(t)$, $0 \leqslant t \leqslant 1$, is actually a closed geodesic. Indeed, put $c(0) = c(1) = p$, $c(1/4) = q$, $c(3/4) = r$. Since $d(q, C(q)) \geqslant \pi/\sqrt{K_1}$, $c|[1/4, 3/4]$ is a minimizing geodesic from q to r. But then, the curve $c(1-t)$, $3/4 \leqslant t \leqslant 1$, followed by $c(1-t)$, $0 \leqslant t \leqslant 1/4$, from q to r cannot have a corner at p.

From the proof of (2.6.5) we know that $c = c_0$ is the limit of a 1-parameter family $c_s = c_s(t)$, $0 \leqslant t \leqslant 1$, of parallel curves to c, $0 \leqslant s \leqslant \varepsilon$, with $L(c_s) < L(c)$, if $s > 0$. Put $c_s(0) = p_s$. Then there exists a uniquely determined minimizing geodesic from p_s to $c_s(t)$, provided $s > 0$, for every $t \in [0, 1]$. The angle between $\dot{c}_s(0)$ and the initial direction of this minimizing geodesic varies between 0 and π, as t varies from 0 to 1.

Fix now α, $0 < \alpha < \pi$. For every $s > 0$ let $c_{s,\alpha}: [0, 1] \to M$ be a minimizing geodesic from $c_s(0) = p_s$ to a point on c_s which with $\dot{c}_s(0)$ forms the angle α. As $s \to 0$, the geodesics $c_{s,\alpha}$ have a limit set which contains a minimizing geodesic $c_\alpha: [0, 1] \to M$ from $p = p_0$ to a point q_α on c such that $\dot{c}(0)$ and $\dot{c}_\alpha(0)$ form the angle α. The existence of such a c_α follows from a compactness argument, using the fact that $\dot{c}_{s,\alpha}(0)$, $0 \leqslant s \leqslant \varepsilon$, belongs to a compact subset of TM, cf. also (1.6.11).

Since the distance from p to its cut locus $C(p)$ is $\pi/\sqrt{K_1}$, q_α must be equal to q. We have therefore shown that every geodesic c' with starts from $p = c(0)$, with $\dot{c}'(0)$ not linearly dependent on $\dot{c}(0)$, meets c again at distance $\pi/\sqrt{K_1}$ in the same point $q = c(1/2)$. Hence, all geodesics through p have their first conjugate point at distance $\pi/\sqrt{K_1}$. Since $K \leqslant K_1$, this implies $K = K_1$, cf. the proof of (2.6.1). □

Following Thorbergsson [1] we can sharpen (3.3.17) for a convex surface as follows:

3.4.9 Theorem. *Let M be a convex surface with $0 < K_1/4 \leqslant K \leqslant K_1$. Then there exists on M a non-hyperbolic simple closed geodesic of index 1 and length $L(c) \in [2\pi/\sqrt{K_1}, 4\pi/\sqrt{K_1}]$.*

Proof. It suffices to consider the case $K_1 = 1$. As in the proof of (3.3.17), we show the existence of a closed geodesic c of length $L(c) \in [2\pi, 4\pi]$ and index $c \leqslant 1$. If $L(c) = 2\pi$, (2.4.6) or (3.4.8) imply that M is isometric to the sphere of constant curvature 1. Thus, c is simple and non-hyperbolic, with index $c = 1$. If $L(c) > 2\pi$, c contains a conjugate point in its interior, hence, index $c = 1$. If c were hyperbolic, index $c^m = m$ index $c = m$, for all coverings c^m of c. On the other hand, for large m, $L(c^m) > (m+1)2\pi$ which implies index $c^m \geqslant m+1$ – thus, c is non-hyperbolic.

If $L(c) < 4\pi$, c is simple since otherwise c would contain a geodesic loop of length $< 2\pi$ which is impossible according to (3.4.8). It remains to show that a closed geodesic c of length $L(c) = 4\pi$ and index 1 is simple.

Indeed, c would otherwise contain a geodesic loop c_1 of length $\leqslant 2\pi$, i. e., actually $L(c_1) = 2\pi$. But then, according to (3.4.8), M is isometric to the sphere of constant curvature 1 and on such a surface, a closed geodesic c of length 4π has index > 1 – a contradiction. □

We continue with a counterpart of (3.4.8), due to Toponogov [1].

3.4.10 Theorem. *On a convex surface M, a simple closed geodesic has length $\leqslant 2\pi/\sqrt{\min K}$. Here, equality holds only if M is isometric to the sphere of constant curvature $K = \min K$.*

Note. Compare the last part with (2.8.1).

Proof. Put $\min K = K_0$. A simple closed geodesic c on M partitions M into two domains. Each of these domains is relatively convex. That is, any two points p, q in one of these domains, say D, can be joined by a geodesic which is not longer than any other curve from p to q inside D. This follows from the fact that the boundary ∂D of D is the image of the simple closed geodesic c.

Let c_1 be a minimizing geodesic from $c(0)$ to $c(1/2)$. If $L(c_1) \geqslant \frac{1}{2} L(c)$ we are done. Indeed, the length of a minimizing geodesic is $\leqslant \pi/\sqrt{K_0}$, thus, $L(c) \leqslant 2\pi/\sqrt{K_0}$.

If $L(c_1) < L(c)/2$ we denote by D the relativity convex domain bounded by $\{\text{image } c\}$ and containing c_1. Let $c(0)c(1/4)c(1/2)$ and $c(0)c(3/4)c(1/2)$ be triangles in D where the side from $c(0)$ to $c(1/2)$ is c_1 and the other sides are relatively minimizing geodesics in D. Consider on M_0 = sphere of constant curvature K_0 the associated Alexandrov triangles. We put these triangles together and obtain a convex quadrangle. It is convex because, according to (2.7.12), the angles in the Alexandrov triangles are $\leqslant$ the angles of the triangles on M. A convex triangle on M has length $\leqslant 2\pi/\sqrt{K_0}$.

Hence, if the sides of the triangles $c(0)c(1/4)c(1/2)$ and $c(0)c(3/4)c(1/2)$ coincide with the arcs on c between these points, we are done.

Otherwise, we take triangles in D with corners the points $c(i/8)$ $c((i+1)/8)c((i+2)/8)$, $i = 0, 2, 4, 6$. From the previously constructed Alexandrov quadrangle and the Alexandrov triangles associated with these four triangles we construct an Alexandrov octagon. It is again convex, hence, its length is $\leqslant 2\pi/\sqrt{K_0}$. If the sides of the newly constructed triangles coincide with the arcs on c between the corners we are done. Otherwise, we continue to subdivide further the arcs on c. Eventually we will reach a stage where the arcs on c are the unique relatively minimizing curves between their endpoints. At that stage it is clear that $L(c)$ must be $\leqslant 2\pi/\sqrt{K_0}$.

The proof of (2.7.12) also shows that here $=$ can hold only if $K = K_0$ everywhere in D and then also in the complement of D on M □.

Remark. In our proof we have used a version of the Toponogov Triangle Theorem (2.7.12) which works with relatively minimizing geodesics in a relatively convex domain instead with minimizing geodesics. That, for surfaces at least, the proofs for the absolute and the relative case are the same was observed already by Alexandrov [1].

We continue with a result on the existence of relatively short closed geodesics on convex surfaces M which permit a fixed point free isometric involution.

Equivalently, we may say that we have a non-simply connected surface $\bar{M}$ of strictly positive curvature. From (2.6.7) we then know that the universal covering M of $\bar{M}$ is a 2-fold covering. Thus, M can be realized as a convex surface in $\mathbb{R}^3$ which is symmetric with respect to the origin of $\mathbb{R}^3$.

3.4.11 Theorem. *Let M be a convex surface which is symmetric with respect to the origin. Denote by $\bar{M}$ the quotient of M w. r. t. the symmetry at the origin. Then there exist on $\bar{M}$ at least two not null-homotopic closed geodesics $\bar{c}_1, \bar{c}_2$ with $E(\bar{c}_1) \leqslant E(\bar{c}_2)$. Moreover, $\bar{c}_1$, is non-hyperbolic of index 0 while $\bar{c}_2$ has index plus nullity $\geqslant 1$. If $\bar{c}_2$ is non-degenerate, $\bar{c}_2$ is hyperbolic of index 1 and hence not a covering of $\bar{c}_1$.*

The coverings c_1, c_2 of $\bar{c}_1, \bar{c}_2$ on M therefore give a pair of closed geodesics where c_1 has index 1 while c_2 in general has index 2 and is hyperbolic.

Remark. An example of such a pair c_1, c_2 is given by the shortest ellipse and the ellipse of medium length on an ellipsoid M with three pairwise different axes, cf. also (3.5.16).

Proof. Denote by Λ' the connected component of $\Lambda\bar{M}$ which is formed by the non-null-homotopic closed curves. According to (2.4.19), there exists in Λ' a critical point $\bar{c}_1$ of $E: \Lambda' \to \mathbb{R}$ where E assumes its infimum. Clearly, index $\bar{c}_1 = 0$. Denote by c_1 the 2-fold covering of $\bar{c}_1$ in M. From (2.6.7) we know index $c_1 > 0$. Therefore, $\bar{c}_1$ and c_1 cannot be hyperbolic, as follows from (3.2.15). Also, $\bar{c}_1$ cannot have multiple points because then $E|\Lambda'$ would not be minimal on $\bar{c}_1$.

If $\bar{c}_1$ is non-degenerate, it is elliptic. In this case, we have index $c_1 = 1$. Indeed, otherwise index $c_1 \geqslant 3$, cf. (3.4.2). But then, $c_1 = c_1(t), 0 \leqslant t \leqslant 1$, must have at least two conjugate points in its interior and $\bar{c}_1 = \bar{c}_1(t), 0 \leqslant t \leqslant 1$, then has at least one conjugate point in its interior. But this contradicts index $\bar{c}_1 = 0$.

Assume now that c_1 is degenerate. Consider first the case when nullity $c_1 = 2$. Then we have from (3.4.2) by the same arguments that index $c_1 = 1$.

Only the case nullity $\bar{c}_1 =$ nullity $c_1 = 1$ remains to be discussed. By changing, if necessary, the initial point of c_1 we can assume that a Jacobi field $Y(t)$ along $c_1(t)$ with $Y(0) = 0$, $\nabla Y(0) \neq 0$, $\langle Y(t), \dot{c}(t) \rangle = 0$ is not periodic. Since index $\bar{c}_1 = 0$, $Y(t)$ can have at most one zero in the t-interval $]0, 1[$. But it also has at least one zero in this interval, since there exists a periodic Jacobi field along c_1. Hence, according to (3.4.2), index $c_1 = 1$.

For the definition of $\bar{c}_2$ we consider continuous mappings

$$f: (I, \partial I) \to (\Lambda\bar{M}, \{\bar{c}_1, \theta\bar{c}_1\}).$$

Here $\theta\bar{c}_1$ is the oppositely oriented curve of $\bar{c}_1$. Define a ϕ-family $\mathscr{A}$ by taking as its elements $A \in \mathscr{A}$ the compact sets $A = f(I), f$ as above, cf. (2.4.17, i). Denote by $\varkappa_2$ the critical value of $\mathscr{A}$. Then clearly $\varkappa_2 \geqslant \varkappa_1 = E(\bar{c}_1) = E(\theta\bar{c}_1)$.

The set $Cr_1\varkappa_2$ of critical points of index $\leqslant 1$ and E-value $\varkappa_2$ is not empty. Indeed, (2.5.16) implies that there must be critical points of index $\leqslant 1$ in every interval $[\varkappa_2, \varkappa_2$

$+1/m]$. Consider a sequence $\{\bar{c}(m), \text{index}\,\bar{c}(m) \leqslant 1, E(\bar{c}(m)) - \bar{\varkappa}_2 \leqslant 1/m\}$. According to condition (C), cf. (2.4.9), $\{\bar{c}(m)\}$ has a convergent subsequence with limit $\bar{c}_2$, a critical point of E-value $\bar{\varkappa}_2$, index $\bar{c}_2 \leqslant 1$.

If $\bar{\varkappa}_2 = \bar{\varkappa}_1$ then $Cr\bar{\varkappa}_1$ cannot exist of the $\mathbb{O}(2)$-orbit $S^1 . \bar{c}_1 \cup \bar{S}^1 . \theta \bar{c}_1$ of $\bar{c}_1$ alone. To see this we choose disjoint neighborhoods $\mathcal{U}_1$ and $\theta \mathcal{U}_1$ of $S^1 . \bar{c}_1$ and $S^1 . \theta \bar{c}_1 = \theta S^1 . \bar{c}_1$. Put $\mathcal{U}_1 \cup \theta \mathcal{U}_1 = \mathcal{U}$. (2.4.18) then would imply the existence of a curve f from $\bar{c}_1$ to $\theta \bar{c}_1$ with image in $\mathcal{U} \cup \Lambda'^{-} = \mathcal{U}$ which is clearly impossible. Thus, if $\bar{\varkappa}_1 = \bar{\varkappa}_2$ there must be a critical point $\bar{c}_2$ outside the $\mathbb{O}(2)$-orbit of $\bar{c}_1$. $\bar{c}_2$ is simple, of course. Index $\bar{c}_2 = 0$ while nullity $\bar{c}_2 > 0$, since there exists a 1-parameter family of curves of E-value differing arbitrarily little from $\bar{\varkappa}_1 = \bar{\varkappa}_2$. Also, $\bar{c}_1$ and $\theta \bar{c}_1$ have nullity > 0. An example where this occurs is given by a flattened ellispoid of revolution, cf. (3.5) for more of this.

On the other hand, if $\bar{c}_2$ is non-degenerate, $\bar{\varkappa}_2 > \bar{\varkappa}_1$. There are nearby shorter curves passing through $\bar{c}_2$, hence, index $\bar{c}_2 = 1$ and, according to (2.4.2), since $\bar{c}_2$ is non-orientable, it must be hyperbolic. □

We conclude with the description of a special deformation processs due to Bangert [1] which applies to a homotopy in the space of multiply covered curves. In (3.4.14) we will give one of the many applications.

3.4.12 Lemma. *Let c_-, c_+ be elements in ΛM with $E(c_-) = E(c_+) = \varkappa_0$. Assume that we have a homotopy on M between c_- and c_+,*

$$f: (\bar{B}^1, \{-1, +1\}) \to (\Lambda M, \{c_-, c_+\}),$$

with $\sup E | f(\bar{B}_1) = \lambda_1 > \varkappa_0$, $\bar{B}^1 = [-1, +1]$. For every $m \in \mathbb{N}^*$ consider the m-fold covering of f:

$$f^m: (\bar{B}^1, \{+1, -1\}) \to (\Lambda M, \{c_-^m, c_+^m\}),$$

is given by $f^m(s) = f(s)^m$. Then, for every $\varepsilon > 0$ with $\varepsilon < \lambda_1 - \varkappa_0$ there exists an integer $m(\varepsilon) > 0$ such that, whenever $m \geqslant m(\varepsilon)$, f^m is homotopic, with fixed end points, to a

$$f_m: (\bar{B}^1, \{-1, +1\}) \to (\Lambda M, \{c_-^m, c_+^m\})$$

with $\sup E | f_m(\bar{B}^1) < \sup E | f^m(\bar{B}^1) - m^2 \varepsilon = m^2(\lambda_1 - \varepsilon)$.

Proof. For a given f and $m > 1$ we define the homotopy

$$F_m: \bar{B}^1 \times [0, 1] \to \Lambda M$$

with $F_m | \bar{B}^1 \times \{0\} = f^m$, $F_m | \bar{B}^1 \times \{1\} \equiv f_m$

$$F_m(-1, \tau) = f^m(-1) = c_-^m;\ F_m(+1, \tau) = f^m(+1) = c_+^m$$

as follows: Put $F_m | \bar{B}^1 \times \{\tau\} = F_{m,\tau}$. Then $F_{m,\tau} \equiv f^m$ on $[-1, -\tau]$ and $[\tau, 1]$. For $s \in [-\tau, -\tau + 2\tau/m]$, $F_{m,\tau}(s)$ shall be the element in ΛM formed by 'segment' $f(\sigma)(0)$, $-\tau \leqslant \sigma \leqslant ms + (m-1)\tau$, followed by the 'loop' $f(ms + (m-1)\tau)(\sigma')$, $0 \leqslant \sigma' \leqslant 1$, followed by the inverse of the 'segment', i. e., $f(\sigma'')(0)$, $ms + (m-1)\tau \geqslant \sigma'' \geqslant -\tau$, followed by the $(m-1)$ 'loops' $f^{m-1}(\tau)(\sigma''') = f(\tau)((m-1)\sigma'''), 0 \leqslant \sigma''' \leqslant 1$. The parameterization of $F_{m,\tau}(s)$ shall be such as to take equal time $1/m$ on each of the $(m-1)$ 'loops' $f^{m-1}(\tau)$ as well as on the initial 'loop with tail', formed by the 'segment', the 'loop' $f(s)$ and the 'inverse segment'. Figuratively speaking, $F_{m,\tau}(s)$, $+\tau \leqslant s \leqslant -\tau + 2\tau/m$ is the pulling-over of the first loop of $f^m(-\tau) = f(-\tau)^m$ into $f(\tau)$, with two 'segments' to join the pulled-over 'loop'.

We define $F_{m,\tau}(s)$, for $-\tau + 2\tau/m \leqslant s \leqslant -\tau + 4\tau/m$, similarly, i. e., as to be the pulling-over of the first loop of $f^{m-1}(-\tau)$ into the second loop of $f^2(\tau)$. After m such pulling-over constructions, all m loops of $f^m(-\tau)$ are carried into the m loops of $f^m(\tau)$. That is, $F_{m,\tau}(\tau) = F_m(\tau, \tau) = f^m(\tau)$.

To get an upper bound for E on the elements of the curve $f_m = F_{m,1} : \bar{B}^1 \to \Lambda M$ from c_-^m to c_+^m we observe that, for each $s \in \bar{B}^1, f_m(s)$ contains $(m-1)$ loops of the form $f(-1) = c_-$ or $f(+1) = c_+$, each run through in the time $1/m$. Therefore, they contribute to $E(f_m(s))$ the amount $m(m-1)\varkappa_0$. The contribution of the "loop with tail" is bounded from above by a number $m\alpha$ where α is the supremum of the E-values for the "loops with tail", run through in the time 1.

Hence, we have

$$E(f_m(s)) \leqslant m(m-1)\varkappa_0 + m\alpha$$

while $$\sup_{s \in \bar{B}^1} E(f^m(s)) = m^2 \lambda_1$$

Thus, if $\varepsilon < \lambda_1 - \varkappa_0, m^2(\lambda_1 - \varepsilon)$ dominates $m^2\varkappa_0 + m(\alpha - \varkappa_0)$ for sufficiently large m. □

We want to give an application of (3.4.12). For that purpose we introduce the concept of a *waist* on an orientable surface M. By this we mean a closed geodesic c which is a local minimum for E. That is, $E(c') \geqslant E(c)$, for all $c' \in \Lambda M$ sufficiently near c.

3.4.13 Proposition. *If c is a waist, then also all multiple coverings c^m of c are a waist.*

Proof. Clearly, index $c = 0$. If c is non-degenerate, it must be hyperbolic, cf. (3.4.2). In this case, index $c^m = m$ index $c = 0$ and nullity $c^m = 0$, i. e., c^m also is a local minimum for E.

If c is degenerate we know from (3.4.4), (3.4.5) that index $c^m = 0$, nullity $c^m = 1$, all m. Thus, in particular, there exists a periodic Jacobi field $Y(t) \neq 0, \langle \dot{c}(t), Y(t) \rangle = 0$, along $c(t)$, and each periodic Jacobi field along $c^m(t) = c(mt)$ is of the form $bY(mt)$. We derive a contradiction from the assumption that there exists a variation c_s', $-\varepsilon < s < \varepsilon$, of c^m with $c_0' = c^m$, such that $E(c_s') < E(c^m)$, for $s \neq 0$.

Let $X(t) = \partial c_s'(t)/\partial s|_0$ be the variation vector field along $c^m(t)$ of the variation c_s', cf. (1.12.11). Since index $c^m = 0$, $X(t)$ is a linear combination $a\dot{c}^m(t) + bY(mt)$ of the base elements of the null space of $D^2E(c^m)$, with $b \neq 0$. This shows that the curves $c_s'(t)$, for $s \neq 0$ sufficiently small, do not intersect the geodesic $c^m(t)$. We use this to construct from c_s' by 'cutting and pasting' m closed curves near c of total E-value $= E(c_s')$. Thus, since each of the m closed curves has E-value $\geqslant E(c)$, $E(c_s') \geqslant E(c^m)$ – the desired contradiction.

For this construction we start with a point on c_s' nearest c and follow c_s' once around c, thereby always switching to that arc of c_s' which is nearest to c. After one such turn we have got a closed curve near c. Proceed in the same manner with the remainder of c_s'. □

We now can prove the

3.4.14 Theorem. *Let M be a surface of the topological type of S^2. Assume that M possesses a waist. Then there exist on M infinitely many geometrically distinct closed geodesics.*

Remarks. 1. This result is due to Bangert [1]. For the case that all closed geodesics on M are non-degenerate, the existence of infinitely many closed geodesics on a surface of the type S^2 with a waist was observed by Thorbergsson (unpublished).

2. The hypothesis that M possesses a waist will only be used to derive the existence of a closed geodesic c_1 with (index plus nullity) $c_1^m \leqslant 1$, all m, and c_1 not a waist. Thus, the conclusion also holds under this weaker assumption.

3. By geometrically distinct we mean that the geodesics do not just differ by the parameterization or by their multiplicity. The existence of infinitely many such closed geodesics on M also holds without the hypothesis that M possesses a waist, e. g., if M is a convex surface. In this case, the proof is much more complicated, however. Either one uses the theory of the Morse complex, cf. Klingenberg [10], [14], or else – and this is more elementary – one works with a refinement of the proof of the Lusternik-Schnirelmann Theorem on the Three Closed Geodesics (cf. (3.7)) described in Klingenberg [12].

Proof. We are going to derive a contradiction from the assumption that there exist on M only finitely many orbits $S^1 . c$ where c is a prime closed geodesic. From (3.4.5) we then know that, for all sufficiently large m, (index plus nullity) c^m is $\neq 2$, for every prime c on M.

Let c_0 be a waist on M. Since M is simply connected, ΛM is connected. Therefore there exists a continuous mapping

$$f: (\bar{B}^1, \{-1, +1\}) \to (\Lambda M, \{c_-, c_+\}), \bar{B}^1 = [-1, +1],$$

joining $c_- = c_0$ and the oppositely oriented geodesic $c_+ = \theta c_0$. We define the ϕ-family $\mathscr{A}_1$ as to consist of the sets $\{f'(\bar{B}^1)\}$ where f' is homotopic to f with fixed end points. Let $\varkappa_1$ be the critical value of $\mathscr{A}_1$. Then $\varkappa_1 > \varkappa_0 = E(c_0) = E(\theta c_0)$. This follows from the fact that the critical set $Cr\varkappa_0$ consists of only finitely many critical S^1-orbits. Indeed, in this case $Cr\varkappa_0$ possesses open neighborhoods $\mathscr{U}$ formed by small tubular pairwise disjoint neighborhoods of these orbits. If we had $\varkappa_0 = \varkappa_1$, there would exist $f'(\bar{B}^1) \subset \mathscr{A}_1$ contained in $\mathscr{U} \cup \Lambda^{\varkappa_0 -}$, cf. (2.4.18). That is, outside every neighborhood of c_0 we would have $c' \in f'(\bar{B}^1)$ with $E(c') < E(c_0)$ which clearly is impossible since c_0 is a waist.

From (3.4.13) we know that with c_0 also $c_0^l, l \geqslant 1$, is a waist. Thus, we may consider for every l the ϕ-family $\mathscr{A}_l$, consisting of the sets $f_l'(\bar{B}^1)$ where f_l' is a mapping homotopic to the l-fold covering f^l of the mapping f. Again, the critical value $\varkappa_l$ of $\mathscr{A}_l$ will be $> E(f_l'(c_0)) = E(c_0^l) = l^2 \varkappa_0$.

Among the finitely many prime orbits there must be one, say $S^1 . c_1$, such that, for an infinite sequence $\{l_j\}$, an element of the ϕ-family $\mathscr{A}_{l_j}$ remains hanging in the sense of (2.4.18) at elements of the tower $\{S^1 . c_1^m, m = 1, 2, \ldots\}$. But then, (index plus nullity) $(c_1^m) \leqslant 1$, for all m, since a 1-dimensional relative homotopy can be deformed below any critical orbit of index $\geqslant 2$, cf. (2.5.16).

We thus are led to the following situation: There exist arbitrarily high multiple coverings of the waist c_0 and homotopies from such coverings to the oppositely oriented coverings which remain hanging at a critical orbit with index plus nullity $\leqslant 1$. We choose a covering c_0^l of c_0 so large that no critical point with E-value $> E(c_0^l)$ has index 2, cf. the remark at the beginning of the proof. We write again c_0 for such a covering. We then have a mapping

$$f: (\bar{B}^1, \{-1, +1\}) \to (\Lambda M, \{c_-, c_+\})$$

with $c_- = c_0, c_+ = \theta c_0$ and a (not necessarily prime) closed geodesic c_1 with (index plus nullity) $c_1^m \leqslant 1$, all m, such that, for every neighborhood $\mathscr{U}$ of $S^1 . c_1$, $\phi_s \circ f(\bar{B}^1)$ meets $\mathscr{U}$ and $\sup E | \phi_s \circ f(\bar{B}^1)$ goes to $\varkappa_1 = E(c_1)$, for $s \to \infty$.

From (3.4.12) we know that, for sufficiently large m, we have a mapping $F_m: \bar{B}^1 \times [0, 1] \to \Lambda M$ with $F_m | \bar{B}^1 \times \{0\} = f^m = m$-fold covering of f and $F_m | \bar{B}^1 \times \{1\} \equiv f_m: \bar{B}^1 \to \Lambda M$ with $\sup E | f_m(\bar{B}^1) < m^2 \varkappa_1$, where we chose the ε in (3.4.12) as $\sup E | f(\bar{B}^1) - \varkappa_1$.

Consider now the ϕ-family $\mathscr{B}_m$, formed by $F_m'(\bar{B}^1 \times [0, 1])$ where F_m' is homotopic to $\phi_s \circ F_m$ with $F_m' = \phi_s \circ F_m$ on $\partial(\bar{B}^1 \times [0, 1])$. We claim that the critical value μ_m of $\mathscr{B}_m$ is $> m^2 \varkappa_1$. Indeed, since $\sup E | F_m'(\bar{B}^1 \times \{1\}) < m^2 \varkappa_1$, $\mu_m \leqslant m^2 \varkappa_1$ would imply (index plus nullity) $c_1^m \geqslant 2$.

Among the elements in $Cr \mu_m$ there must be one, say c_2, with index $c_2 \leqslant 2$, cf. (2.5.16). In our case, this implies (index plus nullity) $c_2 \leqslant 1$. But this contradicts the fact that the mappings F_m' used for the definition of the elements of $\mathscr{B}_m$ have $\bar{B}^1 \times [0, 1]$ as domain, with $F_m' \partial(\bar{B}^1 \times [0, 1]) \subset \Lambda^{m^2 \varkappa_1}$. □

3.5 Geodesics on the Ellipsoid

We want to give a rather detailed exposition of the structure of the geodesic flow on the 2-dimensional ellipsoid. The study of the geodesics on the ellipsoid goes back to Jacobi [1]. He proved with the help of the so-called elliptic coordinates that the differential equation of the geodesics can be solved by quadratures.

By this we mean that for the geodesic flow on the ellipsoid there exists, besides the kinetic energy E, a second function F, which is constant along the flow lines (such functions are called integrals of the geodesic flow) and for which $DE \wedge DF \neq 0$ on an open dense subset of the tangent bundle. A thorough exposition of these results can be found in Darboux [1]. For additional results see Bianchi [1]. Cf. also Klingenberg [9] for the most basic material on second order confocal surfaces.

In our exposition we will treat the geodesic flow on the ellipsoid by geometric-qualitative methods rather than by the extensive use of hyperelliptic integrals, as it is customary in the classical texts.

We conclude with a brief discussion of the geodesic flow for surfaces of revolution.

The structure of the geodesic flow on the n-dimensional ellipsoid (as well as on the other n-dimensional surfaces of second order for $n > 2$) also has been studied extensively, beginning with Jacobi l.c. For an up-to-date presentation, including several new results, cf. Thimm [1]. See also Pars [1].

Recall from (1.10.16) that on the 2-dimensional ellipsoid

$$M = \{x_0^2/a_0 + x_1^2/a_1 + x_2^2/a_2 = 1\}$$

we have the three simple closed geodesics given by the intersection of M with one of the three coordinate planes. We call these also the *basic closed geodesics* or *principal ellipses* on M.

In general we will assume from now on $0 < a_0 < a_1 < a_2$. In the case $a_0 = a_1 < a_2$ we get an elongated ellipsoid of revolution. If $a_0 < a_1 = a_2$, M becomes a flattened ellipsoid of revolution. If $a_0 = a_1 = a_2$, we get the sphere.

For M a surface in $\mathbb{R}^3$, there is defined the concept of an *umbilic point*, cf. Klingenberg [9]. M in general has four umbilics. These are the points with coordinates

$$(x_0, x_1, x_2) = (\pm\sqrt{a_0}\sqrt{a_1 - a_0}/\sqrt{a_2 - a_0}, 0, \pm\sqrt{a_2}\sqrt{a_2 - a_1}/\sqrt{a_2 - a_0}).$$

The special rôle which these points play will soon become clear.

3.5.1 Proposition. *Let* $\varrho \notin \{a_0, a_1, a_2\}$. *Consider the function*

$$A_\varrho\colon (x_0, x_1, x_2) \in \mathbb{R}^3 \mapsto \frac{x_0^2}{a_0 - \varrho} + \frac{x_1^2}{a_1 - \varrho} + \frac{x_2^2}{a_2 - \varrho} \in \mathbb{R}.$$

Whenever $\varrho < a_2$, 1 *is a regular value of* A_ϱ.

More precisely,

$$\{A_\varrho = 1\} \text{ is an } \begin{Bmatrix} \text{ellipsoid} \\ \text{1-sheeted hyperboloid} \\ \text{2-sheeted hyperboloid} \end{Bmatrix} \text{ if } \begin{cases} \varrho < a_0 \\ a_0 < \varrho < a_1 \\ a_1 < \varrho < a_2 \end{cases}$$

Proof. From (1.3.9, i) we know that 1 is a regular value of F_ϱ (if it is a value at all; and this is the case if and only if $\varrho < a_2$). The type of the sets $A_\varrho^{-1}(1)$ then is well known. □

3.5.2 Complement. (i) *The limit of* $\{A_\varrho = 1\}$ *for* $\varrho \to a_0, \varrho < a_0$, *is the* degenerate ellipsoid $\{x_1^2/(a_1 - a_0) + x_2^2/(a_2 - a_0) \leqslant 1; x_0 = 0\}$. *For* $\varrho \to a_0, a_0 < \varrho$, *we obtain the* degenerate 1-sheeted hyperboloid of the first kind $\{x_1^2/(a_1 - a_0) + x_2^2/(a_2 - a_0) \geqslant 1; x_0 = 0\}$.

(ii) *The limit of* $\{A_\varrho = 1\}$ for $\varrho \to a_1, \varrho < a_1$, *is the* degenerate 1-sheeted hyperboloid of the second kind $\{-x_0^2/(a_1 - a_0) + x_2^2/(a_2 - a_0) \leqslant 1; x_1 = 0\}$. *For* $\varrho \to a_1, a_1 < \varrho$, *we obtain the* degenerate 2-sheeted hyperboloid $\{-x_0^2/(a_1 - a_0) + x_2^2/(a_2 - a_0) \geqslant 1; x_1 = 0\}$.

(iii) *The limit of* $\{A_\varrho = 1\}$ *for* $\varrho \to a_2, \varrho < a_2$, *is the* double covered plane $\{x_2 = 0\}$.

(iv) *The* focal hyperbola *is given by*

$$\{-x_0^2/(a_1 - a_0) + x_2^2/(a_2 - a_0) = 1; \; x_1 = 0\}.$$

It intersects the ellipsoid $M = \{A_0 = 1\}$ *in the four umbilics.* □

Through every point p of M outside the coordinate planes there passes exactly one 1-sheeted and one 2-sheeted hyperboloid, $\{A_{u_1} = 1\}$ and $\{A_{u_2} = 1\}$, $(u_1, u_2) \in]a_0, a_1[\times]a_1, a_2[$. We may take (u_1, u_2) as coordinates of p. However, besides p, the seven other points on M which are obtained from p under reflections on the various coordinate planes, all have the same (u_1, u_2)-coordinates.

In the subsequent definition we therefore substitute for (u_1, u_2) parameters (ψ_1, ψ_2) on the torus.

3.5.3 Definition. The ellipsoid M can be described by the so-called *elliptic coordinates* $(u_1, u_2) \in [a_0, a_1] \times [a_1, a_2]$ in the form

$$(*) \qquad \begin{aligned} x_0(u_1, u_2)^2 &= a_0(u_1 - a_0)(u_2 - a_0)/(a_1 - a_0)(a_2 - a_0) \\ x_1(u_1, u_2)^2 &= a_1(u_1 - a_1)(u_2 - a_1)/(a_0 - a_1)(a_2 - a_1) \\ x_2(u_1, u_2)^2 &= a_2(u_1 - a_2)(u_2 - a_2)/(a_0 - a_2)(a_1 - a_2). \end{aligned}$$

To make this description one-to-one, at least outside the four umbilics given by $(u_1, u_2) = (a_1, a_1)$, we let (u_1, u_2) depend on (ψ_1, ψ_2) by

$$(**) \qquad \begin{aligned} u_1 &= a_0 \cos^2\psi_1 + a_1 \sin^2\psi_1, \\ u_2 &= a_1 \sin^2\psi_2 + a_2 \cos^2\psi_2, \end{aligned} \qquad \text{with} \quad (\psi_1, \psi_2) \in S^1 \times S^1.$$

Consider now the ψ_i as functions on the projective line $P^1 = S^1/\mathbb{Z}_2$.

Remark. The mapping above defined from $S^1 \times S^1$ onto M is a *conformal (i.e., angle preserving) covering* of the ellipsoid by a torus, with four branch points over the umbilics. This is the well-known Riemann surface of a function of the form $\sqrt{(z - a_1)(z - a_2)(z - a_3)(z - a_4)}$, that is, a so-called elliptic Riemann surface.

We now write down the line element $ds^2 = \sum_{i,k} g_{ik}(u_1, u_2)\, du_i\, du_k$.

3.5.4 Proposition. *The representation of the scalar product on M in elliptic coordinates is given by*

$$(*)\qquad ds^2 = (-u_1 + u_2)(U_1\, du_1^2 + U_2\, du_2^2)$$

with

$$U_i = U_i(u_i) = (-1)^i u_i / f(u_i);\ f(u_i) = 4(a_0 - u_i)(a_1 - u_i)(a_2 - u_i).$$

Note. A line element of the form given in (3.5.4) (*), where the U_i are arbitrary differentiable functions of u_i alone, is called a *Liouville line element*. A *Liouville surface* is a surface M which permits on an open dense subset $M' \subset M$ an atlas consisting entirely of Liouville charts, i.e., charts where the line element is Liouville. In most of our subsequent arguments, the ellipsoid could be replaced by a general Liouville surface, at least if the underlying topological surface is of the type of the 2-sphere.

Proof. Recall the definition of the $g_{ik}(u_1, u_2)$ of an embedded surface, cf. Klingenberg [9], i.e.,

$$g_{ik}(u_1, u_2) = \sum_{r=0}^{2} \frac{\partial x_r}{\partial u_i}(u_1, u_2) \frac{\partial x_r}{\partial u_k}(u_1, u_2).$$

From (*) (3.5.3) we get (3.5.4) by a simple computation. □

Remark. The u_1-parameter lines $u_2 = \text{const} \in]a_1, a_2[$ consist of a pair of simple closed curves, i.e., the intersection of M with the 2-sheeted hyperboloid $\{A_{u_2} = 1\}$. The two curves belong to the half ellipsoids $M \cap \{x_2 > 0\}$ and $M \cap \{x_2 < 0\}$ and are symmetric with respect to the plane $\{x_2 = 0\}$.

Each of these curves can be viewed as an ellipse on M in the following sense: Take e.g. the curve $u_2 = \text{const}$ on $M \cap \{x_2 > 0\}$. Then this curve consists of those points on $M \cap \{x_2 > 0\}$ for which the sum of the distance from the two umbilics on $M \cap \{x_2 > 0\}$, measured on M, is constant.

Similarly, the set $u_1 = \text{const}$ consists of two closed curves on $M \cap \{x_0 > 0\}$ and $M \cap \{x_0 < 0\}$, respectively which are symmetric with respect to the plane $\{x_0 = 0\}$. Again, these curves are ellipses on M, each with a pair of umbilics as focal points.

The (u_1, u_2)-parameter lines are so-called *curvature lines*, cf. Klingenberg [9].

We now can write down a second integral for the geodesic flow on TM.

3.5.5 Theorem. *In the elliptic coordinates (u_1, u_2) on M, the geodesics are characterized by*

$$(*)\qquad \frac{\sqrt{U_1}}{\sqrt{-u_1 + \gamma}}\, \dot{u}_1 \mp \frac{\sqrt{U_2}}{\sqrt{u_2 - \gamma}}\, \dot{u}_2 = 0,$$

together with the condition $E(u, \dot{u}) = \text{const.}$ *Here, γ is a constant with value in $]a_0, a_1[$ or $]a_1, a_2[$.*

We call the constant γ the *parameter of the geodesic*.

Proof. Choose $\gamma \in]a_0, a_1[$ or $\gamma \in]a_1, a_2[$. On the subdomain of those $(u_1, u_2) \in]a_0, a_1[\times]a_1, a_2[$ which satisfy $u_1 < \gamma < u_2$ we introduce new coordinates u_1', u_2' by

$$du_1' = \sqrt{-u_1+\gamma}\sqrt{U_1}\,du_1 \pm \sqrt{u_2-\gamma}\sqrt{U_2}\,du_2$$

$$du_2' = \sqrt{U_1}\,du_1/\sqrt{-u_1+\gamma} \mp \sqrt{U_2}\,du_2/\sqrt{u_2-\gamma}$$

In these coordinates, the line element is given by

$$ds^2 = du_1'^2 + (-u_1+\gamma)(u_2-\gamma)\,du_2'^2$$

(u_2', u_2') therefore are *geodesic parallel coordinates*; the u_1'-curves are geodesics and the u_2'-curves meet these curves orthogonally. They cut out pieces of equal length from the geodesics and therefore are called parallel. Cf. Klingenberg [9], where the computations for the Christoffel symbols also in the case $g_{11}=1, g_{12}=0, g_{21}=0, g_{22}>0$ are carried out in detail.

Our claim is now equivalent to the equation $du_2'(t)/dt = 0$, together with the condition that the parameter is proportional to arc length. □

In general, we will consider the restriction of the geodesic flow to the unit tangent bundle T_1M only. That is, we will assume $2E=1$. From (3.1.19) we know that this restriction is not essential.

(3.5.5) allows us to read off a *second integral* for the geodesic flow.

3.5.6 Corollary. *Denote by $(T_1M)'$ the open and dense subset of T_1M formed by those unit tangent vectors which are tangent to a geodesic with parameter $\gamma, \gamma \in]a_0, a_1[$ or $\gamma \in]a_1, a_2[$. Define $F: (T_1M)' \to \mathbb{R}$ in elliptic tangent coordinates $(u, \dot u)$ by*

$$F(u, \dot u) = (-u_1+u_2)(u_2 U_1 \dot u_1^2 + u_1 U_2 \dot u_2^2).$$

Then $F(u(t), \dot u(t)) = const = \gamma$, if $u(t) = (u_1(t), u_2(t))$ is a geodesic with parameter γ, parameterized by arc length.

If we denote by $\mu(X)$ the angle between $X \in (T_1M)'$ and the u_1-parameter line through $\tau_M X$ then we may also write

$$F(X) = u_1(\tau_M X)\sin^2\mu(X) + u_2(\tau_M X)\cos^2\mu(X).$$

Remark. That F is constant along a flow line $\phi_t X_0$, where $\tau_M \phi_t X_0$ is a geodesic with parameter γ in $]a_0, a_1[$ or $]a_1, a_2[$, is known as *Liouville's Theorem*.

Actually, it is possible to define F by the same formula on the larger set consisting of all $X \in T_1M$ which do not project onto one of the four umbilics. In this case, the value $F = a_1$ occurs. For $F(X) = a_1$, the geodesic $\tau_M \circ \phi_t X$ will pass through a pair of opposite umbilics, cf. (3.5.16) below. Thus, we will then reach elements $\phi_t X$ where F is not defined by the above formula. One can show, however, that F possesses a differentiable extension to all of T_1M by putting $F(X) = a_1$ whenever $\tau_M X$ is an umbilic.

Proof. We resolve (*) (3.5.5) for γ and observe that X is a unit tangent vetor, i.e., $(-u_1+u_2)\,U_1\dot{u}_1^2+(-u_1+u_2)\,U_2\dot{u}_2^2=1$. The two summands in this equation are $\cos^2\mu$ and $\sin^2\mu$, respectively. □

For the further study of the geodesic flow it is preferable to go to the co-geodesic flow on the total space T^*M of the cotangent bundle. The cotangent coordinates (u, v) are related to the tangent coordinates $(u, \dot{u})$ by

$$\dot{u}_i = g^{ii}(u)\,v_i = v_i/(-u_1+u_2)\,U_i,\ i=1,2.$$

The functions E, F on $(TM)'$ become functions E^*, F^* on $(T^*M)'$, i.e.,

$$E^*(u,v)=\frac{1}{2(-u_1+u_2)}\left(\frac{1}{U_1}v_1^2+\frac{1}{U_2}v_2^2\right),$$

$$F^*(u,v)=\frac{1}{(-u_1+u_2)}\left(\frac{u_2}{U_1}v_1^2+\frac{u_1}{U_2}v_2^2\right).$$

3.5.7 Theorem. *For* $\gamma\in]a_0,a_1[$ *or* $\gamma\in]a_1,a_2[$, *the* ϕ_t*-invariant set* $\{F^*=\gamma\}$ *in the total unit cotangent space* T_1^*M *consists of two disjointly* embedded 2-dimensional tori *which we will denote by* $T_\gamma^\pm$.

We distinguish the cases $\gamma\in]a_1,a_2[$ *and* $\gamma\in]a_0,a_1[$ *as* types I and II, *respectively.*

The flow lines on the tori $T_\gamma^\pm$ *of type I correspond, under the projection* $\tau_M^*: T_1^*M\to M$, *to geodesics which monotonely* wind around *the* x_2*-axis and* oscillate *between the two* u_1*-parameter lines* $\{u_2=\gamma\}$ *on* M, *i.e., the intersection of the ellipsoid with the 2-sheeted hyperboloid* $\{A_\gamma=1\}$. *The flow lines on the two tori are distinguished by the sense in which their projections wind around the* x_2*-axis. As* γ *goes to* a_2, *the tori converge towards the degenerate tori, consisting of the two embedded circles formed by the unit tangent vectors to the ellipse in the* (x_0,x_1)*-plane.*

The flow lines on the tori of type II correspond, under the projection $\tau_M^*: T_1^*M\to M$, *to geodesics which monotonely* wind around *the* x_0*-axis and* oscillate *between the two* u_2*-parameter lines* $\{u_1=\gamma\}$ *on* M, *i.e., the intersection of the ellipsoid with the 1-sheeted hyperboloid* $\{A_\gamma=1\}$. *Again, the two tori* $T_\gamma^\pm$ *correspond to the two senses in which a curve can wind around the* x_0*-axis. As* γ *goes towards* a_0, *the tori* $T_\gamma^\pm$ *become degenerate, i.e., we get the two embedded circles given by the unit tangent vectors to the ellipse in the* (x_1,x_2)*-plane.*

Note. The existence of the invariant tori will be proved again in (3.5.10) by a different method.

Proof. The equations $\{2E^*=1\}, \{F^*=\gamma\}$ yield

$$v_1^2=U_1(\gamma-u_1);\ v_2^2=U_2(u_2-\gamma).$$

Let (ψ_1,ψ_2) be the coordinates introduced in (3.5.3) (**). For the corresponding cotangent coordinates (Ψ_1,Ψ_2) we get, with $\partial u_i(\psi_1,\psi_2)/\partial\psi_j=0$ for $i\neq j$,

$$\Psi_1 = \frac{\partial u_1}{\partial \psi_1} v_1 = 2(a_1 - a_0) \sin \psi_1 \cos \psi_1 \, v_1$$

$$\Psi_2 = \frac{\partial u_2}{\partial \psi_2} v_2 = 2(a_1 - a_2) \sin \psi_2 \cos \psi_2 \, v_2$$

Hence, with $U_i = (-1)^i u_i / 4(a_0 - u_i)(a_1 - u_i)(a_2 - u_i)$,

$$\Psi_1^2 = 4(u_1 - a_0)(a_1 - u_1) v_1^2 = (\gamma - u_1) u_1 / (a_2 - u_1)$$

and

$$\Psi_2^2 = 4(u_2 - a_1)(a_2 - u_2) v_2^2 = (u_2 - \gamma) u_2 / (u_2 - a_0).$$

With $u_i = u_i(\psi_i)$ as in (**) (3.5.3), we get $\Psi_i = \Psi_i(\psi_i)$, i.e., Ψ_i is a function of $\psi_i \in S^1$ only, $i = 1, 2$.

Consider now type I, i.e., $a_1 < \gamma < a_2$. Then $\Psi_2 = \Psi_2(\psi_2)$ describes a simple closed curve in the (ψ_2, Ψ_2)-plane and a fortiori a closed curve in $T_1^* M$. $\Psi_1 = \Psi_1(\psi_1)$ yields two non-closed curves in the (ψ_1, Ψ_1)-plane, one with $\Psi_1 > 0$, the other with $\Psi_1 < 0$, since $\Psi_1(\psi_1)$ is always $\neq 0$. However, in $T_1^* M$, $\Psi_1 = \Psi_1(\psi_1)$, $\psi_1 \in S^1$, describes two closed curves, since the (u, v) are periodic in ψ. Thus, $T_1^* M \cap \{F^* = \gamma\}$ consists of two embedded tori indeed.

The behaviour of the geodesics with parameter γ, as described in the theorem, can now be read off from Liouville's Theorem (3.5.6).

The discussion of type II, i.e., $a_0 < \gamma < a_1$, is similar. □

As an important Corollary we immediately get the

3.5.8 Theorem. *On an ellipsoid M with three different axes, the shortest and the longest ellipse represent stable closed geodesics.*

Proof. Consider e.g. the shortest ellipse, i.e., $\{x_0^2/a_0 + x_1^2/a_1 = 1\}$. If we call this a closed geodesic we implicitely assume the choice of some parameterization $c(t), 0 \leqslant t \leqslant 1$, of this curve.

Let $\{\phi_t X_0\}$ be the corresponding flow line, i.e., $X_0 = \dot{c}(0)/|\dot{c}(0)|$. For every $\gamma_0, a_0 < \gamma_0 < a_1$, denote by T_{γ_0} that one of the two tori $T_{\gamma_0}^{\pm}$ for which the τ_M-projections of the flow lines wind around the x_2-axis in the same sense as c does. Then the union of the $T_\gamma, a_0 \leqslant \gamma \leqslant \gamma_0$ is a flow invariant tubular neighborhood of $\{\phi_t \gamma_0\}$. Here we include T_{a_0} as the degenerate torus formed by the $\{\phi_t X_0\}$. Since we may choose γ_0 arbitrarily near a_0, we obtain in this way a basis for the neighborhoods of $\{\phi_t X_0\}$.

The case of the longest ellipse is treated in the same way, using the tori of type II. □

We will now show that the geodesic flow, when restricted to one of the invariant tori $T_\gamma^{\pm}$ with parameter γ, is equivalent to a linear flow on a flat torus in such a way that thereby the geodesic flow parameter is being preserved. We begin by introducing abbreviations for the periods of certain hyperelliptic integrals. For this cf. e.g. Hadamard [2].

3.5.9 Definition. With the previous notation, put $P(t,\gamma)=(-t)(\gamma-t)(a_0-t)(a_1-t)(a_2-t)$.

For $\gamma\in]a_0,a_1[$ define $\omega_i=(\omega_{1i},\omega_{2i})$ with

$$\omega_{11}=\pm 4\int_{a_1}^{a_2}\frac{(-t)(\gamma-t)}{\sqrt{P(t,\gamma)}}dt;\quad \omega_{12}=4\int_{a_0}^{\gamma}\frac{(-t)(\gamma-t)}{\sqrt{P(t,\gamma)}}dt$$

$$\omega_{21}=\mp 4\int_{a_1}^{a_2}\frac{(-t)}{\sqrt{P(t,\gamma)}}dt;\qquad \omega_{22}=4\int_{a_0}^{\gamma}\frac{(-t)}{\sqrt{P(t,\gamma)}}dt$$

For $\gamma\in]a_1,a_2[$, $\omega_i=(\omega_{1i},\omega_{2i})$ shall be given by

$$\omega_{11}=4\int_{a_0}^{a_1}\frac{(-t)(\gamma-t)}{\sqrt{P(t,\gamma)}}dt;\qquad \omega_{12}=\pm 4\int_{a_1}^{\gamma}\frac{(-t)(\gamma-t)}{\sqrt{P(t,\gamma)}}dt$$

$$\omega_{21}=4\int_{a_0}^{a_1}\frac{(-t)}{\sqrt{P(t,\gamma)}}dt;\qquad \omega_{22}=\mp 4\int_{a_1}^{\gamma}\frac{(-t)}{\sqrt{P(t,\gamma)}}dt.$$

In each case, put $-\omega_{21}:\omega_{22}=\omega(\gamma)$.

Remark. The function $\gamma\to\omega(\gamma)$ is called *period mapping*. In (3.5.14) we will investigate its behaviour. As we are now going to show, $\omega(\gamma)$ measures the ratio between oscillation and winding for a geodesic with parameter γ, when we represent the flow on the flat torus..

3.5.10. Theorem. *The geodesic flow on each of the invariant tori $T_\gamma^\pm$ in appropriate coordinates, is equivalent to the linear flow of slope $\omega(\gamma)$ on the flat square torus.*

It follows that a geodesic c with parameter γ is closed if and only if $\omega(\gamma)$ is rational. In that case, all geodesics of parameter γ are closed and have the same length.

Remarks. 1. The coordinates on $T_\gamma^\pm$ which make the geodesic flow to a linear flow are the integral curves of the two vector fields ζ_{E^*} and $\zeta_{F^*-2\gamma E^*}=\zeta_{F^*}-2\gamma\zeta_{E^*}$, restricted to $T_\gamma^\pm$. Here, ζ_{G^*} is defined as the Hamiltonian vector field associated to the function G^* in the manner of (3.1.9). Actually, the existence of invariant tori and the equivalence of the flow on such tori to a linear flow is a general phenomenon for so-called completely integrable Hamiltonian systems, cf. Arnold and Avez [1] for the details of this theorem, due to V. Arnold.

2. Let c be a closed geodesic on M with parameter γ such that $\omega(\gamma)$ is rational. As a member of a 1-parameter family of closed geodesics of the same length, c is degenerate, i.e., nullity $c\geqslant 1$. That nullity $c=1$ and not 2 is a consequence of the fact that $\omega(\gamma)$ is non-constant as function of γ, cf. (3.5.14).

Proof. Let $\gamma\in]a_0,a_1[$. The differentials du_1', du_2' in the proof of (3.5.5) determine functions $u_1'(u_1,u_2)$, $u_2'(u_1,u_2)$ on $T_\gamma^\pm$, i.e.,

$$u_1' = \int_{a_0}^{u_1} \sqrt{-u_1+\gamma}\sqrt{U_1}\,du_1 \pm \int_{a_1}^{u_2} \sqrt{u_2-\gamma}\sqrt{U_2}\,du_2\,;$$

$$u_2' = \int_{a_0}^{u_1} \frac{\sqrt{U_1}}{\sqrt{-u_1+\gamma}}\,du_1 \mp \int_{a_1}^{u_2} \frac{\sqrt{U_2}}{\sqrt{u_2-\gamma}}\,du_2\,.$$

Here we have to replace the du_i by the $du_i(\psi_i)$, with $u_i = u_i(\psi_i)$, $i = 1, 2$, as in (3.5.3). Moreover, when forming the integrals along curves, we have to subdivide the u_1-intervals and u_2-intervals at their boundary values a_0, γ and a_1, a_2, respectively. The values (u_1', u_2') are determined by the points on $T_\gamma^\pm$ only up to multiples of the periods ω_1, ω_2 in $\mathbb{R}^2$, cf. (3.5.9).

Denote by Ω the matrix with columns (ω_1, ω_2). If we write T_Ω for the flat torus obtained from $\mathbb{R}^2$ modulo the lattice $\mathbb{Z}\,\omega_1 \times \mathbb{Z}\,\omega_2$, the functions $u' = u'(u)$ give a transformation from $T_\gamma^\pm$ onto T_Ω. The geodesic flow lines thereby go into the u_1'-parameter lines. As a final step take the linear transformation Ω^{-1} which carries the lattice $\mathbb{Z}\,\omega_1 \times \mathbb{Z}\,\omega_2$ into the standard lattice $\mathbb{Z} \times \mathbb{Z}$. This then yields the desired representation of $T_\gamma^\pm$ by $T_E = \mathbb{R}^2/\mathbb{Z}^2$. The geodesic flow is given by the vector in the first column of Ω^{-1}. I.e., its slope is of the form $-\omega_{21}:\omega_{22} = \omega(\gamma)$.

The case $\gamma \in \,]a_1, a_2[$ is treated in exactly the same manner.

To see the geometric meaning of $\omega(\gamma)$, take e.g. $\gamma \in \,]a_0, a_1[$. Then we have from $du_2'(t) = 0$ for a geodesic on $T_\gamma^\pm$

$$\int_{t_0}^{t_1} \frac{-u_1}{\sqrt{P(u_1,\gamma)}}\,du_1 = \pm \int_{t_0}^{t_1} \frac{-u_2}{\sqrt{P(u_2,\gamma)}}\,du_2\,.$$

That is, if we write the left hand side as $\mu\omega_{22}$ and the right hand side as $-\nu\omega_{21}$, $\mu:\nu = \omega(\gamma)$ measures the ratio of oscillation to winding of the geodesic c along the t-interval $[t_0, t_1]$. □

For further investigation of the geodesic flow we need the Gauss curvature on M as function of (u_1, u_2).

3.5.11 Lemma. *The Gauss curvature of the ellipsoid in elliptic coordinates is given by*

$$K(u_1,u_2) = \frac{a_0 a_1 a_2}{u_1^2 u_2^2}\,; \quad (u_1,u_2) \in [a_0,a_1] \times [a_1,a_2]\,.$$

Remark. In the above formula, we have also included the umbilics by giving them the coordinates $(u_1, u_2) = (a_1, a_2)$. The ellipse in the (x_0, x_1)-plane has coordinates (u_1, a_2) while the ellipse in the (x_1, x_2)-plane has coordinates (a_0, u_2). The ellipse in the (x_0, x_2)-plane has coordinates (u_1, a_1) and (a_1, u_2).

Proof. For the definition of $K(u_1, u_2)$ see (1.11.3), (1.5.3), (1.8.12). It follows that

$$K(u_1,u_2) = \frac{1}{g_{11}}\left(\Gamma_{11,2}^2 - \Gamma_{12,1}^2 + \sum_l \left(\Gamma_{11}^l \Gamma_{l2}^2 - \Gamma_{12}^l \Gamma_{l1}^2\right)\right)(u_1,u_2)\,.$$

With the $g_{ik}(u_1, u_2)$ as given in (3.5.4) we find for the Christoffel symbols $(i \neq k)$:

$$\Gamma^i_{ii}(u_1, u_2) = \frac{2u_i - u_k}{2u_i(u_i - u_k)} - \frac{f'(u_i)}{2f(u_i)},$$

$$\Gamma^k_{ik}(u_1, u_2) = \frac{1}{2(u_i - u_k)},$$

$$\Gamma^k_{ii}(u_1, u_2) = -\frac{u_i}{2u_k(u_i - u_k)} \cdot \frac{f(u_k)}{f(u_i)}.$$

A straightforward if tedious computation yields the formula for $K(u_1, u_2)$. □

3.5.12 Corollary. *The curvature $K(u_1, u_2)$ of the ellipsoid satisfies the relations*

$$\min K = \frac{a_0}{a_1 a_2} = K(a_1, a_2) \leqslant K(u_1, u_2) \leqslant K(a_0, a_2) = \frac{a_1}{a_0 a_2} \leqslant$$

$$\leqslant K(a_0, u_2) \leqslant K(a_0, a_1) = \frac{a_2}{a_0 a_1} = \max K.$$

In particular, $\min K : \max K = a_0^2/a_2^2$ *does not depend on* a_1. □

We use these results to give a qualitative description of the so-called *conjugate locus* $\operatorname{Conj}(p)$ of a point $p \in M$. This is the set of first conjugate points on the geodesics starting from p. For a detailed discussion, using the theory of hyperelliptic integrals, see v. Mangoldt [1]. Cf. also (2.1.13).

We restrict ourselves to the point $p = (0, \sqrt{a_1}, 0) \in M$. In $T_p M$ we choose the orthonormal basis $e_1 = (1, 0, 0) \in T_p M \subset T_p \mathbb{R}^3 = \mathbb{R}^3$ and $e_2 = (0, 0, 1)$. We parameterize the unit vectors in $T_p M$ by $\phi \in \mathbb{R} \bmod 2\pi$, i.e., $X_\phi = \cos\phi\, e_1 + \sin\phi\, e_2$. Let c_ϕ be the geodesic with $\dot{c}_\phi(0) = X_\phi$. $t_\phi > 0$ shall be the t-parameter of the first conjugate point of $p = c_\phi(0)$ on c_ϕ. Clearly, the symmetries of M imply $t_\phi = t_{-\phi} = t_{\pi+\phi} = t_{\pi-\phi}$. Therefore, we can restrict ourselves to the interval $0 \leqslant \phi \leqslant \pi/2$.

3.5.13 Lemma. *With the previously introduced notation, $\phi \in \mathbb{R} \bmod 2\pi \mapsto t_\phi \in \mathbb{R}$ assumes its maximum for $\phi = 0$ and its minimum for $\phi = \pi/2$. In between, it is monotonic decreasing.*

The curve $c_\phi(t_\phi)$, $0 \leqslant \phi \leqslant \pi/2$, formed by the first conjugate points on the family of geodesics c_ϕ, has the following properties: $c_0 | [0, t_0]$ lies on the smallest ellipse $M \cap \{x_2 = 0\}$ and crosses (at least once) the (x_1, x_2)-plane in the point $(0, -\sqrt{a_1}, 0)$ opposite p. For $0 < \phi < \pi/2$, $c_\phi | [0, t_\phi]$ – with the exception of the initial point $p = c_\phi(0)$ – lies entirely on $M \cap \{x_2 > 0\}$ and crosses (at least once) the (x_1, x_2)-plane. $c_{\pi/2} | [0, t_{\pi/2}]$ lies on the ellipse $M \cap \{x_0 = 0\}$ of maximal length and, with the exception of the initial point $c_{\pi/2}(0)$, lies entirely in the region $M \cap \{x_2 > 0\}$.

Proof. At $\phi = 0$, the function $\phi \to t_\phi$ has a (local) maximum. Indeed, for $\phi =$

$\pm \varepsilon, \varepsilon > 0$ small, the geodesic $c_\phi(t), 0 \leqslant t \leqslant t_0$, will run through a domain where K has become bigger than on $c_0(t), 0 \leqslant t \leqslant t_0$. This can be read off from (3.5.11), (3.5.12) which shows $K(u_1, a_2) < K(u_1, a_2 - \delta)$, for some small $\delta > 0$. Recall that $M \cap \{x_2 = 0\}$ is given by $u_2 = a_2$.

As a consequence, $c_\phi, |\phi|$ small, will intersect $c_0|[0, t_0]$ shortly before its end point. $c_\phi|]0, t_\phi]$ for small $\phi > 0$ therefore will lie entirely in $M \cap \{x_2 > 0\}$.

To see that $c_0|[0, t_0]$ actually crosses the $\{x_1, x_2\}$-plane at an interior point we compare it with the corresponding geodesic c_0^* on the ellipsoid of revolution $M^* = \{x_0^2/a_1 + x_1^2/a_1 + x_2^2/a_2 = 1\}$. The curvature along the shortest ellipse of M^* is constant $= 1/a_2$, cf. (3.5.12). The first conjugate point on c_0^* occurs at distance $\pi\sqrt{a_2}$, i.e., after having crossed the shortest ellipse $=$ circle of radius $\sqrt{a_1}$. If we now let the length $\sqrt{a_1}$ of the first axis in M^* decrease towards the value $\sqrt{a_0}$ of the first axis in M, the distance of the first conjugate point on the shortest ellipse will increase, and the half length of the shortest ellipse will decrease. Thus, $c_0|[0, t_0]$ covers more than half of the shortest ellipse.

Now let ϕ increase towards $\pi/2$. $c_\phi|[0, t_\phi]$ thereby moves into a domain where K increases. This shows that t_ϕ monotonic decreases and $c_\phi|[0, t_\phi]$ remains in the domain $M \cap \{x_2 > 0\}$.

$\phi = \pi/2$ is a minimum of $\phi \to t_\phi$, since the geodesics $c_{\pi/2 \pm \varepsilon}|[0, t_{\pi/2}]$, for small $\varepsilon > 0$, run through a region on M where K becomes smaller than on the ellipse $M \cap \{x_0 = 0\}$ carrying $c_{\pi/2}$, cf. (3.5.12) and note that this ellipse has parameter $u_1 = a_0$. It follows that $c_{\pi/2 \pm \varepsilon}|[0, t_{\pi/2 \pm \varepsilon}]$ meets $c_{\pi/2}$ only after $t = t_{\pi/2}$. Thus, in particular, the geodesic $c_{\pi/2-\varepsilon}|[0, t_{\pi/2-\varepsilon}]$, for small $\varepsilon > 0$, crosses the $\{x_1, x_2\}$-plane shortly before its end point. If we let ϕ decrease further, t_ϕ increases, and always, $c_\phi|[0, t_\phi]$ will cross the $\{x_1, x_2\}$-plane at some $t, 0 < t < t_\phi$.

It remains to be shown that $c_{\pi/2}|[0, \pi/2]$ does not meet the point $(0, -\sqrt{a_1}, 0)$ opposite p, i.e., $c_{\pi/2}|]0, \pi/2]$ lies on $M \cap \{x_2 > 0\}$. This follows from our previous arguments. But it can also be seen by comparing $c_{\pi/2}$ on M with the corresponding geodesic $c_{\pi/2}^{**}$ on the ellipsoid of revolution $M^{**} = \{x_0^2/a_0 + x_1^2/a_1 + x_2^2/a_1 = 1\}$. Here, the curvature along the ellipse $\{x_0 = 0\}$ is constant $= 1/a_0$. Hence, $c_{\pi/2}^{**}$ has its first conjugate point at distance $\pi\sqrt{a_0} < \pi\sqrt{a_1} =$ half length of the circle $M^{**} \cap \{x_0 = 0\}$. The increase of the third half axis of M^{**} from $\sqrt{a_1}$ to $\sqrt{a_2}$ only strengthens this inequality. □

We are now in a position to investigate the dependence of $\omega(\gamma)$ from γ. Recall from (3.5.10) that $\omega(\gamma)$ measures the ratio of oscillation to winding for a geodesic with parameter γ. Classically, this investigation is done by using relations between hyperelliptic integrals, cf. Alkier [1]. Here we will use our qualitative description of the conjugate locus given in (3.5.13).

3.5.14 Theorem. *Let $\gamma \in]a_0, a_1[\cup]a_1, a_2[$. The ratio $\omega(\gamma)$, considered as function of γ, is strictly monotonely decreasing. If $\gamma \in]a_0, a_1[$, then $\omega(\gamma) > 1$ and $\lim_{\gamma \to a_1} \omega(\gamma) = 1$. If $\omega(\gamma) \in]a_1, a_2[$, then $\omega(\gamma) < 1$ and $\lim_{\gamma \to a_1} \omega(\gamma) = 1$.*

Proof. Let $\gamma \in]a_0, a_1[$. Consider the geodesic c on M with parameter γ which starts from $p = (0, \sqrt{a_1}, 0)$ with initial direction $X_\phi = \cos\phi\, e_1 + \sin\phi\, e_2$, with $\phi = \phi_\gamma \in]0, \pi/2[$, cf. the notation introduced before (3.5.13).

Let $\gamma' = \gamma + \varepsilon \in]a_0, a_1[$, be slightly bigger than γ. Then the two parameter lines $\{u_1 = \gamma'\}$ on M are slightly further apart from each other than the two parameter lines $\{u_1 = \gamma\}$ – recall that these parameter lines are symmetric w.r.t. the (x_1, x_2)-plane. Let c' be the geodesic with parameter γ' which starts from p with initial direction $X_{\phi'}$, $\phi' = \phi_{\gamma'} \in]0, \pi/2[$. Then the position of the respective u_1-parameter lines implies $\phi_{\gamma'} < \phi_\gamma$. Hence, we know from (3.5.13) that the first conjugate point on c' occurs at distance $t_{\phi'} > t_\phi$ = distance of the first conjugate point on c. The first intersection of c and c' takes place after both curves have crossed the (x_1, x_2)-plane at least once.

Now let t_γ be the first t-value > 0 where c touches the parameter line $u_1 = \gamma$, i.e., $u_1(c(t_\gamma)) = \gamma$. $c|[0, t_\gamma]$ then has run through one quarter of an oscillation.

Denote by $c(t_\gamma^*)$ the point where c meets again the (x_1, x_2)-plane. Along $[0, t_\gamma^*]$, c has completed one half oscillation. Since t_γ^* occurs before c crosses again the (x_0, x_1)-plane, c has along $[0, t_\gamma^*]$ completed less than one half winding. Thus, the ratio between oscillation and winding is > 1.

Similarly, let $t_{\gamma'}$ be the first t-value > 0 when c' has completed one quarter oscillation while at $c'(t_{\gamma'}^*)$, c' shall meet the (x_1, x_2)-plane again, i.e., c' has completed one half oscillation when running through $[0, t_{\gamma'}^*]$. Since c, c' meet again only after having crossed the (x_1, x_2)-plane, and since along $[t_\gamma, t_\gamma^*]$ and $[t_{\gamma'}, t_{\gamma'}^*]$, the geodesics c and c' each accrue the same amount of winding as they do along $[0, t_\gamma]$ and $[0, t_{\gamma'}]$, respectively, we see that the amount of winding of c' along $[0, t_{\gamma'}]$ is bigger than the amount of winding of c along $[0, t_\gamma]$. Since in each case, the amount of oscillation is one quarter, we get $\omega(\gamma') < \omega(\gamma)$.

As γ moves towards a_1, the parameter lines $\{u_1 = \gamma\}$ on $M \cap \{x_0 > 0\}$ move towards the arc on the ellipse $M \cap \{x_1 = 0\}$ between the umbilics with $x_0 > 0$, run through back and forth. The geodesics c, starting from p with parameter γ and initial angle $\phi_\gamma \in]0, \pi/2[$ move towards the geodesic from p to the umbilic on $M \cap \{x_0 > 0\} \cap \{x_2 > 0\}$. When this geodesic reaches the umbilic, it has completed one quarter winding as well as one quarter oscillation, i.e., $\lim \omega(\gamma) = 1$, for $\gamma \to a_1, \gamma \in]a_0, a_1[$.

The case $\gamma \in]a_1, a_2[$ is treated similarly. Indeed, let $\gamma' = \gamma + \varepsilon \in]a_1, a_2[, \varepsilon > 0$ small. The parameter lines $\{u_2 = \gamma'\}$ will then be closer to the (x_0, x_1)-plane than the parameter lines $\{u_2 = \gamma\}$. Let c, c' be the geodesics with parameter γ, γ' which start from p with initial angle $\phi_\gamma, \phi_{\gamma'}$ in $]0, \pi/2[$. Then $\phi_{\gamma'} < \phi_\gamma$. And since c and c' meet again before they meet the (x_0, x_1)-plane, i.e., before they have completed one half of an oscillation, the amount of winding of c along $[0, t_\gamma]$, when c has completed one quarter of an oscillation, is less than the amount of winding of c' along $[0, t_{\gamma'}]$, when c' has completed one quarter of an oscillation. Thus, $\omega(\gamma') < \omega(\gamma) < 1$. □

From (3.5.14) follows the density of periodic orbits. Actually, this result also holds for arbitrary completely integrable systems of generic type, the genericity referring to the non-constancy of the period mapping $\gamma \mapsto \omega(\gamma)$, cf. Arnold and Avez [1] for details.

3.5.15 Theorem. *The periodic flow lines of the geodesic flow on the ellipsoid are dense. More precisely, if we denote by* Per $T_1 M$ *the set of those unit tangent vectors* X_0 *for which* $\{\phi_t X_0\}$ *is periodic, the closure of* Per $T_1 M$ *is all of* $T_1 M$.

Remark. Here we consider only the case of an ellipsoid with three different axes. But also on the ellipsoid of revolution the periodic flow lines are dense and in particular on the sphere, of course.

For the sphere we even have Per $T_1 M = T_1 M$. There exist other surfaces where Per $T_1 M = T_1 M$, the most important being the so-called Zoll surfaces. For an up-to-date exposition on everything what is known on manifolds M for which Per $T_1 M$ is equal to $T_1 M$ see Besse [1].

There also exist examples of surfaces M where the measure of the closure of Per $T_1 M$, compared to the measure of $T_1 M$, is less than any precribed constant, cf. (3.5.25) below.

While Per $T_1 M$ can be dense, generically it will have measure zero. In fact, as our proof will show, for the general ellipsoid, Per $T_1 M$ locally is of the form $\mathbb{R}^2 \times \mathbb{Q}$. Generically, Per $T_1 M$ will be even thinner, i.e., it will locally look like $\mathbb{R} \times A \subset \mathbb{R} \times \mathbb{Q}^2$ where the factor $\mathbb{R}$ comes from the tangent vectors to a flow line. This reflects the fact that generically the closed geodesics are non-degenerate and hence the critical points $\neq$ const in ΛM are isolated.

Proof. The flow-invariant tori $T_\gamma^\pm, \gamma \in]a_0, a_1[\cup]a_1, a_2[$ fill an open dense set of $T_1 M$. On each $T_\gamma^\pm$, the flow is equivalent to a linear flow with slope $\omega(\gamma)$, cf. (3.5.10). The flow on $T_\gamma^\pm$ is periodic if and only if $\omega(\gamma)$ is rational. Since $\omega(\gamma)$ is a strictly monotonic decreasing function of γ and since the rational numbers are dense in the reals, the closure of the $T_\gamma^\pm$ with $\omega(\gamma)$ rational is all of $T_1 M$. □

We complement (3.5.14) by giving a description of the set of flow lines with γ-value a_1.

3.5.16 Theorem. *The flow-invariant set* $T_1^* M \cap \{F^* = a_1\}$ *is formed by those flow lines which, when projected into* M, *yield the geodesics which pass through an umbilic.*

This set contains in particular the tangent vectors to the middle ellipse, i.e., the ellipse in the (x_0, x_2)*-plane.*

A flow line covering this ellipse is the only periodic flow line in the set $T_1^* M \cap \{F^* = a_1\}$. *For any other flow line, we get a geodesic which passes through one of the two pairs* $\{q, q'\}$ *and* $\{r, r'\}$ *of diametrical umbilics time and again at fixed intervals. This interval is the same for all flow lines and measures the distance from* q *to* q' *or, what is the same, the distance from* r *to* r'.

A parameterization of the middle ellipse yields a hyperbolic closed geodesic of index 2.

More precisely, let $c(t), 0 \leqslant t \leqslant 1$, *be the parameterization of the middle ellipse which starts at* $c(0) = (\sqrt{a_0}, 0, 0)$ *and where* $\dot{c}(0)$ *points into the half space* $\{x_2 > 0\}$. *Let* $0 < t_1 < t_2 < t_3 < t_4 < 1$ *be the parameter values where* c *passes through the four umbilics. Then a stable Jacobi field* $Y_s(t)$ *along* $c(t)$ *has its zeros at* t_2, t_4 *and an unstable Jacobi field* $Y_u(t)$ *along* $c(t)$ *has its zeros at* t_1, t_3.

Finally, if $t_0 \in]t_2, t_3[$ *or* $\in]t_4, t_1 + 1[$, $c|[t_0, t_0 + 1]$ *has two conjugate points in its interior. On the other hand, if* $t_0 \in]t_1, t_2[$ *or* $\in]t_3, t_4[$, $c|[t_0, t_0 + 1]$ *has only one conjugate point in its interior.*

Note. That for a hyperbolic closed geodesic of index > 0 the number of conjugate points depends on the choice of the initial point is a general phenomenon.

Proof. Remove from $T_1^*M \cap \{F^* = a_1\}$ the set of cotangent vectors to the ellipse in the (x_0, x_2)-plane. Then the cotangent coordinates (Ψ_1, Ψ_2), cf. the proof of (3.5.7), read

$$\Psi_1^2 = \frac{u_1(a_1 - u_1)}{a_2 - u_1}; \quad \Psi_2^2 = \frac{u_2(u_2 - a_1)}{u_2 - a_0}.$$

Here, $u_i = u_i(\psi_i)$ as in (**), (3.5.3). Hence, for every $p \in M$ outside the (x_0, x_2)-plane there are four unit vectors belonging to $\{F^* = a_1\}$. From (3.5.14) we have that these are the initial vectors of those geodesics starting from p which go to one of the four umbilics. The four vectors decompose into two pairs of oppositely oriented vectors.

This shows that a geodesic starting from such a $p \in M$ and going to an umbilic, say q, also passes through the opposite umbilic q' and then goes through q again, and so on. This happens at fixed intervals. Indeed, if we write down $du_2' = 0$, (3.5.5), for $\gamma = a_1$, then we have

$$du_1' = \frac{1}{2}\left(\sqrt{\frac{u_1}{(u_1 - a_0)(a_2 - u_1)}}\, du_1 + \sqrt{\frac{u_2}{(u_2 - a_0)(a_2 - u_2)}}\, du_2\right).$$

The length of a geodesic from q to q' is given by integrating the first summand over $[a_0, a_1]$ and the second summand over $[a_1, a_2]$. The resulting value is independent of the choice of the geodesic from q to q'.

Now let $c(t), 0 \leqslant t \leqslant 1$, be the parameterized middle ellipse as in the Theorem. The Poincaré mapping P_ω of the associated periodic orbit in $T_1 M$ has two real eigen values. Indeed, let $Y_s(t)$ be the Jacobi field generated by the geodesics passing through the umbilics $\{c(t_2), c(t_4)\}$ and $Y_u(t)$ the Jacobi field generated by the geodesics passing through $\{c(t_1), c(t_3)\}$. $\tilde{Y}_s(0) = (Y_s(0), \nabla Y_s(0))$ and $\tilde{Y}_u(0) = (Y_u(0), \nabla Y_u(0))$ clearly are eigenvectors of P_ω.

We claim that the eigenvalues of P_ω are > 0 and $\neq 1$. Then c must be hyperbolic of index 2.

To prove our claim we observe that on the segment $c|[t_1, t_3]$ the curvature $K(t)$ during the first half of the interval is bigger than during the second half. Thus, the value $t' \in]t_1, t_3[$ where $\nabla Y_s(t') = 0$ occurs before $(t_3 + t_1)/2$, i.e., $|\nabla Y_s(t_3)| < |\nabla Y_s(t_1)|$ and, a fortiori, $|\nabla Y_s(t_1 + 1)| < |\nabla Y_s(t_1)|$. This shows that P_ω

on $\tilde{Y}_s(0)$ has a positive eigenvalue < 1 and that $Y_s(t)$ is indeed a stable Jacobi field. The above defined field $Y_u(t)$ must then be the unstable field.

Next we show that none of the geodesics passing through a diametral pair of umbilics and not lying in the (x_0, x_2)-plane is closed. Let $\{c(t_1), c(t_3)\} = \{q, q'\}$ be such a diametral pair. Choose an orientation of M. If there were a closed geodesic through $\{q, q'\}$ not in the (x_0, x_2)-plane let c' be the one which starts at q and forms with c a smallest angle $\alpha \in]0, \pi[$. For symmetry reasons, c and c' then also in q' form the angle α.

Denote by c^* the geodesic $c^*(t) = c'(1/2 - t)$. Apply to c^* the reflections on the (x_0, x_1)-plane and on the (x_1, x_2)-plane. This yields a closed geodesic c'' starting from q with angle $\pi - \alpha$.

Observe that the Jacobi field $Y'(t)$ along $c'(t)$ which is generated by the field of geodesics through $\{q, q'\}$ must be unstable, since the correspondingly generated field $Y_s(t)$ along $c(t)$ is stable – this follows by a continuity argument. By the same argument, the correspondingly generated Jacobi field $Y''(t)$ along $c''(t)$ must be stable, since it forms smallest angle with the geodesic $c(-t)$ in q for which $Y_s(-t)$ is unstable. But $Y'(t)$ and $Y''(t)$ are obtained by isometries from each other, thus, there can be no closed geodesic c' through $\{q, q'\}$ outside the (x_0, x_2)-plane – and clearly also none which closes after multiple passings through the pair $\{q, q'\}$.

Finally, we discuss the distribution of conjugate points along $c|[t_0, t_0 + 1]$. If $t_0 \in]t_2, t_3[$ or $\in]t_4, t_1 + 1[$, a geodesic c' starting at $c(t_0)$ and forming with c a small angle $\neq 0$ has its γ-parameter in the interval $]a_1, a_2[$. Thus, $\omega(\gamma) < 1$. That is, when c' meets c again, it has completed one half winding but not yet one half oscillation which means that this meeting point occurs on c before $t_0 + 1/2$. Thus, $c|[t_0, t_0 + 1]$ has two conjugate points in its interior. A similar argument works for $t_0 \in]t_1, t_2[$ or $\in]t_2, t_3[$ with $\omega(\gamma) > 1$. □

We can now also say something about the index of the shortest and the longest ellipse on M .For the index of the longest ellipse see also (3.5.20).

3.5.17 Lemma. *On an ellipsoid M with three different axes, the shortest basic ellipse is the shortest closed geodesic and it has index* 1 *while the largest basic ellipse has index* $\geqslant 3$.

Proof. Let c_0 be a parameterization of the ellipse in the (x_0, x_1)-plane. There are no shorter closed geodesics on M since any other closed geodesic either is winding around the x_0-axis or the x_2-axis or else, is the ellipse of middle length. (3.5.8) then implies that index $c_0 = 1$.

Let now $c_2 = c_2(t)$, $0 \leqslant t \leqslant 1$, be a parameterization of the longest basic ellipse on M. With the arguments from the end of the proof of (3.5.13) we see that c_2 has passed a conjugate point before it reaches the t-value 1/2. According to (3.5.8), c_2 is elliptic. If it is non-degenerate, we have from (3.4.2) index $c_2 \geqslant 3$.

Assume now that c_2 is degenerate. We may choose the initial point of c_2 such that a non-constant periodic Jacobi field $Y(t)$ along $c_2(t)$ vanishes for $t = 0$. $Y(t)$ must have an odd number of zeros in $]0, 1[$, and this number is $\geqslant 2$, thus again, index $c_2 \geqslant 3$. □

We continue with a discussion of the question whether on the ellipsoid there exist simple closed geodesics which are different from the three principal ellipses. A closed curve is called *simple* if it has no self-intersections.

From (3.4.5) we know that the length $L(c)$ of a simple closed geodesic c, different from one of the three ellipses c_1, c_2, c_3, can become arbitrarily large, compared to $L(c_i)$, $i = 1, 2, 3$, if only $a_1 - a_0$ and $a_2 - a_1$ are sufficiently small. Our subsequent argument will show that such a long geodesic c will not be simple. In fact, for a closed geodesic c to be simple, c is allowed to have completed only a single winding when it closes.

We now describe relations between the pinching $\min K : \max K$ of the ellipsoid M and the existence of simple closed geodesics. Our material is largely taken from Viesel [1].

Besides the formula for $K(u_1, u_2)$ in (3.5.11) and (3.5.12) we will need the

3.5.18 Proposition. *$\omega(a_0) = \omega(a_0; \varrho, \sigma)$ is of the form*

$$(*) \qquad \omega(a_0; \varrho, \sigma) = \frac{2}{\pi} \frac{\sqrt{1-\varrho}}{\sqrt{\varrho}} \sqrt{\sigma - \varrho} \cdot \int_0^{\pi/2} \frac{\sqrt{\sigma \cos^2\phi + \sin^2\phi}}{(\sigma - \varrho) + (1 - \sigma)\sin^2\phi} \, d\phi .$$

Here, $\varrho = a_0/a_2$; $\sigma = a_1/a_2$. Thus, $0 < \varrho < \sigma < 1$.

For ϱ fixed, $\omega(a_0; \varrho, \sigma)$ is a monotonic increasing function of σ with upper limit given by $\sqrt{1/\varrho} = \sqrt{a_2/a_0}$. The lower limit, for $\sigma \to \varrho$, is 1.

Proof. Recall from the definition (3.5.9) $\omega(\gamma) = -\omega_{21} : \omega_{22}$, where ω_{21}, ω_{22} also depend on γ. We find

$$\lim_{\gamma \to a_0} \pm \omega_{21}/4 = \int_{a_1}^{a_2} \frac{1}{(t - a_0)} \frac{\sqrt{t}}{\sqrt{(t - a_1)(a_2 - t)}} \, dt .$$

Substitute here $t = a_1 \cos^2\phi + a_2 \sin^2\phi$. Then

$$\lim_{\gamma \to a_0} \pm \omega_{21}/4 = 2 \int_0^{\pi/2} \frac{1}{(a_1 - a_0) + (a_2 - a_1)\sin^2\phi} \sqrt{a_1 + (a_2 - a_1)\sin^2\phi} \, d\phi .$$

To determine $\lim_{\gamma \to a_0} \omega_{22}/4$ we substitute in (3.5.9) $t = \gamma \sin^2\phi + a_0 \cos^2\phi$ and find

$$\lim_{\gamma \to a_0} \omega_{22}/4 = \frac{\pi}{\sqrt{a_2}} \frac{\sqrt{a_0}}{\sqrt{(a_2 - a_0)(a_1 - a_0)}}$$

This yields $(*)$.

We put $\sqrt{\sigma - \varrho} = h(\sigma)$, $\sigma \sin^2\phi + \cos^2\phi = x(\sigma, \phi)$ and

$$2 \int_0^{\pi/2} \frac{\sqrt{x(\sigma, \phi)}}{x(\sigma, \phi) - \varrho} \, d\phi = g(\sigma) .$$

Then

$$g'(\sigma) = -\int_0^{\pi/2} \frac{(x+\varrho)\sin^2\phi}{(x-\varrho)^2\sqrt{x}}\,d\phi,$$

hence, $d\omega(a_0;\varrho,\sigma)/d\sigma$ reads

$$h(\sigma)g'(\sigma)+h'(\sigma)g(\sigma) = \frac{1-\varrho}{\sqrt{\sigma-\varrho}}\int_0^{\pi/2}\frac{\cos^2\phi\,.\sqrt{x}}{(x-\varrho)^2}\,d\phi - $$

$$-\varrho\sqrt{\sigma-\varrho}\int_0^{\pi/2}\frac{\sin^2\phi}{(x-\varrho)^2\sqrt{x}}\,d\phi\,.$$

Since $\sigma \leqslant 1$, this is

$$\geqslant (1-\varrho)\sqrt{\frac{\sigma}{\sigma-\varrho}}\int_0^{\pi/2}\frac{\cos^2\phi}{(x-\varrho)^2}\,d\phi - \varrho\sqrt{\frac{\sigma-\varrho}{\sigma}}\int_0^{\pi/2}\frac{\sin^2\phi}{(x-\varrho)^2}\,d\phi\,.$$

With $t = \operatorname{tg}\phi$, these two integrals can be computed. One finds

$$\frac{\pi}{4(\sigma-\varrho)^2}\left(\frac{\sigma-\varrho}{1-\varrho}\right)^{1/2}\left(\sqrt{\sigma(\sigma-\varrho)}-\varrho\sqrt{(\sigma-\varrho)/\sigma}\right) \geqslant 0\,.$$

The limit $\omega(a_0;\varrho,\sigma)$ for $\sigma \to 1$ is clearly $\sqrt{1/\varrho}$. The limit for $\sigma \to \varrho$ can be determined from the observation that $\omega(a_1) = 1$ and that $\omega(a_0)$ differes arbitrarily little from $\omega(a_1)$ if $a_0 \to a_1$. See also Viesel [1] for an analytic proof. □

3.5.19 Theorem. *On an ellipsoid* $M = \{\sum_0^2 x_i^2/a_i = 1\}$ *there exist non-standard simple closed geodesics, i.e., simple closed geodesics different from the three principal ellipses, if and only if* $\omega(a_0;\varrho,\sigma)$ *is* >2. *Here,* $\varrho = a_0/a_2, \sigma = a_1/a_2$.

More precisely, for each integer value $\omega(\gamma), 1 < \omega(\gamma) < \omega(a_0;\varrho,\sigma)$, *the flow on the two tori* $T_\gamma^\pm$ *is periodic. The projection of these flow lines yields closed geodesics which wind once around the* x_0*-axis while performing* γ *many oscillations. Their length is greater than the length of the middle ellipse in the* (x_0, x_2)*-plane.*

For fixed pinching $\min K : \min K = \varrho^2$, *the greatest number of 2-tori carrying simple closed geodesics occurs for the flattened ellipsoid of revolution, i.e., when* $\omega_0(a_0;\varrho,\sigma)$ *goes to* $\sqrt{1/\varrho}$. *In particular, there are no non-standard simple closed geodesics if* $\min K : \max K \geqslant 1/16$

As M *approaches an elongated ellipsoid of revolution, i.e., as* $\sigma = a_1/a_2$ *goes to* $\varrho = a_0/a_2$, *there will be no nonstandard simple closed geodesics.*

3.5.20 Complement. *The index of the longest ellipse, i.e., the ellipse in the* (x_1, x_2)*-plane, is odd* $\geqslant 3$. *For fixed pinching* $\min K : \max K$, *it becomes maximal for the flattened ellipsoid of revolution. In this case, the index is given by the maximal integer*

$< 2\sqrt{a_2/a_0}$, *plus* 1, *if this integer is even. In particular, if* $\min K : \max K \geq 1/16$, *the longest ellipse has index* 3.

Proof. For a geodesic c with parameter γ to be closed, $\omega(\gamma)$ must be rational. To be simple closed, only a single winding is allowed. Hence $\omega(\gamma)$ must be an integer > 1. Since $\omega(\gamma)$ increases with $\gamma \to a_0$, the families of non-standard simple closed geodesics correspond to the integers in the interval $]1, \omega(a_0)[$.

From (3.5.18) we know that $\omega(a_0)$ becomes maximal, for fixed pinching, in the case that σ goes towards 1. Thereby, the umbilics move towards $(\pm\sqrt{a_0}, 0, 0)$. The longest ellipse c_3 approaches the circle of radius $\sqrt{a_2}$ and the curvature along c_3 becomes constant $= 1/a_0$. Thus, the period of a Jacobi field along c_3 approaches $2\pi/\sqrt{1/a_0} = 2\pi\sqrt{a_0}$. The number of periods on the limit geodesics is the maximal integer $< 2\pi\sqrt{a_2}/2\pi\sqrt{a_0} = \sqrt{1/\varrho}$, which is the limit of $\omega(a_0; \varrho, \sigma)$ for $\sigma \to 1$. The number of conjugate points on c_3, in that case, is the maximal integer $< 2\sqrt{1/\varrho} = 2\sqrt{a_2/a_0}$. This upper bound is ≤ 4 as long as $\min K : \max K \geq 1/16$.

For $\sigma \to \varrho$, as M approaches an ellipsoid of revolution around the x_2-axis, the umbilics move towards $(0, 0, \pm\sqrt{a_2})$. Geodesics of type II become more and more similar to the longest ellipse c_3 and to closed curves through a pair of diametral umbilics. Thus, $\omega(\gamma)$ for $\gamma \in]a_0, a_1[$ goes towards $1 = \omega(a_1)$.

Whenever we have non-basic simple closed geodesics, the shortest among them occur for $\omega(\gamma) = 2$. Let c be such a geodesic. We may assume that $c(0)$ is in the (x_0, x_2)-plane and $\dot{c}(0)$ is orthogonal to this plane. We compare $L(c)/2$ with the length of a minimizing geodesic c' from the umbilic nearest to $c(0)$ to the diametrically opposite umbilic. Then c' is shorter than $L(c)/2$. $2L(c')$ is the length of the geodesic c_2 in the (x_0, x_2)-plane. □

We continue with a brief discussion of the geodesic flow of a surface of revolution. This is another classic example of a surface for which the geodesic flow permits a second integral F besides the kinetic energy E. See Klingenberg [9] for more on surfaces of revolution, and compare also the exposition in Darboux [1].

We mentioned surfaces of revolution in the example before (1.10.10). Here we will restrict ourselves to compact surfaces.

3.5.21 Definition. *A surface of revolution* M is a compact 2-dimensional Riemannian manifold which is isometrically embedded in the Euclidean (x, y, z)-space $\mathbb{R}^3$ and which is transformed into itself under the isometric S^1-action $\varrho : S^1 \times \mathbb{R}^3 \to \mathbb{R}^3$ which consists of the positive rotations around the z-axis.

Consider the intersection $M \cap \{y = 0\}$ of M with the (x, z)-plane. This is a closed curve, transformed into itself under the reflection on the z-axis. Hence, whenever it meets the z-axis, it must do so orthogonally. This follows from the fact that for every point p on M outside the z-axis, the Killing vector $X_\varrho(p)$ is non-zero and tangent to M.

The intersection of M with a plane containing the axis of revolution $\{x = y = 0\}$ is called *meridian*; the intersection of M with a plane orthogonal to the z-axis is called *parallel circle*. The parallel circles are the orbits of the S^1-action.

For $p \in M$, denote by $r(p)$ the distance of p from the z-axis. $r(p)$ is the radius of the parallel circle through p; also, $r(p)$ is the length of the Killing vector $X_\varrho(p)$ at p.

3.5.22 Proposition. (i) *A parallel circle of radius $r_0 > 0$ is a closed geodesic if and only if it is a critical value of $r: M \to \mathbb{R}$.*

(ii) *All meridians are closed geodesics.*

Proof. (i) was proved in (1.10.10) and (ii) follows from (1.10.15). Indeed, a meridian is the fixed point set of the reflection of $\mathbb{R}^3$ on the 2-plane which contains the meridian. □

We now can prove *Clairaut's Theorem.*

3.5.23 Theorem. *On a surface of revolution M, the function*

$$F: T_1 M \to \mathbb{R};\ X \mapsto \langle X, X_\varrho(\tau_M X)\rangle$$

is constant along the geodesic flow lines.

If we denote by $\mu(X)$ the angle between X and the parallel circle through the base point $\tau_M X$ of X we can also write F in the form

$$F(X) = r(\tau_M X)\cos\mu(X).$$

Remark. $F(X)$ is called *angular momentum*. In the case that $\tau_M X$ is on the z-axis, $\mu(X)$ is not well-defined. But this does not matter for the definition of F since in this case $r(\tau_M X) = 0$.

Proof. Let $X \in T_1 M$. Put $\tau_M \phi_t X = c(t)$, i.e., $\phi_t X = \dot{c}(t)$.
Then

$$F(\phi_t X) = \langle \dot{c}(t), X_\varrho(c(t))\rangle = \langle \frac{\partial}{\partial t}\varrho(s, c(t)), \frac{\partial}{\partial s}\varrho(s, c(t))\rangle|_{s=0}.$$

Here $\varrho: S^1 \times M \to M$ denotes the S^1-action, cf. (1.10.8).

Since $t \to \varrho(s, c(t))$ is a geodesic and in particular $|\frac{\partial}{\partial t}\varrho(s, c(t))| = \text{const}$, we get with the help of (1.5.8, i):

$$\frac{d}{dt}F(\phi_t X_0) = \langle \frac{\partial}{\partial t}\varrho(s, c(t)), \frac{\nabla}{\partial s}\frac{\partial}{\partial t}\varrho(s, c(t))\rangle =$$

$$= \frac{1}{2}\frac{\partial}{\partial s}\langle \frac{\partial}{\partial t}\varrho(s, c(t)), \frac{\partial}{\partial t}\varrho(s, c(t))\rangle = 0. \quad \square$$

The counterpart to the elliptic coordinates on the ellipsoid are on a surface of revolution M the parallel- and meridian-coordinates. We call $M \cap \{y = 0; x > 0\}$ the *profile curve of M*. Let u_1 be an arc length parameter on the profile curve and let u_2 be the angular variable on the parallel circles, with $u_2 = 0$ being the intersection with the profile curve. Then the line element reads

$$ds^2 = du_1^2 + r(u_1)^2\, du_2^2 .$$

These coordinates are valid outside the intersection with the z-axis, i.e., whenever $r > 0$.

Again, as for the ellipsoid, if $F(\phi_t X) = \gamma$, γ is called the *parameter of the geodesic* $c(t) = \tau_M \phi_t X$. This means, in the (u_1, u_2)-coordinates,

$$\dot{u}_1^2 + r(u_1)^2 \dot{u}_2^2 = 1; \quad r(u_1)^2 \dot{u}_2 = \gamma$$

or, if we multiply the first equation with γ^2 and substitute the second,

$$(*) \qquad \gamma\, \dot{u}_1 = \pm\, r(u_1) \sqrt{r(u_1)^2 - \gamma^2}\, \dot{u}_2 .$$

We want to show that – at least in general – the set $\{F = \gamma\}$ on $T_1 M$ consists of embedded tori, just as it was the case for the ellipsoid. For this to be true, we have to restrict γ appropriately. For instance, we exclude those values of γ which F assumes on parallel circles which happen to be closed geodesics. As we saw in (3.5.22), these correspond to the critical values of $r(p)$. In this case, the invariant tori become degenerate.

To simplify the formulation of our result we restrict ourselves to special arcs on the profile curve:

3.5.24 Theorem. *Let M be a surface of revolution. Assume that we have for the profile curve a parameter interval $I = [u_1^-, u_1^+]$ of positive length such that $r(u_1) \geqslant r(u_1^-) = r(u_1^+)$ for $u_1 \in I$, and $r'(u_1^-) > 0$, $r'(u_1^+) < 0$. Put $r(u_1^-) = r(u_1^+) = \gamma$. Then the set $T_1 M \cap \{F = \gamma\}$ consists of the two tori $T_\gamma^\pm$. The geodesic flow on each of these tori is equivalent to a linear flow with slope*

$$\omega(\gamma) = \pm \int_I \frac{\gamma}{r(t)\sqrt{r(t)^2 - \gamma^2}}\, dt : \pi$$

Thus, in particular, the underlying geodesics of parameter γ are closed if and only if $\omega(\gamma)$ is rational.

Proof. To see that $T_\gamma^\pm$ is a torus we introduce cotangent coordinates

$$v_1 = g_{11}(u_1, u_2)\, \dot{u}_1 = \dot{u}_1; \quad v_2 = g_{22}(u_1, u_2)\, \dot{u}_2 = r(u_1)^2 \dot{u}_2 .$$

Then $T_\gamma^\pm$ is given by

$$2E^*(u, v) = v_1^2 + \frac{\gamma^2}{r(u_1)^2} = 1; \quad F^*(u, v) = v_2 = \gamma .$$

u_2 is periodic mod 2π. This shows that the second equation $\{v_2 = \gamma\}$ describes a closed curve. But also the first equation describes a closed curve in the (u_1, v_1)-plane which intersects the v_1-axis in $u_1 = u_1^-$ and $u_1 = u_1^+$.

The underlying geodesic of a flow line on $T_\gamma^\pm$ touches the parallel circles $\{u_1 = u_1^-\}$ and $\{u_1 = u_1^+\}$ alternately, thereby describing half an oscillation, while winding around the z-axis. $\omega(\gamma)$ is the ratio between one half oscillation and one half winding, i.e., $\omega(\gamma)$ is the integral of (*) over the interval I, divided by π.

All this also follows from our claim that the flow on $T_\gamma^\pm$ is equivalent to a linear flow. Similar as in the proof of (3.5.10), we introduce on $T_\gamma^\pm$ new parametss u_1', u_2' by

$$du_1' = \frac{\sqrt{r(u_1)^2-\gamma^2}}{r(u_1)}\,du_1 \pm \gamma\,du_2; \quad du_2' = \frac{du_1}{r(u_1)\sqrt{r(u_1)^2-\gamma^2}} \mp \frac{du_2}{\gamma}.$$

With this we get

$$ds^2 = du_1'^2 + \gamma^2\left(r(u_1)^2-\gamma^2\right)du_2'^2 .$$

Thus, the u_1'-curves are precisely the geodesics with $F(u,\dot u) = \gamma$.

As in the proof of (3.5.10), we get a description of $T_\gamma^\pm$ by the flat torus $T_\Omega^2 = \mathbb{R}^2/\mathbb{Z}\omega_1 \times \mathbb{Z}\omega_2$ with periods

$$\omega_1 = (\omega_{11}, \omega_{12}) = \Bigl(2\int_I \frac{\sqrt{r^2(t)-\gamma^2}}{r(t)}\,dt,\ \pm\int_0^{2\pi}\gamma\,dt\Bigr)$$

$$\omega_2 = (\omega_{21}, \omega_{22}) = \Bigl(2\int_I \frac{dt}{r(t)\sqrt{r^2(t)-\gamma^2}},\ \mp\int_0^{2\pi}\frac{dt}{\gamma}\Bigr).$$

It follows that the slope $\omega(\gamma)$ is given by $-\omega_{21}:\omega_{22}$. □

3.5.25 Example of a convex surface of revolution M with $\omega(\gamma) = \text{const}$ for the major part of T_1M. We first define M', an open subset of M, as the set generated by the profile curve

$$\Bigl(a\cos u_1, 0, \pm\int_0^{u_1}\sqrt{1-a^2\sin^2 t}\,dt\Bigr); \ -\pi/2+\varepsilon \leqslant u_1 \leqslant \pi/2-\varepsilon$$

with $0 < a \leqslant 1$, some small $\varepsilon > 0$. This curve yields the well-known spindle-like surface of constant curvature $K = 1$, cf. Klingenberg [9]. For $a = 1$, it is the sphere with some cap around north and south pole removed, the size of this cap depending on ε.

We compute $\omega(\gamma) = \omega(\gamma; a)$ from (3.5.24) for our case. That is, $r(u_1) = a\cos u_1$, $I = [-\alpha, \alpha] \subset [-\pi/2+\varepsilon, \pi/2-\varepsilon]$ such that $\gamma^2 = a^2\cos^2\alpha$. Since the profile curve is symmetric with respect to the x-axis we can write

$$\omega(\gamma; a) = \left(\int_0^\alpha \frac{a\cos\alpha}{a\cos t\sqrt{a^2\cos^2 t - a^2\cos^2\alpha}}\,dt\right) : \frac{\pi}{2} = a^{-1}\omega(\gamma; 1).$$

$\omega(\gamma; 1)$ is the slope of the flow on the sphere, i.e., $= 1$. We therefore see that $\omega(\gamma; a) = a^{-1}$ only depends on a and not on γ, as long as we choose $\gamma^2 = a^2\cos^2\alpha \geqslant a^2\sin^2\varepsilon$. This means that we only consider $X \in T_1M$ which do not form too steep an angle with the parallel circles in order that the geodesic $\tau_M \phi_t X$ stays on the subset $M' \subset M$.

M is now defined by closing the spindle surface with small convex caps around the north and south poles. And the geodesics running entirely on M' are either all closed or are all non-closed, (with the exception of the equator, given by $\gamma^2 = a^2$) depending on whether a is rational or not. Clearly, area (M'): area (M) can be made to approximate 1 arbitrarily well, if only ε is chosen sufficiently small. $\square$

3.6 Closed Geodesics on Spheres

In (3.4.10) we proved the existence of more than one closed geodesic on certain surfaces of the topological type of the 2-sphere. We now develop additional methods which will be employed in the construction of several closed geodesics on more general manifolds. The principal tool is the space of circles on a sphere S^n. A non-null homotopic mapping of S^n into a manifold M can sometimes be used to prove the existence of a good number of closed geodesics on M. The main results concern the case where M is homeomorphic to S^n.

We begin with the space of circles on S^n with $n \geqslant 2$. We think of S^n as embedded the standard way into the Euclidean space $\mathbb{R}^{n+1}$, i. e., $S^n = \{\sum_0^n x_i^2 = 1\}$.

3.6.1 Definition. (i) *The space* AS^n *of parameterized circles on* S^n consists of the mappings $c: S \to S^n$ where c either is a constant map – in which case we call c also a point circle – or else, c is an embedding of the circle $S = [0, 1]/\{0, 1\}$ with parameter proportional to arc length. The image shall lie in the intersection of S^n and a 2-plane in $\mathbb{R}^{n+1}$ having distance < 1 from the origin $0 \in \mathbb{R}^{n+1}$.

We endow AS^n with the topology induced from the inclusion $AS^n \hookrightarrow \Lambda S^n$.

(ii) *A parameterized great circle* is a circle where the image lies in a 2-plane through $0 \in \mathbb{R}^{n+1}$. That is, it is a simple closed geodesic on S^n with parameter interval $[0, 1]$. By BS^n we denote the *space of great circles*.

3.6.2 Proposition. *The space* BS^n *of parameterized great circles is canonically isomorphic to the unit tangent bundle* $T_1 S^n$. *The standard action of* $\mathbb{O}(n+1)$ *on* S^n *operates transitively on* BS^n *with isotropy group isomorphic to* $\mathbb{O}(n-1)$.

Proof. A great circle $c(t)$, $0 \leqslant t \leqslant 1$, is determined by $X_0 = \dot{c}(0)/|\dot{c}(0)| \in T_1 M$, and vice versa. Thus, $BS^n \cong T_1 S^n$.

From the end of (2.2.29) we know that $S^n = \mathbb{O}(n+1)/\mathbb{O}(n)$. Consider on S^n the point $p_0 = (1, 0, \ldots, 0)$ and in $T_{p_0} S^n$ the vector X_0 with coordinates $(0, 1, \ldots, 0)$. The subgroup of $\mathbb{O}(n+1)$, which leaves X_0 invariant, is the group $\mathbb{O}(n-1)$

$\subset \mathbb{O}(n+1)$ which operates as identity on the (x_0, x_1)-plane of $\mathbb{R}^{n+1}$ and in the standard way on the complementary orthogonal $(n-1)$-dimensional subspace.

3.6.3 Proposition. *The space $\Lambda S^n - \Lambda^0 S^n$ of non-constant circles on S^n is the total space of a D^{n-1}-bundle*

$$\alpha : \Lambda S^n - \Lambda^0 S^n \to B S^n$$

over BS^n. Here, the image under α of a circle c is the parameterized great circle c^* which is obtained by first parallel translating the plane of c into the origin $0 \in \mathbb{R}^{n+1}$ such that the midpoint of c goes into 0 and then blowing up the circle into a great circle c^*.

Note. When we speak of a D^{n-1}-bundle we mean the restriction of a vector bundle with fibre dimension $(n-1)$ to the open unit discs in each fibre.

Proof. Obviously, the mapping α is well-determined and surjective. The counter image $\alpha^{-1}(c^*)$ of a great circle c^*, contained in the 2-plane $\mathbb{R}^{*2} \subset \mathbb{R}^{n+1}$, can be described by the set of mid-points of the circles c with $\alpha(c) = c^*$. The mid-points have distance < 1 from $0 \in \mathbb{R}^{n+1}$ and belong to the $(n-1)$-dimensional space $(\mathbb{R}^{*2})^{\perp}$ in $\mathbb{R}^{n+1}$ which is orthogonal to $\mathbb{R}^{*2}$. That is, $\alpha^{-1}(c^*)$ can be identified with the unit disc in $(\mathbb{R}^{*2})^{\perp}$. α is thus seen to be the restriction of a $(n-1)$-dimensional vector bundle. □

Recall that we consider ΛS^n, $\Lambda^0 S^n$ and BS^n as subspaces of ΛS^n with the induced topology. All these spaces are transformed into themselves by the involution $\theta : \Lambda M \to \Lambda M$ as well as by the S^1-action $\chi^{\sim}$, defined in (2.5.A). The involution θ, given by $(c(t)) \mapsto (c(1-t))$ and the involution consisting of the action of $e^{i\pi} \in S^1$, i. e., $(c(t+1/2))$ commute with each other. We put $e^{i\pi}\theta = \theta e^{i\pi} = \theta'$.

This leads us to consider the following quotient spaces of $\Lambda S^n, \Lambda^0 S^n, BS^n$:

3.6.4 Definition. (i) The quotient spaces of $\Lambda S^n, \Lambda^0 S^n, BS^n$ under the $\mathbb{Z}_2$-action generated by θ are denoted by $\bar{\Lambda} S^n, \bar{\Lambda}^0 S^n, \bar{B} S^n$, respectively. We view them as subspaces of $\bar{\Lambda} S^n = \Lambda S^n /_{\theta} \mathbb{Z}_2$.

(ii) The quotient spaces of $\Lambda S^n, \Lambda^0 S^n, BS^n$ under the $\mathbb{Z}_2$-action generated by θ' are denoted by $\Lambda^* S^n, \Lambda^{*0} S^n, B^* S^n$, respectively. They are subspaces of $\Lambda^* S^n = \Lambda S^n /_{\theta'} \mathbb{Z}_2$.

(iii) The quotient space of $\Lambda S^n, \Lambda^0 S^n, BS^n$ under the $(\mathbb{Z}_2 \times \mathbb{Z}_2)$-action generated by θ, θ' are denoted by $\bar{\Lambda}^* S^n, \bar{\Lambda}^{*0} S^n, \bar{B}^* S^n$, respectively. We view them as subspaces of $\bar{\Lambda}^* S^n = \Lambda S^n \underset{(\theta,\theta')}{/} (\mathbb{Z}_2 \times \mathbb{Z}_2)$.

Remark. Traditionally, cf. Klingenberg [10], one also considers the quotient spaces $\Gamma S^n, \Gamma^0 S^n, \Delta S^n$ of $\Lambda S^n, \Lambda^0 S^n, BS^n$ with respect to the full $\mathbb{O}(2)$-action χ. ΔS^n then is called *space of unparameterized great circles*. Such a great circle can be identified with the 2-plane through $0 \in \mathbb{R}^{n+1}$ which is the carrier of the great circles. Thus, ΔS^n can be identified with the Grassmann manifold $G(2, n-1)$ of 2-planes through $0 \in \mathbb{R}^{n+1}$, cf. the end of (2.2.29).

3.6.5 Proposition. (i) *The space BS^n of parameterized great circles on S^n corresponds to the* Stiefel manifold $V(2, n-1)$ *of pairs $\{e, e'\}$ of orthonormal vectors in $\mathbb{R}^{n+1}$. Take for e the initial point of the great circle and for e' its initial tangent vector, normed to the length 1. Thus, BS^n can be identified with the homogenous space*

$$\mathbb{O}(n+1)/\mathbb{O}(n-1) \cong \mathbb{O}(n+1)/S\mathbb{O}(1) \times S\mathbb{O}(1) \times \mathbb{O}(n-1).$$

Here we have indicated on the right hand side the way in which $\mathbb{O}(n-1)$ is embedded in $\mathbb{O}(n+1)$.

(ii) *The involutions θ', θ generate the subgroups $\mathbb{O}(1) \times S\mathbb{O}(1) \times id_{n-1}$ and $S\mathbb{O}(1) \times \mathbb{O}(1) \times id_{n-1}$ of $\mathbb{O}(n+1)$, while the $\mathbb{O}(2)$-action on BS^n is given by $\mathbb{O}(2) \times id_{n-1}$. Hence,*

$$\bar{B}S^n \cong \mathbb{O}(n+1)/S\mathbb{O}(1) \times \mathbb{O}(1) \times \mathbb{O}(n-1)$$

$$B^*S^n \cong \mathbb{O}(n+1)/\mathbb{O}(1) \times S\mathbb{O}(1) \times \mathbb{O}(n-1)$$

$$\bar{B}^*S^n \simeq \mathbb{O}(n+1)/\mathbb{O}(1) \times \mathbb{O}(1) \times \mathbb{O}(n-1)$$

$$\Delta S^n \simeq \mathbb{O}(n+1)/\mathbb{O}(2) \times \mathbb{O}(n-1).$$

(iii) *Since the projection mapping α commutes with the S^1-action and $\mathbb{Z}_2$-action, we get the D^{n-1}-bundles*

$$\bar{\alpha}\ : \bar{\Lambda}S^n - \bar{\Lambda}^0 S^n \to \bar{B}S^n$$

$$\alpha^*: \Lambda^* S^n - \Lambda^{*0} S^n \to B^* S^n$$

$$\bar{\alpha}^*: \bar{\Lambda}^* S^n - \bar{\Lambda}^{*0} S^n \to \bar{B}^* S^n$$

$$\gamma\ : \Gamma S^n - \Gamma^0 S^n \to \Delta S^n$$

Proof. Clear from the definitions. □

Note. BS^n can be viewed in two ways as the total space of an S^{n-1}-bundle over S^n:

$$S^{n-1} \to BS^n \xrightarrow{\tau} S^n; \quad S^{n-1} \to BS^n \xrightarrow{\sigma} S^n$$

by associating with $\{e, e'\} \in BS^n \cong V(2, n-1)$ the element $e \in S^n$ or $e' \in S^n$, respectively. θ and θ' operate as involutions on the fibre of τ and σ, respectively. Taking the quotient space of these actions, we obtain the projective bundles

$$P^{n-1} \to \bar{B}S^n \xrightarrow{\bar{\tau}} S^n; \quad P^{n-1} \to B^*S^n \xrightarrow{\sigma^*} S^n,$$

which are clearly isomorphic to each other.

One can show that the $\mathbb{Z}_2$-cohomology ring of $\bar{B}S^n \cong B^*S^n$ is given by

$$H^*(\bar{B}S^n) \cong H^*(P^{n-1}) \otimes H^*(S^n).$$

Indeed, the $\mathbb{Z}_2$-action θ on BS^n is a fixed point free isometry, i. e., BS^n is a 2-fold covering of $\bar{B}S^n$. The corresponding 1-dimensional cohomology class is the generator of the ring $H^*(P^{n-1}) = \mathbb{Z}_2[X]/(X^n)$.

We use the n cycles in the fibre P^{n-1} of $\bar{B}S^n$ to define n ϕ-families in $\bar{\Lambda}S^n$. We begin with the

3.6.6 Definition. Denote by $B_k, 0 \leqslant k \leqslant n-1$, the subset of BS^n, formed by those parameterized great circles on S^n which start at $p_0 = (1, 0, \ldots, 0) \in S^n$ and lie on $S^{k+1} = \{\sum_0^{k+1} x_i^2 = 1\}$. Clearly, B_k is θ-invariant

The restriction of the D^{n-1}-bundle α to $B_k \subset BS^n$ is denoted by

$$\alpha_k : A_k - A_k^0 \to B_k .$$

Denote by $A_k, \bar{A}_k$ the closure in $\Lambda S^n, \bar{\Lambda} S^n$ of $A_k - A_k^0, \bar{A}_k - \bar{A}_k^0$. (A_k, A_k^0) or simply A_k represents a $(k+n-1)$-chain in $(\Lambda S^n, \Lambda^0 S^n)$. $(\bar{A}_k, \bar{A}_k^0)$ is a $(k+n-1)$-cycle in $(\bar{\Lambda} S^n, \bar{\Lambda}^0 S^n)$.

Remark. B_k can be identified with the sphere S^k and $\bar{B}_k$ can be identified with P^k, a generator of the k-dimensional $\mathbb{Z}_2$-homology of the fibre P^{n-1} of $\bar{\tau}$ over p_0.

Thus, the $(k+n-1)$-cycle $(\bar{A}_k, \bar{A}_k^0)$ can be viewed as a "thickening" of the base cycle $\bar{B}_k$ by the D^{n-1}-discs over $\bar{B}_k$ in the bundle $\bar{\tau}_k$, modulo its boundary.

We use this to define ϕ-families in the sense of (2.4.17).

3.6.7 Definition. Let M be a Riemannian manifold for which there exists a homeomorphism $\phi: S^n \to M$ with the property that it induces a continuous mapping of the space of circles ΛS^n into ΛM.

(i) Denote by A_k any chain in $(\Lambda S^n, \Lambda^0 S^n)$ which is θ-equivariantly homologons to the θ-invariant chain A_k defined in (3.6.6).

(ii) Let $u_k : A_k \to (\Lambda M, \Lambda^0 M)$ be the θ-equivariant mapping, induced by ϕ. Thus, by going to the quotient modulo the $\mathbb{Z}_2$-action θ, we get a $\mathbb{Z}_2$-cycle $\bar{u}_k : \bar{A}_k \to (\bar{\Lambda} M, \bar{\Lambda}^0 M)$. We define the ϕ-family $\mathscr{A}_k$ so as to consist of the sets $\phi_s u_k(A_k)$. Since ϕ_s commutes with the $\mathbb{Z}_2$-action θ, this is indeed a ϕ-family.

Note. The hypothesis on M holds for every manifold for which the underlying topological manifold is the sphere S^n. One way to see this is that on the n-sphere every differentiable structure is related to the standard differentiable structure by a homeomorphism which is differentiable, with the possible exception of a single point, cf. Milnor [1]. Another way would be to observe that any continuous mapping $\phi: S^n \to M$ is homotopically equivalent to a differentiable $\phi: S^n \to M$, cf. Hirsch [1].

3.6.8 Theorem. *Let M be a manifold satisfying the hypothesis in* (3.6.7). *Denote by $\varkappa_k$, the critical values of the ϕ-families $\mathscr{A}_k, 0 \leqslant k \leqslant n-1$. Then*

$$0 < \varkappa_0 \leqslant \varkappa_1 \leqslant \ldots \leqslant \varkappa_{n-1} .$$

If $\varkappa_l = \varkappa_{l+1} =$ *(briefly)* $\varkappa$, *for some* $l < n-1$, *then the compact set of critical points with E-value* $\varkappa$ *consists of an infinite number of* S^1*-orbits.*

Remarks. 1. This result essentially goes back to Lyusternik [1]. Lyusternik and later Alber [1] considered the space $(\Gamma S^n, \Gamma^0 S^n)$ of unparameterized circles instead of the space $(\bar{\Lambda} S^n, \bar{\Lambda}^0 S^n)$.

The advantage of taking the cycles $\bar{u}_j$ of $\bar{\Lambda} S^n \bmod \bar{\Lambda}^0 S^n$ instead of the cycles $v_j = \tilde{\pi}(\bar{u}_j)$ of $\Gamma S^n \bmod \Gamma^0 S^n$ lies in the fact that $\bar{\Lambda} S^n$ is closer to the manifold ΛS^n than is ΠS^n. Actually, if we remove from $\bar{\Lambda} S^n$ the set $\bar{\Lambda}_\theta S^n$, we get a manifold again, having $\Lambda S^n - \Lambda_\theta S^n$ as its double covering, cf. (2.5.A.14). Alber's claim that the quotient cycles $v_j = \tilde{\pi}(\bar{u}_j)$ in $(\Pi M, \Pi^0 M)$ are non-homologous to zero is not correct. However, one does not need this as our proof will show.

2. (3.6.8) can be generalized to the case when we have a mapping $\phi: S^l \to M$, $2 \leqslant l \leqslant \dim M$ which defines a non-trivial homology class, cf. Klingenberg [10], [13].

Proof. The relation $0 < \varkappa_0$ is proved by the same method which was employed in the proof of (2.4.20). Here we use that the first non-vanishing homotopy group of $(\bar{\Lambda} S^n, \bar{\Lambda}^0 S^n)$ which has dimension $n-1$ coincides with the homology group, cf. Spanier [1]. Also, $\varkappa_k \leqslant \varkappa_{k+1}$ clearly holds since any chain u_{k+1} contains, by restriction, a chain u_k.

The only case that remains to be discussed is the case $\varkappa_l = \varkappa_{l+1} =$ (briefly) $\varkappa$, for some $l < n-1$. Denote by $Cr\varkappa$ the set of critical points with E-value $\varkappa$.

We will derive a contradiction from the assumption that $Cr\varkappa$ consists of only finitely many pairs of critical orbits $(S^1 . c_\iota, S^1 . \theta c_\iota)$, $\iota \in I$, I some finite set.

To do this we choose for each $\iota \in I$ an S^1-invariant neighborhood $\mathcal{U}_\iota$ of $S^1 . c_\iota$, outside the fixed point set $\Lambda_\theta M$ of $\theta: \Lambda M \to \Lambda M$. We can assume: $\mathcal{U}_\iota \cap \theta \mathcal{U}_\iota = \emptyset$, all $\iota \in I$, and $\mathcal{U}_\iota \cap \mathcal{U}_\varkappa = \mathcal{U}_\iota \cap \theta \mathcal{U}_\varkappa = \emptyset$, all $(\iota, \varkappa) \in I \times I$, $\iota \neq \varkappa$. By $\mathcal{U}$ we denote the union of these $\mathcal{U}_\iota, \theta \mathcal{U}_\iota$, $\iota \in I$.

Then $\theta \mathcal{U} = \mathcal{U}$. If we denote by $\bar{\mathcal{U}}$ the image of $\mathcal{U}$ in $\bar{\Lambda} M$ then $\bar{\mathcal{U}} \subset \bar{\Lambda} M - \bar{\Lambda}_\theta M$. On $\bar{\Lambda} M - \bar{\Lambda}_\theta M$ we have the 1-dimensional $\mathbb{Z}_2$-homology class which is represented by the image in $\bar{\Lambda} M - \bar{\Lambda}_\theta M$ of any curve in $\Lambda M - \Lambda_\theta M$ from some c to θc. $\bar{\mathcal{U}}$ does not carry a cycle in this class.

There exists a $u_{l+1}: A_{l+1} \to \Lambda M$ with image in $\mathcal{U} \cup \Lambda^{\varkappa-} M$, cf. (2.4.18). Put $u_{l+1}^{-1}(\mathcal{U}) = \mathcal{O}$ and $u_{l+1}^{-1}(\Lambda^{\varkappa-} M) = \mathcal{N}$. $\mathcal{O} \subset A_{l+1} - A_{l+1}^0$. $\mathcal{O}$ and $\mathcal{N}$ are θ-invariant. $\mathcal{O}$ does not carry a curve from some c_0 to θc_0. That is, $\bar{\mathcal{O}}$ does not carry a cycle representing the 1-dimensional homology class = homotopy class of $\bar{A}_{l+1} - \bar{A}_{l+1}^0 \cong P^{l+1}$. To put it differently, every non-null homotopic curve in $\bar{A}_{l+1} - \bar{A}_{l+1}^0 \cong P^{l+1}$ has points outside $\bar{\mathcal{O}}$. Therefore there exists a cycle $\bar{A}_l \subset \bar{\mathcal{N}}$. But then, $u_l(A_l) \subset \Lambda^{\varkappa-} M$, i.e., $\varkappa_l < \varkappa_{l+1}$. □

Remarks. 1. Another possibility is to use the cap product between the $(l+n)$-dimensional cycle $\bar{A}_{l+1}$ and a 1-dimensional non-trivial cocycle ω on $\bar{A}_{l+1} - \bar{A}_{l+1}^0 \cong \bar{B}_{l+1} \cong P^{l+1}$. This yields a cycle $\bar{A}_l$ inside $\bar{\mathcal{N}}$. We come back to this in the proof of (3.6.12) and (3.7.23).

2. Unless $\varkappa_l = \varkappa_{l+1}$ for some $l < n-1$, (3.6.8) does not imply the existence of n different prime (unparameterized) closed geodesics on a manifold M of the homotopy type of S^n. Indeed, we have not excluded the possibility that among the critical orbits at the n different E-levels $\varkappa_0 < \varkappa_1 < \ldots < \varkappa_{n-1}$ some, or even all, are just multiple coverings of the same underlying prime critical orbit. That this actually is not the case and that the underlying prime closed geodesics of the n closed geodesics with E-value $\varkappa_0 < \ldots < \varkappa_{n-1}$ all are different has been shown in Klingenberg [10] with the help of the so-called Morse complex. Here we prove an even stronger result under the additional assumption that the sectional curvature of M is more than 1/4-pinched. The original proof was given in Klingenberg [5].

3.6.9 Theorem. *Let M be a simply connected compact manifold for which the sectional curvature K satisfies the relation $K_1/4 < K \leqslant K_1$, for some $K_1 > 0$. According to (2.8.5), M then is homeomorphic to $S^n, n = \dim M$. There exist on M n different simple closed unparameterized geodesics with length in the interval $[2\pi/\sqrt{K_1}, 4\pi/\sqrt{K_1}[$.*

Proof. By normalizing the metric, we can restrict ourselves to the case $K_1 = 1$, i. e., $K_0 \leqslant K \leqslant 1$, with some $K_0 > 1/4$. From the proof of (2.8.5) and (2.8.6) we have the existence of a homeomorphism

$$\phi : S^n \to M,$$

$n = \dim M$, with the following properties:

Let (p, q) be points on M with maximal distance $d(p, q) = d(M)$. Then $\pi \leqslant d(M) \leqslant \pi/\sqrt{K_0} < 2\pi$. The set $E = E(p, q)$ of points $r \in M$ with $d(p, r) = d(q, r)$ is an embedding of the $(n-1)$-sphere S^{n-1}. We also call E the *equator* of M w. r. t. (p, q). $d(p, E) \geqslant \pi/2, d(q, E) \geqslant \pi/2$. $M - E$ consists of two disjoint open sets $B(p)$ and $B(q)$ with $\partial \bar{B}(p) = \partial \bar{B}(q) = E$. The open ball $B_\pi(p)$ of radius π around p is entirely in the domain of polar coordinates based at p. $\bar{B}(p) = B(p) \cup E$ belongs to $B_\pi(p)$ and every geodesic emanating from p meets $\partial B(p) = E$ transversally. The corresponding facts hold for $B(q)$.

Denote by M_1 the sphere S^n of constant curvature $K_1 = 1$. On M_1 we choose $p_1 = (1, 0, \ldots, 0)$ and $q_1 = (-1, 0, \ldots, 0)$. Define $E_1 = E(p_1, q_1), B(p_1), B(q_1)$ as before. Hence, $B(p_1) = B_{\pi/2}(p_1), B(q_1) = B_{\pi/2}(q_1)$. The mapping $\phi : M_1 = S^n \to M$ constructed in the proof of (2.8.6) now satisfies the hypotheses necessary for the proof of (3.6.9).

In particular, with a half great circle $c_1(t), 0 \leqslant t \leqslant 1$, from p_1 to q_1 there is associated a curve $c(t) = \phi \circ c_1(t), 0 \leqslant t \leqslant 1$ from p to q where $c|[0, 1/2]$ is a minimizing geodesic from p to a point r on E and $c|[1/2, 1]$ is the minimizing geodesic from r to q. $L(c) < 2\pi$.

With help of ϕ we define a cycle $u_k : (A_k, A_k^0) \to (\Lambda M, \Lambda^0 M)$ of the type considered in (3.6.7), by associating with a circle $c_0(t), 0 \leqslant t \leqslant 1$, in A_k the closed curve $\phi \circ c_0(t), 0 \leqslant t \leqslant 1$. The image of a great circle through (p_1, q_1) has length $< 4\pi$, its E-value is $< 8\pi^2$. We will show how u_{n-1} can be deformed θ-equivariantly such as to have its image entirely below the E-level $< 8\pi^2$.

Once this is done we can complete the proof of the Theorem as follows: First of all, $\varkappa_{n-1} < 8\pi^2$, i. e., all closed geodesics constructed with the help of the ϕ-families $\mathscr{A}_k$, $0 \leqslant k \leqslant n-1$, defined in (3.6.7), have length $< 4\pi$. On the other hand, according to (2.6.10), a non-constant geodesic loop has length $\geqslant 2\pi$. Thus, our closed geodesics all are simple. (3.6.8) then shows that there must be at least n simple closed geodesics.

There remains to define the θ-equivariant deformation of u_{n-1} into some $\tilde{u}_{n-1}(A_{n-1}) \subset \Lambda^{8\pi^2-} M$. To do this observe that each circle in A_{n-1} is completely determined by its intersection with the equator $E_1 = E(p_1, q_1)$ on S^n. This intersection consists of exactly two points, except when the circle is a constant curve, i. e., except when the circle belongs to A^0_{n-1} in which case it consists of a single point on E_1.

Now replace each circle of $A_{n-1} - A^0_{n-1}$ by the closed curve, formed by the half great circle from p_1 to q_1 which passes through the first of the two intersection points of the circle with the equator E_1 and then go back to p_1 through the second intersection point. Replace a degenerate circle in A^0_{n-1} by the half great circle from p_1 to q_1, passing through that degenerate circle, and back the same way. Finally, add the family of closed curves obtained by retracting the latter curves back into their intersection with the equator.

The newly defined set $\tilde{A}_{n-1} \subset \Lambda S^n$ is clearly θ-equivariantly homotopic to A_{n-1}. For every $\tilde{c} \in \tilde{A}_{n-1}$, its image $\phi \circ \tilde{c} \in \Lambda M$ consists of closed curves which are formed by not more than four geodesic segments, each of which has length less than the distance from p (or from q) to a point r on the equator E. Since $d(p, r) = d(q, r) < \pi$, the curves $\phi \circ \tilde{c}$ all have E-value $< 8\pi^2$.

Now use the θ-equivariant homotopy from A_{n-1} into $\tilde{A}_{n-1}$ to define an θ-equivariant homotopy from u_{n-1} into an $\tilde{u}_{n-1}$ with image in $\Lambda^{8\pi^2-} M$. □

We wish to define a larger set of ϕ-families than the one defined in (3.6.7). For that purpose we take into consideration not only the $\mathbb{Z}_2$-action generated by θ but also the one generated by $\theta' = e^{i\pi}\theta$.

Thus, we consider the D^{n-1}-bundle $\bar{\alpha}^*$ from (3.6.5). For such a bundle we have the Thom isomorphism

$$H_*(\bar{B}^* S^n) \cong H_{*+(n-1)}(\bar{A}^* S^n, \bar{A}^{*0} S^n)$$

which is defined as follows, cf. Spanier [1]: If $\bar{y}_k^*$ is a k-cycle in $\bar{B}^* S^n$, take the closure $\bar{z}_k^*$ of the fibres $\bar{\alpha}^{*-1}(\bar{y}_k^*)$ over $\bar{y}_k^*$; this is relative a $(k+n-1)$-cycle of $(\bar{A}^* S^n, \bar{A}^{*0} S^n)$.

Besides the D^{n-1}-bundle $\bar{\alpha}^*$, we also consider the D^{n-1}-bundle γ over ΔS^n, cf. (3.6.7). Here again we have the Thom isomorphism

$$H_*(\Delta S^n) \cong H_{*+(n-1)}(\Gamma S^n, \Gamma^0 S^n).$$

We also need the ring structures of $H^*(\bar{B}^* S^n)$ and $H^*(\Delta S^n)$. For the latter case, it was determined by Chern [1]. For our purposes, it is convenient to use also the description given by Borel [1], which is valid for a general class of homogeneous spaces. According to Borel we can write

$$H^*(\Delta S^n) = S(v_1, v_2) \otimes S(v_3, \ldots, v_{n-1})/S^+(v_1, v_2, v_3, \ldots, v_{n-1}).$$

Here, $S(v_1, \ldots, v_k)$ denotes the ring of symmetric polynomials in $(v_1, \ldots, v_k)$ with $\mathbb{Z}_2$-coefficients and $S^+(v_1, \ldots, v_k)$ is the ideal in $S(v_1, \ldots, v_k)$, formed by the polynomials of degree > 0.

An additive basis for $H^*(\Delta S^n)$ is given by the elements

$$(a, b) = \sum_0^{b-a} v_1^{a+i} \cdot v_2^{b-i};\ 0 \leqslant a \leqslant b \leqslant n-1.$$

The multiplicative structure of $H^*(\Delta S^n)$ is determined by the rules

$$(a_1, a_2) = (0, a_1) \cup (0, a_2) + (0, a_1 - 1) \cup (0, a_2 + 1)$$

$$(0, a) \cup (a_1, a_2) = \sum_i (a_1 + i, a_2 + a - i);\ 0 \leqslant i \leq \min(a, a_2 - a_1).$$

Here we put $(a, b) = 0$, whenever $0 \leqslant a \leqslant b \leqslant n-1$ is not satisfied, cf. Chern [1].

We also need $H^*(\bar{B}^* S^n)$ which according to Borel [1] can be written as

$$H^*(\bar{B}^* S^n) = S(u_1) \otimes S(u_2) \otimes S(u_3, \ldots, u_{n+1})/S^+(u_1, u_2, u_3, \ldots, u_{n+1}).$$

If $\pi: \bar{B}^* S^n \to \Delta S^n$ is the canonical projection given by the quotient mapping of the S^1-action, then the induced homomorphism $\pi^*: H^*(\Delta S^n) \to H^*(\bar{B}^* S^n)$ is an embedding, where the image is formed by the subring of polynomials symmetric in u_1, u_2 – note that all elements of the representation of $H^*(\Delta S^n)$ can be expressed by v_1, v_2 alone. In particular, $\pi^*(v_1 + v_2) \equiv \pi^*(0, 1) = u_1 + u_2$; $\pi^*(v_1^2 + v_1 \cdot v_2 + v_2^2) \equiv \pi^*(0, 2) = u_1^2 + u_1 \cdot u_2 + u_2^2$; $\pi^*(v_1 \cdot v_2) = \pi^*(1, 1) = u_1 \cdot u_2$.

Alber [1] has determined the maximal number of cohomology classes of ΔS^n, all with the exception of one of positive dimension, which together have a product $\neq 0$. If we write $n = 2^k + s$ with $0 \leqslant s < 2^k$, such a product is given by $1 \cdot (v_1 + v_2)^{2n-2s-2} \cdot (v_1 \cdot v_2)^s = v_1^{n-1} \cdot v_2^{n-1}$ or, using our second notation, $1 \cdot (0, 1)^{2n-2s-2} \cdot (1, 1)^s = (n-1, n-1)$. To see this observe that $(v_1 + v_2)^{2^l} = v_1^{2^l} + v_2^{2^l}$. Hence,

$$(v_1 + v_2)^{2^k - 1} = \sum_0^{2^k - 1} v_1^i \cdot v_2^{2^k - i - 1};$$

$$(v_1 + v_2)^{2^{k+1} - 2} = ((v_1 + v_2)^{2^k - 1})^2 = \sum_i v_1^{2i} \cdot v_2^{2^{k+1} - 2i - 2} \neq 0,$$

since there occurs the summand $v_1^{2^k - 2} \cdot v_2^{2^k} + v_1^{2^k} \cdot v_2^{2^k - 2}$. But $(v_1 + v_2)^{2^{k+1} - 1} = 0$, since $2^{k+1} - 1 > 2n - 2$. Multiplying the last equation with $(v_1 \cdot v_2)^s$ yields $v_1^{n-1} \cdot v_2^{n-1}$.

We consider the cap product

$$H_*(\bar{A}^* S^n, \bar{A}^{*0} S^n) \otimes H^*(\bar{B}^* S^n) \xrightarrow{\cap} H_*(\bar{A}^* S^n, \bar{A}^{*0} S^n).$$

On the level of simplices and cochains, this is defined as follows, cf. Spanier [1]: For a $(j+k+n-1)$-simplex $\bar{s}^*_{j+k}$ *of* $\bar{A}^* S^n$ and a k-cochain ω^k of $\bar{B}^* S^n$, $\bar{s}^*_j = \bar{s}^*_{j+k} \cap \omega^k$ is the front $(j+n-1)$-simplex of $\bar{s}^*_{j+k}$, with coefficient given by the value of ω^k on the back k-simplex of $\bar{s}^*_{j+k}$. Thus, a cohomology class $\omega^k \in H^k(\bar{B}^* S^n)$ defines a mapping

$$H_{j+k+n-1}(\bar{A}^* S^n, \bar{A}^{*0} S^n) \xrightarrow{\cap \omega^k} H_{j+n-1}(\bar{A}^* S^n, \bar{A}^{*0} S^n).$$

Similarly, we have the cap product

$$H_*(\Gamma S^n, \Gamma^0 S^n) \otimes H^*(\Delta S^n) \xrightarrow{\cap} H_*(\Gamma S^n, \Gamma^0 S^n).$$

Above, we wrote down a basis $\{(a,b); 0 \leqslant a \leqslant b \leqslant n-1\}$ for $H^*(\Delta S^n)$. Denote the dual basis for $H_*(\Delta S^n)$ by $\{[a,b]; 0 \leqslant a \leqslant b \leqslant n-1\}$. We also use the symbol $[a,b]$ to denote the cycle in the class $[a,b]$ consisting of all unparameterized great circles on $S^{b+1} = \{x_0^2 + \ldots + x_{b+1}^2 = 1\}$ which meet the sphere $S^a = \{x_0^2 + \ldots + x_a^2 = 1\}$.

The cycles $[a,b]$ can be covered by chains $y_{a,b}$ in BS^n which project onto cycles $\bar{y}_{a,b}$ of $\bar{B}^* S^n$ and project further onto the cycles $[a,b]$, in such a way that every unparameterized great circle is covered by exactly one parameterized great circle, with the possible exception of the boundary of the chain.

The chain $y_{a,b}$ is formed by all (parameterized) great circles on S^{b+1} which start on the half sphere $S^a \cap \{x_a \geqslant 0\}$ with initial direction tangent to the half sphere $S^{b+1} \cap \{x_{b+1} \geqslant 0\}$.

Clearly, $\bar{y}^*_{a,b}$ = image of $y_{a,b}$ under the quotient mapping $BS^n \to \bar{B}^* S^n$, is a $(a+b)$-cycle of $\bar{B}^* S^n$ and $\pi \bar{y}^*_{a,b} = [a,b]$.

We agree to denote by $z_{a,b}$, $\bar{z}^*_{a,b}$ and $\{a,b\}$ the sets of circles lying over $y_{a,b}$, $\bar{y}^*_{a,b}$ and $[a,b]$ in the bundles α, $\bar{\alpha}^*$ and γ respectively. That is, $\bar{z}^*_{a,b}$ and $\{a,b\}$ are the cycles of dimension $a+b+n-1$ of $(\bar{A}^* S^n, \bar{A}^{*0} S^n)$ and $(\Gamma S^n, \Gamma^0 S^n)$, respectively, which are the images of $\bar{y}^*_{a,b}$ and $[a,b]$ under the Thom isomorphism.

3.6.10 Definition. Write $n = 2^k + s$, $0 \leqslant s < 2^k$, and put $2n - s - 1 = g(n)$. Consider the sequence of $g(n)$ cohomology classes of $\bar{B}^* S^n$ given by

$$(*) \quad \begin{aligned} &u_1^{n-1} \cdot u_2^{n-1} = (u_1+u_2)^{2n-2s-2} \cdot (u_1 \cdot u_2)^s, \ldots, (u_1+u_2)^{2n-2s-2} \cdot (u_1 \cdot u_2) \\ &(u_1+u_2)^{2n-2s-2}, \ldots, (u_1+u_2), 1. \end{aligned}$$

By taking the cap product between a $(3n-3)$-cycle homologons to $\bar{z}^*_{n-1,n-1}$ and cocycles representing the elements of the sequence $(*)$ we obtain a sequence of $g(n)$ so-called subordinated cycles. Each of these cycles goes under $\pi: (\bar{A}^* S^n, \bar{A}^{*0} S^n) \to (\Gamma S^n, \Gamma^0 S^n)$ into a cycle of $(\Gamma S^n, \Gamma^0 S^n)$. Note that every element in $(*)$ comes from an element in $H^*(\Delta S^n)$ under $\pi^*: H^*(\Delta S^n) \to H^*(\bar{B}^* S^n)$.

For abbreviation, we denote the sequence of such defined sets of $\bar{A}^* S^n$ by

(†) $\bar{A}_0^*, \ldots, \bar{A}_{2s-2}^*, \bar{A}_{2s}^*, \ldots, \bar{A}_{2n-3}^*, \bar{A}_{2n-2}^*$.

Thus, the index l of $\bar{A}_l^*$ indicates that $\bar{A}_l^*$ is a $(l+n-1)$-cycle of $(\bar{A}^* S^n, \bar{A}^{*0} S^n)$.

3.6.11 Definition. Let M be homeomorphic to S^n. This is equivalent to the existence of a homeomorphism $\phi: S^n \to M$ which is differentiable with the possible exception of one point. Then also exists a differentiable homotopy equivalence $\phi: S^n \to M$, cf. Hirsch [1]. Such a ϕ determines a mapping $\Lambda\phi: \Lambda S^n \to \Lambda M$, which commutes with the S^1 and $\mathbb{Z}_2$-actions. Thus in particular, $\Lambda\phi | \Lambda S^n$ induces mappings

$$(\bar{A}^* S^n, \bar{A}^{*0} S^n) \to (\bar{\Lambda} M, \bar{\Lambda}^{*0} M) \text{ and } (\Gamma S^n, \Gamma^0 S^n) \to (\Pi M, \Pi^0 M).$$

Let A_l be the $\mathbb{Z}_2 \times \mathbb{Z}_2$-invariant covering in ΛS^n of the element $\bar{A}_l^*$ in the sequence (†) of (3.6.10). Put $\Lambda\phi | A_l = a_l$. The ϕ-family $\mathscr{A}_l$ is now defined by $\{\phi_s a_l(A_l); s \geqslant 0\}$. The critical value of $\mathscr{A}_l$ shall be denoted by $\varkappa_l$.

3.6.12 Theorem. *Let M be homeomorphic to S^n. For the $g(n) = 2n - s - 1$ critical values $\varkappa_l$ of the ϕ-families $\mathscr{A}_l$, as defined in* (3.6.11), *we have the relations*

$$0 < \varkappa_0 \leqslant \ldots \leqslant \varkappa_{2n-2}$$

If, for a subsequent pair $\{l, l'\}$, equality holds, i. e., $\varkappa_l = \varkappa_{l'} =$ (briefly) $\varkappa$, then the number of critical S^1-orbits of E-value $\varkappa$ is infinite.

Proof. $0 < \varkappa_0$ follows as in the proof of (3.6.8) – observe that $\phi_s a_0: A_0 \to \Lambda M$ determines a homotopy equivalence $S^n \to M$. Note that the $(\mathbb{Z}_2 \times \mathbb{Z}_2)$-equivariant homology of $(\Lambda S^n, \Lambda^0 S^n)$ is the same as the homology of $(\bar{A}^* S^n, \bar{A}^{*0} S^n)$.

Assume $\varkappa_l = \varkappa_{l'} =$ (briefly) $\varkappa$. We derive a contradiction from the assumption that there are only finitely many critical S^1-orbits of E-value $\varkappa$. For that purpose we choose an open neighborhood $\mathscr{U}$ of $Cr\varkappa$, by taking the union of small tubular neighborhoods of the critical S^1-orbits in $Cr\varkappa$. $\mathscr{U}$ can be chosen invariant under the S^1- and $\mathbb{Z}_2$-actions, cf. the proof of (3.6.8). There exists an element $\phi_s a_{l'}(A_{l'}) \subset \mathscr{U} \cup \Lambda^{\varkappa-}$. We also write $a_{l'}$. The counter image of $a_{l'}$ of the open sets $\mathscr{U}$ and $\Lambda^{\varkappa-}$ in $A_{l'}$ shall be denoted by $\mathscr{O}$ and $\mathscr{N}$, respectively. $\{\mathscr{O}, \mathscr{N}\}$ is an open covering of the $(\mathbb{Z}_2 \times \mathbb{Z}_2)$-invariant chain $A_{l'}$, which is invariant under the $(\mathbb{Z}_2 \times \mathbb{Z}_2)$-action. Hence, $\{\bar{\mathscr{O}}^*, \bar{\mathscr{N}}^*\}$ is an open covering of the cycle $\bar{A}_{l'}^*$.

Since $\mathscr{U}$ contains no curve from some c to θc or to $\theta' c$, $\bar{\mathscr{O}}^*$ contains no non-trivial cycle homologous to the cycle $\bar{y}_{0,1}^*$ with $\pi\bar{y}_{0,1}^* = [0, 1]$. That is to say, $(\pi^*(0, 1) = u_1 + u_2) | \bar{\mathscr{O}}^* = 0$.

First consider the case $l' > 2s$, i. e., $\bar{A}_l^* = \bar{A}_{l'}^* \cap (u_1 + u_2)$. $\varkappa_l = \varkappa_{l'} = \varkappa$ implies $(\pi^*(0, 1) = u_1 + u_2) | \bar{\mathscr{O}}^* \neq 0$ – a contradiction.

We also may argue as follows: $\pi\mathscr{U}$ has the homotopy type of finitely many points. Hence, the mapping of any 1-cycle w_1 of $\pi\mathscr{O}$ into $\pi\mathscr{U}$, induced by $\pi \circ a_{l'} | \pi\mathscr{O}$, induces over w_1 only the trivial S^1-bundle (or $\mathbb{O}(2)$-bundle). The Stiefel-Whitney class (0, 1) of this bundle (cf. Steenrod [1] for this concept) therefore vanishes, i. e., $(\pi^*(0, 1) = u_1 + u_2) | \bar{\mathscr{O}}^* = 0$.

Assume $l' \leqslant 2s$, i.e., $\bar{A}_{l'}^* = \bar{A}_{l'}^* \cap (u_1 \cdot u_2)$. $\varkappa_l = \varkappa_{l'} = \varkappa$ implies $(\pi^*(1,1) = u_1 \cdot u_2) | \bar{\mathcal{O}}^* \neq 0$. On the other hand, since $\pi \mathcal{U}$ has the homotopy type of a finite number of points, the mapping of any 2-dimensional, simply connected cycle w_2 of $\pi\mathcal{O}$ (note that $(0,1)|\pi\mathcal{O} = 0$) induces over w_2 the trivial S^1-(or $\mathbb{O}(2)$-) bundle. That is to say, the Stiefel-Whitney class $(1,1)$ of this bundle (cf. Steenrod [1]) vanishes. That means $\pi^*(1,1)|\bar{\mathcal{O}}^* = 0$ – a contradiction. □

We conclude this section with an application of (3.6.12), just as we applied (3.6.8) to the proof of (3.6.9). This application is due to Thorbergsson [1]. His error in working with the space $(\Gamma S^n, \Gamma^0 S^n)$ instead of the space $(\bar{B}^* S^n, \bar{B}^{*0} S^n)$, as it must be, was corrected in Klingenberg [13].

At the basis of the desired application is a refinement of the constructions made for the proof of the Sphere Theorem (2.8.5) under the stronger hypothesis that M is a compact, simply connected (4/9)-pinched manifold. Choose $\tilde{p} \in M$. Then we know from (2.6.10) that the distance sphere $\partial \bar{B}_{\pi/2}(\tilde{p})$ of radius $\pi/2$ around $\tilde{p}$, cf. (1.10.6), is a submanifold N of M of codimension 1. Let $\sigma: N \to N$ be the antipodal mapping on N, i.e., $p = \exp_{\tilde{p}} X, |X| = \pi/2$, is mapped into $\exp_{\tilde{p}}(-X)$.

3.6.13 Lemma. *Let M be compact, simply connected such that the sectional curvature K satisfies $4/9 < K \leqslant 1$. With N and $\sigma: N \to N$ as above, choose $p \in N$ and put $\sigma(p) = q$. Consider*

$$\bar{B}(p) = \{r \in M; d(p,r) \leqslant d(\sigma(p), r)\}.$$

and $E(p,q) = \bar{B}(p) \cap \bar{B}(q)$. Then every geodesic of length π, emanating from p, meets $E(p,q)$ in exactly one point at distance $< \pi$.

Remark. Thus, for every $p \in N$, $E(p,q)$ can play the rôle of an equator on M, with $B(p)$ and $B(\sigma(p))$ as hemispheres, just as in the proof of (2.8.5). What we have gained here over the case $1/4 < K \leqslant 1$ is a full family of such equators and hemispheres on M, for all points p on an equatorlike submanifold N of M.

Proof. We first show that the points $r \in E(p,q) = \bar{B}(p) \cap \bar{B}(\sigma(p))$ have distance $< \pi$ from p; thus, they belong to the domain $B_\pi(p)$ of polar coordinates based at p, as well as to the domain $B_\pi(\sigma(p))$ of polar coordinates based at $\sigma(p) = q$.

Indeed, consider a triangle prq. According to (2.7.12), it has length $< 3\pi$, since $K \geqslant K_0 > 4/9$. Since $d(p, \sigma(p)) = \pi$, $d(p,r) = d(p, \sigma(r)) < \pi$.

Now if $c(t), 0 \leqslant t \leqslant \pi$, is a geodesic of length π with $c(0) = p$, the relation

$$d(p, c(t)) < d(\sigma(p), c(t))$$

for all $0 \leqslant t \leqslant \pi$ would yield a triangle $pr\sigma(p)$, $q = c(\pi)$, of length $\geqslant 3\pi$, which is impossible. Thus, for some $t_0 < \pi, c(t_0) \in E(p,q)$. The uniquenes of t_0 follows an in the proof of (2.8.5). □

3.6.14 Definition. On the sphere S^n of curvature $K = 1$ we define the set $\tilde{\Lambda} S^n \subset \Lambda S^n$

as follows. For each pair p, q of antipodal points consider the set

$$\tilde{A}(p, q) = \tilde{A}(q, p)$$

of parameterized closed curves of the following form. The curves begin in p (or q) and go via a half great circle to q (or p) and return again by a half great circle to p (or q). In addition, $\tilde{A}(p, q) = \tilde{A}(q, p)$ shall contain all curves which start at some interior point of a half great circle from p to q at distance $\delta \leqslant \pi/2$ from p (or from q) and move on that half great circle till distance δ from q (or from p) and return the same way to the inital point.

$\tilde{A}S^n$ shall be the union of all the $\tilde{A}(p, q) = \tilde{A}(q, p)$.

Note. $\tilde{A}S^n$ clearly is transformed into itself by the $(\mathbb{Z}_2 \times \mathbb{Z}_2)$-action with generators θ and θ'. Actually, each subset $\tilde{A}(p, q) = \tilde{A}(q, p)$ is $(\mathbb{Z}_2 \times \mathbb{Z}_2)$-invariant. Also, $\tilde{A}S^n$ contains the space BS^n of great circles. We know have the

3.6.15 Proposition. *The subspace* $(\tilde{A}S^n, \tilde{A}^0 S^n)$ *of* $(\Lambda S^n, \Lambda^0 S^n)$ *can* $(\mathbb{Z}_2 \times \mathbb{Z}_2)$-*equivariantly be deformed into the space* $(AS^n, A^0 S^n)$, *such as to keep the common subspace* BS^n *fixed. The deformation carries the subsets* $\tilde{A}(p, q) = \tilde{A}(q, p)$ *into the subsets* $A(p, q)$ *of circles which are parallel to a great circle starting at* p *or at* q.

Proof. Consider a fixed $\tilde{A}(p, q) = \tilde{A}(q, p)$. Each curve $\tilde{c}$ in this set meets for $t = 1/4$ and $t = 3/4$ the equator $E(p, q)$ of the antipodal pair (p, q) orthogonally. We deform $\tilde{c}$ into the circle c which meets $E(p, q)$ orthogonally at $t = 1/4$ and $t = 3/4$. Note that all $\tilde{c} \in \tilde{A}(p, q)$ which meet the equator just once thereby go into a point curve. □

We are now able to prove the application of (3.6.12) which we mentioned earlier.

3.6.16 Theorem. *Let* M *be compact, simply connected with* $4K_1/9 < K \leqslant K_1$, *some* $K_1 > 0$. *Then there exist on* M *at least* $g(n) = 2n - s - 1$ *different simple closed geodesics with length in the interval* $[2\pi/\sqrt{K_1}, 3\pi/\sqrt{K_1}[$.

Proof. We can assume $K_1 = 1$. Our hypothesis implies the existence of a homeomorphism $\phi: S^n \to M$ with the following properties: If $\tilde{p}_1 = (0, \dots, 0, 1)$ and $\tilde{q}_1 = (0, \dots, 0, -1) \in S^n$, ϕ is a diffeomorphism with the possible exception of $\tilde{q}_1$.. The equator $N_1 = E(\tilde{p}_1, \tilde{q}_1)$ of the antipodal pair $(\tilde{p}_1, \tilde{q}_1)$ is mapped into the distance sphere $N = \partial \bar{B}_{\pi/2}(\tilde{p})$ of radius $\pi/2$ around $\tilde{p} = \phi(\tilde{p}_1)$. $\phi | N_1 : N_1 \to N$ commutes with the antipodal mappings on N_1 and N.

This ϕ is constructed like the ϕ in the Sphere Theorem, cf. also the proof of (3.6.9). More precisely, choose $\tilde{p} \in M$ and fix an isometry between $T_{\tilde{p}_1} S^n$ and $T_{\tilde{p}} M$. With the help of the exponential mapping, we define $\phi: \bar{B}_{\pi/2}(\tilde{p}_1) \to \bar{B}_{\pi/2}(\tilde{p})$. ϕ then commutes with the antipodal mapping on the boundary spheres $N_1 = \partial \bar{B}_{\pi/2}(\tilde{p}_1)$ and $N = \partial \bar{B}_{\pi/2}(\tilde{p})$.

Let now $\tilde{q} \in M$ have maximal distance from $\tilde{p}$. For $r \in N$, consider a triangle $\tilde{p}\tilde{q}r$.

Since $4/9 < K$, $d(\tilde{p}, \tilde{q}) < 3\pi/2$. With $d(\tilde{p}, r) = \pi/2$ we have $d(\tilde{q}, r) < \pi$, cf. the corresponding estimate in the proof of (2.8.6) where we used (2.8.3). Therefore, N belongs to the domain $B_\pi(\tilde{q})$ of normal coordinates based at $\tilde{q}$. We thus can extend ϕ to $\complement \bar{B}_{\pi/2}(\tilde{p})$ and get a homeomorphism from S^n to M with the desired properties, except that ϕ on the equator N_1 and in $\tilde{q}_1$ in general will not be differentiable; more precisely, the image of a half great circle from $\tilde{p}_1$ to $\tilde{q}_1$ on S^n is a once broken geodesic from $\tilde{p}$ to $\tilde{q}$, with the break at N. It is possible to smooth out ϕ at N_1. For our subsequent constructions this is not really necessary, however.

We now define a mapping of the subset of $\tilde{\Lambda} S^n$, formed by the $\tilde{\Lambda}(p_1, q_1)$ with (p_1, q_1) on the equator N_1, into ΛM. It suffices to define the image of each half great circle from p_1 to q_1 such that the resulting map of $\tilde{\Lambda}(p_1, q_1)$ into ΛM is $(\mathbb{Z}_2 \times \mathbb{Z}_2)$-equivariant.

Put $\phi(p_1) = p$, $\phi(q_1) = q$. Then $\sigma(p) = q$. Let $c_1(t)$, $0 \leqslant t \leqslant \pi$, be a half great circle from p_1 to q_1. $c'(t) = \phi \circ c_1(t)$, $0 \leqslant t \leqslant \pi$, is a curve from p to q. We define the geodesic segment c^* from p to $E(p, q)$ as having initial direction $\dot{c}'(0)$ and going till the first intersection point r^* with the equator $E(p, q)$. Similarly, c^{**} shall be the geodesic starting from q with initial direction $-\dot{c}'(\pi)$ to the first intersection point r^{**} with the equator $E(p, q)$. Project the parts of $c' - \{p\}$ which lie on $\bar{B}(p)$ from p into $E(p, q)$ and project the parts of $c' - \{q\}$ which lie in $\bar{B}(q)$ from q into $E(q, p)$. Adding to the image the points r^*, r^{**}, we get a curve $\tilde{c}(t)$, $0 \leqslant t \leqslant \pi$, on $E(p, q)$ from r^* to r^{**}. Put $\tilde{c}(\pi/2) = r$. Now deform the segment c^* from p to r^* into the segment from p to r by moving the end point $r^* = \tilde{c}(0)$ along $\tilde{c}(t)$, $0 \leqslant t \leqslant \pi/2$. Similarly, deform c^{**} into the segment from q to r by moving the end point $r^{**} = \tilde{c}(\pi)$ along $\tilde{c}(\pi - t)$, $0 \leqslant t \leqslant \pi/2$. Now define as the image of c_1 the once broken geodesic from p to q passing through r, with parameter proportional to arc length.

It is seen at once that the above defined mapping of the geodesics from p_1 to q_1, p_1 and q_1 antipodal points on N_1, determines a continuous $(\mathbb{Z}_2 \times \mathbb{Z}_2)$-equivariant mapping of $\tilde{\Lambda}(p_1, q_1)$. This is clear not only for those curves of $\tilde{\Lambda}(p_1, q_1)$ which go from p_1 to q_1 but also for the remaining ones which run back and forth on subarcs of the half great circles from p_1 to q_1.

We thus get a mapping of a cycle, homotopic to the cycle $(\bar{Z}^*_{n-1,n-1}$, into $(\Lambda M, \Lambda^0 M)$, with all curves having length $< 4\pi$. That is, the image is in $\Lambda^{8\pi^2 -} M$. From (3.6.12) we get $g(n)$ critical values $< 8\pi^2$, i. e., $g(n)$ closed geodesics of length $< 4\pi$. According to (2.6.10), such geodesics are simple of length $\geqslant 2\pi$. (3.3.15) finally says that all these geodesics have length $< 3\pi$. □

Remark. Ballmann, Thorbergsson and Ziller [1] were able to carry out a construction similar to the one in (3.6.13) under the weaker assumption $1/4 < K \leqslant 1$. From (3.6.12) we then get again $g(n)$ simple closed geodesics.

3.7 The Theorem of the Three Closed Geodesics

In this section we will prove the celebrated Theorem of Lyusternik and Schnirelmann that on a surface of genus zero there always exist three simple closed geodesics.

The basic ideas for the proof were exhibited in Lyusternik and Schnirelmann [1]. A more detailed proof is contained in Lyusternik [1]. For a recent, quite complete proof along the lines of the original proof see Ballmann [1].

Our proof will differ in several details from the proof mentioned above; we first gave it in the Appendix of Klingenberg [10].

For our problem we cannot employ the methods developed in connection with the Hilbert manifold ΛM and the E-decreasing deformations $\phi_s: \Lambda M \to \Lambda M$, cf. chapter 2. The main difficulty with these deformations is that simple closed curves might thereby be carried into non-simple ones. Therefore, all proofs on the existence of simple closed geodesics use a deformation procedure which is different from the $(-\operatorname{grad} E)$-flow deformation.

At the basis of this procedure is a general classical deformation under which a closed curve is carried into a geodesic polygon with many short sides. These polygons, which are completely determined by their corners, are then deformed further by iterating this process.

The advantage of this method lies in the fact that the space of closed curves of energy less than some fixed constant $\varkappa$ is deformed into a finite dimensional manifold. This allows one to employ the methods and results of the Morse-Lyusternik-Schnirelmann theory for such finite dimensional manifolds – a theory which is considerably simpler than the theory of the Hilbert manifold ΛM with its $(-\operatorname{grad} E)$-flow. The disadvantage of this classical approach is that it does not take into account the natural $\mathbb{O}(2)$-action on the space of curves, as we described it in (2.5.A).

If one only wants the most basic existence theorems, the deformation method using geodesic polygons is very efficient and leads to results quickly.

For an excellent exposition of these methods, due originally to Lyusternik and Schnirelmann as well as Birkhoff and Morse, see Seifert and Threlfall [1]. A modern version is contained in Bott [2].

If one wants deeper results, the theory of the Hilbert manifold ΛM with its functional E and its Riemannian metric is indispensable. The source of the power of this approach, of which we developed the basic facts in chapter 2, lies in the fact that from the very beginning everything is developed in an intrinsic manner. This is in sharp contrast to the classical method of geodesic polygons where many arbitrary choices are involved.

In (3.7.3) we give the definition of the polygonal deformations on the space $\Lambda_\infty M$ of all piecewise differentiable curves on a compact Riemannian manifold M. In (3.7.11) we then describe the modification of this deformation (for the case when the curve is simple and lies on a surface of genus zero) so as to keep the curve simple, or at least no proper self-intersections are created. The proof of the existence of three simple closed geodesics on such a surface is then carried out along similar lines to the proof of (3.6.12).

We have made this section completely independent of everything related to the Hilbert manifold ΛM of closed H^1-curves, even if this occasionally forces us to repeat certain things we did earlier. In principle, this section can be read with only the most basic knowledge on local and semi-local Riemannian geometry, where the latter concerns mainly properties related to the injectivity radius. Nothing after (2.1) will be used here without deriving it anew for the problem at hand.

While preparing this section, I have profited a great deal from discussions with V. Bangert. In particular, he pointed out the necessity of introducing on the closure of the set of simple closed curves a metric which is finer than the usual one, cf. (3.7.10).

We begin with a general compact Riemannian manifold M. By $C^0(I, M)$ we denote the space of continuous mappings c of $I = [0, 1]$ into M, with the metric

$$d_\infty(c, c') = \sup_{0 \leqslant t \leqslant 1} d(c(t), c'(t)),$$

cf. (1.6.11) and (2.3.1). $C^0(I, M)$ is a complete metric space. It contains as closed subspace the space $C^0(S, M)$ of continuous closed curves $c: S = [0, 1]/\{0, 1\} \to M$:

By $C'^\infty(S, M)$ or $\Lambda_\infty M$ or simply Λ_∞, we denote the subspace of *piecewise differentiable curves*. The length L and the energy integral E are defined on $\Lambda_\infty M$:

$$L(c) = \int_S |\dot{c}(t)|\, dt; \quad E(c) = \frac{1}{2}\int_S |\dot{c}(t)|^2\, dt$$

These functions are related by

$$L(c) \leqslant \sqrt{2E(c)}.$$

Here equality holds if and only if c is parameterized proportionally to arc length, i. e., $L(c|[0, t]) = tL(c)$, cf. (1.8.6), (1.8.7).

For $\varkappa \geqslant 0$, we denote by $\Lambda_\infty^\varkappa$ or $\Lambda_\infty^\varkappa M$ the subspace of $\Lambda_\infty = \Lambda_\infty M$, formed by the elements c with $E(c) \leqslant \varkappa$.

From (2.1.10) we have the existence of a positive number $\iota(M)$ – the so-called *injectivity radius* – with the following property: Whenever two points p and q on M have distance $< \iota(M)$, then there exists a unique minimizing geodesic $c_{pq} = (c_{pq}(t), 0 \leqslant t \leqslant 1)$ from p to q of length $d(p, q)$.

3.7.1 Lemma. *Let $\{c_m\}$ be a sequence of piecewise differentiable paths $c_m: [0, 1] \to M$. Put $c_m(0) = p_m$, $c_m(1) = q_m$ and assume $d(p_m, q_m) \leqslant \iota(M)/2$, all m.*

If the sequences $\{E(c_m)\}$ and $\{d(p_m, q_m)^2/2\}$ both are convergent with the same limit then $\{c_m\}$ possesses a convergent subsequence with limit a geodesic segment $c: [0, 1] \to M$.

Proof. Since M is compact, there exists a subsequence of $\{c_m\}$ – which we denote again by $\{c_m\}$ – such that $\{p_m = c_m(0)\}$ and $\{q_m = c_m(1)\}$ are convergent with limit p and q, respectively. It follows from (1.6.12) that the sequence $\{c_{p_m q_m}\}$ of minimizing geodesics from p_m to q_m converges to the minimizing geodesic $c = c_{pq}$ from p to q.

Let $t_0 \in [0, 1]$. To complete the proof we show: For every sequence $\{t_m\}$ on $[0, 1]$ with $\lim t_m = t_0$, the sequence $\{c_m(t_m)\}$ converges to $c(t_0)$, cf. the remark after (1.6.11).

For that purpose we put $c_m(t_m) = r_m$. Denote by $\{r_{m(k)}; k = 1, \dots\}$ a convergent subsequence of $\{r_m\}$, with limit r. Consider the sequences

$$\{c_{p_{m(k)}r_{m(k)}}\}, \{c_{r_{m(k)}q_{m(k)}}\}$$

of the minimizing geodesic from $p_{m(k)}$ to $r_{m(k)}$, parameterized by $[0, t_{m(k)}]$, and the minimizing geodesic from $r_{m(k)}$ to $q_{m(k)}$, parameterized by $[t_{m(k)}, 1]$ respectively.

From our hypothesis we have

$$\begin{aligned} E(c_{pr}) + E(c_{rq}) &= \lim\big(E(c_{p_{m(k)}r_{m(k)}}) + E(c_{r_{m(k)}q_{m(k)}})\big) \\ &\leqslant \lim E(c_{m(k)}) = E(c_{pq}). \end{aligned}$$

Since c_{pq} is the unique minimizing curve from p to q and also c_{pr}, followed by c_{rq}, is a curve from p to q, we see that the latter curve coincides with c_{pq}. In particular, $r = c_{pq}(t_0)$. □

We now come to the classical definition of an E-decreasing deformation on subsets of $\Lambda_\infty = \Lambda_\infty M$ of the form $\Lambda_\infty^\varkappa$, some fixed $\varkappa > 0$. As an auxiliary tool we need the

3.7.2 Proposition. *Fix some* $\varkappa > 0$. *Let* k *be an even integer* > 0 *such that* $4\varkappa/k \leqslant \iota(M)^2/4$. *For every* $c \in \Lambda_\infty^\varkappa$ *and for every pair of points* t_0, t_1 *on* $S = [0, 1]/\{0, 1\}$ *with* $|t_0 - t_1| \leqslant 2/k$ *we then have* $d(c(t_0), c(t_1)) \leqslant \iota(M)/2$.

It follows that there exists a unique minimizing geodesic $c_{t_0t_1}: [t_0, t_1] \to M$ *from* $c(t_0)$ *to* $c(t_1)$. $E(c_{t_0t_1}) \leqslant E(c|[t_0, t_1])$.

Proof. From (1.8.7) we get $d(c(t_0), c(t_1))^2 \leqslant 2E(c|[t_0, t_1])|t_1 - t_0| \leqslant 2\varkappa \cdot 2/k \leqslant \iota(M)^2/4$. $d(c(t_0), c(t_1))^2 = 2E(c_{t_0t_1})|t_1 - t_0|$. □

3.7.3 Definition. Fix some $\varkappa > 0$ and choose an even integer $k > 0$ such as in (3.7.2). Let $\mathscr{D}_0 = id$. For σ, $j/k < \sigma \leqslant (j+2)/k$, with j an even integer, $0 \leqslant j \leqslant k-2$, we define

$$\mathscr{D}_\sigma: \Lambda_\infty^\varkappa M \to \Lambda_\infty^\varkappa M$$

by
$$\mathscr{D}_\sigma c(t) = \begin{cases} c(t), & \text{if } t \in \complement[j/k, \sigma] \\ c_{j/k,\sigma}(t), & \text{if } t \in [j/k, \sigma] \end{cases}.$$

Here, $\complement I$ denotes the complement of I on S.

For $1 + j/k < \sigma \leqslant 1 + (j+2)/k$, we define $\mathscr{D}_\sigma$ on $\Lambda_\infty^\varkappa M$ by

$$\mathscr{D}_\sigma c(t) = \begin{cases} c(t), & \text{if } t \in \complement[(j+1)/k, \sigma - 1 + 1/k] \\ c_{(j+1)/k, \sigma - 1 + 1/k}(t), & \text{if } t \in [(j+1)/k, \sigma - 1 + 1/k]. \end{cases}$$

Define now $\mathscr{D}(\sigma, \ )$, $0 \leqslant \sigma \leqslant 2$, to be the subsequent application of the deformations $\mathscr{D}_{2/k}, \dots, \mathscr{D}_{2l/k}, \mathscr{D}_\sigma$, with $2l$ the even integer determined by $2l \leqslant k\sigma < 2l + 2$.

3.7.4 Proposition. *Fix* $\varkappa > 0$. *Define a deformation* $\mathscr{D}(\sigma, \ ): \Lambda_\infty^\varkappa M \to \Lambda_\infty^\varkappa M$, $0 \leqslant \sigma \leqslant 2$, *as in* (3.7.3). *Then, for every* $c \in \Lambda_\infty^\varkappa M, \sigma \in [0, 2] \mapsto E(\mathscr{D}(\sigma, c)) \in \mathbb{R}$ *is a monotonic decreasing function. It is constant if and only if* c *is either a constant mapping or a closed geodesic.*

Proof. Let $\sigma_0 < \sigma_1$. Since $\mathscr{D}(\sigma_1, c)$ is obtained from $\mathscr{D}(\sigma_0, c)$ by replacing arcs with geodesics, the monotonicity follows from (3.7.2).

Assume now $E(\mathscr{D}(2, c)) = E(c)$. For c constant or a closed geodesic, this certainly is true. On the other hand, this equality implies that all restrictions of c to intervals of the form

$$[j/k, (j+2)/k] \text{ and } [(j+1)/k, (j+3)/k]$$

are geodesics, with parameter proportional to arc length if $c \neq$ const. But then, either $c(t) =$ const or else, c is a closed geodesic. □

We also have the following strong continuity property:

3.7.5 Lemma. *Choose* $\varkappa > 0$ *and define* $\mathscr{D}(\sigma, \ )$ *on* $\Lambda_\infty^\varkappa M$ *as in* (3.7.3). *Then the mapping*

$$\mathscr{D}: [0, 2] \times \Lambda_\infty^\varkappa M \to \Lambda_\infty^\varkappa M; (\sigma, c) \mapsto \mathscr{D}(\sigma, c)$$

is continuous.

Proof. Let $\{c_m\}$ be a convergent sequence in $\Lambda_\infty^\varkappa M$ with limit c and let $\{\sigma_m\}$ be a sequence on $[0, 2]$ with limit σ. Then we claim that $\{\mathscr{D}(\sigma_m, c_m)\}$ converges towards $\mathscr{D}(\sigma, c)$. Indeed, the partition points $c_m(j/k)$ converge to $c(j/k)$, and if m is large and hence σ_m is near σ, $\mathscr{D}(\sigma, c_m)$ and $\mathscr{D}(\sigma_m, c_m)$ will differ arbitrarily little. Finally, the sequence $\{\mathscr{D}(\sigma, c_m)\}$ converges to $\mathscr{D}(\sigma, c)$. □

The essential tool for the proof of the existence of closed geodesics now is the following Theorem. It is the analogue of condition (*C*) for the Hilbert manifold ΛM, cf. (2.4.9).

3.7.6 Theorem. *Fix* $\varkappa > 0$ *and define the deformation* $\mathscr{D}(\sigma, \ ): \Lambda_\infty^\varkappa M \to \Lambda_\infty^\varkappa M$ *as in* (3.7.3). *Instead of* $\mathscr{D}(2, c)$ *we also write* $\mathscr{D}c$.

Let $\{c_m\}$ *be a sequence in* $\Lambda_\infty^\varkappa M$ *such that* $\{E(c_m)\}$ *and* $\{E(\mathscr{D}c_m)\}$ *are both convergent with the same limit* $\varkappa_0 > 0$. *Then* $\{c_m\}$ *possesses a convergent subsequence with limit a closed geodesic* c_0 *of E-value* $\varkappa_0$.

Proof. $\{\mathscr{D}c_m\}$ is a sequence of closed curves, each of which is a geodesic polygon. Each side of such a polygon has length $\leqslant \iota(M)/2$. Hence, the polygon is completely determined by its $k/2$ corners. Since M is compact, $\{\mathscr{D}c_m\}$ has a convergent subsequence, which we again denote by $\{\mathscr{D}c_m\}$. Its limit curve is a geodesic polygon c_0. From (3.7.1) we can conclude that the sequence $\{c_m\}$ – possibly after going to another subsequence for which we keep the same notation – converges with its first segments $c_m | [0, 2/k]$ towards $c_0 | [0, 2/k]$. Further subsequences exist for which this is true for all other segments of c_m of the form $c_m|[j/k, (j+2)/k]$ or $c_m|[(j+1)/k, (j+3)/k]$. Finally, since $E(c_0) = \lim E(c_m) = \lim E(\mathscr{D} c_m)$, and since $\mathscr{D}$ is continuous, we have $E(\mathscr{D} c_0) = E(c_0)$. From (3.7.4) we then find that c_0 is a closed geodesic. □

We now can give an elementary proof of (2.4.20) which is essentially the original proof of Lyusternik and Fet [1]:

3.7.7 Theorem. *On every compact Riemannian manifold M there exists a closed geodesic.*

Proof. Assume first $\pi_1 M \neq 0$. Let c be a non-null homotopic closed curve. Consider the sequence $\{\mathscr{D}^m c\}$ of curves, all of which are homotopic to c. The decreasing sequence $\{E(\mathscr{D}^m c)\}$ has a limit $\varkappa_0 \geqslant 0$. But $\varkappa_0 > 0$ since a curve c^* with $E(c^*) < \iota(M)^2/8$, i.e., $L(c^*) \leqslant \sqrt{2E(c^*)} < \iota(M)/2$ lies entirely in the domain of normal coordinates based at $c^*(0)$. Such a c^* then is contractible. From (3.7.6) with $c_m = \mathscr{D}^m c$ we now have the existence of a closed geodesic c_0 of E-value $\varkappa_0$.

In the case $\pi_1 M = 0$ there exists an $f: S^{k+1} \to M$ which is not homotopic to a constant mapping, cf. Spanier [1]. As in the proof of (2.4.20), we associate with f a mapping

$$F = F(f): (\bar{B}^k, \partial \bar{B}^k) \to (\Lambda_\infty, \Lambda_\infty^0 M)$$

Choose $\varkappa > \sup E|\operatorname{im} F$. Define on $\Lambda_\infty^\varkappa$ the deformation $\mathscr{D}(\sigma, \;), 0 \leqslant \sigma \leqslant 2$, and put $\mathscr{D}(2, \;) = \mathscr{D}$. As we saw in the proof of (2.4.20), the deformed mapping $\mathscr{D}F$ of F determines a mapping $\mathscr{D}f$, homotopic to f, such that $F(\mathscr{D}f) = \mathscr{D}F(f)$.

Now consider the decreasing sequence $\{\max E|\operatorname{im} \mathscr{D}^m F\}$. Let $\varkappa_0$ be its limit. Then $\varkappa_0 > 0$. Indeed, otherwise we would have for large m

$$E|\operatorname{im} \mathscr{D}^m F < \iota(M)^2/8.$$

But this means that all curves of $\operatorname{im} \mathscr{D}^m F$ are shorter than $\iota(M)/2$ and therefore can be retracted continuously on their respective initial point. Then, the corresponding mapping $f(\mathscr{D}^m F): S^{k+1} \to M$ is homotopic to a mapping under which each circle $c_p(t), 0 \leqslant t \leqslant 1$, starting from the point p on the half equator $B^k \subset S^{k+1}$, is mapped on the point $\mathscr{D}^m F \circ c_p(0)$. That is, f is homotopic to a mapping where S^{k+1} is mapped on the image of $\bar{B}^k$. But that means that f is homotopic to a constant mapping – a contradiction.

Let now $\{c_m\}$ be a sequence with $c_m \in \operatorname{im} \mathscr{D}^m F, E(\mathscr{D} c_m) = \max E|\operatorname{im} \mathscr{D}^{m+1} F \geqslant \varkappa_0$. (3.7.6) then yields the existence of a closed geodesic c_0 with $E(c_0) = \varkappa_0$. □

Before we restrict ourselves to surfaces, we describe the canonical S^1-action and $\mathbb{Z}_2$-action on $\Lambda_\infty M$ which we introduced in (2.5.A) for the Hilbert manifold ΛM. Note that $\Lambda_\infty M$ is a subset of ΛM, invariant under these actions. In order to be independent of (2.5.A) we repeat the definition of these actions.

3.7.8 Definition. The S^1-*action* $\chi^\sim$ on $\Lambda_\infty M$ is defined by

$$(z, c) \in S^1 \times \Lambda_\infty M \mapsto z \, . \, c \in \Lambda_\infty M$$

with $(z \, . \, c)(t) = c(t + r)$, where $z = e^{2\pi i r}$.

The *involution*

$$\theta : \Lambda_\infty M \to \Lambda_\infty M$$

is defined by $(\theta c)(t) = c(1 - t)$.

3.7.9 Proposition. *The S^1-action $\chi^\sim$ is continuous. For fixed $z \in S^1$, the mapping*

$$\chi_z^\sim : \Lambda_\infty M \to \Lambda_\infty M ; c \mapsto z \, . \, c$$

is an isometry of the metric space $\Lambda_\infty M$, leaving the functions E and L invariant. The same is also true for the involution θ.

For all $c \in \Lambda_\infty M, z \, . \, \theta c = \theta \bar{z} \, . \, c$. In particular, $z = e^{i\pi}$ and θ commute. θ and $\theta' = e^{i\pi}\theta$ together define an isometric, E-preserving $\mathbb{Z}_2 \times \mathbb{Z}_2$-action on $\Lambda_\infty M$.

Proof. $d_\infty(z \, . \, c, z \, . \, c') = d_\infty(c, c')$ and $d_\infty(\theta c, \theta c') = d_\infty(c, c')$ are clear from the definitions, and so is $z \, . \, \theta = \theta \bar{z}$.

It remains to be shown that if $\{c_m\}$ is a sequence with limit c and $\{z_m = e^{2\pi i r_m}\}$ is a sequence with limit $1 = e^{2\pi i 0}$, then $\{z_m \, . \, c_m\}$ converges towards c. But this follows from the relations

$$d_\infty(c, z_m \, . \, c_m) \leqslant d_\infty(c, z_m \, . \, c) + d_\infty(z_m \, . \, c, z_m \, . \, c_m)$$

where the last term can be replaced by $d_\infty(c, c_m)$. □

From now on we denote by M a *surface of genus zero*. That is to say, there exists a diffeomorphism $f : S^2 \to M$ of the standard sphere $S^2 = \{\sum_0^2 x_i^2 = 1\}$ in $\mathbb{R}^3$ onto M.

With the help of f, the image $(c(t) = f \circ c_0(t))$ of a simple closed curve $(c_0(t))$ on S^2 will be a *simple closed curve* on M. Recall that a simple closed curve on M is an embedding of $S = [0, 1]/\{0, 1\}$. This is true in particular if we take for c_0 a circle on S^2. If we now apply to a simple closed curve $c = (c(t))$ on M the deformation $\mathscr{D}(\sigma, c)$, it may happen that the curve ceases to be simple. We therefore introduce a modification $\tilde{\mathscr{D}}_\sigma$ of the deformation $\mathscr{D}_\sigma$ which keeps a simple curve simple – or almost simple, see (3.7.11) below.

We wish to extend the space of simple closed curves on M by certain curves which are 'almost simple' in the sense that they are limits of simple closed curves and the only time where there are multiple points, these are geodesic arcs. Moreover, we want to preserve for such a non-simple curve a definite way in which it is approximated by simple curves.

To make this latter point clearer consider a curve c with $c(1 - t) = c(t)$ where $c | [0, 1/2]$ is a short non-constant geodesic segment. Let (u, v) be Fermi coordinates based on $c | [0, 1/2]$ such that $u \circ c(t) = t, v \circ c(t) = 0, 0 \leqslant t \leqslant 1/2$. Now, c can be viewed as limit of a sequence $\{c_l\}$ of simple curves where $v \circ c_l(t) > 0$ for $0 < t < 1/2$ and $v \circ c_l(t) < 0$ for $1/2 < t < 1$. But we may also view c as the limit of the sequence

$\{\theta c_l\}$ of oppositely oriented simple curves. The point now is that, while $d_\infty(c_l, \theta c_l)$ goes to 0 for $l \to \infty$, in the space of simple closed curves c_l and θc_l do not get arbitrarily close.

This leads us to the following

3.7.10 Definition. (i) Denote by $\mathring{\Lambda}_\infty M$, or simply $\mathring{\Lambda}_\infty$, the space of simple closed curves on M, considered as a subset of the space $\Lambda_\infty M$.

(ii) A curve in $\mathring{\Lambda}_\infty$ from c_0 to c_1 is defined as a continuous mapping $F: S \times [0, 1] \to M$ such that, for each $\tau \in [0, 1]$, $c_\tau \equiv F|S \times \{\tau\}$ is an element of $\mathring{\Lambda}_\infty$ and such that there exists a subdivision of $[0, 1]$ into intervals I with the property that $F|S \times I$ is differentiable.

(iii) The length $L(F)$ of such a curve is defined as $\sup_{t \in S} L(F|\{t\} \times [0, 1])$.

(iv) The distance $\tilde{d}_\infty(c_0, c_1)$ is defined as the infimum of the length of curves from c_0 to c_1 in $\mathring{\Lambda}_\infty$.

(v) We extend $\mathring{\Lambda}_\infty$ by elements c which are defined as follows:

(a) c stands for an equivalence class of Cauchy sequences of $\{\mathring{\Lambda}_\infty, \tilde{d}_\infty\}$, where two such sequences $\{c_m\}, \{c'_m\}$ are called equivalent if $\{\tilde{d}_\infty(c_m, c'_m)\}$ converges to zero.

(b) c, when considered as continuous mapping $c: S \to M$, allows locally a description of the following form: There exist subintervals I of S and Fermi coordinates (u, v) (i. e., $\{u, 0\}$ is a geodesic) such that, for $t \in I, u \circ c(t) = t, v \circ c(t) = \max\{0, v^*(t)\}$ where $v^*: I \to \mathbb{R}$ is piecewise differentiable. We also allow curves which are obtained by finitely many iterations of this procedure.

(vi) The extension of $\mathring{\Lambda}_\infty$ by the elements c defined in (v) is denoted by $\tilde{\Lambda}_\infty$.

Remarks. For this definition to make sense we have assumed a few simple facts which can easily be verified. For example, we have used that $\tilde{d}_\infty$ defines a distance on $\mathring{\Lambda}_\infty$.

Clearly, $d_\infty(c, c') \leqslant \tilde{d}_\infty(c, c')$; If c, c' are sufficiently close to each other in the $\tilde{d}$-metric, equality holds.

Note that an element in $\tilde{\Lambda}_\infty - \mathring{\Lambda}_\infty$ is not just a closed curve. And the underlying closed curve need not be piecewise differentiable, since $(t, v^*(t)), t \in I$, can have an infinite number of t-values where $v^*(t)$ changes its sign. Still, the modified curve $(t, \sup(0, v^*(t)))$ in a certain sense is 'better' since general arcs are replaced by geodesic arcs. In particular, our 'curves' always possess length and energy.

We now come to the definition of the deformation $\tilde{\mathscr{D}}_\sigma$ of the space $\tilde{\Lambda}_\infty M$ into itself.

3.7.11 Definition. Fix a $\varkappa > 0$ and choose $k > 0$ even such as in (3.7.2). Let j be even, $0 \leqslant j \leqslant k-2$. For c a simple closed curve we then define $\tilde{\mathscr{D}}_0 c = c$ and $\tilde{\mathscr{D}}_\sigma c, j/k < \sigma \leqslant (j+2)/k$, as follows:

For small $\sigma - j/k$, $\mathscr{D}_\sigma c$ as defined in (3.7.3) will be simple. However, there might be a $\sigma_0, j/k < \sigma_0 < (j+2)/k$, where $\mathscr{D}_\sigma c$ ceases to be simple. That is to say, $\mathscr{D}_{\sigma_0} c|[j/k, \sigma_0]$ and its complement $\mathscr{D}_{\sigma_0} c|\complement[j/k, \sigma_0]$ have a point in common.

In this case we modify the complement by 'sweeping aside' those parts which come to lie on the 'wrong' side of the geodesic segment $\mathscr{D}_\sigma c|[j/k, \sigma]$, such as to cause self-intersections. More precisely, we consider polar coordinates based at $c(j/k)$. Whenever the radial geodesic from $c(j/k)$ to $c(\sigma)$ meets a point on the complement we keep the radial coordinate of that point constant and change the angular coordinate so as to avoid proper intersections as σ increases. Only in the case (and this can happen only after modifications of this type have taken place) when a point of the complement has the same angular and the same radial coordinate as $c(\sigma)$, do we also push away the radial coordinate of the point on the complement so as to avoid self-intersections. Again, this shall be done only in so far as it is necessary for our purpose.

While modifying $\mathscr{D}_\sigma c$ on $\complement[j/k, \sigma]$ in this manner, we view it as element of $\tilde{\Lambda}_\infty M$ by describing it as the limit of a sequence of simple closed curves in the $\tilde{d}_\infty$-metric.

Whenever, with further increase of σ, the geodesic segment from $c(j/k)$ to $c(\sigma)$ changes the sense in which it moves we allow the points on the complement to move back in the direction of their original position as far as possible without leaving the set $\tilde{\Lambda}_\infty M$. In this way, the complementary arc recaptures part or possibly all of its original position.

In particular, for the final value $\sigma = (j+2)/k$, the deformed curve $\tilde{\mathscr{D}}_\sigma c$, when restricted to $[j/k, (j+2)/k]$, is the minimizing geodesic from $c(j/k)$ to $c((j+2)/k)$. Thus, $\tilde{\mathscr{D}}_{(j+2)/k} c|[j/k, (j+2)/k]$ and $\mathscr{D}_{(j+2)/k} c|[j/k, (j+2)/k]$ coincide, while on $\complement[j/k, (j+2)/k]$, $\mathscr{D}_{(j+2)/k} c$ and $\tilde{\mathscr{D}}_{(j+2)/k} c$ can be different in so far as in the latter case some arcs have been replaced by arcs lying on the minimizing geodesic from $c(j/k)$ to $c((j+2)/k)$.

Now if c is the limit of a $\tilde{d}_\infty$-Cauchy sequence $\{c_m\}$ in $\mathring{\tilde{\Lambda}}_\infty M$ we define $\tilde{\mathscr{D}}_\sigma c$, for $j/k < \sigma \leqslant (j+2)/k$ as the limit of the $\tilde{d}_\infty$-Cauchy sequence $\{\tilde{\mathscr{D}}_\sigma c_m\}$.

In the same way we define $\tilde{\mathscr{D}}_\sigma c$ for $1 + j/k < \sigma \leqslant 1 + (j+2)/k$, cf. (1.7.3).

Finally, $\tilde{\mathscr{D}}(\sigma, c), 0 \leqslant \sigma \leqslant 2$, is defined as the subsequent application of $\tilde{\mathscr{D}}_{2/k}, \ldots, \tilde{\mathscr{D}}_{2l/k}, \tilde{\mathscr{D}}_\sigma$ where $2l$ is the even integer determined by $2l \leqslant k\sigma < 2l+2$.

Remark. In this definition we have used the fact that with a $\tilde{d}_\infty$-Cauchy sequence $\{c_m\}$ in $\mathring{\tilde{\Lambda}}_\infty M$, the sequence $\{\tilde{\mathscr{D}}_\sigma c_m\}$ is also a $\tilde{d}_\infty$-Cauchy sequence. This is an immediate consequence of the continuity of the operation $\mathscr{D}_\sigma$ and of the continuity of the modifications by which we let $\tilde{\mathscr{D}}_\sigma$ differ from $\mathscr{D}_\sigma$.

In analogy with (3.7.4) we have the

3.7.12 Proposition. *Fix $\varkappa > 0$ and define*

$$\tilde{\mathscr{D}}(\sigma, \ ): \tilde{\Lambda}^\varkappa_\infty M \to \tilde{\Lambda}^\varkappa_\infty M, 0 \leqslant \sigma \leqslant 2,$$

as in (3.7.11). *Then, for every $c \in \tilde{\Lambda}^\varkappa_\infty M$, $E(\tilde{\mathscr{D}}(\sigma, c)) \leqslant E(c)$ and $E(\tilde{\mathscr{D}}(2, c)) = E(c)$ can hold if and only if c is constant or c is a simple closed geodesic.*

Proof. The continuity of $\tilde{\mathscr{D}}$ is a consequence of the continuity of $\mathscr{D}$. Clearly, $E(\tilde{\mathscr{D}}(\sigma, c)) \leqslant E(\mathscr{D}(\sigma, c)) \leqslant E(c)$. Finally, if $\tilde{\mathscr{D}}\, c$ and c have the same E-value, we find from (3.7.4) that $c = \text{const}$ or c a closed geodesic. The only closed geodesics in $\tilde{\Lambda}_\infty M$ are the simple ones. □

We can now prove the analogue of (3.7.5):

3.7.13 Lemma. *Choose $\varkappa > 0$ and define $\tilde{\mathscr{D}}(\sigma, \ )$ on $\tilde{\Lambda}_\infty^\varkappa M$ as in* (3.7.11). *Then the mapping*

$$\tilde{\mathscr{D}} : [0, 2] \times \tilde{\Lambda}_\infty^\varkappa M \to \tilde{\Lambda}_\infty^\varkappa M$$

is continuous.

Proof. Let $\{c_m\}$ be a sequence in $\tilde{\Lambda}_\infty^\varkappa$ which converges in the $\tilde{d}_\infty$-distance to c and let $\{\sigma_m\}$ be a sequence on $[0, 2]$ which converges towards σ. Then we have to show that $\{\tilde{\mathscr{D}}(\sigma_m, c_m)\}$ converges in the $\tilde{d}_\infty$-distance to $\tilde{\mathscr{D}}(\sigma, c)$.

Now, as we just saw, $\{\tilde{\mathscr{D}}(\sigma_{m_0}, c_m)\}$ converges to $\tilde{\mathscr{D}}(\sigma_{m_0}, c)$, for each fixed m_0. And obviously, $\{\tilde{\mathscr{D}}(\sigma_m, c)\}$ converges to $\tilde{\mathscr{D}}(\sigma, c)$. □

The analogue of the fundamental Theorem (3.7.6) now reads:

3.7.14 Theorem. *Fix $\varkappa > 0$ and define*

$$\tilde{\mathscr{D}}(\sigma, \ ) : \tilde{\Lambda} M \to \tilde{\Lambda}^\varkappa M, 0 \leqslant \sigma \leqslant 2,$$

as in (3.7.11). *Instead of $\tilde{\mathscr{D}}(2, c)$ we also write $\tilde{\mathscr{D}} c$.*

Let $\{c_m\}$ be a sequence of $\tilde{\Lambda}_\infty^\varkappa M$ such that $\{E(c_m)\}$ and $\{E(\tilde{\mathscr{D}} c_m)\}$ are both convergent with the same limit $\varkappa_0 > 0$. Then $\{c_m\}$ possesses a convergent subsequence with limit a simple closed geodesic c of E-value $\varkappa_0$.

Proof. Proceed exactly as in the proof of (3.7.6). Note that according to (3.7.12) our sequence satisfies the hypotheses made in (3.7.6). From (3.7.12) we then find that the limit geodesic c is simple. □

We will later need a certain analogue of (2.4.16), (2.4.18):

3.7.15 Lemma. *Denote by $\tilde{C}r\tilde{\varkappa}$ the set of simple closed geodesics of E-value $\tilde{\varkappa} > 0$. For some $\varkappa > \tilde{\varkappa}$, define $\tilde{\mathscr{D}}(\sigma, \ ), 0 \leqslant \sigma \leqslant 2$, on $\tilde{\Lambda}_\infty^\varkappa M$ as in* (3.7.11). *Instead of $\tilde{\mathscr{D}}(2, c)$ also write $\tilde{\mathscr{D}} c$.*

Choose an open neighborhood $\tilde{\mathscr{U}}$ of $\tilde{C}r\tilde{\varkappa}$ in $\tilde{\Lambda}^\varkappa M$. Then there exists a small $\varepsilon = \varepsilon(\tilde{\mathscr{U}}) > 0, \tilde{\varkappa} + \varepsilon < \varkappa$, such that $\tilde{\mathscr{D}}$ maps $\tilde{\Lambda}_\infty^{\tilde{\varkappa}+\varepsilon}$ into $\tilde{\mathscr{U}} \cup \tilde{\Lambda}_\infty^{\tilde{\varkappa}-\varepsilon}$.

Proof. If there were no such $\varepsilon = \varepsilon(\tilde{\mathscr{U}}) > 0$, it would mean the existence of a sequence $\{c_m\}$, with $\lim E(c_m) = \lim E(\tilde{\mathscr{D}} c_m) = \tilde{\varkappa}$, $\tilde{\mathscr{D}} c_m \notin \tilde{\mathscr{U}}$. But this is impossible since we know from (3.7.14) that such a sequence has a convergent subsequence with limit in $\tilde{C}r\tilde{\varkappa}$. □

3.7.16 Definition. We denote by BS^2 the space of parameterized great circles on S^2. As a homogeneous space, BS^2 is of the form $\mathbb{O}(3)/S\mathbb{O}(1) \times S\mathbb{O}(1) \times \mathbb{O}(1)$.

By $AS^2 - A^0 S^2$ we denote the space of all non-constant parameterized circles on S^2. These are the embeddings $c : S \to S^2$, with parameter proportional to arc length, for which the image lies in the intersection of S^2 with a 2-plane of $\mathbb{R}^3$ having distance <1 from $0 \in \mathbb{R}^3$.

By $\alpha: AS^2 - A^0S^2 \to BS^2$ we denote the D^1-bundle where the fibre over $c^* \in BS^2$ consists of those circles which have their mid-point on the line orthogonal to the 2-plane carrying c^*.

If we complete $AS^2 - A^0S^2$ by the set A^0S^2 of piont circles we obtain the space AS^2.

The $\mathbb{O}(2)$-action on $\Lambda_\infty S^2$, generated by the S^1-action and the involution θ (cf. (3.7.8)) carries the subspaces AS^2, BS^2, A^0S^2 into itself. This action also commutes with the mapping α.

Besides the full group $\mathbb{O}(2)$, we consider the subgroup $\mathbb{Z}_2 \times \mathbb{Z}_2$, generated by θ and $\theta' = e^{i\pi}\theta = \theta e^{i\pi}$. This yields the following quotient spaces:

$$\bar{B}^*S^2 = BS^2/\mathbb{Z}_2 \times \mathbb{Z}_2 \cong \mathbb{O}(3)/\mathbb{O}(1) \times \mathbb{O}(1) \times \mathbb{O}(1)$$

$$\Delta S^2 = BS^2/\mathbb{O}(2) \cong \mathbb{O}(3)/\mathbb{O}(2) \times \mathbb{O}(1)$$

Putting $(AS^2, A^0S^2)/(\mathbb{Z}_2 \times \mathbb{Z}_2) = (\bar{A}^*S^2, \bar{A}^{*0}S^2), (AS^2, A^0S^2)/\mathbb{O}(2) = (\Gamma S^2, \Gamma^0 S^2)$

we obtain the D^1-bundles

$$\bar{\alpha}^*: \bar{A}^*S^2 - \bar{A}^{*0}S^2 \to \bar{B}^*S^2$$

$$\gamma: \Gamma S^2 - \Gamma^0 S^2 \to \Delta S^2.$$

The space ΔS^2 of unparameterized great circles is isomorphic to the real projective plane P^2: Simply associate with an unparameterized great circle the line through $0 \in \mathbb{R}^3$ which is orthogonal to the plane carrying this great circle. This shows that the $\mathbb{Z}_2$-cohomology ring of ΔS^2 has a multiplicative generator in dimension 1.

The three homology classes of ΔS^2 can be represented by the following three cycles $\Gamma_{j,k}, 0 \leqslant j \leqslant k \leqslant 1$: $\Gamma_{0,0}$ is the great circle in the plane $\{x_2 = 0\}$. $\Gamma_{0,1}$ consists of the great circles passing through $(\pm 1, 0, 0)$ and $\Gamma_{1,1}$ is formed by all great circles on S^2.

Note. While Lusternik and Schnirelmann [1] work with the bundle γ, we will use the bundle $\bar{\alpha}^*$. Once and for all we agree to identify $(\mathbb{Z}_2 \times \mathbb{Z}_2)$-invariant sets in AS^2 with their images in $\bar{A}^*S^2$. Since $\mathbb{Z}_2 \times \mathbb{Z}_2$ acts freely on $AS^2 - A^0S^2$, the $(\mathbb{Z}_2 \times \mathbb{Z}_2)$-equivariant homology of (AS^2, A^0S^2) coincides with the homology of $(\bar{A}^*S^2, \bar{A}^{*0}S^2)$.

3.7.17 Definition. We define $\bar{B}^*_{j,k}, 0 \leqslant j \leqslant k \leqslant 1$, as follows: $\bar{B}^*_{0,0}$ consists of the four great circles in $S^2 \cap \{x_2 = 0\}$, starting from $(\pm 1, 0, 0)$. $\bar{B}^*_{0,1}$ is formed by the great circles on S^2, starting from $(\pm 1, 0, 0)$, and $\bar{B}^*_{1,1}$ is formed by the great circles on S^2 starting from a point on $S^2 \cap \{x_2 = 0\}$.

The counter image of $\bar{B}^*_{j,k}$ under $\bar{\alpha}^*$ is denoted by $\bar{A}^*_{j,k} - \bar{A}^{*0}_{j,k}$. Thus, we obtain the D^1-bundle

$$\bar{\alpha}^*_{j,k}: \bar{A}^*_{j,k} - \bar{A}^{*0}_{j,k} \to \bar{B}^*_{j,k}.$$

When taking the closures $\bar{D}^1$ of D^1 in each fibre we obtain a $\bar{D}^1$-bundle. Denote by $\bar{A}^*_{j,k}$ the total space of this bundle. Put $\partial \bar{A}^*_{j,k} = \bar{A}^{*0}_{j,k}$. Thus, $(\bar{A}^*_{j,k}, \bar{A}^{*0}_{j,k})$ is a relative $(j+k+1)$-dimensional cycle of $(\bar{A}^* S^2, \bar{A}^{*0} S^2)$.

The following proposition gives some idea of the $\mathbb{Z}_2$-homology of the space $(\bar{A}^*_{1,1}, \bar{A}^{*0}_{1,1})$. A similar result holds for $(\Gamma S^2, \Gamma^0 S^2)$.

3.7.18 Proposition. (i) *The* 1-*cycle* $(\bar{A}^*_{0,0}, \bar{A}^{*0}_{0,0})$ *is homologous to* $\bar{B}^*_{0,1}$.

(ii) *The* 2-*cycle* $(\bar{A}^*_{0,1}, \bar{A}^{*0}_{0,1})$ *is homologous to* $\bar{B}^*_{1,1}$.

Proof. Define a mapping

$$h: [0, 1] \times [0, \pi] \times [0, \pi] \to \bar{A}^* S^2$$

as follows: $h(r, 0, 0) =$ circle parallel to the (x_0, x_2)-plane with mid-point $(0, r, 0)$.

$h(r, \phi, 0) = \sigma_\phi h(r, 0, 0)$, where σ_ϕ is the positive rotation by the angle ϕ around the x_0-axis.

$h(r, \phi, \psi) = \tau_\psi h(r, \phi, 0)$, where τ_ψ is the positive rotation by the angle ψ around the x_2-axis.

We put $h|[0, 1] \times [0, \pi] \times \{0\} = g$. Then, with $\mathbb{Z}_2$-coefficients, $\partial g = g|\{0\} \times [0, \pi] + g|[0, 1] \times \{0, \pi\}$ mod $\bar{A}^{*0} S^2$. The first summand yields the cycle $\bar{B}^*_{0,1}$, while the second yields $(\bar{A}^*_{0,0}, \bar{A}^{*0}_{0,0})$.

Similarly, modulo $\bar{A}^{*0} S^2, \partial h = h|\{0\} \times [0, \pi] \times [0, \pi] + h|[0, 1] \times \{0, \pi\} \times [0, \pi] + h|[0, 1] \times [0, \pi] \times \{0, \pi\}$. The first summand yields $\bar{B}^*_{1,1}$. The second summand yields $(\bar{A}^*_{0,1}, \bar{A}^{*0}_{0,1})$, modulo a rotation, while the third summand yields twice the chain g and thus is zero mod 2. □

We use the cycles $(\bar{A}^*_{j,k}, \bar{A}^{*0}_{j,k})$ to define certain families of sets in $(\tilde{\Lambda}_\infty M, \tilde{\Lambda}^0_\infty M)$.

3.7.19 Definition. Choose a diffeomorphism $f: S^2 \to M$. Fix a pair (j, k) and write briefly $(\bar{A}^*, \bar{A}^{*0})$ instead of $(\bar{A}^*_{j,k}, \bar{A}^{*0}_{j,k})$.

(i) We denote by $(\bar{A}^*, \bar{A}^{*0})$ also any $(\mathbb{Z}_2 \times \mathbb{Z}_2)$-invariant $\mathbb{Z}_2$-cycle in $(\bar{A}^* S^2, \bar{A}^{*0} S^2)$ which ist $(\mathbb{Z}_2 \times \mathbb{Z}_2)$-equivariantly homologous to $(\bar{A}^*, \bar{A}^{*0})$.

(ii) By

$$\tilde{u}: (\bar{A}^*, \bar{A}^{*0}) \to (\tilde{\Lambda}_\infty M, \tilde{\Lambda}^0_\infty M)$$

we denote any $(\mathbb{Z}_2 \times \mathbb{Z}_2)$-equivariant mapping which is $(\mathbb{Z}_2 \times \mathbb{Z}_2)$-equivariantly homotopic to the mapping $u_0: (\bar{A}^*, \bar{A}^{*0}) \to (\tilde{\Lambda}_\infty M, \tilde{\Lambda}^0_\infty M)$, induced by $f: S^2 \to M$.

(iii) Denote by $\tilde{\mathscr{A}}$ the family of compact subsets of $\tilde{\Lambda}_\infty M$ of the form $\tilde{u}(\bar{A}^*)$.

(iv) Define the critical value $\tilde{\varkappa} = \tilde{\varkappa}_{\tilde{\mathscr{A}}}$ of the family $\tilde{\mathscr{A}}$ by

$$\tilde{\varkappa} = \inf_{\tilde{A} \in \tilde{\mathscr{A}}} \sup E|\tilde{A}.$$

Remark. The families $\tilde{\mathscr{A}}$ which we have defined in (3.7.19) carry a certain resemblance to the ϕ-families defined in (3.6), for the case $n = 2$.

However, one can show by examples that the $(-\operatorname{grad} E)$-deformations ϕ_s do not always carry a simple closed curve into such a curve. Therefore, we have to replace the deformations ϕ_s by the deformations $\tilde{\mathscr{D}}(\sigma,\ )$ defined in (3.7.11). These deformations are not defined $(\mathbb{Z}_2 \times \mathbb{Z}_2)$-equivariantly, however, even when restricted to a set $\tilde{A} \in \tilde{\mathscr{A}}$. We therefore introduce the following modification of $\tilde{\mathscr{D}}(\sigma,\ )$:

3.7.20 Definition. Consider a covering of the space AS^2 of circles by two finite families $(C_i)_{1 \leqslant i \leqslant m}$ and $(D_i)_{1 \leqslant i \leqslant m}$ of open sets with $C_i \supset \bar{D}_i$ such that the projection mapping by the $(\mathbb{Z}_2 \times \mathbb{Z}_2)$-action,

$$AS^2 \to \bar{A}^* S^2,$$

when restricted to C_i, is a diffeomorphism onto its image $\bar{C}_i^*$, for all i, $1 \leqslant i \leqslant m$. Such coverings obviously exist.

Let $\psi_i: AS^2 \to \mathbb{R}$ be smooth functions with values in the interval $[0, 1]$ such that $\psi_i | D_i = 1$ and $\psi_i | \complement C_i = 0$, $1 \leqslant i \leqslant m$.

For any $(\mathbb{Z}_2 \times \mathbb{Z}_2)$-equivariant mapping

$$\tilde{u}: (AS^2, A^0 S^2) \to (\tilde{\Lambda}_\infty M, \tilde{\Lambda}^0_\infty M)$$

we now define an E-decreasing homotopy

$$\tilde{\mathscr{D}}_{\tilde{u}}: [0, 2] \times (AS^2, A^0 S^2) \to (\tilde{\Lambda}_\infty M, \tilde{\Lambda}^0_\infty M):$$

First choose $\varkappa > 0$ so large that $\operatorname{im} \tilde{u} \subset \tilde{\Lambda}^\varkappa$. Consider a mapping

$$\tilde{\mathscr{D}}: [0, 2] \times (\tilde{\Lambda}^\varkappa_\infty M, \tilde{\Lambda}^0_\infty M) \to (\tilde{\Lambda}^\varkappa_\infty M, \tilde{\Lambda}^0_\infty M)$$

as in (3.7.11). Define now $\tilde{\mathscr{D}}_{\tilde{u}}$ as follows. $\tilde{\mathscr{D}}_{\tilde{u}}(0,\ ) = id$. If $\sigma \in](2i-2)/m, 2i/m]$ we put

$$\tilde{\mathscr{D}}_{\tilde{u}}(\sigma, c_0) = \begin{cases} \tilde{\mathscr{D}}\big((m\sigma - 2i)\psi_i(c_0), \tilde{\mathscr{D}}_{\tilde{u}}((2i-2)/m, c_0)\big), & \text{if } c_0 \in C_i, \\ \theta \tilde{\mathscr{D}}_{\tilde{u}}(\sigma, c_0), & \text{if } \theta c_0 \in C_i, \\ e^{i\pi} . \tilde{\mathscr{D}}_{\tilde{u}}(\sigma, c_0), & \text{if } e^{i\pi} . c_0 \in C_i, \\ \theta e^{i\pi} . \tilde{\mathscr{D}}_{\tilde{u}}(\sigma, c_0), & \text{if } \theta e^{i\pi} . c_0 \in C_i, \\ \tilde{\mathscr{D}}_{\tilde{u}}\big((2i-2)/m, c_0\big), & \text{otherwise.} \end{cases}$$

From the properties (3.7.13), (3.7.14) of $\tilde{\mathscr{D}}(\sigma,\ )$ we immediately obtain the following analogous statement for $\tilde{\mathscr{D}}_{\tilde{u}}(\sigma,\ )$:

3.7.21 Lemma. *Let $\tilde{u}$ and $\tilde{\mathscr{D}}_{\tilde{u}}(\sigma,\ )$, $\sigma \in [0, 2]$, be as in (3.7.20). Instead of $\tilde{\mathscr{D}}_{\tilde{u}}(2, c_0)$ we write $\tilde{\mathscr{D}}_{\tilde{u}} c_0$. Then*

(i) *$\tilde{\mathscr{D}}_{\tilde{u}}(\sigma, c_0)$ is continuous in both arguments.*

(ii) *If $\tilde{\mathscr{D}}_{\tilde{u}} c_0 = \tilde{u}(c_0)$ and $E(\tilde{u}(c_0)) = \tilde{\varkappa} > 0$ then $\tilde{u}(c_0)$ in a simple closed geodesic.*

Proof. Observe that for every $c_0 \in AS^2$ there exists at least one i with $c_0 \in D_i$, thus, $\psi_i(c_i) = 1$. Then (i) and (ii) are immediate consequences of (3.7.13) and (3.7.12). □

We now can prove the analogue of (2.4.18). The main tool for the proof is (3.7.15).

3.7.22 Theorem. *Let $\tilde{\varkappa}$ bw the critical value of one of the families $\tilde{\mathscr{A}}$ defined in* (3.7.19). *Assume $\tilde{\varkappa} > 0$. Denote by $\tilde{C}r\tilde{\varkappa}$ the set of simple closed geodesics of E-value $\tilde{\varkappa}$. Let $\tilde{\mathscr{U}}$ be an open $\tilde{d}_\infty$-neighborhood of $\tilde{C}r\tilde{\varkappa}$ in $\tilde{\Lambda}_\infty M$. Then there exists a $\varepsilon = \varepsilon(\tilde{\mathscr{U}}) > 0$ and an $\tilde{A} \in \tilde{\mathscr{A}}$ which belongs to $\tilde{\mathscr{U}} \cup \tilde{\Lambda}^{\tilde{\varkappa}-\varepsilon} M$.*

In particular, since for $\tilde{C}r\tilde{\varkappa} = \emptyset$ we could have chosen $\tilde{\mathscr{U}} = \emptyset$, $\tilde{C}r\tilde{\varkappa}$ is not empty.

Proof. Fix $\tilde{\mathscr{U}}$. From the continuity of $\tilde{\mathscr{D}}(\sigma, \ )$ we then have the existence of a neighborhood $\tilde{\mathscr{U}}'$ of $\tilde{C}r\tilde{\varkappa}$, $\tilde{\mathscr{U}}' \subset \tilde{\mathscr{U}}$, such that the image of $\tilde{\mathscr{U}}'$ under the subsequent application $\tilde{\mathscr{D}}(\sigma_1, \ )$, $\tilde{\mathscr{D}}(\sigma_2, \ ), \ldots, \tilde{\mathscr{D}}(\sigma_m, \ )$, with m as in the definition of (3.7.20), belongs to $\tilde{\mathscr{U}}$. Here, the σ_i, $1 \leqslant i \leqslant m$, can be chosen arbitrarily in the interval $[0, 2]$.

It follows that an $c = \tilde{u}(c_0) \in \tilde{\mathscr{U}}'$ under $\tilde{\mathscr{D}}_{\tilde{u}}(2, \ )$ is carried into $\tilde{\mathscr{U}}$.

From (3.7.15) we have an $\varepsilon > 0$ such that $\tilde{\Lambda}_\infty^{\tilde{\varkappa}+\varepsilon}$ is carried under $\tilde{\mathscr{D}}(2, \ )$ into $\tilde{\mathscr{U}}' \cup \tilde{\Lambda}_\infty^{\tilde{\varkappa}-\varepsilon}$.

The definition of $\tilde{\varkappa}$ as critical value of the family $\tilde{\mathscr{A}}$ implies the existence of a $\tilde{u}(\bar{A}^*) \in \tilde{\mathscr{A}}$ which lies in $\tilde{\Lambda}_\infty^{\tilde{\varkappa}+\varepsilon}$. We claim that the element $A = \tilde{\mathscr{D}}_{\tilde{u}}(2, \bar{A}^*) \in \tilde{\mathscr{A}}$ belongs to $\tilde{\mathscr{U}} \cup \tilde{\Lambda}_\infty^{\tilde{\varkappa}-\varepsilon}$.

Indeed, let $c = \tilde{u}(c_0) \in \tilde{u}(\bar{A}^*)$. If we consider $\tilde{\mathscr{D}}_{\tilde{u}}(\sigma, c_0)$ for $0 \leqslant \sigma \leqslant 2$, it always will remain in $\tilde{\Lambda}^{\tilde{\varkappa}+\varepsilon}$. But for at least one i, $1 \leqslant i \leqslant m$, $\psi_i(c_0) = 1$, and thus, by the deformation $\tilde{\mathscr{D}}_{\tilde{u}}(\sigma, c_0)$, $2i/m \leqslant \sigma \leqslant (2i+2)/m$, c will be brought into $\tilde{\mathscr{U}}' \cup \tilde{\Lambda}_\infty^{\tilde{\varkappa}-\varepsilon}$. Now, this latter set is transformed into $\tilde{\mathscr{U}} \cup \tilde{\Lambda}_\infty^{\tilde{\varkappa}-\varepsilon}$ for the remaining $\sigma \geqslant (2i+2)/m$ of the deformation $\tilde{\mathscr{D}}_{\tilde{u}}(\sigma, \ )$. □

We now have everything ready for the proof of the Theorem of the Three Closed Geodesics, cf. Lusternik and Schnirelmann [1].

3.7.23 Theorem. *On every surface M of genus zero there exist three simple closed geodesics.*

More precisely, if $f: S^2 \to M$ is a diffeomorphism of the standard sphere S^2 in $\mathbb{R}^3$ onto M, then there exist three simple closed geodesics on M all of which have length $\leqslant \sup \sqrt{2E(f \circ c_0)}$ where c_0 runs through the set AS^2 of circles on S^2.

Remark. This result is optimal in the sense that there exist examples of surfaces where there are no more than three simple closed geodesics. Indeed, as we saw in (3.5.19), on an ellipsoid with axes of different lengths, but these lengths not too different from each other, the three principal ellipses are the only simple closed geodesics.

Proof. The proof goes along lines similar to the proof of (3.6.8) and (3.6.12) for the case $n = 2$.

We consider the families $\tilde{\mathscr{A}}_{j,k}$, $0 \leqslant j \leqslant k \leqslant 1$ of (3.7.19). Since a cycle $\bar{A}^*_{1,1} \bmod \bar{A}^{*0}_{1,1}$ contains a cycle $\bar{A}^*_{0,1} \bmod \bar{A}^{*0}_{0,1}$ and since such a cycle in turn contains a cycle $\bar{A}^*_{0,0} \bmod \bar{A}^{*0}_{0,0}$, we have for the critical values of these three families the relation

$$\tilde{\varkappa}_{0,0} \leqslant \tilde{\varkappa}_{0,1} \leqslant \tilde{\varkappa}_{1,1}.$$

Moreover, $\tilde{\varkappa}_{0,0} > 0$, as follows by the same argument as we employed in the proof of (3.7.7). Therefore, we have from (3.7.22) the existence of simple closed geodesics with E-value $\tilde{\varkappa}_{j,k}$, $0 \leqslant j \leqslant k \leqslant 1$. Thus, if $\tilde{\varkappa}_{0,0} < \tilde{\varkappa}_{0,1} < \tilde{\varkappa}_{1,1}$, we are done.

It remains to us to discuss the cases $\tilde{\varkappa}_{0,0} = \tilde{\varkappa}_{0,1}$ and $\tilde{\varkappa}_{0,1} = \tilde{\varkappa}_{1,1}$. We will show that in either case we get an infinite number of simple unparameterized closed geodesics of E-value $\tilde{\varkappa}_{0,0} = \tilde{\varkappa}_{0,1}$ or $\tilde{\varkappa}_{0,1} = \tilde{\varkappa}_{1,1}$. Actually, the set of simple unparameterized closed geodesics will carry a 1-cycle in these cases.

Assume first $\tilde{\varkappa}_{0,0} = \tilde{\varkappa}_{0,1} =$ (briefly) $\tilde{\varkappa}$. Assume that the set $\tilde{C}r\tilde{\varkappa}$ of simple closed geodesics of E-value $\tilde{\varkappa}$ is formed by a finite number of S^1-orbits $S^1 . c_\iota, \iota \in I$. We will see that this leads to a contradiction.

Indeed, under our assumption we can choose an open $(\mathbb{Z}_2 \times \mathbb{Z}_2)$-invariant neighborhood $\tilde{\mathscr{U}}$ of $\tilde{C}r\tilde{\varkappa}$ which is formed by small open S^1-invariant neighborhoods $\tilde{\mathscr{U}}_\iota$ of the $S^1 . c_\iota$, with $\tilde{\mathscr{U}}_\iota \cap \tilde{\mathscr{U}}_\varkappa = \emptyset$ for $\iota \neq \varkappa$. From (3.7.22) we have the existence of an $\tilde{A} = \tilde{u}(\bar{A}^*_{0,1})$ contained in $\tilde{\mathscr{U}} \cup \tilde{\Lambda}^{\tilde{\varkappa}-}_\infty$. Put $\tilde{u}^{-1}(\tilde{\mathscr{U}}) = \bar{\mathscr{O}}^*, \tilde{u}^{-1}(\tilde{\Lambda}^{\tilde{\varkappa}-}_\infty) = \bar{\mathscr{N}}^*$.

$\{\bar{\mathscr{O}}^*, \bar{\mathscr{N}}^*\}$ forms an open covering of $\bar{A}^*_{0,1}$. Since no $c \in \tilde{\mathscr{U}}$ can be joined inside $\tilde{\mathscr{U}}$ to θc or $\theta' c$, $\bar{\mathscr{O}}^*$ does not carry a 1-cycle homologous to $\bar{B}^*_{0,1}$. A cycle in the class of $\bar{A}^*_{0,0}$ and a cycle in the class of $\bar{B}^*_{0,1}$, both contained in $\bar{A}^*_{0,1}$ and in general position, have intersection number 1 mod 2. One may also interpret this by saying that $\bar{B}^*_{0,1}$ is a cocycle in $\bar{A}^*_{0,1} - \bar{A}^{*0}_{0,1} \approx \bar{B}^*_{0,1}$. From (3.7.18) we know that $\bar{B}^*_{0,1}$ is homologous to $\bar{A}^*_{0,0}$ in $(\bar{A}^*_{0,1}, \bar{A}^{*0}_{0,1})$ and thus, $\bar{A}^*_{0,0}$ has self-intersection number 1 mod 2.

We can assume that the simplices in the 1-dimensional chain complex of $\bar{A}^*_{0,1}$ are chosen so small that they lie entirely in $\bar{\mathscr{O}}^*$ or entirely in $\bar{\mathscr{N}}^*$. Now, since $\bar{\mathscr{O}}^*$ does not carry a 1-cycle, we can find a cocycle $\bar{B}^*_{0,1}$ which has value 0 mod 2 on all 1-simplices which lie in $\bar{\mathscr{O}}^*$. Thus, if we define the cap product $\bar{A}^*_{0,1} \cap \bar{B}^*_{0,1}$ by taking from each 2-simplex σ of $\bar{A}^*_{1,0}$ the front 1-simplex with coefficient the value of $\bar{B}^*_{0,1}$ on the back 1-simplex of σ, we get a cycle in the class $\bar{A}^*_{0,0}$ which lies in $\bar{\mathscr{N}}^*$. Its image under $\tilde{u}$ belongs to $\tilde{\Lambda}^{\tilde{\varkappa}-}_\infty$, i. e., $\tilde{\varkappa}_{0,0} < \tilde{\varkappa}_{0,1}$ – a contradiction.

The case $\tilde{\varkappa}_{0,1} = \tilde{\varkappa}_{1,1} =$ (briefly) $\tilde{\varkappa}$ is treated in the same way: Assume that $\tilde{C}r\tilde{\varkappa}$ consists of only finitely many S^1-orbits. Then choose a neighborhood $\tilde{\mathscr{U}}$ of $\tilde{C}r\tilde{\varkappa}$ as above. There exists $\tilde{u}(\bar{A}^*_{1,1}) \subset \tilde{\mathscr{U}} \cup \tilde{\Lambda}^{\tilde{\varkappa}-}_\infty$. According to (3.7.18), $\bar{A}^*_{0,1}$ is homologous to $\bar{B}^*_{1,1}$. We interpret a cycle in $\bar{B}^*_{1,1}$ as a 1-cocycle on $\bar{A}^*_{1,1}$ by taking the intersection number mod 2 between this cycle and a 1-simplex in a general position. Since $\bar{\mathscr{O}}^*$ carries no 1-cycle in the class $\bar{B}^*_{0,1}$, we may choose this 1-cocycle (which we denote by $\bar{B}^*_{1,1}$) such that it has the value 0 mod 2 on all 1-simplices lying in $\bar{\mathscr{O}}^*$.

We can assume that the 3-simplices σ in $\bar{A}^*_{1,1}$ are chosen so small that they either lie entirely in $\bar{\mathscr{O}}^*$ or entirely in $\bar{\mathscr{N}}^*$. By defining the cap product $\bar{A}^*_{1,1} \cap \bar{B}^*_{1,1}$ in taking the front 2-simplex of each such σ with coefficient the value of $\bar{B}^*_{1,1}$ on the back 1-simplex of σ, we obtain an element in the class $\bar{A}^*_{0,1}$ lying entirely in $\bar{\mathscr{N}}^*$. But then, its image under $\tilde{u}$ lies in $\tilde{\Lambda}^{\tilde{\varkappa}-}_\infty$ – a contradiction. □

Remarks. 1. The E-decreasing deformations $\tilde{\mathscr{D}}$, (3.7.11), can also be applied to closed curves with self-intersection number $\leqslant q$. They will not increase this self-intersection number and thus yield a proof of the existence of an infinite number of prime closed geodesics on a surface of genus 0, cf. Klingenberg [12].

2. Simple closed geodesics on compact surfaces of a type different from S^2 have

been studied by Ballmann [3]. With the exception of the Klein bottle (where there exist at least five simple closed geodesics) and the projective plane (where there exist at least three simple closed geodesics), there always exist infinitely many simple closed geodesics.

3.8 Manifolds of Non-Positive Curvature

In this paragraph we study compact manifolds $\bar{M}$ with sectional curvature $K \leqslant 0$. As we know from (2.6.2), (2.7.2) and (2.7.9), there are neither conjugate points nor focal points on such manifolds. The universal covering M of $\bar{M}$ is diffeomorphic to the tangent space $T_p M$ at one of its points p.

The concept of asymptotes in M allows us to define an intrinsically defined compactification $\mathrm{cp}(M)$ of M by adding the so-called points at infinity.

The action of the fundamental group $\pi_1 \bar{M}$ of $\bar{M}$ as group Γ of deck transformations on M yields information on the structure of $\pi_1 \bar{M}$ and also allows us to construct closed geodesics on $\bar{M}$.

We wish to point out here that most of these concepts and constructions are much richer in structure and content if one assumes M to have strictly negative (bounded) curvature. This we will see in (3.9).

All manifolds are again supposed to be complete and of finite dimension. Unless stated otherwise, geodesics are parameterized by arc length.

We begin with a useful observation on the distance function between geodesics.

3.8.1 Proposition. *Let $c_0(t)$, $c_1(t)$ be two geodesics on a simply connected manifold with sectional curvature $K \leqslant 0$. Here we allow $c_1(t)$ to be constant, whereas $|\dot{c}_0(t)| = 1$. There shall be at most one t-value, say $t = t^*$, where $c_0(t^*) = c_1(t^*)$. Consider the distance function $f(t) = d(c_0(t), c_1(t))$. For all t with $f(t) \neq 0, f$ is differentiably convex, i.e., $\ddot{f}(t) \geqslant 0$. The case $\ddot{f}(t) = 0$ for $t = t_1$ has a special meaning, cf. the end of the proof.*

Proof. We first consider the function $g(t) = \frac{1}{2} f(t)^2$.

Denote by $c_t(s)$, $0 \leqslant s \leqslant 1$, the unique geodesic from $c_0(t)$ to $c_1(t)$. Note that in general $|c_t'(s)| \neq 1$. We consider the differentiable mapping

$$F: (s, t) \in I \times \mathbb{R} \mapsto c_t(s) \in M.$$

Thus, $F | \{0\} \times \mathbb{R} = c_0$, $F | \{1\} \times \mathbb{R} = c_1$, $|\partial F(s, t)/\partial s| = d(c_0(t), c_1(t))$.

$\partial F(s, t)/\partial s = c_t'(s)$ while $s \mapsto \partial F(s, t)/\partial t$ is a Jacobi field along $c_t(s)$ which we also denote by $Y_t(s)$. Note: $Y_t(0) = \dot{c}_0(t)$, $Y_t(1) = \dot{c}_1(t)$.

We now find the derivatives of $g(t) = \langle \partial F/\partial s, \partial F/\partial s \rangle (s, t)/2$ with the help of the formula (1.5.8):

$$\dot{g}(t) = \langle \frac{\nabla}{\partial t} \frac{\partial F}{\partial s}, \frac{\partial F}{\partial s} \rangle (s, t) = \langle \frac{\nabla}{\partial s} \frac{\partial F}{\partial t}, \frac{\partial F}{\partial s} \rangle (s, t)$$

and

$$\ddot{g}(t) = \langle \frac{\nabla}{\partial s} \frac{\partial F}{\partial t}, \frac{\nabla}{\partial s} \frac{\partial F}{\partial t} \rangle (s, t) + \langle R\left(\frac{\partial F}{\partial t}, \frac{\partial F}{\partial s}\right) \frac{\partial F}{\partial t}, \frac{\partial F}{\partial s} \rangle (s, t) +$$

$$+ \frac{d}{ds} \langle \frac{\nabla^2 F}{\partial t^2}, \frac{\partial F}{\partial s} \rangle (s, t).$$

The last term vanishes for $s = 0$ and for $s = 1$ and we therefore get

$$\ddot{g}(t) = \langle \frac{\nabla}{\partial s} Y_t(t), \frac{\nabla}{\partial s} Y_t(s) \rangle - \langle R(Y_t(s), c_t'(s)) c_t'(s), Y_t(s) \rangle > 0.$$

Now consider $\sqrt{2g(t)} = f(t) = d(c_0(t), c_1(t))$. Assume $f(t) \neq 0$. Then

$$\ddot{f}(t) = \frac{1}{f(t)^3} \left\{ \langle \frac{\nabla}{\partial s} Y_t(s), \frac{\nabla}{\partial s} Y_t(s) \rangle \langle c_t'(s), c_t'(s) \rangle - \right.$$

$$\left. - \langle \frac{\nabla}{\partial s} Y_t(s), c_t'(s) \rangle^2 - \langle R_{c_t'(s)} Y_t(s), Y_t(s) \rangle \langle c_t'(s), c_t'(s) \rangle \right\} \geqslant 0.$$

If we have $= 0$, then $\nabla Y_t(s)/\partial s$ must be proportional to $c_t'(s) \neq 0$ and $K = 0$ on the 2-plane spanned by $c_t'(s)$ and $Y_t(s)$. □

We recall from (2.6.2), (2.7.2) and (2.7.9) that a manifold of curvature $K \leqslant 0$ has neither conjugate nor focal points because this is true for the Euclidean space of curvature $K = 0$. The following Lemma should be viewed as a global version of these facts for the case of simply connected manifolds with $K \leqslant 0$.

3.8.2 Lemma. *Let M be simply connected of sectional curvature $K \leqslant 0$. Then for every $p \in M$, the convexity radius $\varkappa(p)$ in the sense of* (1.9.9) *is infinite. More precisely,*

(i) *For every triple p, q, r of points on M, there exists a uniquely determined triangle pqr in the sense of* (2.7.4). *The sides c_{pq}, c_{qr}, c_{pr} are the unique geodesics between these points. The sum of the angles in pqr is $\leqslant \pi$ and actually $< \pi$ if $K < 0$ everywhere. Moreover, there holds a cosine inequality*

$$L(c_{pq})^2 + L(c_{pr})^2 - 2L(c_{pq})L(c_{pr}) \cos \sphericalangle p \leqslant L(c_{qr})^2$$

with $<$ if $K < 0$ everywhere.

(ii) *Let $c(t)$, $t \in \mathbb{R}$, be a geodesic and $p \in M$, p not on c. Then there exists a unique geodesic $c_p(t)$ with $c_p(0) = p$ which meets c orthogonally. c_p is called the* perpendicular from p towards c.

Proof. From (2.6.6) we know that under our hypotheses two points can be joined by exactly one geodesic. If we consider for pqr an associated Alexandrov triangle $p^* q^* r^*$ in the Euclidean space M^* then the angles in $p^* q^* r^*$ are at least as big as the corresponding angles in pqr, cf. (2.7.6). This shows that the angle sum in pqr is $\leqslant\pi$. We also get from this the cosine inequality. Thus we have proved (i).

For the proof of (ii), we note that $f(t) = d(p, c(t))$ is a differentiable convex function which never vanishes, cf. (3.8.1). It is bounded from below, thus, it assumes its infimum, say at $t = t_0$. The second derivative of the function in $t = t_0$ is > 0, as one sees from the proof of (3.8.1). Thus, the minimum is unique and there is no $t_1 \neq t_0$ where $\dot{f}(t_1) = 0$. □

The following Theorem about spaces of curvature $\leqslant 0$ is due to Cartan [2].

3.8.3 Theorem. *Let M be simply connected with $K \leqslant 0$. Then every compact group of isometries possesses a fixed point.*

Proof. This follows with (3.8.2) from (1.10.17). □

We now introduce the concept of asymptotic geodesics. Originally, asymptotic geodesics had been considered for hyperbolic spaces only. Take, e.g., the Poincaré model of the hyperbolic plane, i.e., the unit disc $B^2(0) \subset \mathbb{R}^2$ with the metric $4du^2/(1-u^2)^2$, $u = (u_1, u_2)$. Here asymptotic geodesics are represented by circles which meet $\partial B^2(0)$ orthogonally in the same point. See (3.8.7) for details.

A generalization of the concept of asymptotes to surfaces of variable curvature was considered and used extensively by E. Hopf [1]. The extension to manifolds of arbitrary dimension (and even to spaces with a somewhat more general structure than a manifold, so-called G-spaces) is due to Busemann [1] and his school. Here we mention only the following papers: Bishop and O'Neill [1], Eberlein and O'Neill [1] and Ballmann [2].

The last two papers mainly present results on manifolds of curvature $\leqslant 0$ which have additional properties, such as to be able to carry over a good number of the results on manifolds of strictly negative curvature which will be presented in (3.9).

Another closely related class of manifolds are those without conjugate points and – even stronger – without focal points. For the numerous results in this direction see, besides Busemann [1], Morse and Hedlund [1], Green [1] and Eschenburg [1]. In Eschenburg one also finds additional references.

3.8.4 Definition. Let M be simply connected with $K \leqslant 0$. Two geodesics c, c' on M are called *(positively) asymptotic* if there exists a constant $a = a(c, c') \geqslant 0$ such that $d(c(t), c'(t)) \leqslant a$, all $t \geqslant 0$.

3.8.5 Lemma. *Let M be simply connected, $K \leqslant 0$.* (i) *The relation between geodesics to be asymptotic is an equivalence relation.*

(ii) *Given a geodesic c and a point p on M, there exists exactly one geodesic c' which is asymptotic to c with $c'(0) = p$. Actually, if we denote by c_t the geodesic from p to $c(t)$, t so large that $d(p, c(t^*)) > 0$ for all $t^* > t$, then $\dot{c}'(0)$ is the limit for $t \mapsto \infty$ of the family $\dot{c}_t(0)$.*

Proof. (i) is immediate from the definition (3.8.4).

To prove (ii) we first prove the uniqueness of an asymptote through p: If c_1', c_2' are geodesics with $c_1'(0) = c_2'(0) = p$, c_1', c_2' of bounded distance from c for $t \to \infty$, then $d(c_1'(t), c_2'(t))$ is also bounded for $t \to \infty$. Compare the angle at p of the triangle $pc_1'(t)\,c_2'(t)$ with the angle at p^* of an associated Alexandrov triangle in the Euclidean space. Now, as the latter goes to 0, for $t \to \infty$, so does the first, cf. (2.7.6).

For the existence of an asymptote c' to c through p we may assume $p \notin c$. Let $t_0 \in \mathbb{R}$ be such that the geodesic from p to $c(t_0)$ is perpendicular to c. Put $d(p, c(t_0)) = d$. For each integer $n > t_0$ consider the geodesic c_n' from p to $c(n)$ with $c_n'(0) = p$. From (3.8.1) we have, with $t_n = d(p, c(n)) \leqslant n - t_0 + d$,

$$d(c(t), c_n'(t)) \leqslant d(c(t), c_n'(t_n - n + t)) + d(c_n'(t_n - n + t), c_n'(t)) \leqslant 2d - t_0 + d,$$

for $t_0 \leqslant t \leqslant n$.

The sequence $\{\dot{c}_n'(0)\}$ has an accumulation point, say X'. If c' denotes the geodesic with $\dot{c}'(0) = X'$, our above estimates show that $d(c(t), c'(t))$ is bounded for all $t \geqslant 0$. □

Remark. In the Euclidean space, two geodesics are asymptotic if and only if they are parallel. Thus in particular, in this case positively asymptotic geodesics c, c' are also negatively asymptotic, i.e., they have bounded distance not only for $t \to +\infty$ but also for $t \to -\infty$.

This contrasts with the behaviour of asymptotic geodesics in hyperbolic space, cf. (3.8.7) below. In fact, as we will now show, if there exist two geodesics which are asymptotic in both directions (and do not just differ by a change of parameter) then they bound a flat strip.

3.8.6 Lemma. *Let c_0, c_1 be geodesics with $d(c_0(t), c_1(t)) \leqslant$ some constant, for all $t \in \mathbb{R}$. Let c_1 not be of the form $c_1(t) = c_0(t + t_0)$, some $t_0 \in \mathbb{R}$. Then there exists a so-called flat strip in M, bounded by c_0 and c_1. By this we mean an isometric totally geodesic immersion*

$$F\colon [0, d] \times \mathbb{R} \to M$$

of the flat strip $[0, d] \times \mathbb{R} \subset \mathbb{R} \times \mathbb{R}$, $d > 0$, into M such that $F|\{0\} = c_0$, $F|\{d\} = c_1$. As always, F shall be the restriction to $[0, d] \times \mathbb{R}$ of a differentiable mapping $\tilde{F}\colon]-\varepsilon, d+\varepsilon[\times \mathbb{R} \to M$, some $\varepsilon > 0$.

Proof. We can assume that the perpendicular from $c_1(0)$ towards c_0 meets c_0 in $c_0(0)$. As in the proof of (3.8.1), we consider the geodesics $s \mapsto c_t(s) = F(t, s)$ from $c_0(t)$ to $c_1(t)$. $Y_t(s) = \partial F(t, s)/\partial t$ is a Jacobi field along $c_t(s)$. The function $f(t) = |\partial F(t, s)/\partial s| = d(c_0(t), c_1(t))$ has no zeros and is convex and bounded. Hence, $f = \text{const} = d > 0$. The formula in the proof of (3.8.1) yields $\nabla Y_t(s)/\partial s = 0$, i.e., $s \in [0, d] \mapsto F(t, s) \in M$ is the geodesic from $c_0(t)$ to $c_1(t)$. Since $\langle R_{c_t'(s)} Y_t(s), Y_t(s)\rangle = 0$, the sectional curvature $K(\sigma(s, t))$ for the 2-plane $\sigma(s, t)$ spanned by $\partial F(s, t)/\partial s$ and $\partial F(s, t)/\partial t$ is zero.

It follows from the proof of (1.6.3) that every vector tangent to im F in $F(0, 0)$

possesses a unique extension to a parallel tangential vector field along F. Thus, F is an isometric immersion. □

We now discuss asymptotes in the hyperbolic space. This will give us some idea of what to expect in more general manifolds of strictly negative curvature.

3.8.7 Example. *The hyperbolic space.* Recall from (1.11.7) and (1.11.8) the definition and the main properties of the real hyperbolic space. Here we consider the space of n dimensions and radius $\varrho > 0$ and write H^n_ϱ or also briefly H. H has constant negative curvature $-1/\varrho^2$.

We had two models for H. One was the set

$$\{-x_0^2 + |x|^2 = -\varrho^2; x_0 > 0\}$$

in the space $\mathbb{R} \times \mathbb{R}^n$ with the Lorentz metric $-dx_0^2 + dx^2$. This we will call the *hyperboloid model.* The other consisted of the open ball

$$B^n_\varrho(0) = \{|u|^2 < \varrho\}$$

with the metric $4du^2/(1-|u|^2/\varrho^2)^2$. This we will call the *ball model.*

(i) *The (carriers of the) geodesics in the ball model consist of the circles which meet* $\partial \bar{B}^n_\varrho(0) = S^{n-1}_\varrho(0)$ *orthogonally, in as far as these circles belong to* $B^n_\varrho(0)$.

To see this, recall that in the hyperboloid model the geodesics are the intersection of the hyperboloid with a 2-plane through the origin of $\mathbb{R} \times \mathbb{R}^n$. We use the subgroup $id \times \mathbb{O}(n)$ of the isometry group $\mathbb{O}(1, n)$ of the hyperboloid model to bring such a 2-plane in the position

$$\{\sin\theta x_0 - \cos\theta x_1 = 0; u_3 = \ldots = u_n = 0\},$$

with $0 \leqslant \theta < \pi/2$. This is possible since $id \times \mathbb{O}(n)$ operates by isometries on the ball model. From the mapping of the ball model into the hyperboloid model, cf. (1.11.7), we therefore find for the carrier of a geodesic in $B^n_\varrho(0)$ the formula

$$\{(u_1^2 + u_2^2 + \varrho^2)\sin\theta - 2\varrho u_1 \cos\theta = 0; u_3 = \ldots = u_n = 0\}.$$

This clearly is the equation of a circle, possibly degenerate, i.e., a straight line, if $\theta = 0$.

The intersection points $(u_1^0, \pm u_2^0)$ of this circle with $\partial \bar{B}^n_\varrho(0)$ are given by $\varrho \sin\theta = u_1^0 \cos\theta, u_2^0 = \pm\sqrt{\varrho^2 - u_1^{0\,2}}$. The tangents to the circle in these points read

$$\{u_1(u_1^0 \sin\theta - \varrho\cos\theta) + u_2 u_2^0 \sin\theta = 0\},$$

i.e., they pass through the origin of $B^n_\varrho(0)$.

(ii) *Two geodesics* $c(t)$ *and* $c'(t)$ *in the ball model are asymptotic to each other if and only if they pass through the same point of* $\partial \bar{B}^n_\varrho(0)$, *when* $t \to \infty$.

In fact, the latter property is equivalent to saying that in the hyperboloid model the two 2-planes which carry the geodesics intersect the cone $\{-x_0^2 + |x|^2 = 0\}$ in the same line. Now, if this is the case, the distance between the two geodesics for $|x| \to \infty$ goes to zero even in the Euclidean metric $x_0^2 + |x|^2$ of $\mathbb{R} \times \mathbb{R}^n$. A fortiori, this is true for the Lorentz metric $-x_0^2 + |x|^2$.

Observe now that, for a given geodesic c in $B^n_\varrho(0)$, through every $p \in B^n_\varrho(0)$ there passes one (exactly one) geodesic c' which meets, for $t \to \infty$, $\partial \bar{B}^n_\varrho(0)$ in the same point as c does.

(iii) *Given two non-asymptotic geodesics $c(t)$, $c'(t)$, there exists (up to parameterization) exactly one geodesic $c^*(t)$ which, for $t \to \infty$, is asymptotic to $c(t)$ and which, for $t \to -\infty$, is asymptotic to $c'(t)$.*

Indeed, in the ball model $c(\infty)$ and $c'(\infty)$ determine points ω and ω' on $\partial \bar{B}^n_\varrho(0)$, cf. (i). In the 2-plane through $0 \in \mathbb{R}^n$ which contains ω and ω' there is exactly one circle which meets $\partial \bar{B}^n_\varrho(0)$ orthogonally and passes through $c(\infty)$ and $c'(\infty)$.

(iv) *Let $p \in H$ and consider two geodesics $c(t)$, $c'(t)$, starting from p with an angle 2γ, $0 < 2\gamma < \pi$. Let c^* be the geodesic, asymptotic for $t \to \infty$ and $t \to -\infty$ to c and to c' respectively. Then the distance $d(\gamma)$ from p to c^* is given by*

$$(*) \qquad \operatorname{tgh}(d(\gamma)/\varrho) = \cos\gamma .$$

Since everything is defined intrinsically, we can assume that in the ball model p is the origin 0 while c and c' are given by $\{u_2 = \pm \operatorname{tg}\gamma\, u_1;\ u_i = 0 \text{ for } i > 2\}$. Thus, the intersections with $\partial \bar{B}^n_\varrho(0)$ are $u_1^0 = \varrho \cos\gamma$, $u_2^0 = \pm \varrho \sin\gamma$, $u_i^0 = 0$, for $i > 2$. From the formula in the proof of (i) we get for c^* the circle $\{(u_1^2 + u_2^2 + \varrho^2)\cos\gamma - 2\varrho u_1 = 0\}$. The intersection with the ray $\{0 < u_1 < \varrho\}$ is given by $u_1^* = \varrho(1 - \sin\gamma)/\cos\gamma$.

We find for the length $d = d(\gamma)$ of the segment $[0, u_1^*]$

$$\int_0^{u_1^*} \frac{2\,du_1}{1 - u_1^2/\varrho^2} = 2\varrho \operatorname{arc\,tgh}\{(1 - \sin\gamma)/\cos\gamma\} .$$

With the formula $\operatorname{tgh}(d/\varrho) = 2\operatorname{tgh}(d/2\varrho) : (1 + \operatorname{tgh}^2(d/2\varrho))$ we get $(*)$. □

We not want to extend the properties of the hyperbolic space – at least qualitatively – to spaces of variable negative curvature. We begin by introducing a topology on the classes of asymptotes which will play the rôle of the boundary points of the ball model of the hyperbolic space.

3.8.8 Lemma. *Let M be simply connected, $K \leqslant 0$. Fix $p \in M$. Call a sequence $\{p_n\}$ of points in M* convergent to infinity, with limit $c(\infty)$, *if there exists a geodesic c with $c(0) = p$ such that*

(i) *the sequence $\{d(p, p_n)\}$ goes to ∞.*

(ii) *the sequence $\{\alpha_n\}$ of angles between the initial direction $X_n = \dot{c}_n(0)$ of the geodesic c_n from p to p_n (assume here $p_n \neq p$) and the initial vector $X = \dot{c}(0)$ of c goes to zero.*

If now $p' \in M$ is an arbitrary point on M and c' the asymptote to c, starting at p', then the sequence $\{p_n\}$ also converges to $c'(\infty)$ in the above sense.

Proof. We can restrict ourselves to the case where $p_n \neq p$, $p_n \neq p'$, all n. Let c'_n be the geodesic from p' to p_n. Put $\dot{c}'_n(0) = X'_n$. We then can find a convergent subsequence of $\{X'_n\}$, which we denote again by $\{X'_n\}$, with limit X'. We have to show that then the geodesic c' with $\dot{c}'(0) = X'$ is the asymptote to c through p'.

To see this we note that, for any $t_0 > 0$, any $\varepsilon > 0$, we will have $d(c_n'(t_0), c'(t_0)) < \varepsilon$, $d(c_n(t_0), c(t_0)) < \varepsilon$, all sufficiently large n depending on t_0. The function $d(c_n(t), c_n'(t))$ is convex. For $t = 0$ it is $= d(p, p')$ and it will decrease (weakly) until we reach the value t_n where $c_n(t_n) = p_n$ or $c_n'(t_n) = p_n$. Thus $d(c(t_0), c'(t_0)) < d(p, p') + 2\varepsilon$, i.e., c and c' are asymptotes. □

We now come to the compactification of M by its 'points at infinity', which we mentioned earlier.

3.8.9 Definition. Let M be simply connected with $K \leqslant 0$.

(i) *A point at infinity* is a class of (positively) asymptotic geodesics. If c is a representative of such a class, we denote this class by $c(+\infty)$ or ω_c or simply ω, if there is no danger of confusion.

(ii) For a given geodesic $c = c(t)$, $t \in \mathbb{R}$, we denote by $c(-\infty)$ or α_c or simply α the point at infinity determined by the equivalence class of geodesics asymptotic to $c(-t)$, $t \in \mathbb{R}$. We also say that c is a geodesic from the *α-point* α_c of the geodesic c to the *ω-point* ω_c of the geodesic c.

(iii) By $M(\infty)$ we denote the points of infinity of M.

We will denote the union $M \cup M(\infty)$ also by $\mathrm{cp}(M)$ and call it the *compactification of M*.

(iv) We define a *topology on* $\mathrm{cp}(M)$ by calling a sequence $\{p_n\}$ on M convergent to $\omega \in M(\infty)$ if it satisfies the condition given in (3.8.8). If $\{\omega_n\}$ is a sequence in $M(\infty)$ we call it convergent with limit ω if, for some $p \in M$, the geodesics c_n from p to ω_n (i.e., $c_n(0) = p$, $c_n(\infty) = \omega_n$) converge to the geodesic c from p to ω.

3.8.10 Proposition. *Under the hypotheses of* (3.8.9), *the topology on* $\mathrm{cp}(M) = M \cup M(\infty)$ *is well defined.*

For any $p \in M$ consider the mapping

$$h_p \colon \bar{B}_1(0_p) \subset T_p M \to \mathrm{cp}(M);\ X \mapsto \begin{cases} \exp_p(X/(1-|X|)), & \text{if } |X| < 1, \\ c_X(\infty), & \textit{if } |X| = 1. \end{cases}$$

Here, c_X is the geodesic with $\dot{c}_X(0) = X$. Then this is a homeomorphism from the closed unit ball in $T_p M$ onto $\mathrm{cp}(M)$.

Proof. Due to (3.8.8), we only need to show that the definition of the convergence of a sequence $\{\omega_n\}$ of points at infinity does not depend on the choice of the point $p \in M$. But clearly, we can associate with $\{\omega_n\}$ a sequence $\{p_n = c_n(t_n)\}$, $t_n \to \infty$, in M which converges to $\omega = c(\infty)$. This convergence is independent of the choice of p by (3.8.8).

The very definition of the topology on $\mathrm{cp}(M)$ shows that h_p is a homeomorphism. □

We now consider the extension of an isometry from a simply connected manifold M with $K \leqslant 0$ to its compactification $\mathrm{cp}(M)$. This will be particularly useful when M is the universal covering of a compact manifold of curvature $\leqslant 0$. In this case we have on M the action of the fundamental group as the group of deck transformations.

3.8.11 Lemma. *Let M be simply connected with $K \leqslant 0$. If $\mu: M \to M$ is an isometry, define its extension to $M(\infty)$ by putting $\mu\omega = (\mu c)(\infty)$, if $\omega = c(\infty)$.*

This definition does not depend on the choice of the geodesic c in its class of asymptotic geodesics. $\mu: \mathrm{cp} M \to \mathrm{cp} M$ is a homeomorphism.

If Γ is a group of isometries on M the extension of the action of the $\mu \in \Gamma$ to $\mathrm{cp} M$ yields an action of Γ on $\mathrm{cp} M$.

Proof. Since an isometry carries asymptotic geodesics into asymptotic geodesics, $\mu: M(\infty) \to M(\infty)$ is well defined. It is a bijection with μ^{-1} as inverse.

If $\{p_n\}$ is a sequence on M with limit $\omega \in M(\infty)$ $\{\mu p_n\}$ converges to $\mu\omega$, thus, $\mu: \mathrm{cp} M \to \mathrm{cp} M$ is a homeomorphism. Finally, our definitions yield $(\mu \circ \mu')(\omega) = \mu(\mu'(\omega))$. □

3.8.12 Definition. Let $\mu: M \to M$ be an isometry of a Riemannian manifold M.

(i) The *displacement function* $f_\mu: M \to \mathbb{R}$ is defined by $f_\mu(p) = d(p, \mu p)$.

(ii) A geodesic $c(t)$, $t \in \mathbb{R}$, is called an *axis of* μ if it is transformed into itself by μ, i.e., $\mu c(t) = c(t + d_\mu(c))$, all $t \in \mathbb{R}$, with some $d_\mu(c) \geqslant 0$, only depending on μ and c.

(iii) The *minimal set of* μ, D_μ, consists of the $p \in M$ where $f_\mu(p) = \inf f_\mu$. We denote $\inf f_\mu$ by d_μ.

The following theorem is fundamental:

3.8.13 Theorem. *Let $\bar{M}$ be a compact manifold with $K \leqslant 0$. Let $\Gamma \cong \pi_1 \bar{M}$ be the group of deck transformations, operating on the universal covering M of $\bar{M}$. Let $\mu \in \Gamma, \mu \neq id$.*

Then μ possesses an axis. On each axis, μ operates with the minimal translation $d_\mu = \inf f_\mu > 0$. The minimal set D_μ of μ is the union of the axes of μ. Any two axes have bounded distance for all $t \in \mathbb{R}$ and hence bound a flat strip. All axes therefore have the same negative and positive point at infinity, which we denote by α_μ and ω_μ, respectively. D_μ is convex.

Γ contains no elements of finite order.

Let c be an axis of μ. Then the image $\bar{c}$ in $\bar{M}$ of a segment $c|[t_0, t_0 + d_\mu]$ of length d_μ is a closed geodesic $\bar{c}$ on $\bar{M}$ of length d_μ. $\bar{c}$ is a critical point of $E: \Lambda\bar{M} \to \mathbb{R}$ of index 0; it is an element where E assumes its infimum in the connected component of $\Lambda\bar{M}$ containing $\bar{c}$. There are no other critical points $\neq$ const of $E: \Lambda\bar{M} \to \mathbb{R}$ than the ones obtained in this manner.

Note. Here and in (3.8.14) we have indentified $\bar{c}(t)$ and its reparameterization $\bar{c}(t|\dot{\bar{c}}|)$.

Remark. If $\bar{M}$ has strictly negative curvature we get the following Theorem of Cartan [1]. In the case $\dim \bar{M} = 2$ it was proved by Hadamard [1]. Cartan's proof is not quite complete, as was pointed out by Berger [5].

3.8.14 Theorem. *Let $\bar{M}$ be a compact manifold with strictly negative curvature. Then every class of freely homotopic closed curves on $\bar{M}$, not containing the point curves, contains – up to parameterization – exactly one closed geodesic $\bar{c}$. $\bar{c}$ is the element of* *minimal length in its free homotopy class. $\bar{c}$ is hyperbolic (and hence, in particular, non-degenerate) of index 0. All closed geodesics on $\bar{M}$ are of this type.*

Remark. (3.8.13) and its subsequent application (3.8.17) to the structure of the fundamental group of a compact manifold with non-positive curvature was proved by Gromoll and Wolf [1] and Lawson and Yau [1]. The same result under weaker assumptions was proved by Eschenburg [1]. One also finds an exposition in Cheeger and Ebin [1].

Proof of (3.8.13), (3.8.14). Let $\{p_n\}$ be a sequence on M with $\lim f_\mu(p_n) = d_\mu = \inf f_\mu$. Since $\bar{M}$ is compact, there exists a finite $d > d(\bar{M})$ = diameter of $\bar{M}$. Fix $q \in M$. Then, for every $p \in M$, there exists a $\lambda \in \Gamma$ with $d(q, \lambda p) < d$. Simply observe that the images of q and p in $\bar{M}$ can be joined by a geodesic of length $\leqslant d(\bar{M})$.

Let now $\{\lambda_n\}$ be such that $p_n' = \lambda_n p_n$ has distance $< d$ from q, all n. For an appropriate subsequence of $\{p_n'\}$, which we denote again by $\{p_n'\}$, this sequence has a limit p'.

Put $\lambda_n \mu \lambda_n^{-1} = \mu_n$. Then $f_\mu(p_n) = f_{\mu_n}(p_n')$,

$$f_{\mu_n}(p') \leqslant 2d(p', p_n') + f_{\mu_n}(p_n') = 2d(p', p_n') + f_\mu(p_n),$$

i.e., $\{f_{\mu_n}(p')\}$ is bounded. The Γ-orbit of p' possesses only a finite number of different elements within bounded distance from p'. Therefore, there exists an infinite subsequence of $\{\lambda_n\}$ (which we denote again by $\{\lambda_n\}$) such that λ_n is equal to some λ, all n, hence $\mu_n = \lambda \mu \lambda^{-1}$ = (briefly) μ', all n. $\lim f_{\mu_n}(p') = f_{\mu'}(p') = d_\mu$. Put $\lambda^{-1} p' = p$. Then $f_\mu(p) = d_\mu$.

Clearly, $d_\mu > 0$.

For any p with $f_\mu(p) = d_\mu$, we consider the geodesic c with $c(0) = p$, $c(d_\mu) = \mu p$. Then this is an axis. Indeed, take t_0 with $0 \leqslant t_0 < d_\mu$. $c|[t_0, d_\mu]$, followed by $\mu c|[0, t_0]$ is a curve of length d_μ from $c(t_0)$ to $\mu c(t_0)$. These two points have distance $\geqslant d_\mu$, hence, our curve is the geodesic from $c(t_0)$ to $\mu c(t_0)$. It has no corner at $c(d_\mu)$.

Let c' be an arbitrary axis of μ, i.e., $\mu c'(t) = c'(t + d_\mu')$. Then $d_\mu \leqslant d_\mu'$. The relations

$$d_\mu' = \lim_{n \to \infty} \frac{1}{n} d(c'(0), c'(nd_\mu')) \leqslant$$

$$\leqslant \lim_{n \to \infty} \frac{1}{n} \{d(c'(0), c(0)) + d((c(0), c(nd_\mu)) + d(c(nd_\mu), c'(nd_\mu))\} =$$

$$= d_\mu + \lim_{n \to \infty} \frac{1}{n} d(\mu^n c(0), \mu^n c'(0)) = d_\mu$$

show that $d_\mu = d_\mu'$.

Since $d(c(t), c'(t))$ assumes all possible values in the compact interval $[0, d_\mu]$, $c(t)$ and $c'(t)$ have bounded distance for all $t \in \mathbb{R}$. According to (3.8.6), c and c' bound a flat, totally geodesic strip which is translated into itself by μ. The geodesic between any two points of c and c' lies on that strip and therefore belongs entirely to D_μ. Hence, D_μ is convex.

No $\mu \in \Gamma$, $\mu \neq id$, can have finite order, because if, say $\mu^m = id$, then $c(md_\mu) = c(0)$ for an axis c, which is impossible.

If $\mu c(t) = c(t + d_\mu)$, the image $\bar{c}$ of $c|[0, d_\mu]$ is a closed geodesic $\bar{c}$ in $\bar{M}$ of length d_μ. Any closed curve homotopic to $\bar{c}$ possesses a lift into M consisting of a curve going from some $p \in M$ to $\mu p \in M$. Thus, $\bar{c}$ has minimal length in its class of freely homotopic curves, therefore, index $\bar{c} = 0$. This one also reads off directly from the formula (2.5.1) for $D^2E(\bar{c})$, using $K \leqslant 0$.

If $K < 0$, a closed geodesic is hyperbolic, see (3.2.17), and there can be no flat strip. From (2.6.6) we know that there are no closed geodesics on M. Hence, $\bar{M}$ possesses no closed geodesics homotopic to a point curve. □

3.8.15 Remark. We wish to give still another proof of (3.8.14) which does not use the operation of the fundamental group $\pi_1 \bar{M}$ of $\bar{M}$ as group of deck transformation Γ on the universal covering M. Rather, it employs the theory of the space $\Lambda \bar{M}$ of closed H^1-curves.

First observe that a class of freely homotopic closed curves on $\bar{M}$ constitutes a connected component Λ' of $\Lambda \bar{M}$. As we showed in (2.4.19), whenever Λ' does not contain the set $\Lambda^0 \bar{M}$ of point curves, there exists on Λ' a critical point $\bar{c} \neq$ const with $E(\bar{c}) \leqslant E|\Lambda'$.

As we saw above, all closed geodesics on $\bar{M}$ have index 0 and are hyperbolic and hence, in particular, non-degenerate. Thus, all that remains to be shown is that a $\bar{c} \in \Lambda'$, $\Lambda^0\bar{M} \not\subset \Lambda'$, with minimal E-value on Λ' is essentially unique. More precisely, the only critical points of $E|\Lambda'$ are the elements $z.\bar{c}$ on the orbit $S^1.\bar{c}$ of $\bar{c}$.

The reason for this is that, for a non-degenerate closed geodesic $\bar{c}$ of index 0, E has a strict local minimum on $S^1.\bar{c}$. If there were two different such local minima for $E|\Lambda'$, say $S^1.\bar{c}_0$ and $S^1.\bar{c}_1$ with $E(\bar{c}_0) = \varkappa_0, E(\bar{c}_1) = \varkappa_1$, we would consider a curve $f: I \to \Lambda'$ from $f(0) = \bar{c}_0$ to $f(1) = \bar{c}_1$. The ϕ-family $\{\phi_s f(I); s \geqslant 0\}$ will remain hanging (cf. (2.4.20) for this terminology) at a critical set $Cr\varkappa^*$. Since $S^1.\bar{c}_0$ and $S^1.\bar{c}_1$ are strict local minima, $\varkappa^* > \varkappa_0, \varkappa_1$. But also every $S^1.\bar{c}^* \in Cr\varkappa^*$ is a local minimum which contradicts the existence of a curve $\phi_s f$ from $\bar{c}_0$ to $\bar{c}_1$ where $\sup E|\phi_s f(I)$ comes arbitrarily close to $\varkappa^*$.

That E has a strict local minimum on $S^1.\bar{c}$, if $\bar{c}$ is non-degenerate of index 0, means the following, cf. also (2.5.10). $D^2E(\bar{c})$ is positive definite on the space orthogonal to $T_{\bar{c}}(S^1.\bar{c})$, i.e., $T_{\bar{c}}^- \Lambda = 0$. For an $\bar{e}$ near $\bar{c}$, let $\xi = \xi(\bar{e})$ be its coordinate in the natural chart based at $\bar{c}$. Choose α positive and less than one half of the smallest positive eigen value of $D^2E(\bar{c})$. Then, when ξ^+ is the component of ξ in $T_{\bar{c}}^+ \Lambda$, we have

$$E(\bar{e}) \geqslant E(\bar{c}) + \alpha \|\xi^+\|_1^2$$

whenever $\|\xi^+\|_1 > 0$ and sufficiently small. This means that for all $\bar{e}$ on the boundary of a sufficiently small tubular neighborhood of $S^1 . \bar{c}$, $E(\bar{e}) - E(\bar{c})$ is bounded away from 0. □

As a preparation for the next Theorem we show

3.8.16 Proposition. *Under the hypothesis of* (3.8.13), *let* $\mu \in \Gamma$, $\mu \neq id$ *and let* D *be a closed, convex, non-empty subset of* M *with* $\mu D = D$.

Then D *contains an axis of* μ.

More precisely, if c *is an axis of* μ *such that* $c \cap D = \emptyset$, *and if* $q \in D$ *has minimal distance from* c *then the geodesic* c' *through* q *and* μq *is an axis of* μ.

Proof. Let c be an axis of μ, $c \cap D = \emptyset$. $d(c(t), D)$, $t \in \mathbb{R}$, already assumes all its values in the compact interval $[0, d_\mu]$, $d_\mu = \inf f_\mu$. Let $p = c(0)$ have minimal distance from D and let $q \in D$ be such that $d(p, q) = d(p, D)$. q is uniquely determined. If there were two different q, q' in D with $d(p, q) = d(q, q')$, consider the geodesic segment c^* from q to q'. $c^* \subset D$ and (3.8.1) then yields $d(p, c^*(t)) < d(p, q)$, for some interior point $c^*(t)$ of c^*.

Now let c_n be the geodesic from $\mu^{-n} q$ to $\mu^n q$. $c_n \subset D$ and since the end points of c_n have equal distance from c, (3.8.1) implies $d(c, c_n(t)) = d(p, q)$, for all t. Thus, all c_n meet a compact set and therefore there exists a geodesic $c'(t)$, $t \in \mathbb{R}$, which is the limit, when restricted to a compact interval $I \subset \mathbb{R}$, of a subsequence of $\{c_n | I\}$. Since $c' \subset D$ and $d(c'(t), c(t)) = d(p, q)$, $d(c'(t), \mu' c(t)) = d_\mu$, i.e., c' is an axis containing q. □

The announced consequence for the structure of the fundamental group of a compact manifold with $K \leqslant 0$ now reads, cf. Gromoll and Wolf [1] and Lawson and Yau [1]:

3.8.17 Theorem. *Let* $\bar{M}$ *be compact with* $K \leqslant 0$. *Let* $\Gamma \cong \pi_1 \bar{M}$ *be the group of deck transformations, operating on the universal covering* M *of* $\bar{M}$.

If $\Delta \subset \Gamma$ *is an abelian subgroup of rank* k, *then there exists a flat, totally geodesic* k*-dimensional submanifold* E^k *in* M *on which* Δ *operates as discrete subgroup.*

Hence, $\bar{M}$ *contains an isometrically immersed flat,* k*-dimensional torus* $\bar{E}^k = E^k/\Delta$.

It follows that the rank of an abelian subgroup of Γ *can at most be equal to* dim M.

Remark. The Theorem can easily be extended to solvable subgroups of Γ, cf. the references given above.

Proof. Let $\{\mu_1, .., \mu_k\}$ be a set of generators of the group Δ of rank k. Then

$$D = D_{\mu_1} \cap \ldots \cap D_{\mu_k} \neq \emptyset .$$

This will follow from (3.8.16) if we show that $\mu_i D_{\mu_j} = D_{\mu_j}$, all $i, j \in \{1, \ldots, k\}$.

Now, if $p \in D_{\mu_j}$, D_{μ_j} is μ_i-invariant indeed:

$$f_{\mu_j}(\mu_i p) \doteq d(\mu_i p, \mu_j \mu_i p) = d(\mu_i p, \mu_i \mu_j p) = d(p, \mu_j p) = f_{\mu_j}(p) = d_{\mu_j}.$$

We first consider the case $k = 2$. That is, let μ, μ' be elements of Γ, $\mu\mu' = \mu'\mu$, which generate a group Δ' of rank 2. If $p \in D_\mu \cap D_{\mu'}$ then there exists a 2-dimensional totally geodesic flat submanifold E' containing the μ-axis c through $p=c(0)$ and the μ'-axis c' through p. Moreover, for every $\mu^* \neq id$ in Δ', the geodesic c^* through p and $\mu^* p$ is an μ^*-axis.

To prove this we define E' as the union of the μ'-axes through $c(t)$, $t\in\mathbb{R}$. Fix $t_0\in\mathbb{R}$ and choose $m>0$ so large that $-md_\mu < t_0 < md_\mu$, $d_\mu = d(p,\mu p)$. Then we known from (3.8.13) that $\mu^{-m}c'$ and $\mu^m c'$ are the μ'-axes through $c(-md_\mu)$ and $c(md_\mu)$, respectively, which bound a flat strip containing the μ'-axis through $c(t_0)$. Thus, E' is a totally geodesic flat surface in M through p.

Now let $\mu^* = \mu^a \mu'^{a'}$, $a \neq 0$. We claim that the geodesic $c^* \subset E'$ through p and $\mu^* p$ is an axis of μ^*. To see this we note that $\mu^*(D_\mu \cap D_{\mu'}) = D_\mu \cap D_{\mu'}$. Therefore, $D_\mu \cap D_{\mu'} \cap D_{\mu^*}$ contains an element, say $\tilde{p}$, cf. (3.8.16). Let $\tilde{E}'$ be the totally geodesic flat surface which contains the μ-axis $\tilde{c}$ through $\tilde{p}$ and the μ'-axis $\tilde{c}'$ through $\tilde{p}$. On such a Δ'-invariant flat surface lies the μ^*-axis $\tilde{c}^*$ through $\tilde{p}$; $\tilde{c}^* \subset \tilde{E}'$. Since $d(p,\tilde{p}) = d(\mu^b\mu'^{b'}p, \mu^b\mu'^{b'}\tilde{p})$, all $(b,b')\in\mathbb{Z}\times\mathbb{Z}$, E' and $\tilde{E}'$ have bounded distance from each other, and, in particular, c^* and $\tilde{c}^*$ also have bounded distance from each other. But then they bound a flat strip and with $\tilde{c}^*$ also c^* must be a μ^*-axis.

We now can conclude the proof of the Theorem for the general case $k\geq 2$ as follows – for $k=1$ nothing needs to be proved. Let $p\in D$ and denote by T_pE^k the k-dimensional subspace of T_pM spanned by the tangents to the μ_i-axis through p, $1\leq i\leq k$. For each $\mu \neq id$ in Δ, there passes an μ-axis through p. The tangent vectors to these axes are dense in T_pE^k. Moreover, every 2-plane $\sigma\subset T_pE^k$ occurs as limit of 2-planes σ', where σ' is tangent to a totally geodesic flat surface, invariant under a subgroup of rank 2 of Δ. This shows that $\exp_p T_pE^k$ is a totally geodesic flat k-dimensional submanifold of M. □

As an immediate corollary we get the following result, first proved by Preissmann [1].

3.8.18 Theorem. *Let $\bar{M}$ be a compact manifold with $K<0$. Then every abelian subgroup of the fundamental group $\pi_1\bar{M}$, different from the identity, is infinite cyclic.*

Proof. Clearly, under our hypothesis, $\bar{M}$ cannot contain flat tori of dimension $k>1$, thus (3.8.17) applies. □

Note. This special case of (3.8.17) can also be deduced directly and much simpler from (3.8.14) and (3.8.6).

We continue with a result on manifolds where the hypothesis $K\leq 0$ is weakened to the hypothesis that there are no conjugate points. As we know from (2.6.6), such manifolds have a universal covering which is diffeomorphic to any of its tangent spaces. Despite this, not much is known about the finer geometric structure of such spaces. Mainly in the case of surfaces, the implications of the non-existence of conjugate points are well understood, see Morse and Hedlund [1], Green [1] and Eschenburg [1]. In (3.8.20) we will prove a result of E. Hopf on such surfaces. Cf. also (3.9.17).

We begin with a Theorem of Busemann [1].

3.8.19 Theorem. *Let $\bar{M}$ be a compact manifold without conjugate points. Assume that the fundamental group $\pi_1 \bar{M}$ of $\bar{M}$ is abelian. Then $\bar{M}$ can be simply covered by a $(\dim \bar{M} - 1)$-parameter family of closed geodesics without self-intersections, all having equal length. Such coverings correspond to the prime elements $\neq id$ in $\pi_1 \bar{M}$.*

Proof. We consider on the universal covering M of $\bar{M}$ the group $\Gamma \cong \pi_1 \bar{M}$ of deck transformations. Let $\mu \in \Gamma$ be prime, $\mu \neq 1$, i.e., there shall not exist $\mu_0 \in \Gamma$ and $m > 1$ such that $\mu = \mu_0^m$. We will show that through every point $p \in M$ there passes exactly one geodesic c, $c(0) = p$, with $\mu c(t) = c(t+a)$, $a > 0$, all $t \in \mathbb{R}$. This a does not depend on p.

If we now consider the images in $\bar{M}$ of the geodesics $c|[0, a]$, we get closed geodesics $\bar{c}$ of length a, not having self-intersection points. This is the desired covering of $\bar{M}$ by closed geodesics.

Let $f_\mu : M \to \mathbb{R}$ be the displacement function, i.e., $f_\mu(p) = d(p, \mu p)$. Since $f_\mu(\mu' p) = f_\mu(p)$, for all $\mu' \in \Gamma$, f_μ assumes all its values already in a compact subset of M. Thus, there exists a $p \in M$ where $f_\mu(p) = a = \max f_\mu$.

Let c be the geodesic with $c(0) = p$, $c(a) = \mu p = q$. Put $c(2a) = r$. Consider the triangle $qr\mu r = \mu pr\mu r$. The sides are uniquely determined. The side from μp to μr is given by $\mu c|[0, 2a]$ and has length $2a$. The side from $\mu p = q$ to r is $c|[a, 2a]$ and has length a. Hence $a \geqslant f_\mu(r) = d(r, \mu r) \geqslant - d(r, \mu p) + d(\mu p, \mu r) = a$. That is to say, $r = c(2a)$ is the midpoint of $\mu c|[0, 2a]$, $c(2a) = \mu^2 c(0)$, thus, $\mu c(t) = c(t+a)$, all $t \in \mathbb{R}$.

Let now $q \in M$ be arbitrary. Then

$$na = d(p, \mu^n p) \leqslant d(p, q) + d(q, \mu q) + \ldots + d(\mu^n q, \mu^n p) =$$
$$= 2d(p, q) + nd(q, \mu q) \leqslant 2d(p, q) + na\,.$$

Dividing by n and taking the limit for $n \to \infty$ we see $f_\mu(q) = d(q, \mu q) = a$. Thus, $f_\mu = \text{const}$ on M. For every $q \in M$, the geodesic c with $c(0) = q$, $c(a) = \mu q$ is invariant under μ and μ operates on c as the translation by a.

The above proof shows: If $\mu c(t) = c(t+a)$ then every $\mu' \in \Gamma$ which carries some point $c(t_0)$ of c into some point $c(t_0 + a')$ has c as axis und thus, μ' belongs to the cyclic group generated by μ. Hence, the projection $\bar{c}(t)$, $0 \leqslant t \leqslant a$, of the segment $c|[0, a]$ into $\bar{M}$ is a simple closed geodesic. □

We conclude this section with a Theorem due to E. Hopf [2]. It is an open problem whether it can be extended to arbitrary dimensions.

3.8.20 Theorem. *Let $\bar{M}$ be the 2-dimensional torus with a Riemannian metric without conjugate points. Then $\bar{M}$ is flat flat, i.e., $K \equiv 0$.*

Proof. According to (3.8.19) we can cover $\bar{M}$ by a 1-parameter family of closed geodesics without self-intersections. Let $c(t)$, $0 \leqslant t \leqslant a$, be one such geodesic. Index $c = 0$ and nullity $c > 0$, since c is minimizing in its free homotopy class and since we get a non-constant periodic Jacobi field $Y(t)$ along $c(t)$ as variation vector field of the 1-parameter family of closed geodesics covering M.

Y has no zeros, since the closed geodesics covering $\bar{M}$ do not cross each other. A zero of Y would also mean that there are conjugate points on c.

As in the proof of (3.4.2), we write $Y(t) = y(t)\,E_1(t)$ with a parallel unit vector field along $c(t)$, orthogonal to $E_0(t) = \dot{c}(t)$.

Put $\dot{y}(t)/y(t) = s(t)$. $s(t)$ corresponds to the self-adjoint operator $S(t)$ with $S(t)\,Y(t) = \nabla Y(t)$ in the proof of (3.2.17). $s(t)$ is periodic and satisfies the Riccati equation

$$(*) \qquad \dot{s}(t) + s(t)^2 + K(t) = 0\,.$$

We consider a second, transversal family of closed geodesics covering $\bar{M}$. Forming the integral of $(*)$ over $\bar{M}$ in the coordinates given by these two families of geodesics we get

$$\int_{\bar{M}} s^2\,do + \int_{\bar{M}} K do = 0\,.$$

But since $\bar{M}$ is a torus, according to the Gauss-Bonnet formula, the second integral is zero, cf. Klingenberg [9]. Hence $s(t) = \dot{y}(t) = K(t) = 0$. □

3.9 The Geodesic Flow on Manifolds of Negative Curvature

We now restrict ourselves to compact manifolds of strictly negative curvature K. Then we have positive numbers $0 < k_1 \leqslant k_0$ such that $-k_0^2 \leqslant K \leqslant -k_1^2 < 0$. Occasionally, we also consider non-compact manifolds satisfying such a curvature condition. This then is called a manifold of bounded negative curvature. In particular, the universal covering of a compact manifold of negative curvature will be of this type.

All results of (3.8) concerning manifolds with $K \leqslant 0$ also hold here, of course. But the hypothesis $K < 0$ has a great many additional implications which do not hold e.g. for flat manifolds. At the basis of this additional structure is the presence of the so-called stable and unstable manifolds for the geodesic flow for manifolds of negative curvature. These objects are closely related to the horospheres (horocycles in the case of dimension 2) of the classical hyperbolic geometry. The existence of similar objects in manifolds of variable negative curvature is – at least locally – intimately related to the question of the structural stability of a system of differential equations. This problem has a long history, cf. the introduction to Anosov's paper cited below.

The full impact of these concepts for the global structure of the geodesic flow on an arbitrary negatively curved manifold has been recognized only relatively recently by Anosov and Sinai, cf. the paper of Anosov and Sinai [1] and Anosov [1]. Cf. also Arnold and Avez [1].

All manifolds considered shall be complete; geodesics shall be parameterized by arc length.

We begin with a global version of (3.2.17). Recall from (3.2.1) the definition of the bundle

$$\bar{\tau}_{T_1 M}\colon \bar{T}T_1 M \to T_1 M .$$

If $\dim M = n+1$, the fibre $\bar{T}_{X_0} T_1 M$ over $X_0 \in T_1 M$ is the sum of the two n-dimensional spaces

$$\bar{T}_{X_0 h} T_1 M = \{(X, 0) \in T_{X_0 h} T_1 M; \langle X, X_0 \rangle = 0\} .$$

$$\bar{T}_{X_0 v} T_1 M = \{(0, X) \in T_{X_0 v} T_1 M; \langle X, X_0 \rangle = 0\} .$$

If M has strictly negative curvature, we get an additional and flow-invariant splitting of $\bar{\tau}_{T_1 M}$ as follows:

3.9.1 Theorem. *Let M be a complete manifold of bounded negative curvature, say $-k_0^2 \leqslant K \leqslant -k_1^2 < 0$. Then the bundle $\bar{\tau}_{T_1 M}$ splits continuously into the sum of two $T\phi_t$-invariant subbundles*

$$\bar{\tau}_{T_1 M} = \bar{\tau}_{T_1 M,s} \oplus \bar{\tau}_{T_1 M,u}\colon \bar{T}_s T_1 M \oplus \bar{T}_u T_1 M \to T_1 M ,$$

called stable *and* unstable bundle, *respectively. Each of these bundles has fibre dimension n. $T\phi_t | \bar{T}_s T_1 M$ operates as contraction while $T\phi_t | \bar{T}_u T_1 M$ operates as expansion. More precisely, for all $t_1 \geqslant t_0$,*

$$\frac{k_1}{k_0}\, e^{-k_0(t_1 - t_0)} \leqslant \left| T\phi_{t_1} \circ T\phi_{t_0}^{-1} | \bar{T}_s T_1 M \right| \leqslant \frac{k_0}{k_1}\, e^{-k_1(t_1 - t_0)}$$

and

$$\frac{k_0}{k_1}\, e^{k_0(t_1 - t_0)} \geqslant \left| T\phi_{t_1} \circ T\phi_{t_0}^{-1} | \bar{T}_u T_1 M \right| \geqslant \frac{k_1}{k_0}\, e^{k_1(t_1 - t_0)} .$$

Note. The splitting $\bar{\tau}_{T_1 M}$ into the stable and the unstable sub-bundle in general is not differentiable. Thus, it is not a splitting in the category of smooth vector bundles in the sense of (1.2.10).

Proof. This follows immediately from (3.2.17), where we proved the above relations for the bundle $\upsilon = \upsilon_{X_0}\colon V_{X_0} \to \mathbb{R}$, induced from $\bar{T}T_1 M$ by the immersion $t \mapsto \phi_t X_0$, cf. (3.2.2). We define $\bar{T}_{X_0 s} T_1 M$ and $\bar{T}_{X_0 u} T_1 M$ by $V_{0,s}$ and $V_{0,u}$, using the canonical isomorphism of the fibre V_0 of υ with $\bar{T}_{X_0} T_1 M$.

The continuity of the splitting can be read off from the proof of (3.2.17): Recall that e.g. $V_{t,u} \cong T_{\phi_t X_0 u} T_1 M$ was defined with the help of the self-adjoint operator

$$S_u(t)\colon V_{t,h} \cong \bar{T}_{X_0 h} T_1 M \to V_{t,h} \cong \bar{T}_{X_0 h} T_1 M$$

where $S_u(t)$ is obtained as the uniform limit of operators $S_\tau(t)$, for $\tau \to -\infty$. $S_\tau(0)$ is chosen solely subject to the condition that the spectrum of S_τ belongs to $[k_1, k_0]$. If we let depend such operators smoothly on $X \in T_1 M$, X in a neighborhood of X_0, then the resulting $S_u(t)$ will depend continuously on X. □

We now show how the splitting of $\bar{\tau}_{T_1 M}$ into a stable and an unstable bundle yields the stable and the unstable manifolds.

The local existence theorem, in the form we are going to use it, was proved by Perron [1]. Anosov [1] uses a different method which is an extension and an elaboration of a method invented by Hadamard [3].

Finally, we point out that stable and unstable manifolds occur under much more general situations than the one we are encountering here. All that is needed is a certain hyperbolic structure for a dynamical system, i.e., for the integral curves of a vector field on a manifold. See Anosov [1] for more on this.

3.9.2 Theorem. *Let M be a Riemannian manifold of bounded negative curvature. Then there exists, for every $X_0 \in T_1 M$, immersions*

$$W^{ss}_{X_0}: (\bar{T}_{X_0 s} T_1 M, 0_{X_0}) \to (T_1 M, X_0)$$

$$W^{uu}_{X_0}: (\bar{T}_{X_0 u} T_1 M, 0_{X_0}) \to (T_1 M, X_0)$$

which are tangent to the stable and unstable fibres, respectively. They are called (strong) stable *and* (strong) unstable manifolds *at X_0, respectively.*

The images of these immersions are ϕ_t-invariant, i.e.,

$$\operatorname{im} \phi_t W^{ss}_{X_0} = \operatorname{im} W^{ss}_{\phi_t X_0}; \ \operatorname{im} \phi_t W^{uu}_{X_0} = \operatorname{im} W^{uu}_{\phi_t X_0}.$$

Every orbit $\phi_t X$ of an $X = W^{ss}_{X_0}(\xi)$ on the strong stable manifold at X_0 is (positively) asymptotic to X_0. More precisely, $d_{T_1 M}(\phi_t X, \phi_t X_0) \leqslant a^* e^{-b^* t}$ for $t \geqslant 0$, with constants $a^* \geqslant 1, b^* > 0$, only depending on M.

From $W^{ss}_{X_0}$, $W^{uu}_{X_0}$ we get the immersions

$$W^{s}_{X_0}: T^1_{X_0} T_1 M \oplus \bar{T}_{X_0 s} T_1 M \to T_1 M; \ (t, \xi_s) \mapsto \phi_t W^{ss}_{X_0}(\xi_s)$$

$$W^{u}_{X_0}: T^1_{X_0} T_1 M \oplus \bar{T}_{X_0 u} T_1 M \to T_1 M; \ (t, \xi_u) \mapsto \phi_t W^{uu}_{X_0}(\xi_u),$$

called stable *and* unstable *manifolds (at X_0), respectively. Here we have identified $T^1_{X_0} T_1 M$ with $\mathbb{R}$ via the canonical basis $(X_0, 0)$ of $T^1_{X_0} T_1 M$.*

Proof. Since we can view $W^{uu}_{X_0}$ as $W^{ss}_{-X_0}$, we can restrict ourselves to the stable case.

We introduce Fermi coordinates, based on the geodesic $\phi_t X_0$ in $T_1 M$. A sufficiently small tubular neighborhood of the 0-section of the induced bundle $\upsilon = \upsilon_{X_0}: V = V_{X_0} \to \mathbb{R}$, cf. (3.2.2), is, via these coordinates, immersed into $T_1 M$, although in general it is not embedded. Still, using sections in υ with values in this neighborhood, we get a description of the flow near $\phi_t X_0$ for all $t \in \mathbb{R}$.

From (3.1.17) we know already the linear approximation of the flow ϕ_t near the orbit $\phi_t X_0$. If $\xi = (\xi_s, \xi_u) \in V_{0,s} \oplus V_{0,u}$, this approximation reads

$$t \in \mathbb{R} \mapsto (t, P_t \,.\, \xi_s, P_t \,.\, \xi_u) \in \mathbb{R} \oplus V_{t,s} \oplus V_{t,u}.$$

Here, $P_t: V_0 \to V_t$ corresponds under the bundle morphism $F: V_{X_0} \to \bar{T} T_1 M$ to the transformation $T\phi_t: \bar{T}_{X_0} T_1 M \to \bar{T}_{\phi_t X_0} T_1 M$, cf. (3.2.2).

We write the vector field on V, given by the linear approximation P_t of the flow, in the form

$$(1, A_s(t) \,.\, \xi_s, A_u(t) \,.\, \xi_u) \in T_{(t, \xi)} V.$$

The actual vector field ζ_E of the geodesic flow now is viewed as a perturbation of this vector field, i.e., we write the coordinates of ζ_E in the form

$$(*) \qquad \big(1 + C_0(t, \xi), A_s(t) . \xi_s + C_s(t, \xi), A_u(t) . \xi_u + C_u(t, \xi)\big) \in T_{(t,\xi)} V$$

with $C_*(t, \xi) = o(|\xi|)$, $* \in \{0, s, u\}$.

We are interested in flow lines $\phi_t X$ near $\phi_t X_0$ which, for $t \to \infty$, become asymptotically close to $\phi_t X_0$. From (3.9.1) we have the estimates

$$(**) \qquad \begin{aligned} a^{-1} e^{-c(t-\tau)} &\leqslant |P_t \circ (P_\tau | V_{0,s})^{-1}| \leqslant a e^{-b(t-\tau)}; \; t \geqslant \tau \\ a^{-1} e^{-c(\tau-t)} &\leqslant |P_t \circ (P_\tau | V_{0,u})^{-1}| \leqslant a e^{-b(\tau-t)}; \; \tau \geqslant t \end{aligned}$$

with $a = k_0/k_1 \geqslant 1$, $b = k_1 \leqslant c = k_0$. Therefore, the right hand side of the following integral equation for $\xi(t) = (\xi_s(t), \xi_u(t)) \in V_{t,s} \oplus V_{t,u}$ is well defined, for every bounded $\xi(t)$, $t \geqslant 0$, every $\xi_s \in V_{0,s}$:

$$(\dagger) \qquad \begin{aligned} \xi_s(t) &\equiv \xi_s(t; \xi_s) = P_t . \xi_s + P_t . \int_0^t P_\tau^{-1} . C_s(\tau, \xi(\tau))\, d\tau \\ \xi_u(t) &\equiv \xi_u(t; \xi_s) = P_t . \int_\infty^t P_\tau^{-1} . C_u(\tau, \xi(\tau))\, d\tau . \end{aligned}$$

As we will see below, there exist a solution $\xi(t; \xi_s)$ of $(\dagger)$, depending differentiably on ξ_s. Moreover,

$$|\xi(t; \xi_s)| \leqslant 2a |\xi_s| e^{-b't},$$

with $0 < b' < b < 2b'$ and $a \geqslant 1$ as in $(**)$.

Therefore, we can write

$$\tilde{\xi}(t; \xi_s) = (\xi_0(t; \xi_s), \xi_s(t; \xi_s), \xi_u(t; \xi_s)) \in \mathbb{R} \oplus V_{t,s} \oplus V_{t,u}$$

with

$$\xi_0(t; \xi_s) = \int_\infty^t C_0(\tau, \xi(\tau))\, d\tau .$$

The tangent vector of the curve $t \mapsto t + \tilde{\xi}(t; \xi_s)$ is precisely the vector $(*)$ in the point $(t, \xi(t; \xi_s)) \in V$. That is, $t + \tilde{\xi}(t; \xi_s)$ is the coordinate representation of an orbit $\phi_t X$.

We now define $W^{ss}_{X_0}$ – at least on a neighborhood of the origin of $\bar{T}_{X_0} T_1 M$ – in coordinates by

$$\xi_s \in V_{0,s} \mapsto \tilde{\xi}(0; \xi_s) = \Big(\int_\infty^0 C_0(\tau, \xi(\tau))\, d\tau, \xi_s,$$

$$\int_\infty^0 P_\tau^{-1} . C_u(\tau, \xi(\tau))\, d\tau \in \mathbb{R} \oplus V_{0,s} \oplus V_{0,u} .$$

For $t \to \infty$, $\tilde{\xi}(t; \xi_s) \to 0$, i.e., the corresponding orbit $\phi_t X$ approaches the orbit $\phi_t X_0$ for $t \to \infty$. Therefore, ϕ_t carries $W^{ss}_{X_0}$ into $W^{ss}_{\phi_t X_0}$. Moreover, this shows that for $X = W^{ss}_{X_0}(\xi_s)$, $|\xi_s|$ sufficiently small,

$$W_X^{ss} \text{ near } 0 \in \bar{T}_{Xs} T_1 M \text{ and } W_{X_0}^{ss} \text{ near } \xi_s \in \bar{T}_{X_0 s} T_1 M$$

have the same image. Since $\bar{T}_{X_0 s} T_1 M \cong \mathbb{R}^n$, $W_{X_0}^{ss}$ can be extended uniquely to an immersion to all of $T_{X_0 s} T_1 M$.

All that remains to be done to complete the proof is to solve the integral equation (†), for a given $\xi_s \in V_{0,s} \, |\xi_s|$ small if necessary. But this is a consequence of the implicit function theorem for Banach spaces, cf. Dieudonné [1].

Indeed, consider the Banach space $\mathscr{B}$ of bounded differentiable sections $\xi(t) = (\xi_s(t), \xi_u(t))$ of υ_{X_0}, defined for $t \geqslant 0$, with the maximum norm. We abbreviate the right hand side of (†) by $G(\xi, \xi_s)$. Then we have a differentiable mapping

$$G: \mathscr{B} \times V_{0,s} \to \mathscr{B}$$

with $G(0,0) = 0$, and

$$\{D_1 G(0,0): T_0 \mathscr{B} \to T_0 \mathscr{B}\} = 0 .$$

The implicit function theorem therefore applies to the equation $F(\xi, \xi_s) \equiv \xi - G(\xi, \xi_s) = 0$. That is, for ξ_s in some small ball around $0 \in V_{0,s}$, there exists a differentiable mapping $\xi_s \to \xi(\xi_s) \in \mathscr{B}$ with $G(\xi(\xi_s), \xi_s) = \xi(\xi_s)$. □

Remark. The existence of a solution of the integral equation (†), $\xi = G(\xi, \xi_s)$, is proved by an iteration process. We write down the simple estimates. With a, b as in (∗∗), choose b' with $0 < b' < b < 2b'$ and put $(b - b')/4a = \varepsilon$. Let $\delta > 0$ be so small that

$$|C_*(t, \xi) - C_*(t, \xi')| \leqslant \varepsilon |\xi - \xi'|, \text{ whenever } |\xi| < \delta, |\xi'| < \delta, * \in \{s, u\} .$$

Now fix $\xi_s \in V_{0,s}$ and start with $\xi^{(0)} = (0,0)$. Define $\xi^{(k+1)} \equiv (\xi^{(k+1)}(t; \xi_s)) \in \mathscr{B}$ recursively by $G(\xi^{(k)}, \xi_s)$. Then

$$(\S)^{(k)} \qquad |\xi^{(k+1)}(t; \xi_s) - \xi^{(k)}(t; \xi_s)| \leqslant a 2^{-k} |\xi_s| e^{-tb'} .$$

Thus, the sequence $\{\xi^{(k)}(t; \xi_s)\}$ is uniformly convergent with a limit $\xi(t; \xi_s)$ such that $|\xi(t; \xi_s)| \leqslant 2a |\xi_s| e^{-tb}$.

We prove $(\S)^{(k)}$ by induction on k. Since $\xi^{(0)}(t; \xi_s) = (0,0)$, $\xi^{(1)}(t; \xi_s) = (P_t \,.\, \xi_s, 0)$, i.e., $(\S)^{(0)}$ does hold.

Using $(\S)^{(k-1)}$ we get

$$|\xi_s^{(k+1)}(t; \xi_s) - \xi_s^{(k)}(t; \xi_s)| \leqslant$$

$$\leqslant \int_0^t a e^{-(t-\tau)b} \varepsilon a 2^{-(k-1)} |\xi_s| e^{-\tau b'} d\tau \leqslant a 2^{-k} |\xi_s| e^{-tb'}/2 ,$$

$$|\xi_u^{(k+1)}(t; \xi_s) - \xi_u^{(k)}(t; \xi_s)| \leqslant$$

$$\leqslant \int_t^\infty a e^{-(\tau-t)b} \varepsilon a 2^{-(k-1)} |\xi_s| e^{-\tau b'} d\tau \leqslant a 2^{-k} |\xi_s| e^{-tb'}/2 ,$$

i.e., by summing up, we get $(\S)^{(k)}$. □

We now come to the relation between asymptotic geodesics and stable manifolds.

3.9.3 Theorem. *Let M be a simply connected manifold of bounded negative curvature. Then, for every $X_0 \in T_1 M$, the composed mapping*

$$\tau_M \circ W^s_{X_0}: T^1_{X_0} T_1 M \oplus \bar{T}_{X_0 s} T_1 M \to M$$

is a diffeomorphism. In particular, the stable manifold $W^s_{X_0}$ is an injective immersion.

Put $\tau_M \phi_t X_0 = c_0(t)$. Then a geodesic $c(t), |\dot{c}(t)| = 1$, is asymptotic to $c_0(t)$ if and only if we have a $t_0 \in \mathbb{R}$ and an $X = W^{ss}_{X_0}(\xi)$ such that $\dot{c}(t) = \phi_{t+t_0} X$.

Hence, the ϕ_t-orbits on the stable manifold $W^s_{X_0}$ correspond, under the projection $\tau_M | W^s_{X_0}$, precisely to the class of asymptotic geodesics determined by c_0. Those among these geodesics c for which $\lim_{t \to \infty} d(c_0(t), c(t)) = 0$ are characterized by $\dot{c}(0) = X \in \operatorname{im} W^{ss}_{X_0}$.

Proof. The tangent space in $X = W^s_{X_0}(t_0, \xi)$ to $W^s_{X_0}$ consists of the subspace $T^1_X T_1 M \oplus \bar{T}_{Xs} T_1 M$. The angle between this space and the vertical space $\bar{T}_{Xv} T_1 M = \ker(T\tau_M | \bar{T}_X T_1 M)$ is bounded away from zero. Indeed, while $T^1_X T_1 M$ is orthogonal to the vertical space, we known from the proof of (3.2.17) that the stable subspace at X can be described by

$$\{\xi_h, S_s . \xi_h\} \in \bar{T}_{Xh} T_1 M \oplus \bar{T}_{Xv} T_1 M;$$

Here, S_s is a self-adjoint operator with spectrum in $[-k_0, -k_1]$. This shows that the angle γ between $\xi_s \in \bar{T}_{Xs} T_1 M$ and its vertical component satisfies $\cos\gamma \leq k_0 / \sqrt{1 + k_0^2}$.

Therefore, $\tau_M \circ W^s_{X_0}$ is not only of maximal rank everywhere, but it is also surjective. Since both, the domain and the range, are simply connected, $\tau_M \circ W^s_{X_0}$ is in fact a diffeomorphism.

We know that for $X = W^{ss}_{X_0}(\xi)$, $d_{T_1 M}(\phi_t X_0, \phi_t X) \leq a^* e^{-b^* t}$, some $a^* \geq 1, b^* > 0$, all $t \geq 0$. Thus, the geodesics $c(t) = \tau_M \phi_t X$ with $X \in \operatorname{im} W^{ss}_{X_0}$ satisfy $d(c_0(t), c(t)) \leq a^* e^{-b^* t}$, for $t \to \infty$. Every $p \in M$ occurs exactly once as the image $\tau_M X$ of a $X \in \operatorname{im} W^s_{X_0}$. If $X = \phi_{t_0} W^{ss}_{X_0}(\xi)$, we have for $c(t) = \tau_M \phi_t X$ the relation $\lim_{t \to \infty} d(c(t), c_0(t)) = |t_0|$. □

As a consequence we get a generalization of the classical concept of a horosphere (horocycle, if $\dim M = 2$). Such a generalization, even to spaces of much more general character, has been considered mainly by Busemann [1]. The function $b_{X_0}: M \to \mathbb{R}$ which we define below therefore is called the *Busemann function.*

3.9.4 Corollary. *Let $X_0 \in T_1 M$. Then*

$$h_{X_0} = \tau_M \circ W^s_{X_0}: T^1_{X_0} T_1 M \oplus \bar{T}_{X_0 s} T_1 M \cong \mathbb{R} \times V_s \to M$$

is a diffeomorphism. V_s is an n-dimensional vector space, $n + 1 = \dim M$. For each $t \in \mathbb{R}$, $h_{X_0} | \{t\} \times V_s$ is called horosphere *(with parameters (t, X_0)). As submanifold, we also denote it by $H_{\phi_t X_0}$.*

The family $\{H_{\phi_t X_0}; t \in \mathbb{R}\}$ is a so-called foliation of M of codimension 1, all *leaves* $\cong \mathbb{R}^n$. *In particular, if* t_0 *and* t_1 *are different, then* $H_{\phi_{t_0}X_0}$ *and* $H_{\phi_{t_1}X_0}$ *are disjoint.*

The geodesics $\tau_M \phi_t X$, $X \in \operatorname{im} W^{ss}_{X_0}$, *i.e., the* τ_M*-images of the* ϕ_t*-orbits on* $W^s_{X_0}$, *are the orthogonal trajectories of this foliation.*

Define $b_{X_0}: M \to \mathbb{R}$ *by* $b_{X_0}(q) = t$ *if* $q \in H_{\phi_t X_0}$. *Then the horosphere* $H_{\phi_t X_0}$ *becomes the level set* $\{b_{X_0} = t\}$ *of the function* b_{X_0}. $b_{X_0}(q)$ *measures the distance between* q *and* H_{X_0}. □

Note. Usually, the horospheres are defined directly, not via the strong unstable manifolds. In this case, the hypothesis $K<0$ can be weakened to $K \leqslant 0$, and even less, cf. Eschenburg [1]. K should have a lower bound, however. Under these weaker assumptions, the horospheres may lose in their degree of differentiability. More seriously, under these weaker hypotheses there need not exist a counter part to the projections of the strong unstable manifolds.

The idea of the direct introduction of horosopheres on simply connected manifolds M is to define them as the boundary $\partial \bar{B}_c$ of the set $\bar{B}_c = \bigcup_{t \geqslant 0} \bar{B}_t(c(t))$, for a ray $c(t)$, $t \geqslant 0$, cf. the definition before (2.9.9). If now $K \leqslant 0$, $\partial \bar{B}_c$ is of class C^2. $\bar{B}_c$ itself is totally convex, cf. (2.9.6). Recall that in (2.9.9) we saw that, if $K \geqslant 0$, $\complement B_c$ is totally convex.

In (3.8.9) we had defined the concept of a point at infinity for a simply connected manifold M with $K \leqslant 0$. Also recall that a geodesic c determines its negative and positive point at infinity, denoted by $\alpha_c = c(-\infty)$ and $\omega_c = c(+\infty)$, respectively.

The following result does not alway hold if we only assume $K \leqslant 0$, see e.g. the Euclidean space. Therefore, we restrict ourselves to manifolds of bounded negative curvature. A certain weakening of this hypothesis is possible, cf. Eschenburg [1] and the references given there. Eberlein and O'Neill [1] use the property as an axiom, the so-called visibility axiom.

3.9.5 Theorem. *Let* M *be simply connected of bounded negative curvature. Let* α *and* ω *be two different points at infinity. Then there exists a unique (up to reparameterization) geodesic* c *in* M *with* $c(-\infty) = \alpha$, $c(+\infty) = \omega$.

It follows that in $\operatorname{cp} M = M \cup M(\infty)$, *any pair of different points can be joined by an essentially uniquely determined geodesic.*

3.9.6 Complement. *Assume* $-k_0^2 \leqslant K \leqslant -k_1^2$, *with* $0 < k_1 \leqslant k_0$. *Let* α *and* ω *be different points at infinity. Let* $p \in M$ *and denote by* 2γ *the angle subtended by the geodesics* c' *and* c'' *going from* p *to* α *and* ω, *respectively. Thus in particular* $0 < 2\gamma \leqslant \pi$. *Then the distance* d *from* p *to the geodesic* c *going from* α *to* ω *is bounded by*

$$d_{k_0}(\gamma) \leqslant d \leqslant d_{k_1}(\gamma)\,.$$

Here, $d_{k_0}(\gamma)$ *and* $d_{k_1}(\gamma)$ *denote the distances in the hyperbolic spaces* $H^n_{k_0}$ *and* $H^n_{k_1}$ *for the same geometric situation. According to* (3.8.7, iv), $d_0 = d_{k_0}(\gamma)$ *and* $d_1 = d_{k_1}(\gamma)$ *are given by* $\operatorname{tgh}(d_0/k_0) = \cos\gamma = \operatorname{tgh}(d_1/k_1)$.

Proof. We first prove the complement. Let $\{t_n\}$ be a sequence of positive numbers going to ∞. Put $c'(t_n) = p'_n$, $c''(t_n) = p''_n$. Let c_n be the geodesic from p'_n to p''_n. The perpendicular from p to (the complete) geodesic c_n shall meet c_n in the point q_n. Put $d(p, q_n) = d_n$.

We consider in $M^* = H^n_{k_1}$ Alexandrov triangles $p^* q^*_n p'^*_n$ and $p^* q^*_n p''^*_n$, associated with $pq_n p'_n$ and $pq_n p''_n$, respectively.

In these Alexandrov triangles, the angles have not become smaller. Thus, if we put the two Alexandrov triangles together at their common side $p^* q^*_n$ we get a planar quadrangle where the angle at q^*_n is $\geqslant \pi$. By straightening out this angle we see that the distance d^*_n from p^* to the opposite side in an Alexandrov triangle $p^* p'^*_n p''^*_n$ associated with $pp'_n p''_n$ is $\geqslant d_n$. Since in $p^* p'^*_n p''^*_n$ the angle at p^* is $\geqslant \measuredangle p = 2\gamma$, d^*_n is bounded from above by the number $d_{k_1}(\gamma)$ derived in (3.8.7, iv). Hence, $d_n \leqslant d_{k_1}(\gamma)$.

Similar arguments yield $d_{k_0}(\gamma)$ as lower bound for d_n. Thus, we have proved the Complement.

Let now α and ω on $M(\infty)$ be given. Choose $p \in M$ and let c', c'' be the geodesics from p to α and ω, respectively. $\dot{c}'(0)$ and $\dot{c}''(0)$ form an angle $2\gamma > 0$. Consider, for a sequence $\{t_n\}$ going to infinity, the sequence $\{c_n\}$ of geodesics from $c'(t_n)$ to $c''(t_n)$. As we just saw, $d(p, c_n)$ is bounded. Hence, the sequence $\{X_n\}$ of tangent vectors to c_n in the foot point of the perpendicular from p to c_n belongs to a compact subset. Therefore, $\{X_n\}$ possesses a convergent subsequence with limit X. The geodesic c with $\dot{c}(0) = X$ then goes from α to ω.

c is essentially unique. Indeed, if $\tilde{c}$ with $\dot{\tilde{c}}(0) = \tilde{X}$ is a geodesic from α to ω then it is the τ_M-projection of an orbit $\phi_t \tilde{X}$ on the stable manifold W^s_X which remains in bounded distance from c, also for $t \to -\infty$. But this can only hold if $\tilde{X} = \phi_{t_0} X$, some $t_0 \in \mathbb{R}$. $\square$

The following lemma is fundamental to the subsequent results on the density of periodic orbits and the transitivity of the geodesic flow on a compact Riemannian manifold. For the special case of a compact surface of genus $\geqslant 2$ it can be traced back to F. Klein who investigated discrete groups on the hyperbolic plane. In any case, for surfaces of constant negative curvature it was proved by Koebe [1]. Nielsen [1] extended these results to surfaces of variable negative curvature.

The hypothesis can be weakened to include certain groups Γ with non-compact quotients, but such that the quotient possesses finite volume.

3.9.7 Lemma. *Let Γ be a group of deck transformations on a simply connected manifold M of strictly negative curvature such that $\bar{M} = M/\Gamma$ is compact.*

Then every point ω at infinity belongs to the closure $\overline{\Gamma . p}$ in $\mathrm{cp}M = M \cup M(\infty)$ of the orbit $\Gamma . p$ of an arbitrary point $p \in M$. Moreover, the orbit $\Gamma . \omega$ of every $\omega \in M(\infty)$ is dense in $M(\infty)$.

Proof. Choose $p \in M$, $\omega \in M(\infty)$. Let c be the geodesic with $c(0) = p$, $c(\infty) = \omega$. Since $\bar{M}$ is compact, the diameter $d(\bar{M})$ of $\bar{M}$ is finite. Choose $d > d(\bar{M})$. As we saw in the proof of (3.8.14), for every integer n there exists a $\mu_n \in \Gamma$ with $d(\mu_n p, c(n)) < d$.

For large n, $\mu_n p \neq p$. Let $\{c_n\}$ be the sequence of geodesics from p to $\mu_n p$. $\{c_n\}$ converges towards c. Since

$$d(\mu_n p, p) \geqslant -d(\mu_n p, c(n)) + d(c(0), c(n)) > -d + n,$$

$\{\mu_n p\}$ converges towards ω. Thus $\overline{\Gamma . p} \supset M(\infty)$.

It follows that Γ has no fixed point on $M(\infty)$. Indeed, if there were an $\omega \in M(\infty)$ with $\Gamma . \omega = \omega$, the axes of every $\mu \in \Gamma, \mu \neq id$, would have to pass through ω. Thus, the underlying closed geodesics on $\bar{M}$ would be arbitrarily close, which is impossible.

Let $\mu \in \Gamma - \{id\}$. Let α_μ and ω_μ be the negative and positive limit points of an axis c_μ of μ. Then $\Gamma . \alpha_\mu \cup \Gamma . \omega_\mu$ is dense in $M(\infty)$. Simply observe that the orbit $\Gamma . p$ of the point $p = c_\mu(0)$ has $M(\infty)$ in its closure.

There exists $\mu^* \in \Gamma - \{id\}$ which moves both, α_μ and ω_μ. The sequences $\{\mu^{*n} \alpha_\mu\}$ and $\{\mu^{*n} \omega_\mu\}$ both converge to ω_{μ^*}. Hence, α_μ and $\omega_\mu \in \overline{\Gamma . \omega_{\mu^*}}$, therefore since $\overline{\Gamma . \alpha_\mu \cup \Gamma . \omega_\mu} = M(\infty)$, $\overline{\Gamma . \omega_{\mu^*}} = M(\infty)$. Similarly, $\overline{\Gamma . \alpha_{\mu^*}} = M(\infty)$. Since, for every $\omega \in M(\infty)$, $\overline{\Gamma . \omega} \in \alpha_{\mu^*}$ or $\overline{\Gamma . \omega} \in \omega_{\mu^*}$, we have $\overline{\Gamma . \omega} = M(\infty)$. □

As a first application of (3.9.7) we now show:

3.9.8 Theorem. *Let $\bar{M}$ be a compact manifold of (strictly) negative curvature. Then the periodic orbits of the geodesic flow on $T_1 \bar{M}$ are dense.*

That is to say, if Per $T_1 \bar{M}$ *denotes the set of $X \in T_1 \bar{M}$ for which the orbit $\{\phi_t X\}$ is periodic, then its closure is all of $T_1 \bar{M}$.*

An equivalent statement is the following: Let M be the universal covering of $\bar{M}$ and $\Gamma \cong \pi_1 \bar{M}$ the group of deck transformations operating on M. Then any pair (α, ω) of different points at infinity of M is the limit of a sequence $\{\alpha_n, \omega_n\}$ where α_n and ω_n are the end points $c_n(-\infty)$, $c_n(+\infty)$ of an axis of an element $\mu_n \in \Gamma$.

Remark. Sometimes one also briefly says that, on a compact manifold $\bar{M}$ of negative curvature, the closed geodesics are dense. Note, however, that this means more than saying that every point on $\bar{M}$ can be approximated by points lying on a closed geodesic; in addition, every tangent vector can be approximated by tangent vectors to closed geodesics.

We also remark that the compact manifolds of negative curvature are the best known examples of Riemannian manifolds where the closed geodesics are dense not only for the given metric, but also for all sufficiently small perturbations of the given metric.

Indeed, such perturbations will preserve the property of the manifold to have negative curvature. The deeper reason for this phenomenon is that the geodesic flow on a manifold of negative curvature possesses the hyperbolic structure described in (3.9.1). Whenever a dynamical system is of this type, the periodic orbits are dense (provided some volume form is kept invariant by the system). Even more is true: Such systems are stable in the sense that any small perturbation of the system will yield a system which is C^0-conjugate to the original one. This is due to Anosov [1]. Cf. also Arnold and Avez [1] and Abraham and Marsden [1].

Proof. We prove the second version of the Theorem. That is, let (α, ω) be a pair of different points at infinity and let c be a geodesic joining α to ω, $\alpha = c(-\infty)$, $\omega = c(+\infty)$. Let A and Ω be arbitrarily small neighborhoods of α and ω in $M \cup M(\infty)$, respectively. We will show that there exists a $\mu \in \Gamma$, $\mu \neq id$, of which the endpoints $\alpha_\mu = c_\mu(-\infty)$, $\omega_\mu = c_\mu(+\infty)$ of its axis c_μ belong to A and Ω, respectively.

For such an axis c_μ choose the parametrization such that $c_\mu(0) \in c_\mu$ has minimal distance from $c(0) \in c$. Denote by $2\gamma_\mu$ the angle subtended by the geodesics from $c_\mu(0)$ to α and ω, respectively. If we let A and Ω shrink to α and ω, we get μ's $= \mu(A, \Omega)$'s such that $2\gamma_\mu$ approaches π and $\dot{c}_\mu(0)$ approaches $c(0)$.

We now construct, for given A and Ω with $A \cap \Omega = \emptyset$ a $\mu \in \Gamma$ with the desired properties. Let $\mu^* \in \Gamma$, $\mu^* \neq id$. Let $\alpha^* = \alpha_{\mu^*}$ and $\omega^* = \omega_{\mu^*}$ be the negative and positive end points of the axis $c^* = c_{\mu^*}$ of μ^*. From (3.9.7) we have the existence of a neighborhood Ω^* of ω^* and a $\mu_+ \in \Gamma$ with $\mu_+(\Omega^*) \subset \Omega$. Also, there exists a $\mu_- \in \Gamma$ such that $\mu_-(A)$ is a neighborhood A^* of α^*.

For all sufficiently high powers μ^{*l} of μ^*, $\mu^{*l}(\complement A^*) \subset \Omega^*$. Here $\complement P$ denotes the complement of P in $M \cup M(\infty)$. Thus, the element $\mu = \mu_+ \mu^{*l} \mu_-$ carries $\complement A$ first into $\complement A^*$, then into Ω^* and then into Ω. μ^{-1} therefore carries $\complement \Omega$ into A. But then, the end points α_μ and ω_μ of the axis of μ must be in A and Ω, respectively, since no element outside these sets is kept fixed unter μ. □

We continue by showing that the geodesic flow is topologically transitive. Actually, much stronger properties are true. E.g., that the geodesic flow on a compact manifold of negative curvature is *mixing*. This means the following: We have a measure on $T_1 \bar{M}$ from the Riemannian metric. Let U and U' be two measurable subsets. Then

$$\lim_{t \to \infty} \operatorname{meas}(\phi_t U \cap U') = \operatorname{meas}(U) \cdot \operatorname{meas}(U')/\operatorname{meas}(T_1 \bar{M})$$

For this and for related results see Anosov [1] and Arnold and Avez [1].

3.9.9 Theorem. *Let $\bar{M}$ be a compact manifold of negative curvature. Then there exists a flow line $\phi_t X_0$, $t \in \mathbb{R}$, on $T_1 \bar{M}$ such that $\{\phi_t X_0;\ t \geq 0\}$ is dense in $T_1 \bar{M}$.*

More precisely, every $X \in T_1 \bar{M}$ is the limit of a sequence $\{\phi_{t_n} X_0\}$ where $\lim t_n = \infty$.

If M denotes the universal covering of $\bar{M}$, then there exists $X_0 \in T_1 M$ such that, for every $X \in T_1 M$, there exist sequences $\{\mu_n\}$ in Γ and $\{t_n\}$ in $\mathbb{R}$ with $\lim t_n = \infty$ such that $\lim d_{T_1 M}(\mu_n X, \phi_{t_n} X_0) = 0$.

If we put $\tau_M X_0 = p_0$, $\tau_M \phi_\infty X_0 = \omega_0$, then this shows that every pair (α, ω), $\alpha \neq \omega$, of points at infinity is the closure of the Γ-orbit of the pair (p_0, ω_0).

Note. It can be shown that the set of $X_0 \in T_1 \bar{M}$ with a transitive orbit $\{\phi_t X_0\}$ is measurable with measure $=$ meas $T_1 \bar{M}$. But not every $X_0 \in T_1 \bar{M}$ has a transitive orbit. Take e.g. an element $X_0 \in T_1 \bar{M}$ for which the orbit $\phi_t X_0$, $t \in \mathbb{R}$, is periodic.

As always, we also denote the action of $\mu: M \to M$ on $T_1 M$ by μ.

Proof. We follow an idea contained in the paper of Morse and Hedlund [1]. We first prove a statement which will also be useful for the proof of (3.9.17).

Claim A. *Choose $p_0 \in M$. Let $\Omega(\infty)$ be an open, non-empty set in $M(\infty)$. Then, for every $X \in T_1 M$, the Γ-orbit $\Gamma . X$ meets the closure of the set of tangent vectors to geodesics from p_0 to the points of $\Omega(\infty)$.*

To prove this we first observe that according to (3.9.8) it suffices to consider $X \in T_1 M$ where the geodesic $c = c_X$ with $\dot{c}_X(0) = X$ is an axis of an element $\mu \neq id$ of Γ. From (3.9.7) we know that there exists a $\mu^* \in \Gamma$ which carries $\omega_c = c(\infty)$ into $\Omega(\infty)$. We may thus assume $\omega_c \in \Omega(\infty)$. That is, c and the geodesic c_0 from p_0 to ω_c are asymptotic. The positive powers μ^l of μ carry c into itself and hence $\{\mu^l X\}$ approaches elements in the set $T_1 c_0$ of tangent vectors to c_0.

Claim B. *From Claim A follows the existence of a positive orbit $\{\phi_t X_0, t \geq 0\}$ with the properties required in the Theorem.*

To see this let $\{T_n\}$ be an enumerated basis for the open sets of $T_1 M$. Thus, any neighborhood of any $X \in T_1 M$ contains a neighborhood in this basis. Let now $\Omega_0(\infty)$ non-empty, open in $M(\infty)$. Then we have from Claim A a non-empty open $\Omega_1(\infty) \subset \Omega_0(\infty)$ such that every geodesic from p_0 to a point in $\bar{\Omega}_1(\infty)$ has a tangent vector in common with the orbit $\Gamma . T_1$.

Assume that we have already defined $\Omega_{n-1}(\infty)$. Then Claim A yields an open, non-empty $\Omega_n(\infty)$ in $\Omega_{n-1}(\infty)$ such that every geodesic from p_0 to a point of $\bar{\Omega}_n(\infty)$ has a tangent vector in common with $\Gamma . T_n$. Now,

$$\bigcap_1^\infty \bar{\Omega}_n(\infty) = \text{(briefly) } \bar{\Omega}(\infty) \neq \emptyset$$

and the geodesic c_0 from p_0 to a point $\bar{\omega} \in \bar{\Omega}(\infty)$ has the property: If $X_0 = \dot{c}_0(0)$, then, for every n there is a $t_n \geq 0$ such that $\phi_{t_n} X_0 \in T_n$. Thus, the projection of $\{\phi_t X_0, t \geq 0\}$ into $T_1 \bar{M}$ is dense in $T_1 \bar{M}$.

It only remains to show that we may assume $\lim t_n = \infty$, for $n \to \infty$. To see this we choose the elements T_n relatively compact. E.g., we take counter images of a basis for $T_1 \bar{M}$. Then, for any $t_0 > 0$, also the set $\{\phi_t X_0; t \geq t_0\}$, will meet $\Gamma . T_n$ since there exists $\mu \in \Gamma$ such that $\tau_M(\mu . T_n)$ lies in any prescribed neighborhood of the final point $\bar{\omega}$ of c_0. □

In the remainder of this section we consider so-called manifolds of hyperbolic type.

3.9.10 Definition. Let $\bar{M}$ be a compact Riemannian manifold. $\bar{M}$ is called of *hyperbolic type* if, besides the given Riemannian metric g on $\bar{M}$, on the underlying differentiable manifold (which we also denote by $\bar{M}$), there also exists a Riemannian metric g^* such that $\bar{M}^* = (\bar{M}, g^*)$ has strictly negative curvature.

Remarks. If $\dim \bar{M} = 2$ and $\bar{M}$ is orientable, then hyperbolic type is the same as genus ≥ 2. This case has been investigated by Morse [1]. For arbitrary dimension, see Klingenberg [6]. More restricted classes of such manifolds had been studied already before, see in particular Busemann [2].

A possibly even more general concept of a manifold of hyperbolic type is the following, cf. Klingenberg [10], Eschenburg [1]: Instead of demanding that there exists, besides the given Riemannian metric g, a Riemannian metric g^* of negative curvature, one requires the existence of a metric g^* such that the geodesic flow ϕ_t: $T_1\bar{M}^* \to T_1\bar{M}^*$ on the universal covering is of *Anosov type*.

By this we mean the following, cf. Klingenberg [7]:
The bundle

$$\bar{\tau}_{T_1\bar{M}^*}: \bar{T}T_1\bar{M}^* \to T_1\bar{M}^*$$

shall possess an $T\phi_t$-invariant continuous splitting

$$\bar{\tau}_{T_1\bar{M}^*} = \bar{\tau}_{T_1\bar{M}^*s} \oplus \bar{\tau}_{T_1\bar{M}^*u} : \bar{T}_s T_1\bar{M}^* \oplus \bar{T}_u T_1\bar{M}^* \to T_1\bar{M}^*$$

such that there exist $a \geqslant 1$, $b > 0$, with

$$|T\phi_t\xi_s| \leqslant a|\xi_s|e^{-bt}, \qquad \text{if} \quad \xi_s \in \bar{T}_s T_1\bar{M}^*$$

$$|T\phi_t\xi_u| \geqslant a^{-1}|\xi_u|e^{bt}, \qquad \text{if} \quad \xi_u \in \bar{T}_u T_1\bar{M}^*,$$

all $t \geqslant 0$.

In the paper quoted above we showed that on a manifold of Anosov type there are no conjugate points. From the work of Anosov [1] it follows that the geodesic flow is ergodic and that the periodic orbits are dense.

The only known examples of manifold of Anosov type are – besides those with strictly negative curvature – manifolds which are obtained from manifolds with negative curvature by certain modifications of the Riemannian metric under which some positive curvature is introduced. For examples see Gulliver [1].

We begin with a preparatory result.

3.9.11 Proposition. *Let M be a simply connected manifold of bounded negative curvature $-k_0^2 \leqslant K \leqslant -k_1^2 < 0$. Let $c(t), t \in \mathbb{R}$, be a geodesic on M. Let $c_1(t), 0 \leqslant t \leqslant a$ be a curve from $p_1 = c_1(0)$ to $q_1 = c_1(a)$ with $d(p_1, c) = d(q_1, c) = \varrho, d(c_1(t), c) \geqslant \varrho$, all $t \in [0, a]$, with some positive number ϱ. Denote by p and q the foot points of the perpendicular from p_1 and q_1 to c. Then $L(c_1) \geqslant d(p, q)\cosh(k_1\varrho)$.*

Proof. $c_1(t)$ can assumed to be differentiable. Denote by $e(t)$ the foot point of the perpendicular g_t from $c_1(t)$ to c, cf. (3.8.2). We parameterize g_t by $g_t(s), 0 \leqslant s \leqslant d_t = d(c_1(t), c)$ with $g_t(0) = e(t) \in c$.

Fix now a $t = t_0$. Consider the normal bundle $v_c: T_c^{\perp} \to \mathbb{R}$ of the embedding $c: \mathbb{R} \to M$. Under our hypotheses, $\exp_{c,t_0}: T_{t_0}^{\perp}c \to M$ is an immersion, cf. (1.12.17) for our notations. Hence, the tangent vector $\dot{c}_1(t_0)$ possesses a well defined horizontal component $\dot{c}_{1h}(t_0)$ which is orthogonal to the immersion $\exp_{c,t_0}$ at $d_{t_0}g'_{t_0}(0)$. The Proposition follows once we show

$$(*) \qquad |\dot{c}_{1h}(t_0)| \geqslant \cosh(k_1 d_{t_0})|\dot{e}(t_0)|.$$

To prove (*) consider the space $M^* = H_{k_1}^n$ of constant curvature $K^* = -k_1^2$. A Jacobi field $Y^*(s)$ along a geodesic $c^*(s)$ in M^*, orthogonal to c^* with $\nabla Y^*(0)$

$= 0$, is of the form $\cosh(k_1 s) A(s)$ with $\nabla A(s) = 0$, $A(0) = Y^*(0)$, cf. (1.12.14). (*) therefore follows from (2.7.9) once we have identified $\dot{c}_{1h}(t_0)$ with the value at $s = d_{t_0}$ of a Jacobi fiel $Y(s)$ along $g_{t_0}(s)$ with $Y(0) = \dot{e}(t_0)$ and $\nabla Y(0) = 0$.

To do so let $X(t)$ be the parallel vector field along $e(t)$ with $X(t_0) = g'_{t_0}(0)$. Consider the variation $F(t, s) = \exp_{e(t)} sX(t)$ of $F(t_0, s) = g_{t_0}(s)$ by geodesics. Then $\partial F(t, s)/\partial t$ at $t = t_0$ yields the desired Jacobi fiel $Y(s)$. □

We now can show that on a manifold of hyperbolic type geodesics globally are comparable with geodesics on the underlying manifold of negative curvature. Cf. also Klingenberg [10]. As for the notation, when we consider two Riemannian metrics g and g^* on a differentiable manifold M, we also abbreviate (M, g) by M and (M, g^*) by M^*. Metric concepts, like distance, length etc., have to be distinguished depending on what metric we take. We agree to put a star on objects belonging to M^*, while otherwise, we mean objects belonging to M.

3.9.12 Lemma. *Let $\bar{M} = (\bar{M}, g)$ be a compact Riemannian manifold. Assume that we also have a metric g^* such that $\bar{M}^* = (\bar{M}, g^*)$ has strictly negative curvature, $-k_0^2 \leqslant K^* \leqslant -k_1^2 < 0$.*

Let $M = (M, g)$ and $M^* = (M, g^*)$ be the universal Riemannian coverings of $\bar{M}$ and $\bar{M}^*$, respectively. Then there exists a constant $\varrho \geqslant 0$ with the following property: Whenever we take two points p, q in M and minimizing geodesics c and c^* from p to q in M and M^*, respectively, then every point $r \in c$ has M-distance $\leqslant \varrho$ from c^*.

Proof. Since the underlying differentiable manifold $\bar{M}$ is compact, there exists a constant $a \geqslant 1$ such that

$$a^{-2} g^*(X, X) \leqslant g(X, X) \leqslant a^2 g^*(X, X)$$

for all $X \in TM$. Thus, for a curve c,

$$a^{-1} L^*(c) \leqslant L(c) \leqslant a L^*(c),$$

and, for the distances between two points p and q,

$$a^{-1} d^*(p, q) \leqslant d(p, q) \leqslant a d^*(p, q).$$

Now let c and c^* be as in the Lemma. Assume that all points p on the subarc $c|[t_0, t_1]$ of c from $p_0 = c(t_0)$ to $p_1 = c(t_1)$ have M^*-distance $\geqslant \varrho'$ from c^*, while $d^*(p_0, c^*) = d^*(p_1, c^*) = \varrho'$, some $\varrho' > 0$. Denote by p_0^* and p_1^* the foot points of the M^*-perpendicular from p_0 and p_1 towards c^*. Then we get from (3.9.11) the estimates

$$(*) \qquad d(p_0, p_1) = L(c|[t_0, t_1]) \geqslant a^{-1} L^*(c[t_0, t_1]) \geqslant a^{-1} d^*(p_0^*, p_1^*) \cosh(k_1 \varrho').$$

On the other hand, the length of the curve from p_0 to p_0^* to p_1^* to p_1 yields an upper bound for $d(p_0, p_1)$:

$$(**) \qquad d(p_0, p_1) \leqslant a d^*(p_0, p_1) \leqslant a(2\varrho' + d^*(p_0^*, p_1^*)).$$

(*) and (**) show that there exists a ϱ'_0 such that, whenever $\varrho' \geqslant \varrho'_0$, the distance $d^*(p_0^*, p_1^*)$ of points constructed as above becomes $\leqslant 1$. We claim that $\varrho = a(3\varrho'_0 + 1)$ is a constant with the property required in the Lemma.

Indeed, let $p \in c$ be a point with maximal M-distance from c^*.

If $d(p, c^*) \leqslant a\varrho'_0$, we are done. Otherwise, $d^*(p, c^*) \geqslant a^{-1} d(p, c^*) > \varrho'_0$. Then there exists an arc $c|[t_0, t_1]$ of c from $p_0 = c(t_0)$ to $p_1 = c(t_1)$, containing p in its interior, with $d^*(p_0, c^*) = d^*(p_1, c^*) = a\varrho'_0$. Thus, we get with (**) the following estimates, using $d^*(p_0^*, p_1^*) \leqslant 1$:

$$d(p, c^*) \leqslant d(p, p_0) + d(p_0, c^*) \leqslant d(p_0, p_1) + d(p_0, c^*)$$
$$\leqslant a(2\varrho'_0 + 1) + a\varrho'_0 = \varrho . \quad \square$$

From (3.9.12) we now immediately obtain

3.9.13 Theorem. *Let $\bar{M} = (\bar{M}, g)$ be a compact manifold of hyperbolic type. That is, there exists a metric g^* on $\bar{M}$ with negative curvature. Consider the compactification* $\mathrm{cp}(M^*) = M^* \cup M^*(\infty)$ *of the universal covering M^* of $\bar{M}^* = (\bar{M}, g^*)$.*

Then any two points on $\mathrm{cp}M \cong \mathrm{cp}M^*$ *can be joined by a minimizing geodesic c. If* c^* *is the M^*-geodesic in* $\mathrm{cp}M^*$*, joining the same points, then $d(c(t), c^*) \leqslant \varrho$, for all t, with a constant $\varrho \geqslant 0$ only depending on $\bar{M}$ and $\bar{M}^*$.*

Proof. If none of the two points is at infinity, this follows from (3.9.12). Let now $p \in M, \omega \in M(\infty)$. Consider the M^*-geodesic from p to ω. For each $n > 0$, let c_n be a minimizing M-geodesic from p to $c^*(n)$. The sequence $\{\dot{c}_n(0)\}$ of initial vectors possesses a convergent subsequence. If X is the limit vector of this sequence, the geodesic c with $\dot{c}(0) = X$ converges, for $t \to \infty$, to ω, since all points on c_n have bounded distance from c^*.

If α and ω are two different points at infinity we consider a M^*-geodesic $c^*(t)$ from α to ω. For each $n > 0$, let c_n be a minimizing M-geodesic from $c^*(-n)$ to $c^*(+n)$. From (3.9.12) we have a compact set in $T_1 M$ such that, for every n, there exists a t_n with $\dot{c}_n(t_n)$ in this set. Let X be in the limit set of the sequence $\{\dot{c}_n(t_n)\}$. The geodesic c with $\dot{c}(0) = X$ then is a minimizing M-geodesic from α to ω. $\square$

Remark. If $M = (M, g)$ is the simply connected covering of a compact manifold of hyperbolic type, then there will be in general also non-minimizing geodesics on M. In fact, choose on M a ball $B_{2\varrho}(p)$ of radius $2\varrho > 0$, contained entirely in the domain of normal coordinates based at p. With the technique of the so-called metric surgery one can modify the given metric on $B_\varrho(p) \subset B_{2\varrho}(p)$ in an arbitrarily prescribed manner, and then smooth out this new metric on $B_{2\varrho}(p) - B_\varrho(p)$ to come back to the original metric on $\complement B_{2\varrho}(p)$. For details cf. Weinstein [1].

We want to see under what conditions a manifold of hyperbolic type still possesses a geodesic flow which is topologically transitive. Some restrictions on the metric will be necessary. One natural such restriction is to exclude conjugate points. However, one does not know whether this suffices, at least if $\dim M > 2$. We therefore require even more.

3.9.14 Definition. Let $M = (M, g)$ be the simply connected covering of a compact manifold $\bar{M}$ of hyperbolic type. For M, the set $M(\infty)$ of points at infinity is well defined, using a Riemannian metric g^* of strictly negative curvature.

We say that $\bar{M}$ or also M has *no conjugate points, including at infinity*, if there are no conjugate points on M and if, moreover, for a (finite) point $p \in M$ and a point at infinity $\omega \in M(\infty)$, there is exactly one geodesic c with $c(0) = p$, $c(\infty) = \omega$.

Remark. In the literature, this property (or a very similar property) usually is called *geodesic instability*. Because it means that, given a half geodesic $c(t)$, $t \geqslant 0$, from $p = c(0)$ with limit $c(\infty) = \omega \in M(\infty)$, any geodesic $c'(t)$, starting from p with $\dot{c}'(0) \neq \dot{c}(0)$ has not bounded distance from $c(t)$, as $t \to \infty$.

We now prove the result of Green [1] that in dimension 2 no conjugate points implies no conjugate points at infinity.

As a preliminary step we show:

3.9.15 Lemma. *Let $c(t)$, $t \in \mathbb{R}$, be a geodesic on a surface. Assume that $c(-\varepsilon)$, some $\varepsilon > 0$, has no conjugate points $c(t)$, along $c|[-\varepsilon, t]$, all $t \geqslant -\varepsilon$. Moreover, the Gauss curvature $K(t)$ in $c(t)$ shall have a lower bound of the form $K(t) > -k^2$, all $t \geqslant -\varepsilon$.*

If $Y(t)$ is a Jacobi field orthogonal to $\dot{c}(t)$ with $Y(0) = 0$, $\nabla Y(0) \neq 0$, then $|Y(t)|$ converges to ∞ for $t \to \infty$.

Proof. By introducing an orthonormal frame $\{E_0(t) = \dot{c}(t), E_1(t)\}$ along $c(t)$ we can write every Jacobi field $Z(t)$ orthogonal to $\dot{c}(t)$ as $Z(t) = z(t) E_1(t)$ with

$$(\dagger) \qquad \ddot{z}(t) + K(t) z(t) = 0 .$$

On the other hand, we have the Jacobi fields $z^*(t)$ on the surface M^* of constant curvature $K^* = -k^2 < 0$, satisfying the equation

$$(\dagger)^* \qquad \ddot{z}^*(t) - k^2 z^*(t) = 0 .$$

Our hypotheses imply that if $z(t)$ is a solution of $(\dagger)$ with $z(t_0) = 0, \dot{z}(t_0) > 0$, then $z(t) \neq 0$, for all $t > t_0$, provided $t_0 \geqslant -\varepsilon$. We can thus form for such a $z(t)$ the quotient $s(t) = \dot{z}(t)/z(t)$, $t > t_0$. $s(t)$ satisfies the Riccati equation

$$(\S) \qquad \dot{s}(t) + s(t)^2 + K(t) = 0, \quad t > t_0 .$$

We know this equation from the proof of (3.2.17).
For a solution $z^*(t)$ of $(\dagger)^*$ with $z^*(t) \neq 0$ for $t > t_0$ we form $s^*(t) = \dot{z}^*(t)/z^*(t)$ and then have the corresponding equation

$$(\S)^* \qquad \dot{s}^*(t) + s^*(t)^2 - k^2 = 0, \quad t > t_0 .$$

The solutions of $(\S)^*$ corresponding to $z^*(t_0) = 0$, $\dot{z}^*(t_0) > 0$ are well known, i.e.,

$$s^*(t) = k \operatorname{ctgh}(k(t - t_0))$$

We claim

(*) $\quad -k \leqslant s(t) \leqslant s^*(t) = k \operatorname{ctgh}(k(t-t_0))$, for all $t > t_0$.

The right hand side of (*) is contained in the proof of the Rauch Comparison Theorem (2.7.2). In fact, in the present case it is simply the Sturm Comparison Theorem of which we give the usual proof: Let $z(t), z^*(t)$ be the solutions of $\ddot{z} + Kz = 0$ and $\ddot{z}^* - k^2 z^* = 0$ with $z(t_0) = z^*(t_0) = 0$, $\dot{z}(t_0) > 0$. Then

$$0 = \int_{t_0}^{t} [z(\ddot{z}^* - k^2 z^*) - z^*(\ddot{z} + Kz)](\tau)\, d\tau =$$

$$(z\dot{z}^* - z^*\dot{z})(t) + \int_{t_0}^{t} [zz^*(-k^2 - K)](\tau)\, d\tau \leqslant z(t)\dot{z}^*(t) - z^*(t)\dot{z}(t)$$

i.e., $s(t) \leqslant s^*(t)$.

If there were a $t_1 \geqslant t_0$ with $s(t_1) < -k$, consider the solution $\tilde{s}^*$ of (§)* with $\tilde{s}^*(t_1) = s(t_1)$. $\tilde{s}^*$ is of the form $k \operatorname{ctgh}(k(t-t_2))$ with an appropriately chosen $t_2 > t_1$. It follows that $\tilde{s}^*(t)$ goes to $-\infty$ as $t < t_2$ goes to t_2. We will show

(**) $\quad \tilde{s}(t) \leqslant \tilde{s}^*(t)$ for $t \geqslant t_1$.

Thus, $s(t_1) < -k$ leads to a contradiction. To prove (**) we note

$$\dot{s}(t) - \dot{\tilde{s}}^*(t) < -s^2(t) + \tilde{s}^{*2}(t)$$

which shows that $\dot{s}(t') < \dot{\tilde{s}}^*(t')$ whenever $s(t') = \tilde{s}^*(t')$. This completes the proof of (*).

Now consider solutions $y(t), z(t)$ of the Jacobi equation with $y(0) = 0$, $\dot{y}(0) = 1$ and $z(-\varepsilon) = 0$, $\dot{z}(-\varepsilon) > 0$ such that $z(0) = 1$. Since $\dot{y}(t)z(t) - \dot{z}(t)y(t) = \text{const} = 1$ we see, for $t > 0$, $d(y(t)/z(t))/dt = z^{-2}(t)$, hence,

$$\frac{y(t)}{z(t)} = \int_0^t z^{-2}(\tau)\, d\tau = \text{(briefly) } h(t).$$

$$\frac{1}{y^2(t)} = \frac{z(t)}{y(t)}\left(\frac{\dot{y}(t)}{y(t)} - \frac{\dot{z}(t)}{z(t)}\right) \leqslant h(t)^{-1}\left(k \operatorname{ctgh}(kt) + k \operatorname{ctgh}(k(t+\varepsilon))\right).$$

For $t \geqslant$ some t_0', the second factor on the right hand side becomes $\leqslant 4k$. Hence

$$h(t) \leqslant 4k\, y^2(t), \quad t \geqslant t_0'.$$

If there were a sequence $\{t_n\}$ going to ∞ with $\lim y(t_n) = a < \infty$, then $a > 0$ and we can also assume $\lim h(t_n) = b < \infty$. Thus, $\lim z(t_n) = y(t_n)/h(t_n)$ exists. But then $\{\dot{h}(t_n) = z^{-2}(t_n)\}$ is bounded away from zero – a contradiction. □

As an immediate consequence we get, cf. Green [1]:

3.9.16 Theorem. *Let $\bar{M}$ be a compact surface of genus $\geqslant 2$, without conjugate points. Then, on the universal covering M of $\bar{M}$, there are no conjugate points, even at infinity.*

Proof. Consider $p \in M$ and assume that there were a point ω at infinity which is the limit $c_0(\infty) = c_1(\infty)$ of two different geodesics c_0, c_1, starting from p. Put $\dot{c}_0(0) = X_0$, $\dot{c}_1(0) = X_1$.

c_0 and c_1 bound a domain on M with the following property: Consider the family X_s, $0 \leqslant s \leqslant 1$, of unit vectors in $T_p M$ tangent to this domain. The geodesics c_s with $\dot{c}_s(0) = X_s$, then also have ω as limit, for $t \to \infty$.

We thus have a mapping $F\colon [0,1] \times [0, \infty] \to M$ with $F(s, t) = c_s(t)$. $t \mapsto \partial F(s, t)/\partial s$ is a Jacobi field $Y_s(t)$ along $c_s(t)$ with $Y_s(0) = 0$. From (3.9.15) we know that $|Y_s(t)| \to \infty$, for $t \to \infty$.

On the other hand, the length of the circular arcs $s \mapsto F(t, s)$ is bounded, for each t, since all c_s have uniformly bounded distance from a geodesic c^* in $M^* = (M, g^*)$ with negative curvature, cf. (3.9.13). Since the length of such an arc is given by the integral from 0 to 1 over $|Y_s(t)|$, which is unbounded as $t \to \infty$, we get the desired contradiction to our assumption that there exist two different geodesics c_0 and c_1 from p to ω. $\square$

We now can prove the extension of (3.9.9) announced above.

3.9.17 Theorem. *Let $\bar{M}$ be a compact manifold of hyperbolic type such that there exist no conjugate points, even at infinity. If $\dim \bar{M} = 2$, it suffices to assume that there are no conjugate points. Then the geodesic flow on $T_1 \bar{M}$ is topologically transitive.*

Proof. We show the validity of Claim A in the proof of (3.9.9). Claim B then follows in exactly the same way.

By M and M^* we denote the universal Riemannian coverings of $\bar{M}$ and $\bar{M}^* = (\bar{M}, g')$, respectively, where $\bar{M}^*$ has negative curvature.

Let $\Omega(\infty)$ be a non-empty open set in $M(\infty) = M^*(\infty)$. The proof of (3.9.11) yields the existence of a transitive geodesic ray c_0^* in M^* from p_0 to a point $\omega_0 \in \Omega(\infty)$. That is to say, if we put $\dot{c}_0^*(0) = X_0^*$ and let $X^* \in T_1 M^*$ be arbitrary, then there exist sequences $\{\mu_n\}$ in Γ and $\{t_n\}$ in $\mathbb{R}$ with $\lim t_n = \infty$ such that $\mu_n X^*$ approaches $\phi_{t_n}^* X_0^*$. Here, we have denoted by ϕ^* the geodesic flow on $T_1 M^*$.

If c^* is the M^*-geodesic with $\dot{c}^*(0) = X^*$ and $\alpha = c^*(-\infty)$, $\omega = c^*(+\infty)$, then it follows that $\{\mu_n \alpha\}$ converges towards $\alpha_0 = c_0^*(-\infty)$, while $\{\mu_n \omega\}$ converges towards ω_0.

Now let $X \in T_1 M$ be given. Let c be the geodesic with $\dot{c}(0) = X$. Put $c(-\infty) = \alpha$, $c(+\infty) = \omega$, $c(0) = p$. By c^* we denote an M^*-geodesic from α to ω. Put $c^*(0) = p^*$, $\dot{c}^*(0) = X^*$. Then, as we have just seen, with appropriately chosen sequences $\{\mu_n\}$ in Γ and $\{t_n\}$ in $\mathbb{R}$, $\lim t_n = \infty$, $\mu_n X^*$ approaches $\phi_{t_n}^* X_0^*$. Thus, in particular, $\{\mu_n p^*\}$ goes to ω_0, $\{\mu_n \alpha\}$ goes to α_0. Since c has bounded distance from c^*, also $\{\mu_n p\}$ goes to ω_0.

Denote by c_n the M-geodesic from p_0 to $\mu_n p = \mu_n c(0)$. $\{c_n\}$ converges to the unique M-geodesic c_0 from p_0 to ω_0. Thus, for all large n, $c_n(\infty)$ will be in $\Omega(\infty)$. Since the $\mu_n c$ and $\mu_n c^*$ have bounded distance from each other and $\{\mu_n c^*\}$ converges to c_0^*, $d(p_0, \mu_n c)$ is bounded. Hence, the tangent vector to c_n in its end point $\mu_n p$ and the tangent vector $\mu_n X$ to $\mu_n c$ for $t = 0$ become arbitrarily close to each other, as n goes to infinity. This shows that the $\mu_n X$ can be approximated by tangent vectors belonging to the set of M-geodesics from p_0 to $\Omega(\infty)$. □

Remark. The geodesic flow on manifolds of negative curvature has served as a model for the study of ergodic systems. A survey article with references to many recent results is Katok [1].

References

Abraham, R. and Marsden, J.
[1] Foundations of Mechanics. Reading, Mass.: Benjamin/Cummings 1978
Abraham, R. and Robbin, J.
[1] Transversal mappings and flows. New York and Amsterdam: Benjamin 1967
Alber, S. I.
[1] On periodicity problems in the Calculus of Variations in the Large. Uspehi Mat. Nauk *12* No. 4, 57–125 (1957). Russian – Amer. Math. Soc. Transl. (2) *14* (1960)
Alexander, S. and Bishop, R.
[1] Prolongations and Completions of Riemannian manifolds. J. Diff. Geom. *6*, 403–410 (1972)
Alexandrov, A. D.
[1] The Intrinsic Geometry of Convex Surfaces. Moscow-Leningrad: Gosudarstv. Izdat Tehn-Teor. Lit. (1948) (Russian). German translation: Die Innere Geometrie der konvexen Flächen. Berlin: Akademie Verlag 1955
Alkier, H.
[1] Über geodätische Linien auf Flächen zweiten Grades. Diss. Leipzig 1925.
Aloff, S. and Wallach, N.
[1] An infinite family of distinct 7-manifolds admitting positively curved Riemannian structures. Bull. Amer. Math. Soc. *81*, 93–97 (1975)
Ambrose, W.
[1] Parallel translation of Riemannian curvature. Ann. of Math. *64*, 337–363 (1956)
Anosov, D.
[1] Geodesic flows on closed riemannian manifolds of negative curvature. Trudy Mat. Inst. Steklov *90* (1967) [Russian]. English translation: Proc. Steklov Inst. Mat., Providence, R. I.: Amer. Math. Soc., 1967
Anosov, D. and Sinai, Ya. G.
[1] Some smooth ergodic systems. Uspehi Mat. Nauk *22*, 107–172 (1967) (Russian); Russian Math. Surveys *22*, 103–167 (1967)
Arnold, V. I.
[1] Critical points of smooth functions. Proc. Int. Congress Math., Vol. 1, 19–39. Vancouver 1974
Arnold, V. I. and Avez, A.
[1] Problèmes ergodiques de la méchanique classique. Paris: Gauthier-Villars 1967. – English translation: Ergodic problems of Classical Mechanics. New York and Amsterdam: Benjamin 1968
Ballmann, W.
[1] Der Satz von Lyusternik und Schnirelmann. Bonn. Math. Schr. Nr. 102, 1978
[2] Einige neuere Resultate über Mannigfaltikeiten nicht-positiver Krümmung. Bonn. Math. Schr. Nr. 113, 1978

[3] Doppelpunktfreie geschlossene Geodätische auf kompakten Flächen. Math. Z. *166*, 41–46 (1978)

Ballmann, W., Thorbergsson, G. and Ziller, W.

[1] Closed Geodesics on Positively Curved Manifolds. Preprint, Bonn and Philadelphia, 1981

Bangert, V.

[1] Closed geodesics on complete surfaces. Math. Ann. *251*, 83–96 (1980)

Berard-Bergery, L.

[1] Les variétés Riemanniennes homogènes simplement connexes de dimension impaire à courbure strictement positive. J. Math. Pures Appl. *55*, 47–68 (1976)

Berger, M.

[1] Sur quelques variétés riemanniennes suffisament pincées. Bull. Soc. Math. France *88*, 57–71, (1960)

[2] Les variétés riemanniennes (1/4)-pincées. Ann. Scuola Norm. Pisa (3) *14*, 161–170 (1960)

[3] Les variétés Riemanniennes homogènes normales simplement connex à courbure strictement positive. Ann. Scuola Norm. Pisa *15*, 179–246 (1961)

[4] An extension of Rauch's metric theorem and some applications. Illinois J. Math. *6*, 700–712 (1962)

[5] Lectures on Closed Geodesics in Riemannian Geometry. Bombay: Tata Institute 1965

Besse, A.

[1] Manifolds all of whose Geodesics are Closed. Berlin–Heidelberg–New York: Springer 1978

Bianchi, L.

[1] Vorlesungen über Differentialgeometrie. Deutsche Übersetzung von M. Lukat. Leipzig: Teubner 1899

Bishop, R. and Crittenden, R.

[1] Geometry of Manifolds. New York and London: Academic Press 1964

Bishop, R. and O'Neill, B.

[1] Manifolds of negative curvature. Trans. Amer. Math. Soc. *145*, 1–49 (1969)

Bishop, R. L.

[1] Decomposition of cut loci. Proc. Amer. Math. Soc. *65*, 133–136 (1977)

Borel, A.

[1] La cohomologie mod 2 des certaines espaces homogènes. Comment. Math. Helv. *27*, 165–197 (1953)

Bott, R.

[1] The stable homotopy of the classical groups. *70*, 313–337 (1954)

[2] On the iteration of closed geodesics and the Sturm intersection theory. Comm. Pure Appl. Math. *9*, 171–206 (1956)

Bourbaki, N.

[1] Topologie Génèrale. Paris: Hermann et Cie, 1953–61

[2] Variétés différientielles et analytiques. Fasc. des résultats. Paris: Hermann et Cie 1967/71

Busemann, H.

[1] The Geometry of Geodesics. New York, N.Y.: Academic Press 1955

[2] Extremals on closed hyperbolic space forms. Tensor *16*, 313–318 (1965)

Buser, P. and Karcher, H.

[1] Gromov's Almost Flat Manifolds. Astérisque *81* (1981)

Cartan, E.

[1] Leçons sur la Géométrie des espaces de Riemann, Paris: Gauthier-Villars 1928

[2] Groupes simples clos et ouverts et géometrie riemannienne, J. Math. Pures Appl. *8*, 1–33 (1929)

Cheeger, J.

[1] Finiteness theorems for riemannian manifolds. Amer. J. Math. *92*, 61–74 (1970)

Cheeger, J. and Ebin, D.

[1] Comparison Theorems in Riemannian Geometry. Amsterdam and Oxford: North-Holland Publ. Comp. 1975

Cheeger, J. and Gromoll, D.

[1] On the structure of complete manifolds of nonnegative curvature. Ann. of Math. *96*, 413–443 (1972)

Chern, S.S.

[1] On the multiplication of the characteristic ring of a sphere bundle. Ann. of Math. *49*, 362–372 (1948)

[2] Studies in Global Geometry and Analysis. Math. Assoc. of America. Englewood Cliffs, N.J.: Prentice Hall 1967

Chevalley, C.

[1] Theory of Lie Groups I. Princeton: Princeton Univ. Press 1946

Cohn-Vossen, S.

[1] Kürzeste Wege und Totalkrümmung auf Flächen. Comp. Math. *2*, 69–133 (1935)

[2] Complete riemannian spaces of positive curvature. Dokl. Akad. Nauk SSSR *3*, 387–389 (1935) (Russian)

[3] Totalkrümmung und geodätische Linien auf einfach zusammenhängenden offenen vollständigen Flächenstücken. Recueil Math. de Moscou *43*, 139–163 (1936)

Darboux, G.

[1] Leçons sur la théorie générale des surfaces. Vol. I à IV. Paris: Gauthier-Villars, 1887–1896. Reprint of 3rd edition, New York: Chelsea 1977

Dieudonné, J.

[1] Foundations of Modern Analysis. New York and London: Academic Press 1960

Dubois, E.

[1] Beiträge zur Riemannschen Geometrie im Großen. Comm. Math. Helv. *41*, 30–50 (1966/67)

Duistermaat, H.

[1] On the Morse Index in Variational Calculus. Advances in Math. *21*, 173–195 (1976)

Eberlein, P. and O'Neill, B.

[1] Visibility manifolds. Pacific J. of Math. *46*, 45–109 (1973)

Ehrlich, P.

[1] Continuity Properties of the Injectivity Radius Function. Comp. Math. *29*, 151–178 (1974)

Eliasson, H.

[1] Die Krümmung des Raumes Sp(2)/SU(2) von Berger. Math. Ann. *164*, 317–323 (1966)

[2] On the geometry of manifolds of maps. J. Diff. Geom. *1*, 165–194 (1967)

Eschenburg, J.H.

[1] Stabilitätsverhalten des geodätischen Flusses Riemannscher Mannigfaltigkeiten. Bonn. Math. Schr. Nr. 87, 1976

[2] New examples of manifolds with strictly positive curvature. Preprint, Math. Inst. Universität Münster, 1980

Flaschel, P. und Klingenberg, W.

[1] Riemannsche Hilbertmannigfaltigkeiten. Periodische Geodätische. Lecture Notes in Mathematics 282, Berlin–Heidelberg–New York: Springer 1972

Fubini, G.

[1] Sulle metriche definite da una forma Hermitiana. Atti Ist. Veneto *6*, 501–513 (1904)

Gauss, C.F.

[1] Disquisitiones generales circa superficies curvas. Commentationes societatis regiae scientiorum Gottingensis recentiores *6*, Göttingen 1828

Gluck, H. and Singer, D.

[1] Scattering of a geodesic field I, II. Ann. of Math. *108*, 347–372 (1978), *110*, 205–225 (1979)
Green, L.W.
[1] Surfaces without conjugate points. Trans. Amer. Math. Soc. *76*, 529–546 (1954)
Greene, R.
[1] Complete metrics of bounded curvature on non-compact manifolds. Arch. Math. *31*, 89–95 (1978)
Greene, R. and Wu, H.
[1] Integrals of Subharmonic Functions on Manifolds of Nonnegative Curvature. Inv. math. *27*, 265–298 (1974)
Gromoll, D., Klingenberg, W. und Meyer, W.
[1] Riemannsche Geometrie im Grossen. Lecture Notes in Mathematics 55, Berlin–Heidelberg–New-York: Springer 1968. 2te Auflage 1975
Gromoll, D. and Meyer, W.
[1] On complete open manfolds of positive curvature. Ann. of Math. *90*, 75–90 (1969)
[2] An exotic sphere with non-negative curvature. Ann. of Math. *100*, 401–406 (1974)
Gromoll, D. and Wolf, J.
[1] Some relations between the metric structure and the algebraic structure of the fundamental group in manifolds of non-positive curvature. Bull. Amer. Math. Soc. *77*, 545–551 (1971)
Gromov, M.
[1] Manifolds of negative curvature. J. Diff. Geom. *13*, 223–230 (1978)
[2] Almost flat manifolds. J. Diff. Geom. *13*, 231–241 (1978)
[3] Curvature, diameter and Betti numbers. Comment. Math. Helv. *56*, 179–195 (1981)
Grove, K.
[1] Condition (C) for the energy integral on certain path spaces and applications to the theory of geodesics. J. Diff. Geom. *8*, 207–223 (1973)
Grove, K., Karcher, H. and Ruh, E.
[1] Jacobi fields and Finsler Metrics on Compact Lie Groups with an Application to Differentiable Pinching Problems. Math. Ann. *211*, 7–21 (1974)
Grove, K. and Shiohama, K.
[1] A generalized sphere theorem. Ann. of Math. *106*, 201–211 (1977)
Gulliver, R.
[1] On the variety of manifolds without conjugate points. Trans. Amer. Math. Soc. *210*, 185–201 (1975)
Hadamard, J.
[1] Les surfaces à courbures opposées et leur lignes géodesiques. J. Math. Pures Appl. (5) *4*, 27–73 (1896)
[2] Sur la forme des lignes géodesique à l'infini et sur les géodesiques des surfaces réglées du second ordre. Bull. Soc. Math. France *26*, 195–216 (1898)
[3] Sur l'iteration et les solutions asymptotiques des équations differentielles. Bull. Soc. Math. France *29*, 224–228 (1901)
Hedlund, G.A.
[1] Poincaré's rotation number and Morse's type number. Trans. Amer. Math. Soc. *34*, 75–97 (1932)
Heintze, E.
[1] The curvature of $SU(5)/(Sp(2) \times S^1)$. Inv. math. *13*, 205–212 (1971)
Helgason, S.
[1] Differential Geometry and Symmetric Spaces. New York and London: Academic Press 1962
Hilbert, D.
[1] Über das Dirichletsche Prinzip. J. Reine Angew. Math. *129*, 63–67 (1905)

Hirsch, M.
[1] Differential Topology. New York–Heidelberg–Berlin: Springer 1976
Hopf, E.
[1] Statistik der Lösungen geodätischer Probleme vom unstabilen Typus, II. Math. Ann. *117*, 590–608 (1940)
[2] Closed surfaces without conjugate points. Proc. Nat. Acad. Sci *34*, 47–51 (1948)
Hopf, H.
[1] Zum Clifford-Kleinschen Raumproblem. Math. Ann. *95*, 313–335 (1925)
Hopf, H. und Rinow, W.
[1] Über den Begriff der vollständigen differentialgeometrischen Fläche. Math. Ann. *116*, 749–766 (1938)
Jacobi, C.G.J.
[1] Note von der geodätischen Linie auf einem Ellipsoid und den verschiedenen Anwendungen einer merkwürdigen analytischen Substitution. Crelles J. *19*, 309–313 (1839). Die kürzeste Linie auf dem dreiaxigen Ellipsoid. Achtundzwanzigste Vorlesung. Vorlesungen über Dynamik, gehalten an der Universität zu Königsberg im Wintersemester 1842–1843. Hrsg. A. Clebsch. Berlin: Reimer 1866
Karcher, H.
[1] Schnittort und konvexe Mengen in vollständigen Riemannschen Mannigfaltigkeiten. Math. Ann. *177*, 105–121 (1968)
Katok, A.
[1] Entropy and Closed Geodesics. Technical Report, U. of Maryland, 1981
Kervaire, M.
[1] A manifold which does not admit any differentiable structure. Comment. Math. Helv. *34*, 257–270 (1960)
Klingenberg, W.
[1] Contributions to Riemannian Geometry in the large. Ann. of Math. *69*, 654–666 (1959)
[2] Neue Ergebnisse über konvexe Flächen. Commment. Math. Helv. *34*, 17–36 (1960)
[3] Über riemannsche Mannigfaltigkeiten mit positiver Krümmung. Comment. Math. Helv. *35*, 47–54 (1961)
[4] Über Riemannsche Mannigfaltigkeiten mit nach oben beschränkter Krümmung. Ann. Mat. Pura Appl. *60*, 49–59 (1962)
[4a] Manifolds with restricted conjugate locus I, II. Anm. of Math. *78*, 527–547 (1963); *80*, 330–339 (1964).
[5] Simple closed geodesics on pinched spheres. J. Diff. Geom. *2*, 225–232 (1968)
[6] Geodätischer Fluss auf Mannigfaltigkeiten vom hyperbolischen Typ. Inv. math. *14*, 63–82 (1971)
[7] Manifolds with geodesic flow of Anosov type. Ann. of Math. *99*, 1–13 (1974)
[8] Der Indexsatz für geschlossene Geodätische. Math. Z. *139*, 231–256 (1974)
[9] A Course in Differential Geometry. New York–Heidelberg–Berlin: Springer 1978
[10] Lectures on Closed Geodesics. Grundlehren Math. Wiss. Bd. 230, Berlin–Heidelberg–New York: Springer 1978
[11] Über den Index geschlossener Geodätischer auf Flächen. Nagoya Math. J. *69*, 107–116 (1978)
[12] Closed Geodesics on Surfaces of Genus 0. Ann. Scuola Norm. Pisa (IV) *6*, 19–38 (1979)
[13] On the existence of closed geodesics on spherical manifolds. Math. Z. *176*, 319–325 (1981)
[14] Über die Existenz unendlich vieler geschlossener Geodätischer. Akad. Wiss. Lit. Mainz. Abh. Math.-Naturw. Kl. Nr. 1, 1981
Klingenberg, W. and Sakai, T.
[1] Injectivity radius for 1/4-pinched manifolds. Arch. Math. *34*, 371–376 (1980)

Klingmann, M.
[1] Das Morse'sche Indextheorem bei allgemeinen Randbedingungen. J. Diff. Geom. *1*, 371–380 (1967)
Kobayashi, S.
[1] Fixed points of isometries. Nagoya Math. J. *13*, 63–68 (1958)
[2] Riemannian manifolds without conjugate points. Ann. Math. Pura Appl. *53*, 149–155 (1961)
[3] On conjugate and cut loci. Studies in Global Geometry and Analysis, Math. Assoc. Amer., 96–122 (1967)
Kobayashi, S. and Nomizu, K.
[1] Foundations of Differential Geometry. Vol. 1 and 2. New York: Interscience 1963/69
Koebe, P.
[1] Riemannsche Mannigfaltigkeiten und nichteuklidische Raumformen. Sitz. Ber. Preuss. Akad. Wiss., 164–196 (1927), 345–442 (1928), 414–457 (1929), 304–364 (1930), 506–584 (1931)
Lang, S.
[1] Differential Manifolds. Reading. Mass.: Addison-Wesley 1972
Lawson, H. B. and Yau, S. T.
[1] On compact manifolds of nonpositive curvature. J. Diff. Geom. *7*, 211–228 (1972)
Levi-Cività, T.
[1] Nozione di parallelismo in una varietà qualunque e consequente spezificazione geometrice della curvature Riemanniana. Rend. Circ. Mat. Palermo *42*, 173–205 (1917)
Lusternik, L. et Schnirelmann, L.
[1] Sur le problème de trois géodésiques fermées sur les surfaces de genre 0. C. R. Acad. Sci. Paris *189*, 269–271 (1929)
Lyusternik, L.
[1] The Topology of Function Spaces and the Calculus of Variations in the Large. Trudy Mat. Inst. Steklov *19*, (Russian) (1947)
Lyusternik, L. and Fet, A. I.
[1] Variational problems on closed manifolds. Dokl. Akad. Nauk SSSR (N.S.) *81*, 17–18 (Russian) (1951)
v. Mangoldt, H.
[1] Über diejenigen Punkte auf positiv gekrümmten Flächen, welche die Eigenschaft haben, daß die von ihnen ausgehenden geodätischen Linien nie aufhören, kürzeste Linien zu sein. J. Reine Angew. Math. *91*, 23–52 (1881)
Meyer, K.
[1] Generic Bifurcation of Periodic Points. Trans. Amer. Math. Soc. *149*, 95–107 (1970)
Milnor, J.
[1] On manifolds homeomorphic to the 7-sphere. Ann. of Math. *64*, 394–405 (1956)
[2] Morse Theory. Ann. Math. Studies No. 51, Princeton, N.J.: Princeton Univ. Press 1963
Morse, M.
[1] A fundamental class of closed geodesics on any closed surface of genus greater than one. Trans. Amer. Math. Soc. *26*, 25–60 (1971)
[2] Calculus of Variations in the Large. Amer. Math. Soc. Colloq. Publ. vol. 18. Providence, R. I.: Amer. Math. Soc. 1934
[3] A generalization of the Sturm separation and comparison theorems in n-space. Math. Ann. *103*, 52–69 (1930)
[4] Generalized concavity theorems. Proc. Nat. Acad. Sci. USA *21*, 359–362 (1935)
Morse, M. and Hedlund, G.
[1] Manifolds without conjugate points. Trans. Amer. Math. Soc. *51*, 363–386 (1942)
Moser, J.

[1] Stable and Random Motions in Dynamical Systems. Ann. Math. Studies No. 77, Princeton, N.Y.: Princeton University Press 1973
Myers, S.B.
[1] Connections between differential geometry and topology I. Duke Math. J. *1*, 376–391 (1935)
[2] Riemannian manifolds in the large. Duke Math. J. *1*, 39–49 (1935)
Myers, S.B. and Steenrod, N.
[1] The group of isometries of a Riemannian manifold. Ann. of Math. *40*, 400–416 (1939)
Nash, J.
[1] The imbedding problem for riemannian manifolds. Ann. of Math. *63*, 20–63 (1956)
Nielsen, J.
[1] Untersuchungen zur Topologie der geschlossenen zweiseitigen Flächen. Acta Math. *90*, 189–358 (1927)
O'Neill, B.
[1] The fundamental equations for a submersion. Michigan Math. J. *13*, 459–469 (1966)
Nomizu, K. and Ozeki, H.
[1] The existence of complete Riemannian metrics. Proc. Amer. Math. Soc. *12*, 889–891 (1961)
Ozols, V.
[1] Largest normal neighbourhoods. Proc. Amer. Math. Soc. *61*, 99–101 (1976)
Palais, R.
[1] Morse Theory on Hilbert manifolds. Topology *2*, 299–340 (1963)
Palais, R. and Smale, S.
[1] A generalized Morse theory. Bull. Amer. Math. Soc. *70*, 165–172 (1964)
Pars, L.A.
[1] A treatise on Analytical Dynamics. London: Heinemann 1965
Perron, O.
[1] Die Stabilitätsfrage bei Differentialgleichungen. Math. Z. *32*, 703–728 (1930)
Poincaré, H.
[1] Les méthodes nouvelles de la mécanique céleste. Vol. I, II, III. Paris: Gauthier-Villars 1892/99
[2] Sur les lignes géodésiques des surfaces convexes. Trans. Amer. Math. Soc. *17*, 237–274 (1905)
Preissmann, A.
[1] Quelques propriétés globales des espaces de Riemann. Comment. Math. Helv. *15*, 175–216 (1943)
Rauch, H.E.
[1] A contribution to differential geometry in the large. Ann. of Math. *54*, 38–55 (1951)
de Rham, G.
[1] Sur la réductibilité d'un espace de Riemann. Comment. Math. Helv. *26*, 328–344 (1952)
[2] Variétés Différentiables. Paris: Hermann et Cie 1955
Sakai, T.
[1] On the cut loci of symmetric spaces. Hokkaido Math. J. *6*, 136–161 (1977)
Schatten, R.
[1] A Theory of Cross-Spaces. Ann. Math. Studies No. 26, Princeton University Press: Princeton, N.J. 1950
Schoenberg, J.J.
[1] Some applications of the calculus of variations to Riemannian geometry. Ann. of Math. *33*, 485–495 (1932)
Seifert, H. and Threlfall, W.
[1] Variationsrechnung im Grossen. Leipzig: Teubner 1938

Serre, J. P.
[1] Homologie singulière des espace fibrés. Ann. of Math. *54*, 425–505 (1951)
Smale, S.
[1] Generalized Poincaré conjecture in dimensions greater than four. Ann. of Math. *74*, 391–406 (1961)
Spanier, E.
[1] Algebraic Topology. New York–London: McGraw Hill 1966
Solá-Morales, J.
[1] On the continuation of the -grad E flow of $H^1(S^1, M)$. Arch. Math. *34*, 140–142 (1980)
Stallings, J.
[1] The piecewise-linear structure of euclidean space. Proc. Cambride Phil. Soc. *58*, 481–488 (1962)
Steenrod, N.
[1] The topology of fibre bundles. Princeton, N. Y.: Princeton Univ. Press 1951
Study, E.
[1] Kürzeste Wege im komplexen Gebiet. Math. Ann. *60*, 312–377 (1905)
Sulanke, R. und Wintgen, P.
[1] Differentialgeometrie und Faserbündel. Berlin: VEB Deutscher Verlag der Wissenschaften 1972
Synge, J.
[1] On the connectivity of spaces of positive curvature. Quart. J. Math. *7*, 316–320 (1936)
Thimm, A.
[1] Integrabilität beim geodätischen Fluss. Bonn. Math. Schr. Nr. 103 (1978)
Thom, R.
[1] Stabilité structurelle et morphogénèse. Reading, Mass.: Benjamin, 1972
Thorbergsson, G.
[1] Non-hyperbolic Closed Geodesics. Math. Scand. *44*, 135–148 (1979)
Toponogov, V. A.
[1] Computation of the length of a closed geodesic on a convex surface. Dokl. Akad. Nauk SSSR, *124*, 282–284 (1959) (Russian)
[2] Dependence between curvature and topological structure of Riemannian spaces of even dimensions. Dokl. Akad. Nauk SSSR *133*, 1031–1033 (1960) (Russian). Soviet Mathematics *1*, 943–945 (1961)
[3] The metric structure of riemannian spaces of nonnegative curvature which contain straight lines. Sibirsk. Mat. Z. *5*, 1358–1369 (1964) (Russian). – Transl. Amer. Math. Soc. (2) *70*, 225–239 (1968)
[4] Riemannian spaces having their curvature bounded below by a positive number. Uspehi Mat. Nauk *14*, 87–130 (1959) (Russian)
Viesel, H.
[1] Über einfach geschlossene Geodätische auf dem Ellipsoid. Arch. Math. *22*, 106–112 (1971)
Wall, C. T. C.
[1] Geometric properties of generic differentiable manifolds. Geometry and Topology, Rio de Janeiro 1976. Ed. R. Palais and M. do Carmo. Lecture Notes in Mathematics 597, 707–774. Berlin–Heidelberg–New York: Springer 1977
Wallach, N.
[1] Compact homogeneous Riemannian manifolds with strictly positive curvature. Ann. of Math. *96*, 277–295 (1972)
Walter, R.
[1] Konvexität in riemannschen Mannigfaltigkeiten. Jahresber. DMV *83*, 1–31 (1981)
Warner, F.

[1] The conjugate locus of a Riemannian manifold. Amer. J. Math. *87*, 575–604 (1965)
[2] Foundations of Differentiable Manifolds and Lie Groups. Glenview, Ill.-London: Scott, Foresmond and Co. 1971

Weinstein, A.
[1] The cut locus and conjugate locus of a riemannian manifold. Ann. of Math *87*, 29–41 (1968)
[2] On the homotopy type of positively pinched manifolds. Arch. Math. *18*, 523–524 (1967)
[3] Distance spheres in complex projective spaces. Proc. Amer. Math. Soc. *39*, 649–650 (1973)

Whitehead, J. H. C.
[1] Convex regions in the geometry of paths. Quart. J. Math. Oxford Ser. *3*, 33–42 (1932)

Whitney, H.
[1] Differentiable manifolds. Ann. of Math. *37*, 645–680 (1936)

Wolf, J.
[1] Spaces of Constant Curvature. New York–Toronto–London: McGraw Hill 1967

Wolter, F. E.
[1] Distance function and cut loci on a complete Riemannian manifold. Arch. Math. *32*, 92–96 (1979)

Yau, S.T.
[1] Problem section, in Seminar on Differential Geometry. Ann. Math. Studies No. 102, Princeton, N.J: Princeton Univ. Press 1982.

Ziller, W.
[1] Geschlossene Geodätische auf globalsymmetrischen und homogenen Räumen. Bonn. Math. Schr. Nr. 85, 1976

Index

de Gruyter Studies in Mathematics

An international series of monographs and textbooks of a high standard, written by scholars with an international reputation presenting current fields of research in pure and applied mathematics.

Editors: Heinz Bauer, Erlangen, and Peter Gabriel, Zürich

Forthcoming titles:

Michel Métivier
Semimartingales
(in press)

Ludger Kaup · Burchard Kaup
Holomorphic Functions of Several Variables

Ulrich Krengel
Ergodic Theorems

Heiner Zieschang · Gerhard Burde
Theory of Knots

Cornelius Constantinescu
Spaces of Measures

Verlag Walter de Gruyter · Berlin · New York

Journal für die reine und angewandte Mathematik

Multilingual Journal

founded in 1826 by
August Leopold Crelle

continued by
C. W. Borchardt, K. Weierstrass, L. Kronecker, L. Fuchs,
K. Hensel, L. Schlesinger, H. Hasse, H. Rohrbach

at present edited by
Otto Forster · Willi Jäger · Martin Kneser
Horst Leptin · Samuel J. Patterson · Peter Roquette

with the cooperation of
M. Deuring, P. R. Halmos, O. Haupt,
F. Hirzebruch, G. Köthe, K. Krickeberg, K. Prachar,
H. Reichardt, L. Schmetterer, B. Volkmann

Walter de Gruyter · Berlin · New York 1982

Frequency of publication: yearly approx. 8 volumes
(1982: volume 330–337)
Price per volume: DM 148,–; $ 74.00

Back volumes: volume 1–250 bound complete DM 35 000,–; $ 17,500.00
Single volume each DM 168,–; $ 84.00

For USA and Canada: Please send all orders to
Walter de Gruyter, Inc., 200 Saw Mill River Road, Hawthorne, N.Y. 10532

Prices are subject to change